Institute for Mathematics and
its Applications
IMA

The **Institute for Mathematics and its Applications** was established by a grant from the National Science Foundation to the University of Minnesota in 1982. The IMA seeks to encourage the development and study of fresh mathematical concepts and questions of concern to the other sciences by bringing together mathematicians and scientists from diverse fields in an atmosphere that will stimulate discussion and collaboration.

The IMA Volumes are intended to involve the broader scientific community in this process.

Avner Friedman, Director
Willard Miller, Jr., Associate Director

* * * * * * * * * *

IMA ANNUAL PROGRAMS

1982–1983	Statistical and Continuum Approaches to Phase Transition
1983–1984	Mathematical Models for the Economics of Decentralized Resource Allocation
1984–1985	Continuum Physics and Partial Differential Equations
1985–1986	Stochastic Differential Equations and Their Applications
1986–1987	Scientific Computation
1987–1988	Applied Combinatorics
1988–1989	Nonlinear Waves
1989–1990	Dynamical Systems and Their Applications
1990–1991	Phase Transitions and Free Boundaries
1991–1992	Applied Linear Algebra
1992–1993	Control Theory and its Applications
1993–1994	Emerging Applications of Probability
1994–1995	Waves and Scattering
1995–1996	Mathematical Methods in Material Science

IMA SUMMER PROGRAMS

1987	Robotics
1988	Signal Processing
1989	Robustness, Diagnostics, Computing and Graphics in Statistics
1990	Radar and Sonar (June 18 - June 29)
	New Directions in Time Series Analysis (July 2 - July 27)
1991	Semiconductors
1992	Environmental Studies: Mathematical, Computational, and Statistical Analysis
1993	Modeling, Mesh Generation, and Adaptive Numerical Methods for Partial Differential Equations
1994	Molecular Biology

* * * * * * * * * *

SPRINGER LECTURE NOTES FROM THE IMA:

The Mathematics and Physics of Disordered Media

Editors: Barry Hughes and Barry Ninham
(Lecture Notes in Math., Volume 1035, 1983)

Orienting Polymers

Editor: J.L. Ericksen
(Lecture Notes in Math., Volume 1063, 1984)

New Perspectives in Thermodynamics

Editor: James Serrin
(Springer-Verlag, 1986)

Models of Economic Dynamics

Editor: Hugo Sonnenschein
(Lecture Notes in Econ., Volume 264, 1986)

The IMA Volumes in Mathematics and its Applications

Volume 68

Series Editors
Avner Friedman Willard Miller, Jr.

Max D. Gunzburger

Editor

Flow Control

With 111 Illustrations

Springer-Verlag
New York Berlin Heidelberg London Paris
Tokyo Hong Kong Barcelona Budapest

Max D. Gunzburger
Department of Mathematics and
   Interdisciplinary Center for Applied Mathematics
Virginia Polytechnic Institute and State University
Blacksburg, VA 24061-0531
USA

Series Editors:
Avner Friedman
Willard Miller, Jr.
Institute for Mathematics and its Applications
University of Minnesota
Minneapolis, MN 55455
USA

Mathematics Subject Classifications (1991): 76MXX, 65P05, 65MXX, 65NXX, 65K10, 49J20, 49K20, 93B05, 93B52

Library of Congress Cataloging-in-Publication Data
Flow control / Max D. Gunzburger, editor.
       p.   cm. — (The IMA volumes in mathematics and its
  applications ; v. 68)
   Includes bibliographical references (p.).
   ISBN 0-387-94445-1
   1. Fluid dynamics—Mathematical models. 2. Numerical
  calculations. I. Gunzburger, Max D. II. Series.
  TA357.F538 1995
  620.1′064—dc20   94-46237

Printed on acid-free paper.

Production managed by Laura Carlson; manufacturing supervised by Jeffrey Taub.
Camera-ready copy prepared by the IMA.
Printed and bound by Braun-Brumfield, Ann Arbor, MI.
Printed in the United States of America.

9 8 7 6 5 4 3 2 1

ISBN 0-387-94445-1 Springer-Verlag New York Berlin Heidelberg

The IMA Volumes
in Mathematics and its Applications

Current Volumes:

Volume 1: Homogenization and Effective Moduli of Materials and Media
Editors: Jerry Ericksen, David Kinderlehrer, Robert Kohn, and
J.-L. Lions

Volume 2: Oscillation Theory, Computation, and Methods of
Compensated Compactness
Editors: Constantine Dafermos, Jerry Ericksen,
David Kinderlehrer, and Marshall Slemrod

Volume 3: Metastability and Incompletely Posed Problems
Editors: Stuart Antman, Jerry Ericksen, David Kinderlehrer, and
Ingo Muller

Volume 4: Dynamical Problems in Continuum Physics
Editors: Jerry Bona, Constantine Dafermos, Jerry Ericksen, and
David Kinderlehrer

Volume 5: Theory and Applications of Liquid Crystals
Editors: Jerry Ericksen and David Kinderlehrer

Volume 6: Amorphous Polymers and Non-Newtonian Fluids
Editors: Constantine Dafermos, Jerry Ericksen, and
David Kinderlehrer

Volume 7: Random Media
Editor: George Papanicolaou

Volume 8: Percolation Theory and Ergodic Theory of Infinite Particle
Systems
Editor: Harry Kesten

Volume 9: Hydrodynamic Behavior and Interacting Particle Systems
Editor: George Papanicolaou

Volume 10: Stochastic Differential Systems, Stochastic Control Theory,
and Applications
Editors: Wendell Fleming and Pierre-Louis Lions

Volume 11: Numerical Simulation in Oil Recovery
Editor: Mary Fanett Wheeler

Volume 67: Mathematics in Industrial Problems, Part 7
by Avner Friedman

Volume 68: Flow Control
Editor: Max D. Gunzburger

Forthcoming Volumes:

1991-1992: *Applied Linear Algebra*

Linear Algebra for Signal Processing

1992 Summer Program: *Environmental Studies*

1992–1993: *Control Theory*

Control and Optimal Design of Distributed Parameter Systems

Robotics

Nonsmooth Analysis & Geometric Methods in Deterministic Optimal Control

Adaptive Control, Filtering and Signal Processing

Discrete Event Systems, Manufacturing, Systems, and Communication Networks

1993 Summer Program: *Modeling, Mesh Generation, and Adaptive Numerical Methods for Partial Differential Equations*

1993-1994: *Emerging Applications of Probability*

Discrete Probability and Algorithms

Random Discrete Structures

Mathematical Population Genetics

Stochastic Networks

Stochastic Problems for Nonlinear Partial Differential Equations

Image Models (and their Speech Model Cousins)

Stochastic Models in Geosystems

Classical and Modern Branching Processes

1994 Summer Program: *Molecular Biology*

FOREWORD

This IMA Volume in Mathematics and its Applications

FLOW CONTROL

is based on the proceedings of a workshop that was an integral part of the 1992–93 IMA program on "Control Theory." Historically, flow control problems have been addressed through experimental investigations. Analytic and computational research had been based on drastically simplified flow models. However, recently, a number of mathematicians and other scientists have been addressing flow control problems without invoking such simplifications. The purpose of the workshop was to bring together these scientists and other mathematicians interested in entering this rapidly growing research area that will have significant impact on applications.

We thank Max D. Gunzburger for organizing the workshop and for editing the proceedings. We also take this opportunity to thank the National Science Foundation and Office of Naval Research whose financial support made the workshop possible.

Avner Friedman

Willard Miller, Jr.

PREFACE

This volume contains the proceedings of the Period of Concentration in Flow Control held at the IMA in November, 1992. This gathering of engineers and mathematicians was especially timely as it coincided with the emergence of the role of mathematics and systematic engineering analysis in flow control and optimization. Since this meeting, this role has significantly expanded to the point where now sophisticated mathematical and computational tools are being increasingly applied to the control and optimization of fluid flows. Thus, these proceedings serve as a valuable record of some important work that has gone on to influence the practical, everyday design of flows. Moreover, they also represent very nearly the state of the art in the formulation, analysis, and computation of flow control problems.

My own article in the proceedings attempts to set the stage for the remaining articles by describing the history of attempts at flow control and optimization and explaining why the time is ripe for the introduction of sophisticated tools from the theory of partial differential equations, from optimization theory, and from computational fluid dynamics into the study of flow control. The remaining articles in the volume show how these tools may be introduced to attack flow control problems. Mathematical issues in optimal control, feedback control, and controllability of fluid flows are treated in the articles by E. Casas, A. Fursikov and O. Imanuvilov, K. Ito, H. Tran and J. Scroggs, S. Sritharan, S. Stojanovic, T. Svobodny, and R. Temam. Computational studies of algorithms and of particular applications are found in the articles by H. Banks and R. Smith, J. Borggaard, J. Borggaard and J. Burns, J. Brock and W. Ng, J. Burkardt and J. Peterson, Y.-R. Ou, G. Strumulo, and A. Taylor, P. Newman, G. Hou, and H. Jones. Among the applications considered in this volume are acoustics, compressible flows and incompressible flows, chemical vapor deposition, turbulent flows, and flows with shock waves.

I would like to express my sincere thanks to all of the participants in the Period of Concentration, and especially to the speakers and those who contributed to these proceedings. Thanks are also due to the staff of the IMA for their help in the production of these proceedings. Of special note in this regard are Patricia V. Brick, Ruth Capp, Stephan J. Skogerboe and Kaye Smith. Finally, I would like to acknowledge the hospitality and help extended to me and the other participants by Avner Friedman and Willard Miller, Jr.. Without them, neither the Period of Concentration in Flow Control nor this volume would have been possible.

Max D. Gunzburger
Blacksburg, 1994

xiii

CONTENTS

ACTIVE CONTROL OF ACOUSTIC PRESSURE FIELDS USING SMART MATERIAL TECHNOLOGIES *

H.T. BANKS[†] AND R.C. SMITH[‡]

Abstract. An overview describing the use of piezoceramic patches in reducing noise in a structural acoustics setting is presented. The passive and active contributions due to patches which are bonded to an Euler-Bernoulli beam or thin shell are briefly discussed and the results are incorporated into a 2-D structural acoustics model. In this model, an exterior noise source causes structural vibrations which in turn lead to interior noise as a result of nonlinear fluid/structure coupling mechanisms. Interior sound pressure levels are reduced via patches bonded to the flexible boundary (a beam in this case) which generate pure bending moments when an out-of-phase voltage is applied. Well-posedness results for the infinite dimensional system are discussed and a Galerkin scheme for approximating the system dynamics is outlined. Control is implemented by using LQR optimal control theory to calculate gains for the linearized system and then feeding these gains back into the nonlinear system of interest. The effectiveness of this strategy for this problem is illustrated in an example.

1. Introduction. The recent development of highly fuel-efficient turboprop and turbofan engines which also produce high levels of interior cabin noise (especially at low frequencies) has stimulated a substantial effort on the development of a comprehensive active control methodology for interior pressure field cavities that have been excited by some primary or external source. In this overview paper, we shall discuss recent approaches and preliminary results in the growing effort to develop "smart" or "adaptive" material concepts (materials that possess the capability for both sensing and actuation are often called "smart" materials) and control strategies for such a comprehensive methodology.

Interior cavity noise in aircraft with turboprop engines is produced primarily through (nonlinear) fluid/structure interaction mechanisms. The turboprop blades produce an external acoustic pressure field which is converted into mechanical vibrations through fluid/structure interactions at the exterior aircraft cabin walls. In turn, these mechanical vibrations produce, through interactions of the interior cabin walls with the air in the cabin cavity, pressure waves or an interior acoustic pressure field.

Our discussion here focuses on a time domain state space approach to active or feedback control of noise in the interior acoustic cavity. We are especially interested in models and methodologies which treat tran-

* The research of H.T.B. was supported in part by the Air Force Office of Scientific Research under grant AFOSR-90-0091. This research was also supported by the National Aeronautics and Space Administration under NASA Contract Numbers NAS1-18605 and NAS1-19480 while H.T.B. was a visiting scientist and R.C.S. was in residence at the Institute for Computer Applications in Science and Engineering (ICASE), NASA Langley Research Center, Hampton, VA 23681.

† Center for Research in Scientific Computation, North Carolina State University, Raleigh, NC 27695.

‡ Department of Mathematics, Iowa State University, Ames, IA 50011.

sient dynamics. There is a substantial literature on active control of noise
in a frequency domain setting (see [18,21,24,26,28] for some examples and
further references to both experimental and analytic efforts) as well as
a growing literature on infinite dimensional state space time domain approaches (e.g. [2,3,8,9,10]). Earlier efforts by most researchers focused on
a control methodology implemented through *secondary source techniques*
with the input or secondary noise based on feedback of noise levels in the
acoustic cavity. In this approach, a system of microphones and speakers is
strategically placed in the interior cavity where one can sense the pressure
field (composed of the primary source plus any secondary sources present).
This information is used as feedback for the actuators or speakers which
produce a (hopefully) optimally interfering signal (secondary noise) to reduce the total noise levels in certain critical zones (related to passenger
comfort). Both frequency and time domain settings have been used in providing not only "proof of concept" analyses but also in designing and and
implementing these ideas (to date, mainly in luxury class automobiles).

More recently, a second approach utilizing *smart materials technology*
has captured the attention of investigators. There are a large number of
classes of smart materials (e.g. electrorheological fluids, magnetostrictives,
shape memory alloys) but we shall restrict our discussions in this paper
to piezoceramic devices such as piezoceramic patches which, when bonded
to a structure such as a beam, plate, or curved cylindrical shell, act as an
electro-mechanical transducer. When excited by an electric field, the patch
induces a strain in the material to which it is bonded and hence can be
employed as an actuator. Moreover, if the host material undergoes a deformation (either bending or extension/contraction), this produces a strain in
the patch which results in a voltage across the patch that is proportional
to the strain and thereby permits the use of the patch as a mechanical
sensor. If constructed and wired with proper circuits, these patches can
be employed as "self-sensing actuators" [20], thereby providing a smart or
adaptive material capability for the structure to which the device is bonded
or in which it is embedded. When combined with a computational adaptive
or feedback control element, the potential for self-controlled or intelligent
structures is enormous.

In our presentation and discussions of active control of noise, we shall
concentrate on actuator aspects of piezoceramics. In the noise suppression
example detailed below, we tacitly assume that acoustic pressure in the
cavity and wall displacements and velocities are sensed for feedback. For a
complete smart material system, one would use piezoceramic (strain) sensors and cavity pressure sensors to construct a state estimator for feedback.

The motivating example we consider consists of an exterior noise source
which is separated from an interior cavity by an active wall or plate. This
plate transmits noise or vibrations from the exterior field to the interior
cavity via fluid/structure interactions thus leading to the formulation of
a system of partial differential equations consisting of an acoustic wave

equation coupled with elasticity equations for the plate. The control is implemented in the example via piezoceramic patches on the plate which are excited in a manner so as to produce pure bending moments. It should be noted that the incorporation of the feedback control in this manner leads to a system with an unbounded input term (in this case, a system with input coefficients involving the Dirac delta "function" and its "derivative"). Experiments are being designed and carried out at NASA Langley Research Center in which the interior cavity is taken to be cylindrical with a circular active plate to which sectorial piezoceramic patches are bonded.

While the motivating structural acoustics applications are three dimensional in nature, many of the theoretical and numerical issues concerning system modeling, the simulation of system dynamics, estimation of physical parameters, and the developments of feasible control strategies can be studied in 2-D geometries. In this work, we consider a 2-D domain $\Omega(t)$ which is bounded on three sides by hard walls and on the fourth by a flexible beam (see Figure 1). A periodic forcing function f, modeling an exterior noise source, causes vibrations in the beam which then lead to unwanted interior noise.

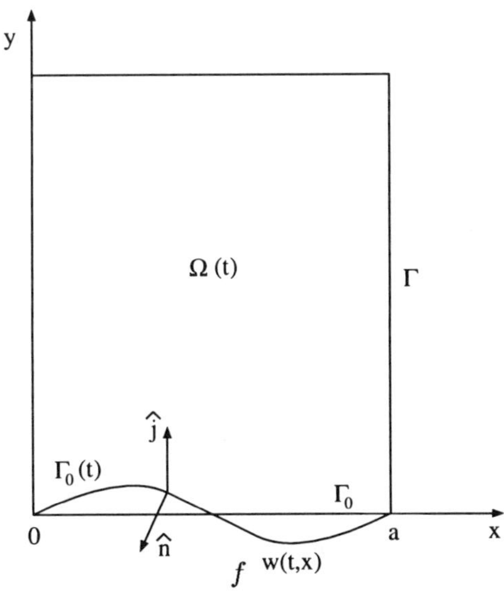

FIG. 1. *The 2-D domain.*

This specific problem was chosen since it is a two dimensional slice from a three dimensional cylindrical domain which models an experimental apparatus consisting of a rigid cylindrical pipe with a clamped aluminum plate at one end.

As a 2-D analogue of the plate, the perturbable boundary $\Gamma_0(t)$ (see Figure 1) is modeled by a fixed-end Euler-Bernoulli beam having Kelvin-Voigt damping. Bonded to the beam are s pairs of piezoceramic patches

4   H.T. Banks and R.C. Smith

which are configured and excited in a manner so as to produce pure bending moments (see Figure 2). We reiterate that it is through the excitation of these patches that the sound pressure levels are controlled.

The acoustic response inside the cavity is modeled by a linear wave equation with zero normal velocity boundary conditions taken on three walls in order to simulate the rigid walls of the experimental pipe. The boundary conditions on the fourth (beam) side of the acoustic cavity result from nonlinear velocity and pressure couplings between the acoustic and structural responses (as discussed in [14], these coupling terms are nonlinear since they take place along the surface of the vibrating beam). Finally, under the assumption of small beam displacements which is inherent in the Euler-Bernoulli theory, the variable domain $\Omega(t)$ is replaced by the fixed domain $\Omega \equiv [0, a] \times [0, \ell]$ as shown in Figure 2.

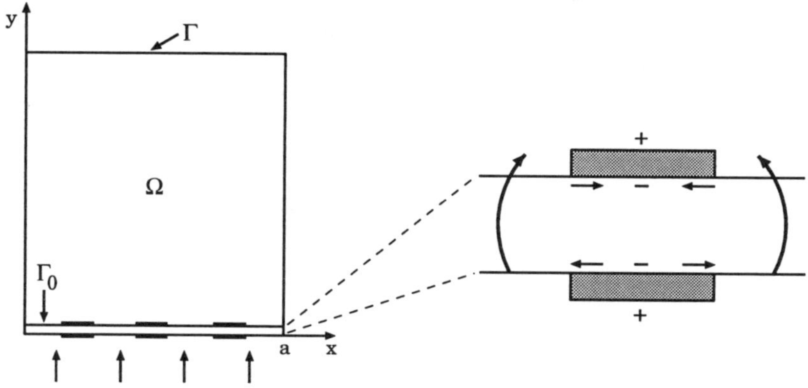

FIG. 2. *Acoustic cavity with piezoceramic patches creating pure bending moments.*

In terms of the velocity potential ϕ (so that $p = \rho_f \phi_t$ is the acoustic pressure) and the transverse beam displacements w, the strong form of the approximate controlled model for the coupled system is then given by

$$\phi_{tt} = c^2 \Delta \phi \quad , \quad (x, y) \in \Omega \ , t > 0 \ ,$$

$$\nabla \phi \cdot \hat{n} = 0 \quad , \quad (x, y) \in \Gamma \ , t > 0 \ ,$$

$$\nabla \phi(t, x, w(t, x)) \cdot \hat{n} = w_t(t, x) \quad , \quad 0 < x < a \ , t > 0 \ ,$$

$$(1.1) \quad \rho w_{tt} + \frac{\partial^2 \mathcal{M}}{\partial x^2} = -\rho_f \phi_t(t, x, w(t, x)) + f(t, x) \quad , \quad \begin{array}{c} 0 < x < a \ , \\ t > 0 \ , \end{array}$$

$$w(t, 0) = \frac{\partial w}{\partial x}(t, 0) = w(t, a) = \frac{\partial w}{\partial x}(t, a) = 0 \quad , \quad t > 0 \ ,$$

$$\phi(0, x, y) = \phi_0(x, y) \quad , \quad w(0, x) = w_0(x)$$

$$\phi_t(0, x, y) = \phi_1(x, y) \quad , \quad w_t(0, x) = w_1(x)$$

(for further details concerning the development of this model, see [14]). Here ρ, ρ_f and c are the beam density, equilibrium density of the atmosphere, and speed of sound in the cavity, respectively. The general beam moment $\mathcal{M}(t, x)$ consists of an internal component, depending on material and geometric properties of the beam and patches, and an external component (the control term) which results from the activation of the patches through an applied voltage. Specific descriptions of these moments in a variety of settings are given in the next section. Finally, the nonlinear coupling between the beam vibrations and the interior acoustic field manifests itself in the velocity term $\nabla\phi(t, x, w(t, x)) \cdot \hat{n} = w_t(t, x)$ and the backpressure $\rho_f \phi_t(t, x, w(t, x))$.

2. Piezoceramic Patch/Structure Interactions. As discussed in the last section, control is implemented in the system through the excitation of piezoceramic patches which are bonded to the beam. This affects the dynamics of the beam in two ways. The first effect is passive and results from the structural changes incurred with the bonding of the patches to the structure. In addition to the patch thickness, there is a nontrivial bonding layer, and both contribute to a moment of inertia which differs from that found in regions of the structure not covered with patches. Moreover, the density, Young's modulus and damping coefficient of the glue and patch differ from those of the beam, and as a result, these parameters must be modeled as piecewise constants in order to accurately match system frequencies (see [17]). The third passive contribution is due to the piezoelectric property which dictates that when the patch is subjected to an in-plane strain, a voltage proportional to the strain is produced. Hence longitudinal and transverse vibrations in the beam lead to the generation of current which provides additional damping in the structure. The final (active) contribution from the piezoceramic patches results from the in-plane strains which are produced when a voltage is applied. This leads to the generation of external moments and forces which enter the equations of motion as external loads.

The initial part of this section contains a discussion concerning the contributions due to patches which are bonded to an Euler-Bernoulli beam. The changes which are necessary for extending these arguments to plates and shells are then outlined in the latter part of the section with further details given in [16].

2.1. Piezoceramic Patch/Beam Interactions. In the discussion which follows, we consider an Euler-Bernoulli beam of length ℓ, width b and thickness h as depicted in Figure 3. The Young's modulus, mass density (in mass per unit volume) and damping coefficient for the homogeneous beam are denoted by E_b, ρ_b and c_{Db}, respectively. Bonded to the beam are piezoceramic patches which can be mounted either individually or in pairs as shown in Figures 3 and 4. In the initial discussion concerning the contribution due to the patch pairs, it is assumed that both patches have thickness T, Young's modulus E_{pe}, density ρ_{pe}, and damping coefficient c_{Dpe}. Moreover, it is assumed that the bonding layers for each patch have the same thickness, Young's modulus, density and damping coefficient, and these parameters are denoted by $T_{b\ell}$, $E_{b\ell}$, $\rho_{b\ell}$ and $c_{Db\ell}$, respectively. We emphasize that these assumptions are made solely for clarity of presentation, and similar results can be obtained in an analogous manner for the more general case in which the patches and bonding layers have differing thicknesses and material properties (see, for example, [16]).

For an Euler-Bernoulli beam having this configuration, force and moment balancing yields the strong form of the dynamic equations

$$(2.1) \qquad \rho(x)\frac{\partial^2 u}{\partial t^2} - \frac{\partial N_x}{\partial x} = \hat{q}_x \;,$$

$$\rho(x)\frac{\partial^2 w}{\partial t^2} + \frac{\partial^2 M_x}{\partial x^2} = \hat{q}_n - \frac{\partial \hat{m}_y}{\partial x}$$

where N_x and M_x are the internal force and moment resultants, respectively (see [12,16]). As depicted in Figure 3, w and u denote the transverse and longitudinal displacements, respectively. The external surface loads $\hat{q}_n, \hat{q}_x$ and $\hat{m}_y$ denote normal forces, in-plane forces and moments, respectively. For patch pairs with edges at x_1 and x_2, the density of the structure is

$$\rho(x) = \rho_b hb + 2b\left(\rho_{b\ell}T_{b\ell} + \rho_{pe}T\right)\chi_{pe}(x)$$

where the characteristic function is given by

$$(2.2) \qquad \chi_{pe}(x) = \begin{cases} 1 & , \quad x_1 \le x \le x_2 \\ 0 & , \quad \text{otherwise} \;. \end{cases}$$

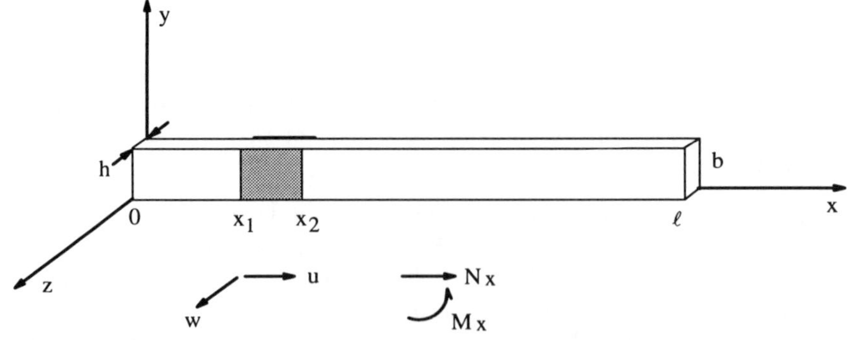

FIG. 3. *Cantilever beam with piezoceramic patches.*

A corresponding weak or variational form of the equations can be determined by choosing $V = H_b^1(\Gamma_0) \times H_b^2(\Gamma_0)$ for the space of trial functions where Γ_0 denotes the beam and the subscript b again denotes the set of functions which must satisfy the essential boundary conditions. Through an energy derivation, one arrives at the variational form for the beam equations

(2.3)
$$\int_0^\ell \left\{ \rho(x)\frac{\partial^2 u}{\partial t^2}\phi_1 + N_x \frac{\partial \phi_1}{\partial x} - \hat{N}_x \frac{\partial \phi_1}{\partial x} \right\} dx = 0$$

$$\int_0^\ell \left\{ \rho(x)\frac{\partial^2 w}{\partial t^2}\phi_3 + M_x \frac{\partial^2 \phi_3}{\partial x^2} - \hat{q}_n \phi_3 - \hat{M}_x \frac{\partial^2 \phi_3}{\partial x^2} \right\} dx = 0$$

for all $\phi_1 \in H_b^1(\Gamma_0)$ and $\phi_3 \in H_b^2(\Gamma_0)$. Here $\hat{N}_x$ and $\hat{M}_x$ are external *line* force and moment resultants. As discussed in [16], the surface loads $\hat{q}_x$ and $\hat{m}_y$ of (2.1) are *locally* related to the forces and moments $\hat{N}_x$ and $\hat{M}_x$ (which are more natural quantities to use in a weak formulation) through the expressions $\hat{q}_x = -\frac{\partial \hat{N}_x}{\partial x}$, $\hat{m}_y = -\frac{\partial \hat{M}_x}{\partial x}$. Global expressions for the specific loads which result from the activation of the patches in both the strong and weak formulations are discussed later in the section.

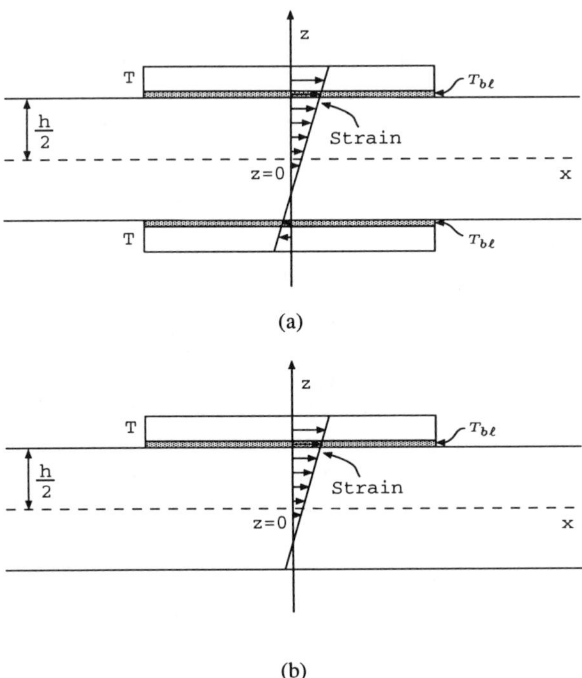

FIG. 4. *Strain distribution for the composite structure undergoing bending and extension; (a) patch pair, and (b) single patch.*

Internal Moment and Force Resultants

In order to determine expressions for the internal force and moment resultants N_x and M_x, the patch pair configuration illustrated in Figure 4a is considered first. Because these resultants depend upon the stresses and ultimately upon the strains occurring in the structure, the description of the resultants begins with a description of the in-plane strains.

In accordance with the Euler-Bernoulli theory, the strain is assumed to be linear and is continuous throughout the combined structure. With ε and κ denoting the midsurface strain and change in curvature, respectively, the strain at an arbitrary point in the beam, bonding layer, or patch is given by $e = \varepsilon + \kappa z$ where z is the distance of the point from the middle surface of the beam (see Figure 4a). Because of the differing Young's moduli and damping coefficients in the beam, bonding layer and patch, the stress slopes will differ in the various layers. Under the assumption that the stress is proportional to a linear combination of strain and strain rate, the stress is given by

$$(2.4) \qquad \sigma = \begin{cases} E_b e + c_{D_b}\dot{e} & , \quad \text{beam} \\ E_{b\ell}e + c_{D_{b\ell}}\dot{e} & , \quad \text{bonding layers} \\ E_{pe}e + c_{D_{pe}}\dot{e} & , \quad \text{patches .} \end{cases}$$

The coefficients c_{D_b} and $c_{D_{b\ell}}$ are the Kelvin-Voigt damping coefficients for the beam and bonding layer while the coefficient $c_{D_{pe}}$ is taken to be a combination of the Kelvin-Voigt damping coefficient for the patch and the damping which results from the production of current when the structure vibrates. This latter contribution to the damping results from the piezoelectric effect of the patches which dictates that a voltage is produced when the patch is subjected to in-plane strains. Under the assumption that the Kelvin-Voigt (material) and electrical damping have approximately the same types of effect in the patch, we have combined the two into the coefficient $c_{D_{pe}}$ which is considered to be unknown and like the other parameters, must ultimately be estimated using data fitting techniques with experimental data when considering actual applications. We also point out that the expression (2.4) can easily be generalized to include the possibility of differing material properties in the two patches or bonding layers.

The force and moment resultants are obtained by integrating the stress across the thickness of the structure thus yielding the expressions

$$(2.5) \qquad N_x = \begin{cases} b \displaystyle\int_{-h/2}^{h/2} \sigma\, dz & , \text{ regions without patches} \\ b \displaystyle\int_{-h/2-T_{b\ell}-T}^{h/2+T_{b\ell}+T} \sigma\, dz & , \text{ regions with patches} \end{cases}$$

and

$$(2.6) \qquad M_x = \begin{cases} b \displaystyle\int_{-h/2}^{h/2} \sigma z\, dz & , \text{ regions without patches} \\[1.2em] b \displaystyle\int_{-h/2-T_{b\ell}-T}^{h/2+T_{b\ell}+T} \sigma z\, dz & , \text{ regions with patches} \end{cases}$$

The substitution of (2.4) into (2.5) and (2.6) yields expressions for the resultants in terms of the midsurface strain ε and change in curvature κ. By considering infinitesimal deformations of the middle surface, ε and κ can be related to the longitudinal and transverse displacements u and w through the strain-displacement equations

$$\varepsilon = \frac{\partial u}{\partial x} \quad , \quad \kappa = \frac{\partial^2 w}{\partial x^2}$$

(see [25], pages 9 and 46). For a beam having two patches bonded to it, the internal (material) force and moment resultants are then given by

$$(2.7) \qquad \begin{aligned} N_x &= Eh(x)\frac{\partial u}{\partial x} + c_D h(x)\frac{\partial^2 u}{\partial x \partial t} \\[0.8em] M_x &= EI(x)\frac{\partial^2 w}{\partial x^2} + c_D I(x)\frac{\partial^3 w}{\partial x^2 \partial t} \end{aligned}$$

where

$$Eh(x) = E_b hb + 2b\left[E_{b\ell}T_{b\ell} + E_{pe}T\right]\chi_{pe}(x)$$

$$EI(x) = E_b \frac{h^3 b}{12} + \frac{2b}{3}\left[E_{b\ell}a_{3b\ell} + E_{pe}a_{3pe}\right]\chi_{pe}(x)$$

$$c_D h(x) = c_{D b} hb + +2b\left[c_{D b\ell}T_{b\ell} + c_{D pe}T\right]\chi_{pe}(x)$$

$$c_D I(x) = c_{D b}\frac{h^3 b}{12} + \frac{2b}{3}\left[c_{D b\ell}a_{3b\ell} + c_{D pe}a_{3pe}\right]\chi_{pe}(x)\ .$$

Here $\chi_{pe}(x)$ again denotes the characteristic function described in (2.2), and the constants $a_{3b\ell}$ and a_{3pe} are given by $a_{3b\ell} = (h/2 + T_{b\ell})^3 - (h/2)^3$ and $a_{3pe} = (h/2 + T_{b\ell} + T)^3 - (h/2 + T_{b\ell})^3$.

The substitution of the force and moment resultants in (2.7) into the dynamic equations (2.1) yields the equations of motion for the combined structure in terms of the transverse and longitudinal displacements w and u. As should be expected for a beam containing a pair of identical patches which are bonded symmetrically about the middle surface, the differential equations (under the first order Euler-Bernoulli assumptions) describing the vibrations in the two coordinate directions are *uncoupled*.

To see how this differs from the case in which a single patch is bonded to the beam, we now consider the case in which a patch of width T is bonded to the beam over the region $x_1 \leq x \leq x_2$ as shown in Figure 4b.

Integrating the stresses through the combined thickness of the structure yields the resultant expressions

$$
\text{(2.8)}\quad
\begin{aligned}
N_x &= Eh(x)\frac{\partial u}{\partial x} + c_D h(x)\frac{\partial^2 u}{\partial x\partial t} + E_2(x)\frac{\partial^2 w}{\partial x^2} + c_{D2}(x)\frac{\partial^3 w}{\partial x^2\partial t}\\[2mm]
M_x &= EI(x)\frac{\partial^2 w}{\partial x^2} + c_D I(x)\frac{\partial^3 w}{\partial x^2\partial t} + E_2(x)\frac{\partial u}{\partial x} + c_{D2}(x)\frac{\partial^2 u}{\partial x\partial t}\ .
\end{aligned}
$$

The parameters in this case are given by

$$
Eh(x) = E_b hb + b\left[E_{b\ell}T_{b\ell} + E_{pe}T\right]\chi_{pe}(x)
$$

$$
EI(x) = E_b\frac{h^3 b}{12} + \frac{b}{3}\left[E_{b\ell}a_{3b\ell} + E_{pe}a_{3pe}\right]\chi_{pe}(x)
$$

$$
E_2(x) = \frac{b}{2}\left[E_{b\ell}a_{2b\ell} + E_{pe}a_{2pe}\right]\chi_{pe}(x)
$$

$$
c_D h(x) = c_{Db} hb + b\left[c_{Db\ell}T_{b\ell} + c_{Dpe}T\right]\chi_{pe}(x)
$$

$$
c_D I(x) = c_{Db}\frac{h^3 b}{12} + \frac{b}{3}\left[c_{Db\ell}a_{3b\ell} + c_{Dpe}a_{3pe}\right]\chi_{pe}(x)
$$

$$
c_{D2}(x) = \frac{b}{2}\left[c_{Db\ell}a_{2b\ell} + c_{Dpe}a_{2pe}\right]\chi_{pe}(x)
$$

with $a_{3b\ell}$ and a_{3pe} defined as before and $a_{2b\ell}$ and a_{2pe} given by $a_{2b\ell} = (h/2 + T_{b\ell})^2 - (h/2)^2$, $a_{2pe} = (h/2 + T_{b\ell} + T)^2 - (h/2 + T_{b\ell})^2$.

When the force and moment expressions in (2.8) are substituted into the dynamic equations (2.1), it is apparent that the longitudinal and transverse vibrations are *coupled* as a result of the asymmetry of the structure due to the single patch. This is in contrast to the case when patch pairs are bonded to the beam and helps to indicate the, in general, nontrivial effect that the patches have on the passive or material properties of the structure.

External Moment and Force Resultants

The second contribution from the piezoceramic patches is the generation of external moments and forces which results from the converse piezoelectric property that when a voltage is applied, in-plane strains are induced in the patch. The magnitude of these induced free strains is given by

$$
e_{pe_1} = \frac{d_{31}}{T}V_1\quad,\quad e_{pe_2} = \frac{d_{31}}{T}V_2
$$

where d_{31} is a piezoceramic strain constant, and V_1 and V_2 are the voltages into the two patches in the pair. We point out that when a voltage is applied to a free patch with edge coordinates x_1 and x_2, the point $\bar{x} = (x_1 + x_2)/2$

will not move whereas the symmetric points on either side will move an equal amount in opposite directions. This motivates the use of the indicator function in several of the following definitions.

The stresses due to the excitation of the patches are given by

$$(\sigma_x)_{pe_1} = -E_{pe}e_{pe_1} \quad , \quad (\sigma_x)_{pe_2} = -E_{pe}e_{pe_2}$$

with the negative signs resulting from conservation of forces when balancing the material and induced stresses in the patch.

The integration of these stresses through the thickness of the patches yields the expressions

(2.9)
$$(M_x)_{pe} = [(M_x)_{pe_1} + (M_x)_{pe_2}]\chi_{pe}(x)$$

$$(N_x)_{pe} = [(N_x)_{pe_1} + (N_x)_{pe_2}]\chi_{pe}(x)S_{1,2}(x)$$

where

$$(M_x)_{pe_1} = -\frac{1}{2}E_{pe}bd_{31}(h + 2T_{b\ell} + T)V_1$$

$$(M_x)_{pe_2} = \frac{1}{2}E_{pe}bd_{31}(h + 2T_{b\ell} + T)V_2$$

$$(N_x)_{pe_1} = -E_{pe}bd_{31}V_1$$

$$(N_x)_{pe_2} = -E_{pe}bd_{31}V_2$$

for the external moments and forces generated by the activation of the patches. The presence of the indicator function

(2.10)
$$S_{1,2}(x) = \begin{cases} 1 & , & x < (x_1 + x_2)/2 \\ 0 & , & x = (x_1 + x_2)/2 \\ -1 & , & x > (x_1 + x_2)/2 \end{cases}$$

results from the fact that for homogeneous patches having uniform thickness, opposite but equal strains are generated about the point $\bar{x} = (x_1 + x_2)/2$.

These expressions can then be substituted directly into the weak equations (2.3) as loads on the beam (with $\hat{q}_n = 0$ and $\hat{N}_x = (N_x)_{pe}, \hat{M}_x = (M_x)_{pe}$). In order to determine the patch loads for the strong form of the beam equations, the corresponding surface moments and forces are found via the relationships

$$\hat{q}_x = -S_{1,2}(x)\frac{\partial(N_x)_{pe}}{\partial x} \quad , \quad \hat{m}_y = -\frac{\partial(M_x)_{pe}}{\partial x}$$

and these latter values are used in (2.1). We point out that this results in the need to differentiate across discontinuities in characteristic and indicator functions (once for the force and twice for the moment) whereas this problem is avoided in the weak formulation since the derivatives are transferred on the test functions. In fact, the effect of the characteristic functions in the latter case is to simply restrict the integrals to the region covered by the patches.

The general moments in the beam component of the structural acoustic system (1.1) can now be described in terms of the internal and external moments just discussed. By combining both the passive and active contributions due to a single pair of patches which are excited out-of-phase, the general moment is given by

$$\mathcal{M} = M_x - (M_x)_{pe}$$

where the internal and external moments are

$$M_x = EI(x)\frac{\partial^2 w}{\partial x^2} + c_D I(x)\frac{\partial^3 w}{\partial x^2 \partial t}$$

$$(M_x)_{pe} = E_{pe} b d_{31}(h + 2T_{b\ell} + T)V\chi_{pe}(x) = \mathcal{K}^B V \chi_{pe}(x)$$

as given in (2.7) and (2.9), respectively (the latter expression is obtained by taking $V = V_1 = -V_2$ in (2.9)). We emphasize that the out-of-phase excitation of the patches produces pure bending moments and hence only transverse vibrations are present in the beam response.

For a system in which s pairs of patches are bonded to a beam and are excited out-of-phase, the beam component of the system (1.1) has the form

$$\rho(x)\frac{\partial^2 w}{\partial t^2} + \frac{\partial^2}{\partial x^2}\left(EI(x)\frac{\partial^2 w}{\partial x^2} + c_D I(x)\frac{\partial^3 w}{\partial x^2 \partial t}\right) + \rho_f \phi_t(w)$$

$$= f + \sum_{i=1}^{s} \mathcal{K}_i^B u_i(t)\frac{d^2}{dx^2}\chi_{pe_i}(x)$$

where $\chi_{pe_i}(x)$ denotes the characteristic function over the ith patch pair and $u_i(t)$ is the voltage into the ith pair. The parameters EI and $c_D I$ are given by

$$EI(x) = E_b \frac{h^3 b}{12} + \sum_{i=1}^{s}\frac{2b}{3}\left[E_{b\ell_i} a_{3b\ell_i} + E_{pe_i} a_{3pe_i}\right]\chi_{pe_i}(x)$$

$$c_D I(x) = c_{Db}\frac{h^3 b}{12} + \sum_{i=1}^{s}\frac{2b}{3}\left[c_{Db\ell_i} a_{3b\ell_i} + c_{Dpe_i} a_{3pe_i}\right]\chi_{pe_i}(x)$$

while the patch parameters are given by $\mathcal{K}_i^B = E_{pe_i} b d_{31_i}(h + 2T_{b\ell_i} + T_i)$ (in these definitions, the bonding layers and patches in the ith pair are considered to have thickness $T_{b\ell_i}$ and T_i, respectively). We note that the discontinuous parameters $\rho, EI, c_D I$ and $\mathcal{K}_i^B$ lead to second derivatives of characteristic functions which causes difficulties in the strong form of the equations. The transfer of these derivatives onto test functions eliminates these problems in the weak form of the equations and is one motivation for using the weak form of the system equations as discussed in the next section.

2.2. Patch Contributions to Plate and Shell Dynamics. In the first part of this section, the contributions from piezoceramic patches to the longitudinal and transverse vibrations of an Euler-Bernoulli beam were examined. It was noted that the patch contributions could be categorized into two types; the first resulted from the structural changes incurred when the patches were bonded to the beam while the second effect was due to the activation of the patches when a voltage was applied. These same types of effects result when piezoceramic patches are bonded to more complex structures such as thin plates or shells.

The motion of a plate differs from that of a beam in that two sets of longitudinal motion are present with the stretching in one coordinate direction related to the contraction in the other through the Poisson ratio ν. In thin shells, the transverse and longitudinal vibrations are coupled due to the underlying curvature of the structure. However, once the underlying dynamic equations in terms of the force and moment resultants are known, the effects due to the presence and activation of the piezoceramic patches can be determined in a manner analogous to that discussed above for thin beams (see [16]).

To illustrate, we consider a thin circular cylindrical shell of radius R, thickness h and having the axial coordinate x as shown in Figure 5. As in the beam discussion, the variable z measures the distance of a point on the structure from the corresponding point on the middle surface ($z = 0$) along the normal to the middle surface.

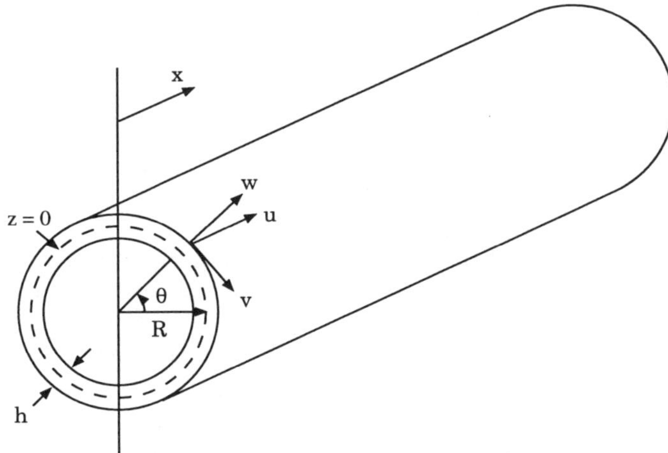

FIG. 5. *The thin cylindrical shell.*

As discussed in [16], the infinitesimal strain relationships for a cylindrical shell are

$$e_x = \varepsilon_x + z\kappa_x$$

$$e_\theta = \frac{1}{1 + z/R}\left(\varepsilon_\theta + z\kappa_\theta\right)$$

$$\gamma_{x\theta} = \frac{1}{1 + z/R}\left[\varepsilon_{x\theta} + z\left(1 + \frac{z}{2R}\right)\tau\right]$$

where e_x and e_θ are normal strains at an arbitrary point within the cylindrical shell and $\gamma_{x\theta}$ is the shear strain. Here ε_x, ε_θ and $\varepsilon_{x\theta}$ are the normal and shear strains in the middle surface and κ_x, κ_θ and τ are the midsurface changes in curvature and midsurface twist (see [23], page 8).

In terms of the axial, tangential and radial displacements u, v and w, respectively, the expressions for the midsurface strains and changes in curvature for the cylindrical shell are

$$\varepsilon_x = \frac{\partial u}{\partial x} \quad , \quad \varepsilon_\theta = \frac{1}{R}\frac{\partial v}{\partial \theta} + \frac{w}{R} \quad , \quad \varepsilon_{x\theta} = \frac{\partial v}{\partial x} + \frac{1}{R}\frac{\partial u}{\partial \theta}$$

$$\kappa_x = -\frac{\partial^2 w}{\partial x^2} \quad , \quad \kappa_\theta = -\frac{1}{R^2}\frac{\partial^2 w}{\partial \theta^2} + \frac{1}{R^2}\frac{\partial v}{\partial \theta} \quad , \quad \tau = -\frac{2}{R}\frac{\partial^2 w}{\partial x \partial \theta} + \frac{2}{R}\frac{\partial v}{\partial x} \; .$$

If a generalized Hooke's law in which stress is assumed to be proportional to a linear combination of strain and strain rate is used as the constitutive relation, the stresses in the shell are given by

$$\sigma_x = \frac{E_s}{1 - \nu_s^2}(e_x + \nu_s e_\theta) + \frac{c_{D_s}}{1 - \nu_s^2}(\dot{e}_x + \nu_s \dot{e}_\theta)$$

$$\sigma_\theta = \frac{E_s}{1 - \nu_s^2}(e_\theta + \nu_s e_x) + \frac{c_{D_s}}{1 - \nu_s^2}(\dot{e}_\theta + \nu_s \dot{e}_x)$$

$$\sigma_{x\theta} = \sigma_{\theta x} = \frac{E_s}{2(1 + \nu_s)}\gamma_{x\theta} + \frac{c_{D_s}}{2(1 + \nu_s)}\dot{\gamma}_{x\theta}$$

where σ_x and σ_θ are normal stresses and $\sigma_{x\theta}$ and $\sigma_{\theta x}$ are tangential shear stresses. The constants E_s, ν_s and c_{D_s} are the Young's modulus, Poisson ratio, and damping coefficient for the shell. Similar relations are found in the bonding layers and patches (see (2.4) for analogous expressions for the beam).

The internal or material moment and force resultants are obtained by integrating the stresses across the thickness of the structure. For patches having thickness T and bonding layers of thickness $T_{b\ell}$, this yields the expressions

$$\left[\begin{array}{c} N_x \\ N_{x\theta} \end{array}\right] = \int_{-h/2-T_{b\ell}-T}^{h/2+T_{b\ell}+T} \left[\begin{array}{c} \sigma_x \\ \sigma_{x\theta} \end{array}\right] \left(1 + \frac{z}{R}\right) dz$$

$$\left[\begin{array}{c} N_\theta \\ N_{\theta x} \end{array}\right] = \int_{-h/2-T_{b\ell}-T}^{h/2+T_{b\ell}+T} \left[\begin{array}{c} \sigma_\theta \\ \sigma_{\theta x} \end{array}\right] dz$$

$$(2.11) \qquad \left[\begin{array}{c} M_x \\ M_{x\theta} \end{array}\right] = \int_{-h/2-T_{b\ell}-T}^{h/2+T_{b\ell}+T} \left[\begin{array}{c} \sigma_x \\ \sigma_{x\theta} \end{array}\right] \left(1 + \frac{z}{R}\right) z\,dz$$

$$\left[\begin{array}{c} M_\theta \\ M_{\theta x} \end{array}\right] = \int_{-h/2-T_{b\ell}-T}^{h/2+T_{b\ell}+T} \left[\begin{array}{c} \sigma_\theta \\ \sigma_{\theta x} \end{array}\right] z\,dz$$

in regions of the structure covered by the patches with similar expressions in those region of the structure consisting solely of shell material (the limits of integration in this latter case are $-h/2$ and $h/2$). Explicit descriptions for these internal moment and force resultants can be found in [16].

In a shell which is excited by the activation of piezoceramic patches, the external moments and forces are due to the in-plane strains

$$e_{pe_1} = (e_x)_{pe_1} = (e_\theta)_{pe_1} = \frac{d_{31}}{T}V_1 \quad , \quad e_{pe_2} = (e_x)_{pe_2} = (e_\theta)_{pe_2} = \frac{d_{31}}{T}V_2$$

which result from the input of the voltages V_1 and V_2 into the outer and inner patches. The resulting external stresses are given by

$$(\sigma_x)_{pe_1} = (\sigma_\theta)_{pe_1} = -\frac{E_1}{1-\nu_1}e_{pe_1} \quad , \quad (\sigma_x)_{pe_2} = (\sigma_\theta)_{pe_2} = -\frac{E_2}{1-\nu_2}e_{pe_2} .$$

For a patch with bounding values x_1, x_2, θ_1 and θ_2 the total external line moments and forces are

$$(M_x)_{pe} = [(M_x)_{pe_1} + (M_x)_{pe_2}]\chi_{pe}(x,\theta)$$

$$(M_\theta)_{pe} = [(M_\theta)_{pe_1} + (M_\theta)_{pe_2}]\chi_{pe}(x,\theta)$$

$$(2.12) \qquad (N_x)_{pe} = [(N_x)_{pe_1} + (N_x)_{pe_2}]\chi_{pe}(x,\theta)S_{1,2}(x)\hat{S}_{1,2}(\theta)$$

$$(N_\theta)_{pe} = [(N_\theta)_{pe_1} + (N_\theta)_{pe_2}]\chi_{pe}(x,\theta)S_{1,2}(x)\hat{S}_{1,2}(\theta)$$

where the indicator function $S_{1,2}(x)$ is defined in (2.10) (with a similar definition for $\hat{S}_{1,2}(\theta)$) and

$$\chi_{pe}(x,\theta) = \left\{\begin{array}{ll} 1 & , \quad x_1 \le x \le x_2 \ , \ \theta_1 \le \theta \le \theta_2 \\ 0 & , \quad \text{otherwise} . \end{array}\right.$$

The individual patch moments are obtained by integrating the external stress distribution through the thickness of the patches in the same manner used in the beam analysis (see [16]).

In order to obtain a strong form of the equations of motion, force and moment balancing can be used to obtain Donnell-Mushtari shell equations

$$R\rho(x,\theta)\frac{\partial^2 u}{\partial t^2} - R\frac{\partial N_x}{\partial x} - \frac{\partial N_{\theta x}}{\partial \theta} = -R\frac{\partial (N_x)_{pe}}{\partial x}S_{1,2}(x)\hat{S}_{1,2}(\theta)$$

$$R\rho(x,\theta)\frac{\partial^2 v}{\partial t^2} - \frac{\partial N_\theta}{\partial \theta} - R\frac{N_{x\theta}}{\partial x} = -\frac{\partial (N_\theta)_{pe}}{\partial \theta}S_{1,2}(x)\hat{S}_{1,2}(\theta)$$

$$R\rho(x,\theta)\frac{\partial^2 w}{\partial t^2} - R\frac{\partial^2 M_x}{\partial x^2} - \frac{1}{R}\frac{\partial^2 M_\theta}{\partial \theta^2} - 2\frac{\partial^2 M_{x\theta}}{\partial x \partial \theta} + N_\theta$$
$$= R\hat{q}_n - R\frac{\partial^2 (M_x)_{pe}}{\partial x} - \frac{1}{R}\frac{\partial (M_\theta)_{pe}}{\partial \theta}$$

(see [16,23] for a more detailed derivation of these equations as well as a discussion concerning the assumptions that are made in obtaining this and other forms of the equations of motion for a thin shell). The contributions due to the patches are incorporated in the internal moments and forces (2.11), the external moments and forces (2.12), and the variable density $\rho(x,\theta)$.

3. Weak Form and Well-Posedness of the Structural Acoustics Model.

As discussed in the last two sections, the incorporation of the piezoceramic patch contributions into the strong form of the modeling system equations leads to first and second derivatives of characteristic functions since both the internal and external moments contain discontinuities at the edges of the patches. This yields an unbounded control input operator and leads to difficulties when approximating the dynamics of the coupled system. To avoid these difficulties, it is advantageous to formulate the problem in weak or variational form (the use of the variational form also permits the use of basis functions having less smoothness than those used when approximating the solution to the strong form of the equations).

3.1. Weak Form of the System Equations.

The state for the second-order form of the 2-D structural acoustics problem is taken to be $z = (\phi, w)$ in the Hilbert space $H = \bar{L}^2(\Omega) \times L^2(\Gamma_0)$ with the energy inner product

$$\left\langle \begin{pmatrix} \phi \\ w \end{pmatrix}, \begin{pmatrix} \xi \\ \eta \end{pmatrix} \right\rangle_H = \int_\Omega \frac{\rho_f}{c^2}\phi\xi \, d\omega + \int_{\Gamma_0} \rho_b w\eta \, d\gamma \ .$$

The choice of the space $\bar{L}^2(\Omega)$, defined as the quotient of $L^2(\Omega)$ over the constant functions, results from the fact that the potentials are determined only up to a constant.

To provide a class of functions which are considered when defining a variational form of the problem, we also define the Hilbert space $V = \bar{H}^1(\Omega) \times H_0^2(\Gamma_0)$ where $\bar{H}^1(\Omega)$ is the quotient space of H^1 over the constant functions and $H_0^2(\Gamma_0)$ is given by $H_0^2(\Gamma_0) = \{\psi \in H^2(\Gamma_0) : \psi(x) = \psi'(x) = 0 \text{ at } x = 0, a\}$. The V inner product is taken as (here and below we use the notation $D = \frac{\partial}{\partial x}$)

$$\left\langle \begin{pmatrix} \phi \\ w \end{pmatrix}, \begin{pmatrix} \xi \\ \eta \end{pmatrix} \right\rangle_V = \int_\Omega \rho_f \nabla \phi \cdot \nabla \xi \, d\omega + \int_{\Gamma_0} EID^2 w D^2 \eta \, d\gamma \, .$$

As discussed in [14], integration in combination with the use of Green's theorem then yields the nonlinear first-order variational form

$$
\begin{aligned}
\text{(3.1)} \quad & \int_\Omega \frac{\rho_f}{c^2} \phi_{tt} \xi \, d\omega \quad + \quad \int_{\Gamma_0} \rho w_{tt} \eta \, d\gamma \\[2ex]
& \qquad + \quad \int_\Omega \rho_f \nabla \phi \cdot \nabla \xi \, d\omega + \int_{\Gamma_0} EID^2 w D^2 \eta \, d\gamma \\[2ex]
& \qquad + \quad \int_{\Gamma_0} \left\{ c_D I D^2 w_t D^2 \eta + \rho_f \left[\phi_t(w)\eta - w_t \xi \right] \right\} d\gamma \\[2ex]
& \qquad = \quad \int_{\Gamma_0} \sum_{i=1}^{s} \mathcal{K}_i^B u_i(t) \chi_{pe_i}(x) D^2 \eta \, d\gamma + \int_{\Gamma_0} f \eta \, d\gamma
\end{aligned}
$$

for all (ξ, η) in V (here $\chi_{pe_i}(x)$ denotes the characteristic function over the ith patch). We note that the nonlinear coupling term can be written as $\phi_t(t, x, w(t, x)) = \phi_t(t, x, 0) + \tilde{\phi}_t(t, x, w(t, x))$ where $\tilde{\phi}_t(t, x, w(t, x)) \equiv \phi_t(t, x, w(t, x)) - \phi_t(t, x, 0)$. We will make use of this decomposition in the abstract formulation of the nonlinear system as a perturbation of a linearized system in our discussion below. Again, a more complete discussion and motivation concerning the formulation of the first-order system in weak form is given in [14].

We point out that in this variational form the derivatives have been transferred from the plate and patch moments onto the test functions. This eliminates the problem of having to approximate the derivatives of the characteristic function and the Dirac delta as is the case with the strong form of the equations.

The system (3.1) can be formally approximated by replacing the state variables by their finite dimensional approximations and constructing the resulting matrix system. Hence it is in a form which is suitable for use in applications. In order to discuss the well-posedness of the model, however, it is advantageous to pose the problem in terms of sesquilinear forms and the bounded operators which they define, and this is the subject of the rest of the section.

3.2. Abstract First-Order Formulation. As motivated by theoretical results in [3,4,6,15], we consider the Gelfand triple $V \hookrightarrow H \simeq H^* \hookrightarrow V^*$ with pivot space H and define sesquilinear forms $\sigma_i : V \times V \to I\!R$, $i = 1, 2$ by

$$\sigma_1(\Phi, \Psi) = \int_\Omega \rho_f \nabla \phi \cdot \nabla \xi d\omega + \int_{\Gamma_0} EID^2 w D^2 \eta d\gamma \ ,$$

$$\sigma_2(\Phi, \Psi) = \int_{\Gamma_0} \{c_D ID^2 w D^2 \eta + \rho_f(\phi \eta - w\xi)\}d\gamma$$

where $\Phi = (\phi, w)$ and $\Psi = (\xi, \eta)$ are in V (see [29] for basic definitions and fundamental functional analysis theory).

As detailed for a similar problem in [15], it is straightforward to show that with these definitions, σ_1 and σ_2 are bounded (there exist c_1 and c_2 such that $|\sigma_1(\Phi, \Psi)| \le c_1 |\Phi|_V |\Psi|_V$ and $|\sigma_2(\Phi, \Psi)| \le c_2 |\Phi|_V |\Psi|_V$), σ_1 is V-elliptic and σ_2 is H-semielliptic (there exist $c > 0$ and $b \ge 0$ such that $\mathrm{Re}\,\sigma_1(\Phi, \Phi) \ge c|\Phi|_V^2$ and $\mathrm{Re}\,\sigma_2(\Phi, \Phi) \ge b|\Phi|_H^2$ for all $\Phi \in V$) and that σ_1 is symmetric ($\sigma_1(\Phi, \Psi) = \sigma_1(\Psi, \Phi)$ for all $\Phi, \Psi \in V$). As a result of the boundedness, we can define operators $A_1, A_2 \in \mathcal{L}(V, V^*)$ by

$$\langle A_i \Phi, \Psi \rangle_{V^*, V} = \sigma_i(\Phi, \Psi)$$

for $i = 1, 2$.

To account for the control contributions, we let U denote the Hilbert space containing the control inputs ($U = I\!R$ in our structural acoustics example), and we define the control operator $B \in \mathcal{L}(U, V^*)$ by

$$\langle Bu, \Psi \rangle_{V^*, V} = \int_{\Gamma_0} \sum_{i=1}^{s} \mathcal{K}_i^B u_i \chi_{pe_i}(x) D^2 \eta d\gamma$$

for $\Psi \in V$, where $\langle \cdot, \cdot \rangle_{V^*, V}$ is the usual duality pairing. Finally, letting $F = (0, f/\rho_b)$ and $G(z, z_t) = (0, -\rho_f \tilde{\phi}_t(w))$ where again, $\tilde{\phi}_t(w) = \tilde{\phi}_t(t, x, w(t, x)) = \phi_t(t, x, w(t, x)) - \phi_t(t, x, 0)$ denotes the nonlinear perturbation to the linear coupling term, we can write the control system in weak or variational form

$$(3.2) \quad \begin{aligned} \langle z_{tt}(t), \Psi \rangle_{V^*, V} &+ \sigma_2(z_t(t), \Psi) + \sigma_1(z(t), \Psi) \\ &= \langle Bu(t) + F(t) + G(z(t), z_t(t)), \Psi \rangle_{V^*, V} \end{aligned}$$

for Ψ in V. This then yields the system

$$z_{tt}(t) + A_2 z_t(t) + A_1 z(t) = Bu(t) + F(t) + G(z(t), z_t(t))$$

in V^*.

To apply infinite dimensional control results for periodic forcing functions to this problem, it is advantageous to write the system in first-order form. This is accomplished by defining the product spaces $\mathcal{H} = V \times H$ and $\mathcal{V} = V \times V$ with the norms

$$|(\Phi, \Psi)|_{\mathcal{H}}^2 = |\Phi|_V^2 + |\Psi|_H^2 \ ,$$

$$|(\Phi, \Psi)|_{\mathcal{V}}^2 = |\Phi|_V^2 + |\Psi|_V^2 \ .$$

We point out that $\mathcal{V} \hookrightarrow \mathcal{H} \simeq \mathcal{H}^* \hookrightarrow \mathcal{V}^*$ again forms a Gelfand triple.

The sesquilinear form $\sigma : \mathcal{V} \times \mathcal{V} \to I\!R$ is then defined by

$$\sigma(\Theta, \chi) = \sigma((\Upsilon, \Lambda), (\Phi, \Psi)) = -\langle \Lambda, \Phi \rangle_V + \sigma_1(\Upsilon, \Psi) + \sigma_2(\Lambda, \Psi)$$

where $\chi = (\Phi, \Psi)$ and $\Theta = (\Upsilon, \Lambda)$.

For the state $\mathcal{Z}(t) = (z(t), z_t(t))$ in $\mathcal{H}$, we can subsequently write the system in the first-order variational form

$$(3.3) \quad \langle \mathcal{Z}_t(t), \chi \rangle_{\mathcal{V}^*, \mathcal{V}} + \sigma(\mathcal{Z}(t), \chi) = \langle \mathcal{B}u(t) + \mathcal{F}(t) + \mathcal{G}(\mathcal{Z}(t)), \chi \rangle_{\mathcal{V}^*, \mathcal{V}}$$

where $\mathcal{F}(t) = (0, F(t)), \mathcal{G}(\mathcal{Z}(t)) = (0, G(z(t), z_t(t)))$ and $\mathcal{B}u(t) = (0, Bu(t))$. As usual, the relation (3.3) must hold for all $\chi \in \mathcal{V}$. Finally, the weak form (3.3) is *formally* equivalent to the system

$$(3.4) \qquad\qquad \mathcal{Z}_t(t) = \mathcal{A}\mathcal{Z}(t) + \mathcal{C}(t, \mathcal{Z}(t))$$

in $\mathcal{V}^*$ where

$$(3.5) \qquad\qquad \mathcal{C}(t, \mathcal{Z}(t)) = \mathcal{B}u(t) + \mathcal{F}(t) + \mathcal{G}(\mathcal{Z}(t))$$

and

$$\operatorname{dom} \mathcal{A} = \{ \Theta = (\Upsilon, \Lambda) \in \mathcal{H} : \Lambda \in V, A_1 \Upsilon + A_2 \Lambda \in H \}$$

$$(3.6) \qquad\qquad \mathcal{A} = \begin{bmatrix} 0 & I \\ -A_1 & -A_2 \end{bmatrix} .$$

3.3. Model Well-Posedness. In the previous discussion, the weak form of the coupled structural acoustic equations was written as an abstract first-order semilinear initial value problem with a state in $\mathcal{H}$. The nonlinear forcing term $\mathcal{C}(t, \mathcal{Z}(t)) = \mathcal{B}u(t) + \mathcal{F}(t) + \mathcal{G}(\mathcal{Z}(t))$ however lies in $\mathcal{V}^*$ rather than $\mathcal{H}$ since the control term $B \in \mathcal{L}(U, V^*)$ defines the product space control term $\mathcal{B}u(t) = (0, Bu(t)) \in \{0\} \times V^* \subset V \times V^* = \mathcal{V}^*$. Hence the standard theory for abstract semilinear Cauchy problems does not apply directly, and the first step in the following discussion is the outline for arguments which can be used to extend the operator $\mathcal{A}$ to a space where the theory does apply. A more extensive discussion concerning the well-posedness of a linear problem of this type can be found in [15] and details for the following arguments can be found in that work.

The first step in determining the well-posedness of the system model is to argue that $\mathcal{A}$ generates a C_0-semigroup on $\mathcal{H}$. As noted earlier, the sesquilinear form σ_1 is V-elliptic, continuous and symmetric while σ_2 is continuous and H-semielliptic. From the Lumer-Philips theorem (with further arguments found in [1] and pages 82-84 of [4]) this then implies that the operator $\mathcal{A}$ defined in (3.6) generates a C_0-semigroup on the state space $\mathcal{H}$. Moreover, the semigroup satisfies the exponential bound $|\mathcal{T}(t)| \leq e^{\omega t}$ for $t \geq 0$ (where in fact, $\omega = 0$ due to the fact that $\mathcal{A}$ is dissipative as shown in [4]).

Since $\mathcal{B}u(t)$ lies in V^* rather than $\mathcal{H}$, the next step is to extend the semigroup $\mathcal{T}(t)$ on $\mathcal{H}$ to a semigroup $\tilde{\mathcal{T}}(t)$ on a larger space $\mathcal{W}^* \supset \{0\} \times V^*$ so as to be compatible with the forcing term (this is accomplished using "extrapolation space" ideas and arguments similar to those presented in [6,7,22]).

As detailed in [15], the space of interest is defined in terms of $\mathrm{dom}\,\mathcal{A}^*$ where

$$\mathrm{dom}\,\mathcal{A}^* = \{\chi = (\Phi, \Psi) \in \mathcal{H} | \Psi \in V, A_1^*\Phi - A_2^*\Psi \in H\}$$

$$\mathcal{A}^*\chi = \begin{pmatrix} -\Psi \\ A_1^*\Phi - A_2^*\Psi \end{pmatrix}.$$

Specifically, the space $\mathcal{W} = [\mathrm{dom}\,\mathcal{A}^*]$ is taken to be $\mathrm{dom}\,\mathcal{A}^*$ with the inner product

$$\langle \Phi, \Psi \rangle_{\mathcal{W}} = \langle (\lambda_0 - \mathcal{A}^*)\Phi, (\lambda_0 - \mathcal{A}^*)\Psi \rangle_{\mathcal{H}}$$

for some arbitrary but fixed λ_0 with $\lambda_0 > \omega$ (recall that the original solution semigroup satisfies the bound $|\mathcal{T}(t)| \leq e^{\omega t}$). As proven in [7], the resulting $\mathcal{W}$ norm is equivalent to the graph norm corresponding to $\mathcal{A}^*$. Moreover, we have that $\{0\} \times V^* \subset \mathcal{W}^* = [\mathrm{dom}\,\mathcal{A}^*]^*$ (see [15] for details).

From the definition of $\mathcal{A}^*$ and the equivalence of the $\mathcal{W}$ norm with the graph norm corresponding to $\mathcal{A}^*$, we can define $\tilde{\mathcal{A}}\Theta \in \mathcal{W}^*$ by

$$\left(\tilde{\mathcal{A}}\Theta \right)(\chi) = \langle \Theta, \mathcal{A}^*\chi \rangle_{\mathcal{H}}$$

for all $\Theta \in \mathcal{H}$, $\chi \in \mathcal{W}$. With this definition and the Riesz representation theorem, it is shown in [15] that $\tilde{\mathcal{A}}$ is an extension of the original operator $\mathcal{A}$ from $\mathrm{dom}\,\mathcal{A} \subset \mathcal{H}$ to all of $\mathcal{H}$. Finally, as proven in [7], the operator $\tilde{\mathcal{A}}$ is the infinitesimal generator of a C_0-semigroup $\tilde{\mathcal{T}}(t)$ on $\mathcal{W}^*$ which is an extension of $\mathcal{T}(t)$ from $\mathcal{H}$ to $\mathcal{W}^*$.

In the corresponding linear problem, under reasonable regularity conditions on $t \mapsto u(t)$ and $t \mapsto F(t)$, one can immediately argue the existence of a unique strong solution to the system in terms of the extended semigroup $\tilde{\mathcal{T}}(t)$. For the semilinear problem of interest, however, the nonlinear nonhomogeneous terms must satisfy certain continuity criteria in order to obtain similar results. For example, if we let X denote the reflexive Banach space $\mathcal{W}^*$ and assume that $\mathcal{C} : [0, T] \times X \to X$ defined in (3.5) is continuous in t on $[0, T]$ and uniformly Lipschitz continuous on X, then the integral equation

$$(3.7) \quad \mathcal{Z}(t) = \tilde{T}(t)\mathcal{Z}_0 + \int_0^t \tilde{T}(t-s) \left(\begin{array}{c} 0 \\ Bu(s) + F(s) + G(\mathcal{Z}(s)) \end{array} \right) ds$$

is well-defined for $Bu + F + G(\mathcal{Z}) \in L^2((0,T), V^*)$. Moreover, for $\mathcal{Z}(0) = \mathcal{Z}_0$, the solution $\mathcal{Z}(t)$ of (3.7) is a unique mild solution to (3.4) (see Theorem 1.2, page 184 of [27]). In addition, if $C : [0, T] \times X \to X$ is Lipschitz continuous in both variables, then it follows from Theorem 1.6, page 189 of [27] that (3.7) provides the strong solution to (3.4) interpreted in the $\mathcal{W}^*$ sense.

The required continuity of the nonhomogeneous terms Bu and F is demonstrated in [15] and hence the remaining question concerns the Lipschitz continuity of the nonlinear coupling term $G(z, z_t) = (0, -\rho_f \tilde{\phi}_t(w))$. If we assume that the input terms F and Bu are sufficiently smooth so as to assure the necessary continuity in $G(z, z_t)$, then our open loop nonlinear system is well-posed.

3.4. Well-Posedness of the Closed Loop System. The arguments leading to the well-posedness results for the linear and nonlinear open loop models can also be extended to the closed loop systems which result when the gains determined for a corresponding LQR problem are fed back into the system. In determining these gains, the perturbing force $\mathcal{F}$ is assumed to be periodic (this is a reasonable assumption since $\mathcal{F}$ models the exterior noise which in this problem is generated by the revolution of turboprop or turbofan blades).

Discussing first the *linearized* problem, the periodic LQR problem consists of finding $u \in L^2(0, \tau; U)$ which minimizes a quadratic cost functional of the form

$$J(u) = \frac{1}{2} \int_0^\tau \left\{ \langle Q\mathcal{Z}(t), \mathcal{Z}(t) \rangle_{\mathcal{H}} + \langle Ru(t), u(t) \rangle_U \right\} dt$$

subject to $\mathcal{Z}_t(t) = \mathcal{A}\mathcal{Z}(t) + \mathcal{B}u(t) + \mathcal{F}(t)$ with $\mathcal{Z}(0) = \mathcal{Z}(\tau)$. Since $\mathcal{Z} = (\phi, w, \phi_t, w_t)$, the operator Q can be chosen so as to emphasize the minimization of particular state variables as well as to create windows that can be used to decrease state variations of certain frequencies. The control space U is taken to be $I\!R^s$ if s patches are used in the model, and it is assumed that the operator $R \in \mathcal{L}(U)$ is an $s \times s$ diagonal matrix where $r_{ii} > 0, i = 1, \cdots, s$ is the weight on the controlling voltage into the ith patch. In the case that $\mathcal{B}$ is bounded on $\mathcal{H}$, a complete feedback theory for this periodic problem can be given as discussed in [19]. This theory can be extended to also include the case of unbounded $\mathcal{B}$, i.e., $\mathcal{B} \in \mathcal{L}(U, V^*)$, of interest here (see [5]). Under usual stabilizability and detectability assumptions on the system as well as standard assumptions on Q, the optimal control is given by

$$(3.8) \qquad u(t) = -R^{-1}\mathcal{B}^*[\Pi\mathcal{Z}(t) - r(t)]$$

where $\Pi \in \mathcal{L}(\mathcal{V}^*, V)$ is the unique nonnegative self-adjoint solution of the algebraic Riccati equation

$$(3.9) \qquad \mathcal{A}^*\Pi + \Pi\mathcal{A} - \Pi\mathcal{B}R^{-1}\mathcal{B}^*\Pi + \mathcal{Q} = 0 \ .$$

Here r is the unique τ-periodic solution of

$$(3.10) \qquad \dot{r}(t) = -(\mathcal{A}^* - \Pi\mathcal{B}R^{-1}\mathcal{B}^*)r(t) + \Pi\mathcal{F}(t)$$

and the optimal trajectory $\mathcal{Z}$ is the solution of

$$(3.11) \qquad \dot{\mathcal{Z}}(t) = (\mathcal{A} - \mathcal{B}R^{-1}\mathcal{B}^*\Pi)\mathcal{Z}(t) + \mathcal{B}R^{-1}\mathcal{B}^*r(t) + \mathcal{F}(t) \ .$$

As discussed in [5] for the case when $\mathcal{B} \in \mathcal{L}(U, \mathcal{V}^*)$, one also finds that the operator $\mathcal{A} - \mathcal{B}R^{-1}\mathcal{B}^*\Pi$ generates an exponentially stable C_0-semigroup $\mathcal{S}(t)$ on the state space $\mathcal{H}$. From Corollary 10.6, page 41 of [27], this implies that $\mathcal{A}^* - \Pi\mathcal{B}R^{-1}\mathcal{B}^*$ generates the corresponding adjoint semigroup $\mathcal{S}^*(t)$ on $\mathcal{H}^* \simeq \mathcal{H}$. The semigroup $\mathcal{S}(t)$ can then be extended through the extrapolation space techniques just discussed to a larger space $\widetilde{\mathcal{W}}^* \supset \{0\} \times V^*$, and with reasonable regularity assumptions on $t \mapsto F(t)$, this implies the existence of solutions to the tracking equation (3.10) and closed loop system (3.11) for $r(0) = r_0$ and $\mathcal{Z}(0) = \mathcal{Z}_0$.

As discussed in greater detail in the next section where the corresponding finite dimensional control problem is considered, an effective strategy for controlling the original nonlinear system is to determine the gains for the linearized model and feed these back into the nonlinear system. This then yields the nonlinear closed loop system

$$\dot{\mathcal{Z}}(t) = (\mathcal{A} - \mathcal{B}R^{-1}\mathcal{B}^*\Pi)\mathcal{Z}(t) + \mathcal{B}R^{-1}\mathcal{B}^*r(t) + \mathcal{F}(t) + \mathcal{G}((\mathcal{Z}(t))$$

where again, $\mathcal{A} - \mathcal{B}R^{-1}\mathcal{B}^*\Pi$ generates the C_0-semigroup $\mathcal{S}(t)$ which can then be extended to $\widetilde{\mathcal{W}}^*$. With the assumption that the input term F is sufficiently smooth so as to assure the necessary continuity in nonhomogeneous terms, the closed loop nonlinear system is also well-posed.

4. System Approximation and the Finite Dimensional Control Problem. The discussion thus far has centered around the infinite dimensional model for the structural acoustic system as well as issues concerning its well-posedness. However, in order to develop viable schemes for approximating the nonlinear system dynamics, estimating physical parameters, and determining control gains, appropriate finite dimensional approximations to the state variables w and ϕ must be developed. For reasons discussed in [3], a Galerkin scheme was chosen and the potential and beam displacement were discretized in terms of spline and spectral expansions, respectively.

4.1. System Approximation. A tensored Legendre basis was used for the discretization of the acoustic velocity potential. Letting $P_i^a(x)$ and $P_i^\ell(y)$ denote the standard Legendre polynomials that have been scaled by transformation to the intervals $[0, a]$ and $[0, \ell]$, respectively, the basis functions $\{B_{ij}^m\}$ for the cavity were then defined as

$$B_{ij}^m(x, y) = P_i^a(x)P_j^\ell(y) \quad \text{for} \quad i = 0, 1, \cdots, m_x, \quad j = 0, 1, \cdots, m_y, \quad i + j \neq 0 ,$$

where $m = (m_x + 1) \cdot (m_y + 1) - 1$. The condition $i + j \neq 0$ eliminates the constant function thus guaranteeing that the set of functions is suitable as a basis for the quotient space. The m dimensional cavity approximating subspace is taken to be $H_c^m = span\,\{B_i^m\}_{i=1}^m$ and the approximate cavity solution is given by

$$\begin{aligned}
\phi^N(t, x, y) &= \sum_{i=1}^{m} \phi_i^N(t)B_i^m(x, y) \\[2mm]
&= \sum_{j=0}^{m_y} \sum_{\substack{i=0 \\ i+j\neq0}}^{m_x} \tilde{\phi}_{ij}^N(t)P_i^a(x)P_j^\ell(y) .
\end{aligned}$$

Cubic splines were used as a basis for the beam displacement since they satisfy the smoothness requirement as well as being easily implemented when adapting to the fixed-end boundary conditions and patch discretizations. Letting $\{B_i^n\}_{i=1}^{n-1}$ denote the cubic splines which have been modified to satisfy the boundary conditions (see [3,14] for details), the corresponding $n - 1$ dimensional beam approximating subspace is given by $H_b^n = span\,\{B_i^n\}_{i=1}^{n-1}$ and the approximate beam solution is taken to be

$$w^N(t, x) = \sum_{i=1}^{n-1} w_i^N(t)B_i^n(x) .$$

The approximating state space was then taken to be $H^N = H_c^m \times H_b^n$ where $N = m + n - 1$, and the product space for the first order system is $\mathcal{H}^N = H^N \times H^N$. By restricting the infinite dimensional system (3.1) to $\mathcal{H}^N \times \mathcal{H}^N$, one obtains the nonlinear finite dimensional system

$$M^N \dot{y}^N(t) = \tilde{A}^N\left(y^N(t)\right) + \tilde{B}^N u(t) + \tilde{F}^N(t)$$
$$M^N y^N(0) = \tilde{y}_0^N$$

or equivalently

$$(4.1) \qquad \begin{aligned}
\dot{y}^N(t) &= A^N\left(y^N(t)\right) + B^N u(t) + F^N(t) \\[2mm]
y^N(0) &= \tilde{y}_0^N .
\end{aligned}$$

Explicit descriptions of the mass and stiffness operators M^N and $\tilde{A}^N(y^N(t))$ as well as detailed definitions of the control matrix $\tilde{B}^N$ and the force vector

$\tilde{F}^N(t)$ can be found in [3,13]. The vector $y^N(t) = (\phi_1^N(t), \cdots, \phi_m^N(t), w_1^N(t),$ $\cdots, w_{n-1}^N(t), \dot{\phi}_1^N(t), \cdots, \dot{\phi}_m^N(t), \dot{w}_1^N(t), \cdots, \dot{w}_{n-1}^N(t))^T$ contains the $2N \times 1$ approximate state coefficients while $u(t) = (u_1(t), \cdots u_s(t))^T$ contains the s control variables. As detailed in [14], the nonlinearity in the operator $\tilde{\mathcal{A}}^N(y^N(t))$ manifests itself in the dependence of the operator on the unknown coefficients $\{w_j(t)\}$.

4.2. The Finite Dimensional Control Problem. Due to the nonlinearity in the infinite dimensional system (3.1) and hence the finite dimensional matrix system (4.1), LQR feedback control results for problems with periodic forcing terms can not be directly applied as there were in [3]. Instead, the following strategy was adopted. The infinite dimensional system was linearized by replacing the nonlinear coupling term $\phi_t(t, x, w(t, x))$ by its linear component $\phi_t(t, x, 0)$ (this is equivalent to taking $G(z(t), z_t(t)) = 0$ in (3.2) or $(\mathcal{G}(\mathcal{Z}(t)) = 0$ in (3.3) or (3.4)). This linearization is motivated by the assumption of small beam displacements which is inherent in the Euler-Bernoulli theory (for physically reasonable input forces, the beam displacements are of the order $10^{-5}m$ for the geometries of interest). The feedback gains for this approximate linearized system were calculated from a periodic LQR theory (see [3]) and were then fed back into the nonlinear problem to create a stable nonlinear closed loop control system.

To illustrate this control strategy, the LQR theory for problems with periodic input terms is briefly outlined. The resulting gains are then applied to the nonlinear problem of interest with the results being illustrated in an example.

Linear Periodic Control Problem

As discussed in [3], the approximation of the nonlinear coupling term $\phi_t(t, x, w(t, x))$ by its linear component, and the projection of the resulting system into the finite dimensional subspace $\mathcal{H}^N \times \mathcal{H}^N$ yields the linear finite dimensional Cauchy equation

$$\dot{y}^N(t) = A^N y^N(t) + B^N u(t) + F^N(t)$$
(4.2)
$$y^N(0) = y_0^N$$

(this system can also be obtained by restricting the infinite dimensional system (3.2) with $G(z(t), z_t(t)) = 0$ to $\mathcal{H}^N \times \mathcal{H}^N$). The components of the linear stiffness matrix can be found in [3].

The periodic finite dimensional control problem is then to find $u \in L^2(0, \tau)$ which minimizes

$$J^N(u) = \frac{1}{2} \int_0^\tau \left\{ \langle Q^N y^N(t), y^N(t) \rangle_{I\!R^N} + \langle Ru(t), u(t) \rangle_{I\!R^s} \right\} dt \, , \quad N = m+n-1$$

where y^N solves (4.2), τ is the period, R is an $s \times s$ diagonal matrix and $r_{ii} > 0, i = 1, \cdots, s$ is the weight or penalty on the controlling voltage into the ith patch.

The nonnegative definite matrix Q^N is chosen in a manner so as to to emphasize the minimization of particular state variables. From energy considerations as discussed in [3], an appropriate choice for Q^N in this case is

$$Q^N = M^N \mathcal{D}$$

where M^N is the mass matrix, and the diagonal matrix $\mathcal{D}$ is given by

$$\mathcal{D} = diag\left[d_1 I^m, d_2 I^{n-1}, d_3 I^m, d_4 I^{n-1}\right] \ .$$

Here I^k, $k = m, n-1$, denotes a $k \times k$ identity and the parameters d_i are chosen to enhance stability and performance of the feedback.

The optimal control is then given by

$$u^N(t) = R^{-1}(B^N)^T \left[r^N(t) - \Pi^N y^N(t)\right]$$

where Π^N is the solution to the algebraic Riccati equation

$$(4.3) \qquad (A^N)^T \Pi^N + \Pi^N A^N - \Pi^N B^N R^{-1}(B^N)^T \Pi^N + Q^N = 0 \ .$$

For the regulator problem with periodic forcing function $F^N(t)$, $r^N(t)$ must satisfy the linear differential equation

$$(4.4) \qquad \begin{aligned} \dot{r}^N(t) &= -\left[A^N - B^N R^{-1}(B^N)^T \Pi^N\right]^T r^N(t) + \Pi^N F^N(t) \\ r^N(0) &= r^N(\tau) \end{aligned}$$

while the optimal trajectory is the solution to the linear differential equation

$$\dot{y}^N(t) = \left[A^N - B^N R^{-1}(B^N)^T \Pi^N\right] y^N(t) + B^N R^{-1}(B^N)^T r^N(t) + F^N(t)$$

$$y^N(0) = y^N(\tau) \ .$$

The finite dimensional optimal control, Riccati solution, tracking equation and closed loop system can be compared with the original infinite dimensional relations given in (3.8), (3.9), (3.10) and (3.11), respectively. In order to guarantee the convergence $\Pi^N \to \Pi$, $r^N \to r$, and hence the convergence of $u^N \to u$, it is sufficient to impose various conditions on the original and approximate systems. These hypotheses include convergence requirements for the uncontrolled problem as well as the requirement that the approximation systems preserve stabilizability and detectability margins uniformly. A fully developed theory (see [5]) is available for the case when $\mathcal{F} \equiv 0$ (in this case the tracking variable r does not appear in the solution) even when $\mathcal{B}$ is unbounded. However, the theory in [5] requires strong damping in the second-order system whereas the only damping in our system is the strong Kelvin-Voigt damping in the beam (damping in the cavity was omitted due to the relatively small dimensions involved).

Although the convergence theory of [5] does not directly apply here, numerical tests indicate that convergence is obtained even though this system contains only weak or boundary damping.

Nonlinear Control Problem

To extend these results to the nonlinear system of interest, the linear gains were calculated and fed back into the nonlinear system (4.1), thus yielding the suboptimal control

$$u^N(t) = R^{-1}(B^N)^T \left[r^N(t) - \Pi^N y^N(t) \right]$$

and the closed loop system

$$\dot{y}^N(t) = \mathcal{A}^N \left(y^N(t) \right) - B^N R^{-1}(B^N)^T \Pi^N y^N(t) + B^N R^{-1}(B^N)^T r^N(t) + F^N(t)$$

$$y^N(0) = y^N(\tau) \ .$$

The Riccati matrix Π^N and tracking vector $r^N(t)$ are solutions to (4.3) and (4.4) which arise when formulating the corresponding LQR problem.

Example: Nonlinear Control

To illustrate the dynamics and effects of feedback control on a nonlinear system modeling a 2-D analogue of a 3-D experimental setup, a .6 m by 1 m cavity with a flexible beam at one end was considered (see Figure 6). The beam was assumed to have width and thickness .1 m and .005 m, respectively, and the Young's modulus and beam density were taken to be $E = 7.1 \times 10^{10} \ N/m^2$ and $\rho_b = 2700 \ kg/m^3$. This yielded the stiffness parameter $EI = 73.96 \ Nm^2$ and linear mass density $\rho = 1.35 \ kg/m$. The damping parameter for the beam was chosen to be $c_D I = .001 \ kg \, m^3/sec$. The speed of sound and atmospheric density inside the cavity were taken to be $c = 343 \ m/sec$ and $\rho_f = 1.21 \ kg/m^3$, respectively.

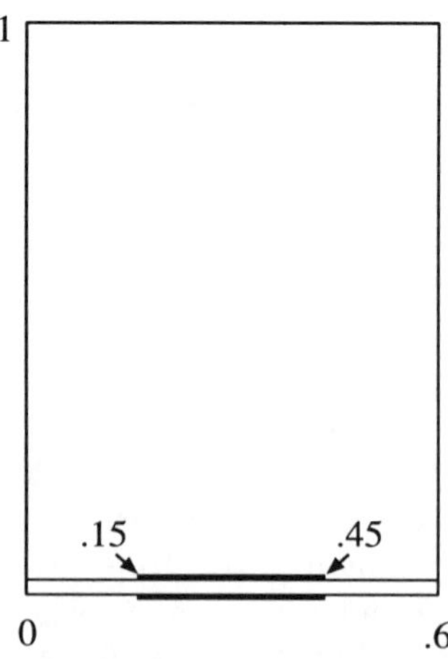

FIG. 6. *Acoustic cavity with one centered 1/2 length patch.*

Several forcing functions modeling uniform (in space) periodic exterior sound sources were considered. In this example, the forcing function was taken to be $f(t, x) = 2.04 \sin(470\pi t)$ which models a periodic plane wave with a root mean square (rms) sound pressure level of 117 dB. The frequency of 235 hertz is approximately halfway between the first and fourth natural frequencies of the system (as shown in [11], these occur at 65.9 hertz and 387.8 hertz, respectively).

The dynamics of the uncontrolled system were approximated using 80 cavity basis functions ($m_x = m_y = 8$) and 11 beam basis functions ($n = 12$). The time interval of interest was taken to be $[0, 16/235]$ which admitted 16 periods of the driving frequency, and time histories of the beam displacement at $X = .3$ and cavity pressure at $X = .3, Y = .1$ on this temporal interval are plotted in Figures 7 and 8.

The frequency plots of the uncontrolled beam displacement and cavity pressure in Figures 9 and 10 exhibit not only the driving frequency but also transient responses at $65.9, 181.6, 345.2, 387.7$ and 519.5 hertz which are due to the natural frequencies of the coupled system (see [11] for a complete discussion of the dynamics and natural frequencies for the corresponding linearized system). In particular, the high energy response at 181.6 hertz indicates a strong excitation of the system at what corresponds to the frequency for the first mode of the uncoupled cavity (care must be taken when describing the dynamics of the system in terms of the undamped beam and cavity modes since the nonlinear coupling and beam damping yield system responses which differ somewhat from those of the isolated components). The presence of the multiple frequencies can also be seen in the time history plots of the uncontrolled beam displacement and cavity pressure in Figures 7 and 8.

Control was then implemented by using Potter's method to calculate the gains for the linearized system and feeding them back into the nonlinear system as discussed previously. The following results were obtained with an out-of-phase single pair (so as to create pure bending moments) of centered patches covering one half of the beam length as shown in Figure 6. The quadratic cost functional parameters were taken to be $d_1 = d_2 = d_4 = 1, d_3 = 10^4$ and $R = 10^{-6}$ with d_3 chosen to have larger magnitude so as to more heavily penalize large pressure variations.

Figure 11 contains a plot of the controlling voltage $u(t)$. As expected, it is periodic, and the magnitude remains below $25\,V$ which is a physically reasonable voltage to apply to the piezoceramic patches.

We point out that the application of the controlling voltage resulted in a high frequency transient response and 168 cavity basis functions ($m_x = m_y = 12$) and 15 beam basis functions ($n = 16$) were needed to resolve the controlled system dynamics.

From Figures 7 and 8, it can be seen that the controlled responses undergo a transient phase of approximately three periods and then are maintained at a low magnitude throughout the rest of the time interval.

By calculating the rms pressure levels, it was determined that at the point $(X, Y) = (.3, .1)$, the uncontrolled sound pressure level is 82.8 dB whereas the controlled sound pressure is reduced 15.7 dB to 67.1 dB. The level of reduction becomes even more significant as one moves deeper into the cavity since the strong cavity excitation in the uncontrolled case yields high magnitude pressure oscillations near the back wall which are uniformly reduced by the application of the controlling voltage. Finally, it is noted that the relative reduction in pressure is more significant than the reduction in beam displacement. This is due to the heavier penalization of pressure fluctuations through the choices $d_2 = 1$ and $d_3 = 10^4$.

The frequency plots of the controlled responses (in Figures 9 and 10) show that the dominant response is now at the driving frequency of 235 hertz. They also demonstrate the presence of high frequency transient responses which are much more significant than those found in the uncontrolled case. This indicates that the interior pressure oscillations are reduced through two mechanisms when the controlling voltage is applied; the first is due to the reduced magnitude of the beam displacements while the second is due to the excitation of high frequency beam oscillations which couple less readily with the interior acoustic field. The combination of the two results in significantly reduced interior sound pressure levels.

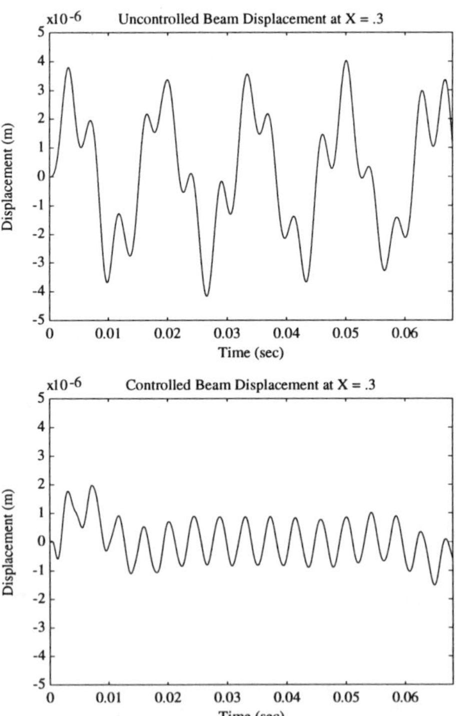

FIG. 7. *Uncontrolled and controlled beam displacement at $X = .3$.*

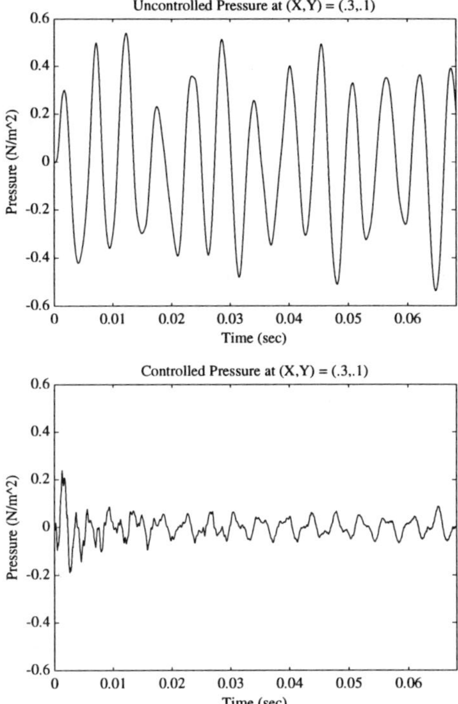

FIG. 8. *Uncontrolled and controlled acoustic pressure at* $(X, Y) = (.3, .1)$

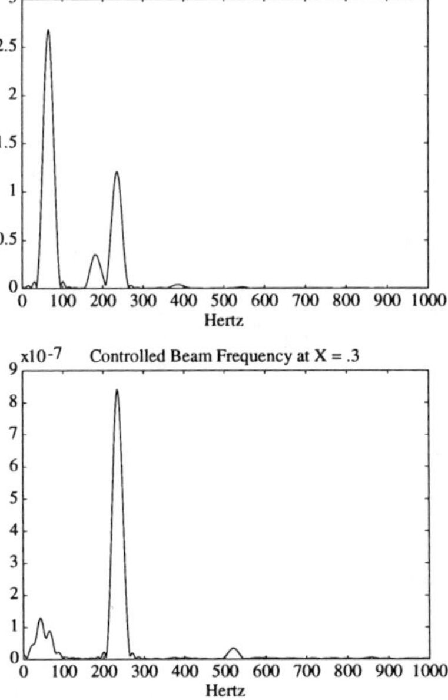

FIG. 9. *Uncontrolled and controlled frequencies calculated on the beam.*

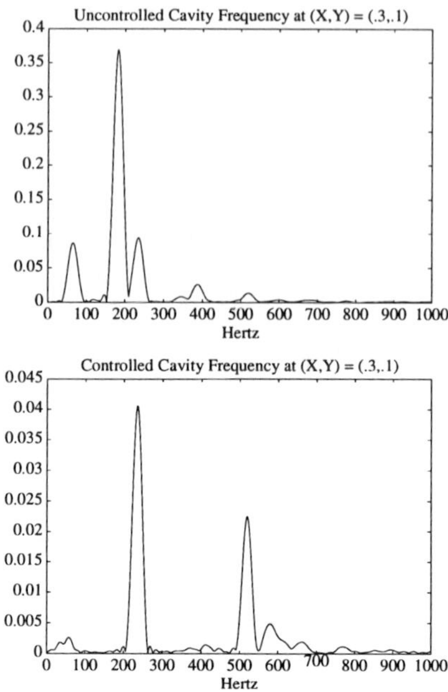

FIG. 10. *Uncontrolled and controlled frequencies calculated in the cavity.*

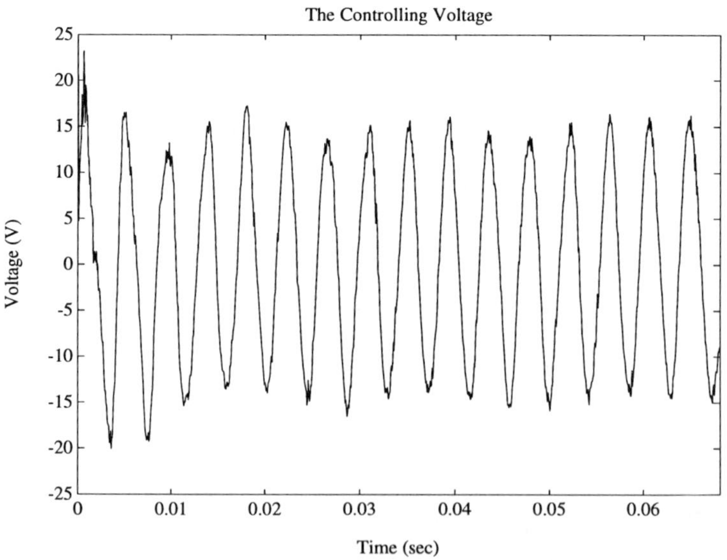

FIG. 11. *The controlling voltage $u(t)$.*

5. Conclusion. In this paper, we have discussed several of the issues which are involved in using piezoceramic patches as actuators in a nonlinear structural acoustics application. The patches affect the dynamics of the coupled system by contributing external forces and moments to the structure when a voltage is applied, and the first part of the discussion is centered around a description of the interactions between the patches and an Euler-Bernoulli beam and a thin cylindrical shell. In this discussion, care was taken to distinguish between the passive (material) contributions, due to the added thickness and differing material properties of the patch and bonding layer, and the active (external) contributions which result from the strains which are produced when a voltage is applied to the patches.

As a result of the differing material properties and presence of the piezoceramic patches, the material and control parameters of the combined structure are piecewise constant in nature and hence lead to discontinuities in the moment and force resultants. This leads to difficulties in the strong form of the system equations when the moments are differentiated and is one motivation for using the weak or variational form where the derivatives are transferred onto the test functions. The weak form is also advantageous for many approximation schemes since it reduces the smoothness requirements for the basis elements. Finally, well-posedness issues were considered by posing the weak form in the context of sesquilinear forms.

Due to the nonlinearities arising in the coupling between the beam vibrations and the interior acoustic field, LQR feedback control results could not be directly applied to the problem. Instead, gains corresponding to the linearized problem were calculated and fed back into the nonlinear system. As demonstrated by the results in the example as well as the more extensive set of examples in [14], this strategy is very effective for this problem. This is partly due to the weakness of the nonlinearity. By comparing the nonlinear results reported here and in [14] with the corresponding linear ones in [11], one can see that qualitatively, the two sets agree closely. This can be explained by the fact that the beam displacements are very small and hence the linearized coupling terms quite accurately approximate the true nonlinear expressions.

ACKNOWLEDGEMENTS:
The authors would like to thank H.C. Lester and R.J. Silcox of the Acoustics Division, NASA Langley Research Center, for numerous discussions concerning the structural acoustic component of this work.

REFERENCES

[1] H.T. Banks and J.A. Burns, *Introduction to Control of Distributed Parameter Systems,* Birkhäuser, to appear.

[2] H.T. Banks and F. Fakhroo, "Legendre-Tau Approximations for LQR Feedback Control of Acoustic Pressure Fields," Center for Research in Scientific Computation Technical Report, CRSC-TR92-5, North Carolina State University, *Journal of Mathematical Systems, Estimation and Control,* submitted.

[3] H.T. Banks, W. Fang, R.J. Silcox and R.C. Smith, "Approximation Methods for Control of Acoustic/Structure Models with Piezoceramic Actuators," *Journal of Intelligent Material Systems and Structures,* 4(1), 1993, pp. 98-116.

[4] H.T. Banks and K. Ito, "A Unified Framework for Approximation in Inverse Problems for Distributed Parameter Systems," in *Control-Theory and Advanced Technology* 4(1), 1988, pp. 73-90.

[5] H.T. Banks and K. Ito, "Approximation in LQR Problems for Infinite Dimensional Systems with Unbounded Input Operators," SIAM Conference on Control in the 90's, San Francisco, May 1989, preprint.

[6] H.T. Banks, K. Ito and Y. Wang, "Well-Posedness for Damped Second Order Systems with Unbounded Input Operators," Center for Research in Scientific Computation Technical Report, CRSC-TR93-10, North Carolina State University, *Differential and Integral Equations,* submitted.

[7] H.T. Banks, F. Kappel and C. Wang, "Weak Solutions and Differentiability for Size Structured Population Models," in *Int. Series of Num. Math.,* Vol. 100, Birkhäuser, 1991, pp. 35-50.

[8] H.T. Banks, S.L. Keeling and R.J. Silcox, "Optimal Control Techniques for Active Noise Suppression," Proc. 27th IEEE Conf. on Decision and Control, Austin, Texas, 1988, pp. 2006-2011.

[9] H.T. Banks, S.L. Keeling, R.J. Silcox and C. Wang, "Linear Quadratic Tracking Problems in Hilbert Space: Applications to Optimal Active Noise Suppression," in Proc. 5th IFAC Symp. on Control of DPS, (A. El-Jai and M. Amouroux, eds.), Perpignan, 1989, pp. 17-22.

[10] H.T. Banks, S.L. Keeling and C. Wang, "Linear Quadratic Tracking Problems in Infinite Dimensional Hilbert Spaces and a Finite Dimensional Approximation Framework," LCDC/CCS Rep. 88-28, October, 1988, Brown University.

[11] H.T. Banks, R.J. Silcox and R.C. Smith, "The Modeling and Control of Acoustic/Structure Interaction Problems via Piezoceramic Actuators: 2-D Numerical Examples," *ASME Journal of Vibration and Acoustics,* to appear.

[12] H.T. Banks and R.C. Smith, "Models for Control in Smart Material Structures," *Identification and Control in Systems Governed by Partial Differential Equations,* SIAM, Philadelphia, 1993, pp. 26-44.

[13] H.T. Banks and R.C. Smith, "Feedback Control of Noise in a 2-D Nonlinear Structural Acoustics Model," *Control-Theory and Advanced Technology,* submitted.

[14] H.T. Banks and R.C. Smith, "Parameter Estimation in a Structural Acoustic System with Fully Nonlinear Coupling Conditions," *Inverse Problems,* submitted.

[15] H.T. Banks and R.C. Smith, "Well-Posedness of a Model for Structural Acoustic Coupling in a Cavity Enclosed by a Thin Cylindrical Shell," ICASE Report 93-10, *Journal of Mathematical Analysis and Applications,* to appear.

[16] H.T. Banks, R.C. Smith and Y. Wang, "Modeling Aspects for Piezoceramic Patch Activation of Shells, Plates and Beams," Center for Research in Scientific Computation Technical Report, CRSC-TR92-12, North Carolina State University, *Quarterly of Applied Mathematics*, to appear.

[17] H.T. Banks, Y. Wang, D.J. Inman and J.C. Slater, "Variable Coefficient Distributed Parameter System Models for Structures with Piezoceramic Actuators and Sensors," Proc. of the 31st Conf. on Decision and Control, Tucson, AZ, Dec. 16-18, 1992, pp. 1803–1808.

[18] A.J. Bullmore, P.A. Nelson, A.R.D. Curtis and S.J. Elliott, "The Active Minimization of Harmonic Enclosed Sound Fields, Part II: A Computer Simulation," *Journal of Sound and Vibration*, **117**(1), 1987, pp. 15-33.

[19] G. Da Prato, "Synthesis of Optimal Control for an Infinite Dimensional Periodic Problem," *SIAM J. Control Opt.*, **25**, 1987, pp. 706-714.

[20] J.J. Dosch, D.J. Inman and E. Garcia, "A Self-Sensing Piezoelectric Actuator for Collocated Control," *Journal of Intelligent Material Systems and Structures*, **3**, 1992, pp. 166-185.

[21] S.J. Elliott, A.R.D. Curtis, A.J. Bullmore and P.A. Nelson, "The Active Minimization of Harmonic Enclosed Sound Fields, Part III: Experimental Verification," *Journal of Sound and Vibration*, **117**(1), 1987, pp. 35-58.

[22] A. Haraux, "Linear Semigroups in Banach Spaces," in *Semigroups, Theory and Applications, II* (H. Brezis, et al., eds.), Pitman Res. Notes in Math, Vol 152, Longman, London, 1986, pp. 93-135.

[23] A.W. Leissa, *Vibration of Shells*, NASA SP-288, 1973.

[24] H.C. Lester and C.R. Fuller, "Active Control of Propeller Induced Noise Fields Inside a Flexible Cylinder," AIAA Tenth Aeroacoustics Conference, Seattle, WA, 1986.

[25] S. Markuš, *The Mechanics of Vibrations of Cylindrical Shells*, Elsevier, New York, 1988.

[26] P.A. Nelson, A.R.D. Curtis, S.J. Elliott and A.J. Bullmore, "The Active Minimization of Harmonic Enclosed Sound Fields, Part I: Theory," *Journal of Sound and Vibration*, **117**(1), 1987, pp. 1-13.

[27] A. Pazy, *Semigroups of Linear Operators and Applications to Partial Differential Equations*, Springer-Verlag, New York, 1983.

[28] R.J. Silcox, H.C. Lester and S.B. Abler, "An Evaluation of Active Noise Control in a Cylindrical Shell," NASA Technical Memorandum 89090, February 1987.

[29] J. Wloka, *Partial Differential Equations*, Cambridge University Press, New York, 1987.

ON THE PRESENCE OF SHOCKS IN DOMAIN OPTIMIZATION OF EULER FLOWS

J.T. BORGGAARD*

Abstract. In this paper we consider a shape optimization problem for a 1-D Euler flow. We show that for problems with shocks, the use of high order CFD schemes can produce artificial local minima in the approximate cost functional. These local minima can cause optimization algorithms to fail. We illustrate this phenomenon, show how hybrid algorithms may be constructed to overcome this problem and speculate on potential difficulties that may occur in more complex situations.

1. Introduction and motivation. The use of domain optimization techniques in the design of fluid flow systems has shown great promise in many areas of application. In this paper, we focus on domain optimization problems which involve shocks. An example of a problem of this type is the optimal forebody-simulator design problem. A forebody-simulator (FBS) is a device that is shaped and attached to a jet engine in order to produce a flow that "simulates" the flow that would result from a full aircraft forebody. The optimization problem is to find the shape of this FBS which will provide flow to the engine inlet which is as close as possible to the flow which the engine would receive in flight. This problem is described in detail in papers by Huddleston [6] and Borggaard, Burns, Cliff and Gunzburger [1].

Fluid flow systems which are modeled by Euler equations are interesting since they are of mixed type, which can lead to discontinuities (or shocks) in their flow solutions. The existence of shocks produces some interesting difficulties in the resulting optimization problem, which is primarily caused by the choice of a numerical approximation. To understand this behavior, we study a simple model which displays the same features as the optimal FBS design problem.

The model problem consists of a steady-state Euler flow in a 1-D duct with variable cross-sectional area. The goal of the optimal design problem is to find the cross-sectional area that minimizes the distance between the flow and a desired flow. With the proper choice of inlet and outlet conditions and constraints on the variation of the duct cross-sectional area, this problem contains a shock. Although this problem is complex enough to capture the difficulties presented by shocks in the flow, the 1-D problem can be solved analytically. Consequently, the cost functional can easily be computed and there are several numerical schemes which can be used to solve this

---

* This work was supported in part by the Air Force Office of Scientific Research under grant AFOSR F49620-92-J-0078. The author would like to express his appreciation to Dr. John A. Burns for his helpful suggestions. Department of Mathematics, Interdisciplinary Center for Applied Mathematics, Virginia Polytechnic Institute and State University, Blacksburg, Virginia 24061.

problem. This model problem has been used by Frank and Shubin [5] in their study of optimal design.

As noted above, steady state Euler equations produce interesting behavior when used in an optimization scheme. The basic problem comes from trying to "match" discontinuous flow solutions. We shall see that in theory the minimum is very distinct, i.e. the cost functional has a clear global minimum caused by the penalty of not matching shock locations. However, using numerical methods to solve the minimization problem can cause local minima in the approximate cost functional. Thus, optimization strategies which are used to solve this class of problems have to account for this phenomena.

The remainder of this paper is organized as follows. We present a description of the model problem in the next section. This 1-D problem may be found in the paper by Frank and Shubin [5]. However, for completeness we present a description of the reduction of the Euler equations to a single ordinary differential equation and give the solution. The optimal design problem is presented and approximated by using numerical schemes to solve the Euler equations. We show how the approximate cost functional is affected by different numerical approximation schemes and demonstrate how using a particular numerical method (which accurately models the shock) can lead to negative results when used in conjunction with an optimization algorithm. A hybrid optimal design algorithm is presented which produces an optimal design and avoids the problem caused by the local minima. Finally, we summarize our findings and discuss the potential applicability to more complex problems, such as the optimal FBS design problem.

2. Model problem description.

2.1. One dimensional Euler equations.
Although the formulation presented below may be found in [5], it is included here for completeness and to introduce notation. Assuming a steady, inviscid flow in a duct (see Figure 2.1) where the flow variables (ρ, u, e and p denoting density, velocity, internal energy and pressure) depend only on the length along the duct, the balance laws produce the following form of the Euler equations;

$$(2.1) \qquad \rho u A = \text{constant} \equiv C$$

$$(2.2) \qquad \left[\rho u^2 A \right]_\xi = -p_\xi A$$

$$(2.3) \qquad \left[\rho e + \frac{1}{2}\rho u^2 + p \right] u A = \text{constant} \equiv L,$$

where $A(\xi)$ is the cross-sectional area as shown in Figure 2.1. This set of equations can be reduced to a single ordinary differential equation for the velocity by substituting equations (2.1), (2.3) and the ideal gas law,

$$(2.4) \qquad p = (\gamma - 1)\rho e, \qquad (\gamma = 1.4 \text{ for air})$$

into the momentum balance equation (2.2). The resulting equation is

$$(2.5) \qquad \left[u + \frac{\bar{H}}{u} \right]_\xi + \frac{A_\xi}{A}\left(\bar{\gamma} u - \frac{\bar{H}}{u} \right) = 0,$$

where $\bar{\gamma} = (\gamma - 1)/(\gamma + 1)$ and $\bar{H} = 2\bar{\gamma}L/C$. Defining

$$(2.6) \qquad f(u) = u + \frac{\bar{H}}{u} \quad \text{and} \quad g(u, A) = \frac{A_\xi}{A}\left(\bar{\gamma} u - \frac{\bar{H}}{u} \right),$$

we describe the state u, given a cross-sectional area $A(\cdot)$ (with $A_\xi > 0$), as the solution to the following two-point boundary value problem:

$$[f(u)]_\xi + g(u, A) = 0$$

$$(2.7) \qquad u(0) = u_{in} \quad \text{and} \quad u(1) = u_{out}.$$

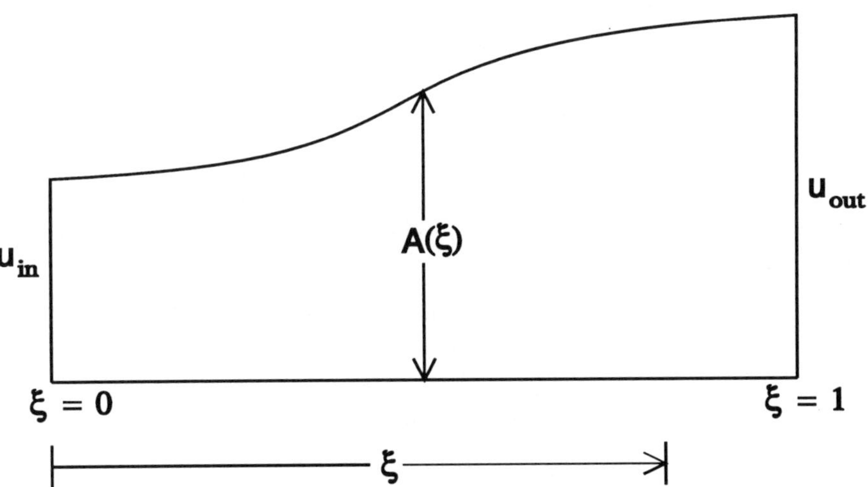

FIG. 2.1. *1-D Duct with Varying Cross-Sectional Area*

When u is smooth, equation (2.5) can be written as

$$(2.8) \qquad Au\left(\frac{\bar{H}}{\bar{\gamma}} - u^2 \right)^{\frac{1}{\gamma-1}} = K,$$

38   J.T. BORGGAARD

where K is a constant. Since $A_\xi > 0$, it follows from the theory of hyperbolic equations ([2] and [7]) that the geometry of a diffusing duct ($A_\xi > 0$) can produce at most one (normal) shock. Thus, equation (2.8) can be used to determine the flow on either side of the shock (by using the boundary conditions to determine the values of the constants K_{in} and K_{out}). All that remains is to determine the location of the shock.

Since the flow is steady, the shock speed a is zero. This implies that

$$(2.9) \qquad a(u_*) \equiv \frac{df}{du}(u_*) = 1 - \frac{\bar{H}}{u_*^2} = 0, \quad (\Longrightarrow u_* = \sqrt{\bar{H}})$$

where u_* is the speed of sound. Applying the Rankine-Hugoniot condition [7] yields

$$(2.10) \qquad u_l u_r = \bar{H},$$

where u_l and u_r represent the limiting velocity from the left and from the right of the shock, respectively. The value of the shock location, ξ_*, can be determined by solving the equation

$$(2.11) \qquad \frac{1}{K_{in}} u_l \left(\frac{\bar{H}}{\bar{\gamma}} - u_l^2\right)^{\frac{1}{\gamma-1}} = A^{-1}(\xi_*) = \frac{1}{K_{out}} u_r \left(\frac{\bar{H}}{\bar{\gamma}} - u_r^2\right)^{\frac{1}{\gamma-1}}$$

along with (2.10), for u_l and u_r. Applying (2.8), $A(\xi_*)$ can be found and since $A_\xi > 0$ it follows that one can solve for ξ_*.

Consider the solution of equation (2.7) with

$$(2.12) \qquad A(\xi) = 1.05 + (1.745 - 1.05)\xi - 0.09\sin(2\pi\xi),$$

$\bar{\gamma} = 0.4/2.4$, $\bar{H} = 1.14$, $u_{in} = 1.299$ and $u_{out} = 0.506$. Using the technique described above, we get $A(\xi_*) = 1.3705$ which leads to $\xi_* = 0.4786$ (using MATLAB to invert equation (2.12)). The resulting solution for $\hat{u}$ is plotted in Figure 2.2.

2.2. Optimization problem. Let $\mathcal{A} = \{A : [0,1] \to \mathbb{R} \mid A \in C^1(0,1), A_\xi(\xi) > 0\}$ and define $\mathcal{J} : \mathcal{A} \to \mathbb{R}$ by

$$(2.13) \qquad \mathcal{J}(A) = \int_0^1 [u(x; A(\cdot)) - \hat{u}(x)]^2 dx$$

where $u(x; A(\cdot))$ is the solution of the boundary value problem (2.7). The optimal design problem is to find $A^* \in \mathcal{A}$ such that

$$\mathcal{J}(A^*) \leq \mathcal{J}(A) \text{ for all } A \in \mathcal{A}.$$

This is now an infinite dimensional optimal control problem and one could use the theory of distributed parameter control to attack this problem.

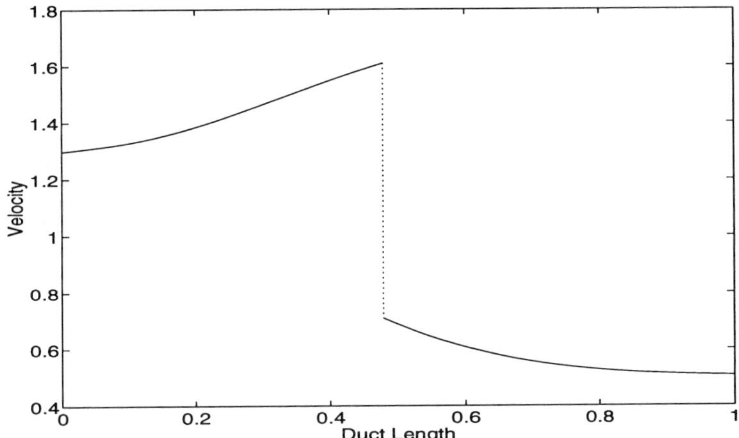

FIG. 2.2. *Exact Solution of* 1-D *Varying Duct Problem*

However, we shall restrict our attention to the case (as is common in practice) where $A(\cdot)$ has been parameterized. This leads to the problem of minimizing $\mathcal{J}$ over a finite dimensional subspace $\mathcal{B} \subseteq \mathcal{A}$.

In our example, we consider a subspace of the Bernstein-Bezier quadratic polynomials, $\mathcal{B}^2$. Bernstein-Bezier polynomials possess several nice properties when used in approximations. The most important for us are the *convex hull* and *endpoint interpolation* properties (see e.g. [4]). These properties allow us to satisfy the monotonicity requirement and match the inflow and outflow cross-sectional area easily. Thus we look at optimization over

$$(2.14) \quad \mathcal{B} \;=\; \big\{ B : [0,1] \to \mathrm{IR} \,|\, B = 1.05 B_0^2 + b_p B_1^2 + 1.745 B_2^2; \\ b_p \in (1.05, 1.745) \big\}$$

where

$$B_i^r(\xi) = \binom{r}{i} \xi^i (1-\xi)^{r-i}, \quad \xi \in [0,1].$$

In particular, we optimize over a one parameter family of C^1 curves.

It is also the case that (as in the optimal FBS problem) given any $A \in \mathcal{B}$, it may not be possible to analytically solve (2.7) for u. Consequently, one must consider numerical approximations, such as finite difference methods, for solving (2.7). In any practical problem one must consider a discrete analog of the optimization problem above.

The discretized optimization problem now becomes: Find $A^* \in \mathcal{B}$ to minimize

$$(2.15) \qquad \mathcal{J}^N(A) = \mathcal{J}^N(b_p) = \frac{1}{N} \sum_{i=1}^{N} |u_i^N(A) - \hat{u}_i^N|^2$$

where $u_i^N(A)$ is a numerical approximation of $u(A)$ at discrete points in $[0,1]$ and $\hat{u}_i^N$ is discrete data. For our problem, u_i^N will come from finite difference solutions of (2.7) discussed in Section 3, and $\hat{u}_i^N$ will correspond to points taken from the curve given in Figure 2.2. Note that we assume that experimental data is given at the finite difference mesh points. If this were not true, then some type of interpolation must be used.

3. Numerical results.

3.1. Introduction. In this section, we study the phenomenon of local minima in the approximate cost functional, $\mathcal{J}^N$. We consider three numerical schemes [5] for finding approximations u^N of the boundary value problem (2.7). These are the Godunov, the Enquist-Osher and the artificial viscosity methods. The methods discretize the interval $[0,1]$ into N cells of length $h \equiv 1/N$, with centers, $\xi_j = h(j - 1/2)$, $j = 1, \ldots, N$. The flow velocity u is modeled as a constant over each of these cells (u_j^N is the constant associated with the jth cell). In all of these methods, u^N is found as the root of a system of nonlinear equations,

$$(3.1) \qquad \frac{f_{j+1/2} - f_{j-1/2}}{h} + g_j = 0 \qquad j = 1, \ldots, N$$

where $f_j = f(u_j^N)$ and $g_j = g(u_j^N, A(\xi_j))$. The three methods differ in how the flux $f_{j+1/2}$ is determined from f_j, f_{j+1} and $f(u_*)$. The Godunov method uses the formula

$$f_{j+1/2} = \begin{cases} f_{j+1} & u_j, u_{j+1} < u_*; \\ f_j & u_j, u_{j+1} > u_*; \\ f(u_*) & u_j < u_* < u_{j+1}; \\ \max(f_j, f_{j+1}) & u_{j+1} < u_* < u_j, \end{cases}$$

the Enquist-Osher method uses

$$f_{j+1/2} = \begin{cases} f_{j+1} & u_j, u_{j+1} < u_*; \\ f_j & u_j, u_{j+1} > u_*; \\ f(u_*) & u_j < u_* < u_{j+1}; \\ f_j + f_{j+1} - f(u_*) & u_{j+1} < u_* < u_j, \end{cases}$$

and the artificial viscosity method uses

$$f_{j+1/2} = 1/2 \left\{ f_{j+1} + f_j - u_{j+1} + u_j \right\}.$$

These three methods were applied to (2.7) where the cross-sectional area was given by equation (2.12). Plots of these solutions for $N = 45$

along with the exact solution are shown in Figure 3.1. Of the three methods considered here, the Godunov method provides the better approximation to the discontinuous solution $\hat{u}$. The Enquist-Osher method gives similar results except there is one more mesh point in the transition region between the supersonic and subsonic flow. The artificial viscosity method, on the other hand, has significant error in this region.

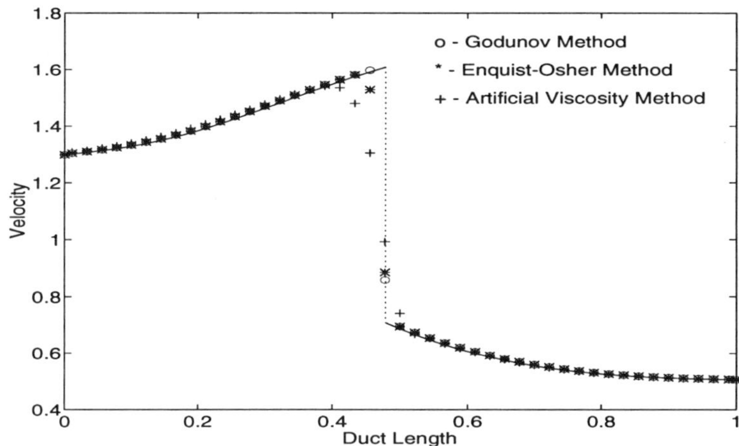

FIG. 3.1. *Numerical Solutions of* 1-*D Varying Duct Problem*

As shown in the next sections, the approximate cost functional $\mathcal{J}^N$, which is obtained using the Godunov method to solve (2.7) for u^N, contains (numerically generated) local minima, while the approximate cost functionals obtained using the Enquist-Osher and artificial viscosity methods do not. Furthermore, an example is presented where cascading the Godunov method into an optimization algorithm does not produce the global minimum. In addition, a hybrid scheme is presented which circumvents the optimization problems caused by the local minima and achieves the global minimum.

3.2. Cost functionals. Two levels of approximation were used to obtain $\mathcal{J}^N$ from $\mathcal{J}$. The first was to replace the integral operator in $\mathcal{J}$ by the sum of terms $[u(\xi_i; A(\cdot)) - \hat{u}(\xi_i)]^2$ weighted by $N(= \frac{1}{h})$. The second was to replace $u(\xi_i; A(\cdot))$ by u_i^N and $\hat{u}(\xi_i)$ by $\hat{u}_i^N$. The accuracy of the second level of approximation is solely determined by the particular choice of the numerical scheme (for this problem, there is no error in replacing $\hat{u}(\xi_i)$ by $\hat{u}_i^N$). However, it is the first level of approximation which can introduce the interesting behavior in the approximate cost functionals.

We demonstrate this fact by plotting the cost functional $\mathcal{J}$ vs. the Bezier parameter b_p in Figure 3.2. The cost functional $\mathcal{J}$ contains a well defined unique minimum due to the large penalty associated with not matching the shock locations. Figure 3.3 contains plots of $\mathcal{J}^N$ vs. b_p (for a few values of N) where u^N consists of (mesh) point evaluations of the exact solution to (2.7). Therefore, only the first level of approximation is used. We see that the resulting approximate cost functionals contain steps which are clearly local minima when b_p is to the left of the global minimum, b_p^*.

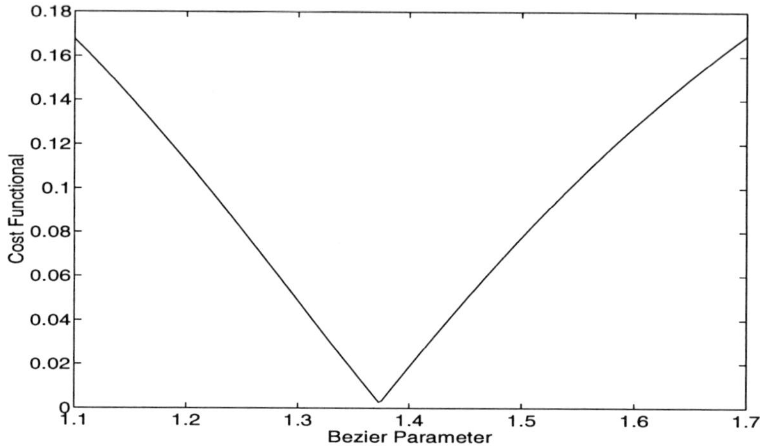

FIG. 3.2. *Cost Functional with Exact Flow Solution*

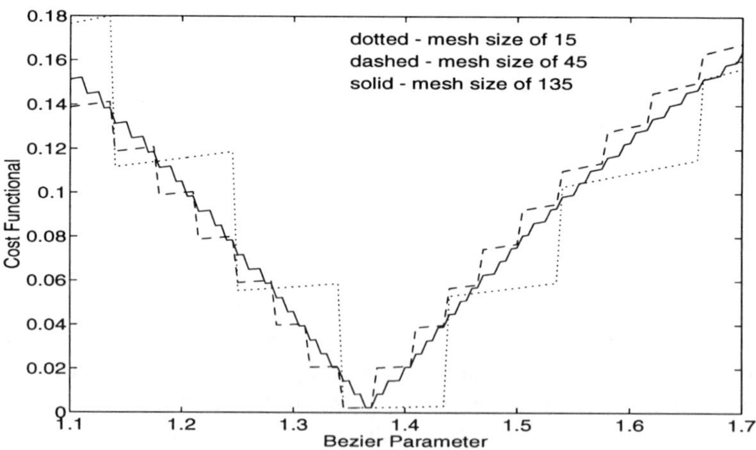

FIG. 3.3. *Approximate Cost Functional with Exact Flow Solution*

The reason for the existence of these steps (and hence the local minima) in the approximate cost functional may be explained as follows. The approximate cost functional $\mathcal{J}^N$ contains a weighted sum of terms of the form $[u(\xi_i; A) - \hat{u}(\xi_i)]^2$. If $u(\xi_i; A)$ and $\hat{u}(\xi_i)$ represent flow on different sides of the shock, i.e. if $u(\xi_i; A)$ is supersonic and $\hat{u}(\xi_i)$ is subsonic, or visa versa, then the ith term in the sum adds a large contribution to $\mathcal{J}^N$. Whereas, if $u(\xi_i; A)$ and $\hat{u}(\xi_i)$ represent flow on the same side of the shock, i.e. either both are supersonic or both are subsonic, then the ith term is not nearly as significant. As the Bezier parameter b_p varies, the location of the shock in u also varies. Hence, as the shock location passes through ξ_i, $u(\xi_i; A)$ will jump from supersonic to subsonic, or visa versa. This will significantly change the value of the ith term in the summation. It is this change which causes the steps.

This reasoning also leads to the conclusion that performing the first level of approximation of $\mathcal{J}$ by a Riemann sum

$$\sum_{i=1}^{N} c_i \left[u(x_i; A) - \hat{u}(x_i) \right]^2 ,$$

where $\{c_i\}$ are weights and $\{x_i\}$ is any "practical" distribution of points in $[0,1]$ (i.e. not all clustered near the endpoints), produces the same behavior in $\mathcal{J}^N$. Remarkably, this includes Gaussian integration rules of arbitrary accuracy.

We turn now to the second level of approximation, where $u(\xi_i; A)$ is replaced by $u_i^N(A)$, one of the numerical approximations to the boundary value problem (2.7). As shown in Figure 3.1, there is some approximation error near the shock using any of these methods. Consequently, there are values of u^N which lie between the supersonic and subsonic flow curves. The terms in the summation which correspond to these points do not have the same dramatic change in value when the shock lies just on either side of them. Therefore, this approximation error in u^N tends to smooth out $\mathcal{J}^N$.

The approximate cost functionals $\mathcal{J}^{15}$ and $\mathcal{J}^{45}$ obtained using the Godunov scheme are shown in Figure 3.4. We note that there are local minima to the left of the global minimum. This is similar to the behavior in Figure 3.3, except the approximate cost functional is much "smoother". For the cases when the Enquist-Osher and artificial viscosity methods are used to compute $\mathcal{J}^{15}$ and $\mathcal{J}^{45}$ (plotted in Figures 3.5 and 3.6, respectively), these artificial local minima do not occur. The fact that there are more mesh points connecting the supersonic and the subsonic flow curves account for the smooth approximate cost functionals.

3.3. Optimization results. In this section, we present optimization results using a "black box" method [5]. This method couples the evaluation of $\mathcal{J}^N$ described in Section 2.2 with a BFGS based quasi-Newton minimization algorithm using finite difference approximations to evaluate the gradient [3]. This method is applied to the optimization problem (2.15).

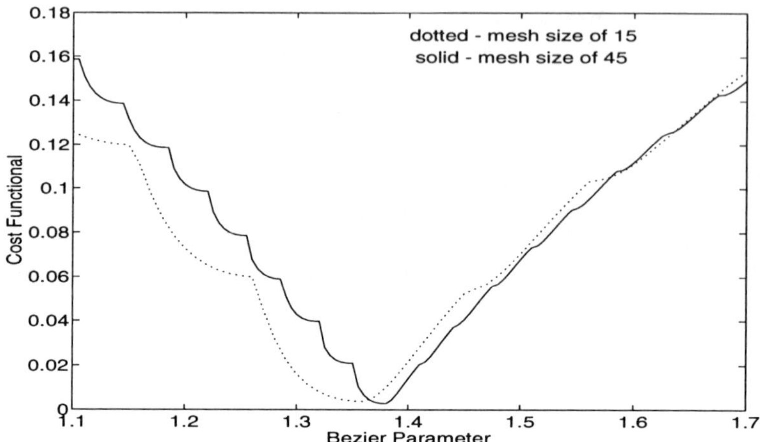

FIG. 3.4. *Approximate Cost Functional - Godunov Scheme*

TABLE 3.1
Minimization using Godunov Method

Iteration	Bezier Parameter	Objective Function	Gradient
0	1.13000	6.24486	-6.92614
1	1.13937	6.20974	-1.19887
2	1.14133	6.20818	-0.40852
3	1.14235	6.20795	-0.04043
4	1.14246	6.20795	-0.00156
5	1.14246	6.20795	-0.00001
6	1.14246	6.20795	0.00000

This "black box" method is used to try to find the Bezier parameter b_p^* which uniquely describes the cross-sectional area $A^* \in \mathcal{B}$. We will compare the convergence of minimization schemes when the cost functional $\mathcal{J}^{45}$ is computed using numerical solutions u_i^{45} obtained using both the Godunov and the Enquist-Osher methods.

We saw in the last section that when the cost functional is computed using the Godunov method, there were a large number of local minima. Observe that when an iteration is started with an initial guess for the Bezier parameter of 1.13 (near a local minimum), the iteration stalls at a local minimum since the gradient goes to zero. This iteration history is presented in Table 3.1 above.

When the iteration is started to the right of this local minima, e.g. an initial Bezier parameter of 1.18, we find that the iteration converges (see Table 3.2) to the global minimum seen in Figure 3.4. However, the

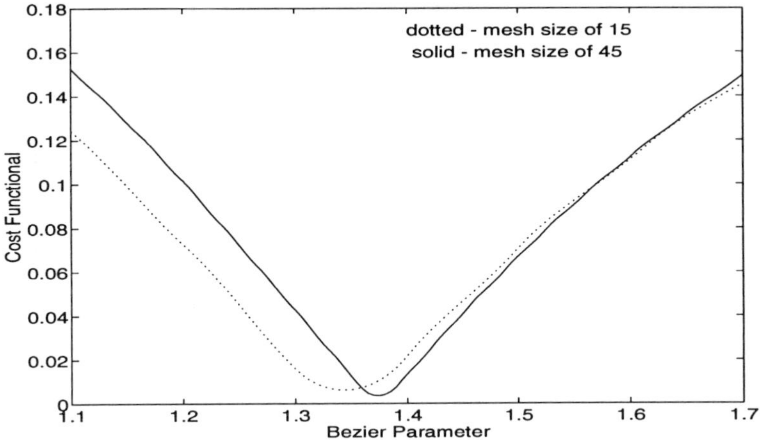

FIG. 3.5. *Approximate Cost Functional - Enquist-Osher Scheme*

TABLE 3.2
Minimization using Godunov Method

Iteration	Bezier Parameter	Objective Function	Gradient
0	1.18000	5.28376	-89.36653
1	1.19309	4.60160	-26.30958
2	1.19855	4.49142	-14.93566
3	1.20572	4.41886	-6.18889
6	1.21580	4.29234	-85.33195
11	1.24494	3.49924	-2.23058
22	1.30412	1.76934	-5.01563
33	1.35864	1.70586	-11.89953
44	1.37125	0.10889	0.43637

convergence is slow, taking 44 steps. The slow convergence is caused by the algorithm having to pass over the "steps" in the approximate cost functional.

Fast convergence is observed, however, when we compute the approximate cost functional using the Enquist-Osher method. This is expected since, as we saw in Figure 3.5, the extra point in the shock region removes the local minima from $\mathcal{J}^N$. Starting at the same initial point of 1.13 which caused the Godunov method to fail, we find that convergence is reached in just 6 iterations. The iteration history using the Enquist-Osher method in our "black box" scheme is presented in Table 3.3 below. It is important to note that although this convergence is rapid, the value of the global minimizer for this approximate cost functional $\mathcal{J}^{45}$ is still not the best estimate

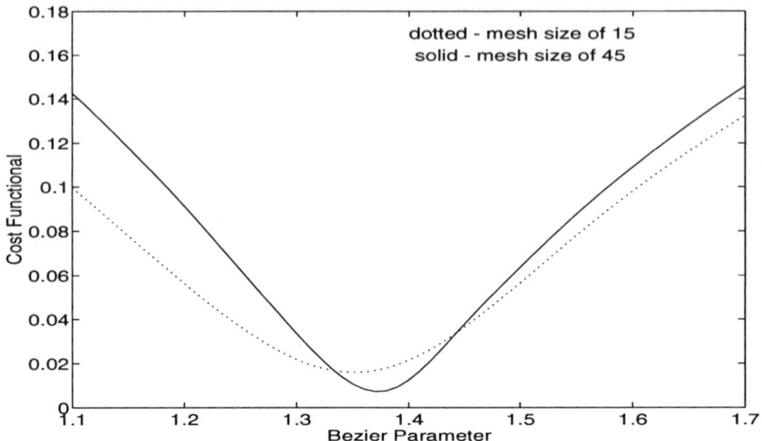

FIG. 3.6. *Approximate Cost Functional - Artificial Viscosity Scheme*

TABLE 3.3
Minimization using Enquist-Osher Method

Iteration	Bezier Parameter	Objective Function	Gradient
0	1.13000	6.15394	-23.80509
1	1.33000	0.90358	-27.61271
2	1.37000	0.15503	5.46034
3	1.36340	0.14660	-2.92231
4	1.36357	0.14329	0.02460
5	1.36568	0.14329	0.00040
6	1.36568	0.14329	0.00040

of b_p^* (the minimum of $\mathcal{J}$) we can achieve. As we saw above (and in Figure 3.4) the Godunov method produces a more accurate approximation of $\mathcal{J}$.

3.4. Hybrid optimization scheme. Examining the approximate cost functionals in Figure 3.3, we see that erroneous results could be obtained from optimization schemes which would use a perfect numerical solution of the boundary value problem (2.7). We have also seen that using the Godunov scheme to evaluate the approximate cost functional can lead to problems if we are unlucky enough to start near a local minimum.

Here we propose a strategy designed to overcome the problem when approximate cost functionals have artificial local minima as those in Figures 3.3 or 3.4. We begin the iteration computing $\mathcal{J}^N$ using a numerical scheme which uses enough dissipation to "smear" the shock over several grid points. This would allow us to converge quickly to a region which is close to the minimum. At this point, we use the more accurate estimate of u^N in $\mathcal{J}^N$

TABLE 3.4

Minimization using Hybrid Method

Iteration	Bezier Parameter	Objective Function	Gradient
0	1.13000	6.15394	-23.80509
1	1.33000	0.90358	-27.61271
2	1.37000	0.15503	5.46034
3	1.36340	0.14660	-2.92231
0	1.36340	0.12991	-5.72003
1	1.37050	0.10929	-0.70580
2	1.37150	0.10917	0.18529
3	1.37077	0.10911	-0.57659
4	1.37094	0.10902	-0.49796
5	1.37138	0.10899	1.21808
6	1.37107	0.10896	-0.44042
7	1.37115	0.10892	-0.40346
8	1.37123	0.10889	-0.36974
9	1.37126	0.10889	0.49282
10	1.37125	0.10889	-0.36076
11	1.37125	0.10889	0.46915
12	1.37125	0.10889	0.16183

and continue the optimization. This hybrid method will work only if (as is the case here) the first method produces a global minimum close enough to the global minimum of the second approximation so that using this as a starting value avoids local minima and converges to a more accurate global minimum.

An example of this hybrid method is presented below. The scheme switches from the Enquist-Osher to the Godunov method when the change in $\mathcal{J}^N$ is less than 0.01. The optimization results are given in Table 3.4 below. We see that starting with an initial guess of 1.13, this algorithm converged to the optimum value achieved in Table 3.2. The first 3 iterations used the Enquist-Osher method and then switched to the Godunov method for the remaining 12 iterations.

4. Comments. Shape optimization problems for systems with shocks can produce unexpected results. We illustrated one of these features by studying a particular Euler flow using a model 1-D steady state duct problem. Numerical approximations of the cost functional coupled with discontinuous solutions (shocks) arising from the Euler equations, produced artificial local minima in the approximate cost functional, as we expected. It was observed that these local minima are more pronounced when the numerical approximation scheme predicts the shock more accurately. Also, the existence of these numerically generated local minima caused a domain optimization scheme to fail to converge to the global minimum. The point of this example is to show that applying optimization loops to "accurate" Euler flow solvers may produce designs that are not optimal.

We demonstrated that Euler flow solvers which have sufficient numerical dissipation do not produce these artificial minima. This was used to construct a hybrid algorithm which used a flow solver containing numerical dissipation to produce a better initial guess for an optimization algorithm based on a more accurate flow solver. This hybrid method requires that one can find an initial guess which is close enough to the minimum so that the numerically generated local minima are not encountered.

However, other strategies for avoiding the local minima problem can be considered. A natural approach is to find a good initial guess by first optimizing on a coarse mesh. In addition to the issue of efficiency, one should encounter fewer local minima. As noted above, introduction of implicit numerical dissipation (present in some numerical schemes) may avoid the local minima problem altogether. In practice, one may be able to increase the artificial dissipation level in those numerical schemes using artificial dissipation for stability. This may be necessary when computing on very fine meshes. Experience with the optimal FBS design problem has shown that the dissipation present for stabilizing the numerical scheme may itself by sufficient to avoid the problem.

A desirable strategy would be to use numerical methods which track the shock locations. If the location of the shocks in the experimental data is also known, then the integration in equation (2.13) could be approximated more carefully and avoid the problem altogether. Unfortunately this is not always feasible. The optimal FBS design problem is an example where the shock location is not known.

REFERENCES

[1] Borggaard, J., Burns, J., Cliff, E. and Gunzburger, M., *Sensitivity Calculations for a 2D, Inviscid, Supersonic Forebody Problem*, in Identification and Control of Distributed Parameter Systems, to appear.

[2] Courant, R. and Friedrichs, K. L., *Supersonic Flow and Shock Waves*, Interscience Publishers, Inc., New York, 1948.

[3] Dennis Jr., J. E. and Schnabel, R. B., *Numerical Methods for Unconstrained Optimization and Nonlinear Equations*, Prentice Hall, Inc., Englewood Cliffs, New Jersey, 1983.

[4] Farin, G., *Curves and Surfaces for Computer Aided Geometric Design: A Practical Guide*, Academic Press, Inc., San Diego, CA, 1988.

[5] Frank, P. and Shubin, G., "A Comparison of Optimization-Based Approaches for a Model Computational Aerodynamics Design Problem," *Journal of Computational Physics*, Vol. 98, (1992), pp. 74–89.

[6] Huddleston, D., *Development of a Free-Jet Forebody Simulator Design Optimization Method*, AEDC-TR-90-22, Arnold Engineering Development Center, Arnold AFB, TN, December, 1990.

[7] Lax, P. D., "Hyperbolic Systems of Conservation Laws and the Mathematical Theory of Shock Waves," Regional Conference Series in Applied Mathematics, Vol. 11, SIAM, Philadelphia, 1973.

A SENSITIVITY EQUATION APPROACH TO SHAPE OPTIMIZATION IN FLUID FLOWS

JEFF BORGGAARD* AND JOHN BURNS[†]

Abstract. In this paper we apply a sensitivity equation method to shape optimization problems. An algorithm is developed and tested on a problem of designing optimal forebody simulators for a 2D, inviscid supersonic flow. The algorithm uses a BFGS/Trust Region optimization scheme with sensitivities computed by numerically approximating the linear partial differential equations that determine the flow sensitivities. Numerical examples are presented to illustrate the method.

1. Introduction. The development of practical computational methods for optimization based design and control often relies on cascading simulation software into optimization algorithms. Black-box methods are examples of this approach. Although the precise form of the overall "optimal design" (OD) algorithm may change, there is an often unstated assumption that properly combining the "best" simulation algorithm with the "best" optimization scheme will produce a good OD algorithm. There are many examples to show that in general this assumption is not valid. However, in many cases it is a valid assumption and often this approach is the only practical way of attacking complex optimal design problems. If one uses this cascading approach, then it is still important to carefully pass information between the simulation and the optimizer. Typically, one uses a simulation code to produce a finite dimensional model and this discrete model is then used to supply approximate function evaluations to the optimization algorithm. Moreover, the approximate functions are then differentiated to supply gradients needed by the optimizer. Although there are numerous variations on this theme, they all may be formulated as "approximate-then-optimize" approaches. There are other approaches that first formulate the problem as an infinite dimensional optimization problem and then use numerical schemes to approximate the optimal design. All-at-once, one-shot and adjoint methods are examples of this "optimize-then-approximate" approach. Regardless of which approach one chooses, some type of approximation must be introduced at some point in the design process.

---

* Supported in part by the Air Force Office of Scientific Research under grant F49620-92-J-0078. Interdisciplinary Center for Applied Mathematics, Department of Mathematics, Virginia Polytechnic Institute and State University, Blacksburg, Virginia 24061.

† Supported in part by the Air Force Office of Scientific Research under grants F49620-92-J-0078 and F49620-93-1-0280, the National Science Foundation under grant INT-89-22490 and by the National Aeronautics and Space Administration under Contract No. NASI-19480 while the author was a visiting scientist at the Institute for Computer Applications in Science and Engineering (ICASE), NASA Langley Research Center, Hampton, VA 23681-0001. Interdisciplinary Center for Applied Mathematics, Department of Mathematics, Virginia Polytechnic Institute and State University, Blacksburg, Virginia 24061.

The sensitivity equation (SE) method is an approach that views the simulation scheme as a device to produce approximations of both the function and the sensitivities. The basic idea is to produce approximations of the infinite dimensional sensitivities and to pass these "approximate derivatives" to the optimizer along with the approximate function evaluations. There are several theoretical and practical issues that need to be considered when this approach is used. For example, there is no assurance that the SE method produces "consistent derivatives." This will depend on the particular numerical scheme used to discretize the problem. However, the SE method allows one the option of using separate numerical schemes for flow solves and sensitivities, so that consistent derivatives can be forced. We shall not address these issues in this short paper. The goal here is to illustrate that a SE based method can be used with standard optimization schemes to produce a practical fast algorithm for optimal design. We concentrate on a particular application (the optimal forebody design problem) and use a specific iterative solver for the flow equations (PARC). Many flow solvers are iterative and for these types of codes, the SE method has perhaps the maximum potential for improving speed and accuracy.

In the next section we describe the forebody design problem and formulate the optimal design problem. In Sections 3 and 4 we review the derivation of the sensitivity equations and in Section 5 we discuss modifications to an existing simulation code that are needed in order to use that code for computing sensitivities. In Section 6, we present numerical results for the optimal design problem and Section 7 contains conclusions and suggestions for future work.

2. Optimal design of a forebody simulator. This problem is a 2D version of the problem described in [1,4,8] . The Arnold Engineering Development Center (AEDC) is developing a free-jet test facility for full-scale testing of engines in various free flight conditions. Although the test cells are large enough to house the jet engines, they are too small to contain the full airplane forebody and engine. Thus, the effect of the forward fuselage on the engine inlet flow conditions must be "simulated." One approach to solving this problem is to replace the actual forebody by a smaller object, called a "forebody simulator" (FBS), and determine the shape of the FBS that produces the best flow match at the engine inlet. The 2D version of this problem is illustrated in Figure 2.1 (see [1],[4], [8] and [9]).

The underlying mathematical model is based on conservation laws for mass, momentum and energy. For inviscid flow, we have that

$$(2.1) \qquad \frac{\partial}{\partial t}Q + \frac{\partial}{\partial x}F_1 + \frac{\partial}{\partial y}F_2 = 0$$

where

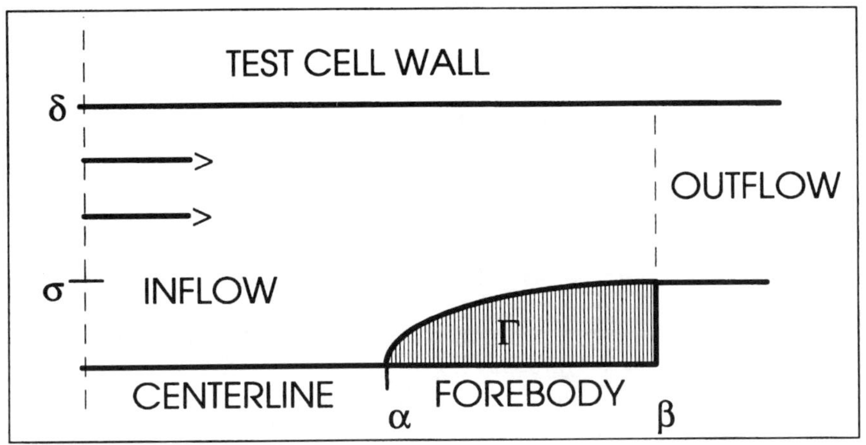

FIGURE 2.1: *2D Forebody Problem*

$$
(2.2) \quad Q = \begin{pmatrix} \rho \\ m \\ n \\ E \end{pmatrix}, \quad F_1 = \begin{pmatrix} m \\ mu + P \\ mv \\ (E+P)u \end{pmatrix} \text{ and } F_2 = \begin{pmatrix} n \\ nu \\ nv + P \\ (E+P)v \end{pmatrix}.
$$

The velocity components u and v, the pressure P, the temperature T, and the Mach number M are related to the conservation variables, i.e., the components of the vector Q, by

$$
u = \frac{m}{\rho}, \qquad v = \frac{n}{\rho}, \qquad P = (\gamma - 1)\left(E - \frac{1}{2}\rho(u^2 + v^2)\right),
$$

$$
(2.3) \quad T = \gamma(\gamma - 1)\left(\frac{E}{\rho} - \frac{1}{2}(u^2 + v^2)\right) \qquad \text{and} \qquad M^2 = \frac{u^2 + v^2}{T}.
$$

At the inflow boundary, we want to simulate a free-jet, so that we specify the total pressure P_0, the total temperature T_0 and the Mach number M_0. We also set $v = 0$ at the inflow boundary. If u_I, P_I and T_I denote the inflow values of the x-component of the velocity, the pressure and the temperature, these may be recovered from P_0, T_0 and M_0 by

$$
T_I = \frac{T_0}{(1 + \frac{\gamma-1}{2}M_0^2)}, P_I = \frac{P_0}{(1 + \frac{\gamma-1}{2}M_0^2)^{\frac{\gamma}{\gamma-1}}} \text{ and } u_I^2 = M_0^2 T_I = \frac{M_0^2 T_0}{(1 + \frac{\gamma-1}{2}M_0^2)}.
$$

(2.4)

The components of Q at the inflow may then be determined from (2.4) through the relations

$$(2.5) \qquad \rho_I = \frac{\gamma P_I}{T_I}, \quad m_I = \rho_I u_I, \quad n_I = 0 \quad \text{and} \quad E_I = \frac{P_I}{\gamma - 1} + \rho_I \frac{u_I^2}{2}.$$

The forebody is a solid surface, so that the normal component of the velocity vanishes, i.e.,

$$(2.6) \qquad u\eta_1 + v\eta_2 = 0 \qquad \text{on the forebody,}$$

where η_1 and η_2 are the components of the unit normal vector to the boundary. Note that we impose (2.6) on the velocity components u and v, and not on the momentum components m and n. Insofar as the state is concerned, it is clear that it does not make any difference whether (2.6) is imposed on m and n or on u and v, since $m = \rho u$ and $n = \rho v$ and $\rho \neq 0$. It can be shown that it does not make any difference to the sensitivities as well.

Assume that at $x = \beta$ the desired steady state flow $\hat{Q} = \hat{Q}(y)$ is given as data on the line (called the Inlet Reference Plane)

$$IRP = \{(x, y) | x = \beta, \sigma \leq y \leq \delta\}.$$

Also, we assume here that the inflow (total) Mach number M_0 can be used as a design (control) variable along with the shape of the forebody. Let the forebody be determined by the curve $\Gamma = \Gamma(x)$, $\alpha \leq x \leq \beta$ and let $p = (M_0, \Gamma(\cdot))$. The problem can be stated as the following optimization problem:

Problem FBS. Given data $\hat{Q} = \hat{Q}(y)$ on the IRP, find the parameters $p^* = (M_0^*, \Gamma^*(\cdot))$ such that the functional

$$(2.7) \qquad \mathcal{J}(p) = \frac{1}{2} \int_\sigma^\delta \| Q_\infty(\beta, y) - \hat{Q}(y) \|^2 \, dy$$

is minimized, where $Q_\infty(x, y) = Q_\infty(x, y, p)$ is the solution to the steady state Euler equations

$$(2.8) \qquad G(Q, p) = \frac{\partial}{\partial x} F_1 + \frac{\partial}{\partial y} F_2 = 0.$$

In the FBS design problem, the data $\hat{Q}$ is generated both experimentally and numerically. In particular, the full airplane forebody (which is longer and larger than the desired FBS) is used to generate the data. Since the FBS is "constrained" to be shorter and smaller, we shall consider the optimization problem illustrated in Figure 2.2. The data $\hat{Q}$ is generated by solving (2.1)-(2.6) for the long forebody in Figure 2.2-(a) and the problem is to find p^* to minimize $\mathcal{J}$ where the shortened FBS is constrained to be one half the length of the "real forebody." This problem provides a realistic test of the optimal design algorithm in that the data can not be fitted exactly. Also, we note that we have a problem with shocks in the flow field. As shown in [2], optimization of flows with shocks can be difficult and requires some understanding of the impact that shocks have on the smoothness of the cost functional.

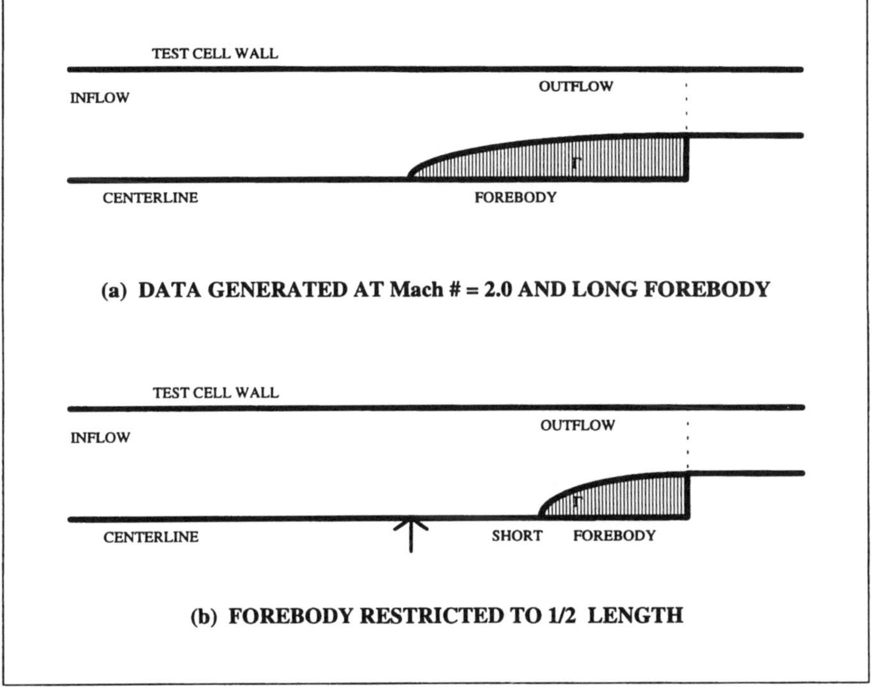

FIGURE 2.2:   *A 2D Optimal forebody Design Problem*

Clearly the statement of the problem is not complete. For example, one should carefully specify the set of admissible curves $\Gamma(\cdot)$ and questions remain about existence, uniqueness and integrability of "the" solution Q_∞. We will not address these issues in this short paper.

Most optimization based design methods require the computation of the derivatives $\frac{\partial}{\partial p} Q_\infty(x, y, p)$. These derivatives are called sensitivities and various schemes have been developed to approximate the sensitivities numerically (see [7], [8], [10] and [11]). A common approach is to use finite differences. In particular, the steady state equation (2.8) is solved for $\tilde{p}$ and again for $\tilde{p} + \Delta p$ and then $\frac{\partial}{\partial p} Q_\infty(x, y, \tilde{p})$ is approximated by $\frac{Q_\infty(x,y,\tilde{p}+\Delta p)-Q_\infty(x,y,\tilde{p})}{\Delta p}$. This method is often costly and can introduce large errors. Another approach is to first derive an equation (the sensitivity equation) for $\frac{\partial}{\partial p} Q_\infty(x, y, p)$ and then numerically solve this equation. We

54 J. T. BORGGAARD AND J. A. BURNS

shall illustrate this approach for the forebody design problem. In the next two sections we derive the sensitivity equations. Although these derivations may be found in [3] we repeat them here for completeness.

3. Sensitivities with respect to the inflow Mach number. First, we consider the design parameter M_0^2. We will derive equations for the sensitivity

$$(3.1) \qquad Q' \equiv \frac{\partial Q}{\partial M_0^2} \equiv \begin{pmatrix} \rho' \\ m' \\ n' \\ E' \end{pmatrix},$$

where

$$(3.2) \qquad \rho' \equiv \frac{\partial \rho}{\partial M_0^2}, \quad m' \equiv \frac{\partial m}{\partial M_0^2}, \quad n' \equiv \frac{\partial n}{\partial M_0^2} \quad \text{and} \quad E' \equiv \frac{\partial E}{\partial M_0^2}.$$

The differential equation system (2.1) has no explicit dependence on the design parameter M_0^2, so that equations for the components of Q' are easily determined by formally differentiating (2.1) with respect to M_0^2. The result is the system

$$(3.3) \qquad \frac{\partial Q'}{\partial t} + \frac{\partial F_1'}{\partial x} + \frac{\partial F_2'}{\partial y} = 0,$$

where

$$(3.4) \quad F_1' = \begin{pmatrix} m' \\ mu' + m'u + P' \\ mv' + m'v \\ (E+P)u' + (E'+P')u \end{pmatrix} \quad \text{and} \quad F_2' = \begin{pmatrix} n' \\ nu' + n'u \\ nv' + n'v + P' \\ (E+P)v' + (E'+P')v \end{pmatrix},$$

and where,

$$(3.5) \quad u' = \frac{\partial u}{\partial M_0^2}, \qquad v' = \frac{\partial v}{\partial M_0^2}, \qquad P' = \frac{\partial P}{\partial M_0^2} \quad \text{and} \quad T' = \frac{\partial T}{\partial M_0^2},$$

and where, through (2.3), the sensitivities (3.2) and (3.5) are related by

$$u' = \frac{1}{\rho} m' - \frac{m}{\rho^2} \rho', \quad P' = (\gamma - 1) \left(E' - \frac{1}{2}\rho'(u^2 + v^2) - \rho(uu' + vv') \right),$$

$$(3.6) \qquad v' = \frac{1}{\rho} n' - \frac{n}{\rho^2} \rho' \quad \text{and} \quad T' = \gamma(\gamma - 1) \left(\frac{1}{\rho} E' - \frac{E}{\rho^2} \rho' - (uu' + vv') \right).$$

Note that (3.3) is of the same form as (2.1), with a different flux vector. In particular, (3.3) is in conservation form. As a result of the fact that (3.3) is *linear* in the primed variables, and that by (3.6) u', v' and P' are linear in the components of Q', (3.3) is a linear system in the sensitivity (3.1), i.e., in the components of Q'.

Now, we need to discuss the boundary conditions for Q'. Except for the inflow conditions, all boundary conditions are independent of the design parameter M_0^2. Thus, the latter may be differentiated with respect to M_0^2 to obtain boundary conditions for the sensitivities. For example, at the forebody where (2.6) holds, we simply would have that

$$(3.7) \qquad u'\eta_1 + v'\eta_2 = 0 \qquad \text{on the forebody.}$$

Similar operations yield boundary conditions for the sensitivities along symmetry lines, other solid surfaces and at the outflow boundary. Note that if instead of (2.6), one interprets the no penetration condition as one on the momentum, i.e., $m\eta_1 + n\eta_2 = 0$ on the forebody, then instead of (3.7) we would have that

$$(3.8) \qquad m'\eta_1 + n'\eta_2 = 0 \qquad \text{on the forebody}$$

which is seemingly different from (3.7). However, (2.6) and (3.6) can be used to show that

$$(3.9) \quad m'\eta_1 + n'\eta_2 = \rho(u'\eta_1 + v'\eta_2) + \rho'(u\eta_1 + v\eta_2) = \rho(u'\eta_1 + v'\eta_2)$$

so that, since $\rho \neq 0$, (3.7) and (3.8) are identical.

The inflow boundary conditions for the sensitivities may be determined by differentiating (2.4) and (2.5) with respect to the design parameter M_0^2. Note that this parameter appears explicitly in the right-hand-sides of the equations in (2.4) and (2.5). Without difficulty, one finds from (2.5) that

$$\rho'_I = \frac{\gamma}{T_I} P'_I - \frac{\gamma P_I}{T_I^2} T'_I, \qquad m'_I = \rho_I u'_I + u_I \rho'_I,$$

$$(3.10) \quad n'_I = 0 \qquad \text{and} \qquad E'_I = \frac{1}{\gamma - 1} P'_I + \frac{1}{2} u_I^2 \rho'_I + \rho_I u_I u'_I,$$

where, from (2.4),

$$T'_I = -\left(\frac{\gamma - 1}{2}\right) \frac{T_0}{\left(1 + \frac{\gamma-1}{2} M_0^2\right)^2}, \qquad P'_I = -\left(\frac{\gamma}{2}\right) \frac{P_0}{\left(1 + \frac{\gamma-1}{2} M_0^2\right)^{\frac{2\gamma-1}{\gamma-1}}}$$

$$(3.11) \text{and} \quad u'_I = \frac{\sqrt{T_I}}{2M_0} + \frac{M_0}{2\sqrt{T_I}} T'_I = \frac{\sqrt{T_0}}{2M_0 \left(1 + \frac{\gamma-1}{2} M_0^2\right)^{3/2}} \left(1 + (\gamma - 1)M_0^2\right).$$

4. Sensitivities with respect to the forebody design parameters. We assume that the forebody is described in terms of a finite number of design parameters which we denote by P_k, $k = 1, \ldots, K$, and that the forebody may be described by the relation

$$(4.1) \qquad y = \Phi(x; P_1, P_2, \ldots, P_K), \qquad \alpha \leq x \leq \beta$$

We express the dependence of the state variable Q on the coordinates and the design parameters by $Q = Q(t, x, y; M_0^2, P_1, P_2, \ldots P_K)$. We have already seen what equations can be used to determine the sensitivity of the state with respect to M_0^2, i.e., for Q'. We now discuss what equations can be used to determine the sensitivities with respect to the forebody design parameters P_k, $k = 1, \ldots, K$, i.e., for

$$(4.2) \qquad Q_k \equiv \frac{\partial Q}{\partial P_k} \equiv \begin{pmatrix} \rho_k \\ m_k \\ n_k \\ E_k \end{pmatrix},$$

where

$$(4.3) \quad \rho_k \equiv \frac{\partial \rho}{\partial P_k}, \; m_k \equiv \frac{\partial m}{\partial P_k}, \; n_k \equiv \frac{\partial n}{\partial P_k} \text{ and } E_k \equiv \frac{\partial E}{\partial P_k}, \; k = 1, \ldots, K.$$

System (2.1) has no explicit dependence on the design parameters P_k, so that equations for the components of Q_k are easily determined by differentiating (2.1) with respect to P_k, $k = 1, \ldots, K$. This produces the systems, $k = 1, \ldots, K$, given by

$$(4.4) \qquad \frac{\partial Q_k}{\partial t} + \frac{\partial F_{k1}}{\partial x} + \frac{\partial F_{k2}}{\partial y} = 0,$$

where

$$F_{k1} = \begin{pmatrix} m_k \\ mu_k + m_k u + P_k \\ mv_k + m_k v \\ (E + P)u_k + (E_k + P_k)u \end{pmatrix} \text{ and } F_{k2} = \begin{pmatrix} n_k \\ nu_k + n_k u \\ nv_k + n_k v + P_k \\ (E + P)v_k + (E_k + P_k)v \end{pmatrix},$$

$$(4.5)$$

and where,

$$(4.6) \qquad u_k = \frac{\partial u}{\partial P_k}, \quad v_k = \frac{\partial v}{\partial P_k}, \quad P_k = \frac{\partial P}{\partial P_k} \quad \text{and} \quad T_k = \frac{\partial T}{\partial P_k}.$$

Moreover, by (2.3), the sensitivities (4.3) and (4.6) are related by

$$u_k = \frac{1}{\rho} m_k - \frac{m}{\rho^2} \rho_k, \quad P_k = (\gamma - 1)\left(E_k - \frac{1}{2}\rho_k(u^2 + v^2) - \rho(uu_k + vv_k) \right),$$

$$(4.7) \quad v_k = \frac{1}{\rho} n_k - \frac{n}{\rho^2} \rho_k \text{ and } T_k = \gamma(\gamma - 1)\left(\frac{1}{\rho} E_k - \frac{E}{\rho^2} \rho_k - (uu_k + vv_k) \right),$$

for $k = 1, \ldots, K$.

All boundary conditions except the one on the forebody also do not depend on the forebody design parameters P_k, $k = 1, \ldots, K$. For example,

consider the inflow boundary conditions (2.4)-(2.5). Differentiating these
with respect to P_k, $k = 1, \ldots, K$ yields that

$$(4.8) \qquad \rho_{kI} = m_{kI} = n_{kI} = E_{kI} = T_{kI} = P_{kI} = u_{kI} = v_{kI} = 0$$

at the inflow boundary. Now consider the boundary condition (2.6) on the
forebody. We have that on the forebody

$$(4.9) \qquad \frac{\eta_1}{\eta_2} = -\frac{\partial \Phi}{\partial x}.$$

Combining (2.6) and (4.9) we have that

$$(4.10) \qquad u\frac{\partial \Phi}{\partial x} - v = 0$$

along the forebody or, displaying the full functional dependence on the coordinates and design parameters, we have at a point (x, y) on the forebody,
and at any time t,

$$u\left(t, x, y = \Phi(x; P_1, P_2, \ldots, P_K); M_0^2, P_1, P_2, \ldots, P_K\right) \frac{\partial \Phi}{\partial x}(x; P_1, P_2, \ldots, P_K)$$

$$(4.11) \quad - v\left(t, x, y = \Phi(x; P_1, P_2, \ldots, P_K); M_0^2, P_1, P_2, \ldots, P_K\right) = 0.$$

We can proceed to differentiate (4.11) with respect to any of the forebody design parameters P_k, $k = 1, \ldots, K$. The result is that, along the
forebody for $k = 1, \ldots, K$,

$$(4.12) \quad u_k\frac{\partial \Phi}{\partial x} - v_k = -\left(\frac{\partial u}{\partial y}\right)\left(\frac{\partial \Phi}{\partial P_k}\right)\left(\frac{\partial \Phi}{\partial x}\right) - u\frac{\partial}{\partial x}\left(\frac{\partial \Phi}{\partial P_k}\right) + \left(\frac{\partial v}{\partial y}\right)\left(\frac{\partial \Phi}{\partial P_k}\right),$$

where u, v and their derivatives are evaluated at the forebody $(x, y = \Phi(x))$.

If an iterative scheme is used to find a steady state solution of this
system $((4.4), (4.8), (4.12))$, then we assume that present guesses for the
state variables u and v and their derivatives $\partial u/\partial y$ and $\partial v/\partial y$ and for
the design parameters M_0^2 and P_k, $k = 1, \ldots, K$, are known. It follows
that the right-hand-side of (4.12) is known as well and equation (4.12) the
boundary conditions along the forebody for the sensitivities with respect
to the forebody design parameters, is merely an inhomogeneous version of
(4.10), the boundary condition along the forebody for the state.

Let us now specialize to the type of forebodies considered by Huddleston, [8,9], i.e.,

$$(4.13) \qquad \Phi(x; P_1, P_2, \ldots, P_K) = \sum_{k=1}^{K} P_k \phi_k(x),$$

where $\phi_k(x)$, $k = 1, \ldots, K$, are prescribed functions, e.g., Bezier curves (see [6]). In this case,

$$(4.14) \qquad \frac{\partial \Phi}{\partial P_k} = \phi_k(x) \qquad \text{and} \qquad \frac{\partial}{\partial x}\left(\frac{\partial \Phi}{\partial P_k}\right) = \frac{d\phi_k}{dx}(x),$$

and

$$(4.15) \qquad \frac{\partial \Phi}{\partial x} = \sum_{k=1}^{K} P_k \frac{d\phi_k}{dx}(x).$$

Combining (4.12)-(4.15), one obtains that, at any point $(x, \Phi(x))$ on the forebody and for each $k = 1, \ldots, K$,

$$(4.16) \quad \left(\sum_{j}^{K} P_j \frac{d\phi_j}{dx}\right) u_k - v_k = -\left(\frac{\partial u}{\partial y}\right)\left(\sum_{j=1}^{K} P_j \frac{d\phi_j}{dx}\right)\phi_k - u\frac{d\phi_k}{dx} + \left(\frac{\partial v}{\partial y}\right)\phi_k.$$

For forebodies of the type (4.13), (4.16) gives the boundary conditions along the forebody for the sensitivities with respect to the forebody design parameters P_k, $k = 1, \ldots, K$. It is now clear that, given guesses for the state variables u and v and their derivatives $\partial u/\partial y$ and $\partial v/\partial y$ and for the design parameters M_0^2 and P_k, $k = 1, \ldots, K$, then the right-hand-side of (4.16) is known.

Consider now the problem of minimizing $\mathcal{J}(p)$ as defined above. Most optimization algorithms use gradient information. In particular, if P_k denotes one of the shape parameters, then the derivative

$$(4.17)\frac{\partial}{\partial P_k}\mathcal{J}(\tilde{p}) = \int_{\sigma}^{\delta} < \left[\frac{\partial}{\partial P_k}Q_\infty(\beta, y, \tilde{p})\right], Q_\infty(\beta, y, \tilde{p}) - \hat{Q}(y) > dy$$

may be required in the optimization loop. The sensitivity $\frac{\partial}{\partial P_k}Q_\infty(x, y, \tilde{p})$ satisfies the steady-state version of the sensitivity equations (4.4). In practice one must construct approximations to $\frac{\partial}{\partial P_k}Q_\infty(x, y, \tilde{p})$ and feed this information into the optimizer.

Assume that one has a particular simulation scheme (finite differences, finite elements, etc.) to approximate the flow $Q_\infty(x, y, \tilde{p})$ on a given grid, i.e.

$$(4.18) \qquad Q_h(x, y, \tilde{p}) \rightarrow Q_\infty(x, y, \tilde{p}).$$

as the "step size" $h \rightarrow 0$. Given the design parameter $\tilde{p}$, one constructs a grid (depending on $\tilde{p}$) and then computes $Q_h(x, y, \tilde{p}) \approx Q_\infty(x, y, \tilde{p})$. This process may require some type of iterative scheme. We will address this issue below. In theory, one could use the same grid and computational scheme to approximate $\frac{\partial}{\partial P_k} Q_\infty(x, y, \tilde{p})$ so that one generates "approximate sensitivities"

$$(4.19) \qquad \left[\frac{\partial}{\partial P_k} Q_\infty(x, y, \tilde{p})\right]_h \to \frac{\partial}{\partial P_k} Q_\infty(x, y, \tilde{p})$$

as $h \to 0$. It is important to note that in general

$$(4.20) \qquad \left[\frac{\partial}{\partial P_k} Q_\infty(x, y, \tilde{p})\right]_h \neq \frac{\partial}{\partial P_k} [Q_h(x, y, \tilde{p})],$$

i.e. this approach may not provide "consistent sensitivities". However, some schemes do provide consistent derivatives and even if (4.20) holds, the error

$$(4.21) \qquad ED_h = \left[\frac{\partial}{\partial P_k} Q_\infty(x, y, \tilde{p})\right]_h - \frac{\partial}{\partial P_k} [Q_h(x, y, \tilde{p})]$$

may be sufficiently small so that the optimization algorithm converges. Trust region methods are particularly well suited for problems of this type, where derivative information may contain (small) errors. As we shall see below, there are certain cases where $\left[\frac{\partial}{\partial P_k} Q_\infty(x, y, \tilde{p})\right]_h$ can be computed fast and accurately. Hence, the SE method provides estimates for sensitivities that may prove "good enough" for optimization and yet relatively cheap to compute. A comparison of $\left[\frac{\partial}{\partial P_k} Q_\infty(x, y, \tilde{p})\right]_h$ and various finite difference approximations of $\frac{\partial}{\partial P_k} [Q_h(x, y, \tilde{p})]$ may be found in [3].

It is important to note that the details of the computations needed to approximate a sensitivity are not the central issue here. For example, the sensitivity equations (3.3) and (4.4) are viewed as independent partial differential equations that must be solved by "some" numerical scheme. This scheme does not necessarily have to be the same scheme used to solve the flow equation (2.1), although as we shall see below, there are cases where using the same scheme is a useful approach.

Also, note that the sensitivity equations are derived for the problem formulated on the "physical" domain. If one uses a computational method that maps the problem to a computational domain (as does PARC), then the SE method does not require derivatives of this mapping. One simply maps the sensitivity equation (including the necessary boundary conditions), grids the computational domain, solves the resulting transformed equations and then maps back to the physical domain. If, on the other hand, one mapped the flow equation (2.1) and derived a sensitivity equation in the computational domain, then to obtain the correct sensitivities one would have to compute the mapping sensitivity. Therefore, it is more efficient to derive the sensitivity equations in the physical domain.

Finally, we note that the SE method described here has one additional benefit. To compute a sensitivity, say $\frac{\partial}{\partial P_k} Q_\infty(x, y, \tilde{p})$, then one first selects the parameter value $\tilde{p}$, constructs a computational grid and solves for $\left[\frac{\partial}{\partial P_k} Q_\infty(x, y, \tilde{p})\right]_h$. There is no need to compute grid sensitivities.

5. Computing sensitivities using an existing code for the state.

Suppose one has available a code to compute the state variables, i.e., to find approximate solutions of (2.1) along with boundary and initial conditions. In principle, it is an easy matter to amend such a code so that it can also compute sensitivities.

First, let us compare (2.1) with (3.3). If one wishes to amend the existing code that can handle (2.1) so that it can treat (3.3) as well, one has to change the definitions of the flux functions from those given in (2.2) to those given in (3.4). Note that the solution for the state is needed in order to evaluate the flux functions of (3.4).

Next, note that (3.3) and (4.4) are identical differential equations. Thus, the changes made to the code in order to treat (3.3) can also be used to treat (4.4). In fact, as long as the differential equation and any other part of the problem specification do not explicitly depend on the design parameters, the analogous relations will be the same for all the sensitivities.

The only changes that vary from one sensitivity calculation to another are those that arise from conditions in which the design parameters appear explicitly. In our example, for the sensitivity with respect to M_0^2, one must change the portion of the code that treats the inflow conditions (2.4)-(2.5) so that it can instead treat (3.10)-(3.11). In the problem considered here, the nature (i.e. what variables are specified) of the boundary conditions at the inflow, and everywhere else, is not affected. Note that for the sensitivity with respect to M_0^2, the boundary condition (3.7) on the forebody is the same as that for the state, given by (2.6).

For the sensitivities with respect to the forebody design parameters, the inflow boundary conditions simplify to (4.8), i.e., they become homogeneous. The boundary condition at the forebody is now given by (4.12) or (4.16). Once again, the nature of the boundary conditions is unchanged from that for the state and only the specified data is different. For the inflow boundary conditions, we may still specify the same conditions for the sensitivities, but now they would be homogeneous. The boundary conditions along the forebody change in that they become inhomogeneous, (compare (4.10) and (4.16)).

In summary, to change a code for the state so that it also handles the sensitivities, one must redefine the flux functions in the differential equations, and the data in the boundary conditions. The changes necessary in the code to account for any particular relation that does not explicitly involve the design parameters are independent of which sensitivity one is presently considering.

The previous remarks are concerned only with the changes one must effect in a state code in order to handle the fact that one is discretizing a different problem when one considers the sensitivities. We have seen that these changes are not major in nature. However, there are additional changes that may be needed when one attempts to solve the discrete equations. In the numerical results presented below we use the finite difference code "PARC" (see [4] and [8]) to solve the state and sensitivity equations. However, the following comments apply equally well to other CFD codes of this type.

Since we are interested in steady design problems, the time derivative in (2.1) is considered only to provide a means for marching to a steady state. Now, suppose that at any stage of a Gauss-Newton, or other iteration, we have used PARC to find an approximate steady state solution of (2.1) plus boundary conditions. In order to do this, one has to solve a sequence of linear algebraic systems of the type

$$(5.1) \quad \left(I + \Delta t A(Q_h^{(n)})\right) Q_h^{(n+1)} = \left(Q_h^{(n)} + \Delta t B(Q_h^{(n)})\right), \quad n = 0, 1, 2, \ldots,$$

where the sequence is terminated when one is satisfied that a steady state has been reached and where $Q_h^{(n)}$ denotes the discrete approximation to the state Q at the time $t = n\Delta t$. We denote this steady state solution for the approximation to the state by Q_h. One problem of the type (5.1) is solved for every time step. In (5.1), the matrix A and vector B arise from the spatial discretization of the fluxes and the boundary conditions. Both of these depend on the state at the previous time level.

Having computed a steady state solution by (5.1), the task at hand is now to compute the sensitivities. We will focus on Q', the sensitivity with respect to the inflow Mach number. Analogous results hold for the sensitivities with respect to the forebody design parameters. Recall that given a state, the sensitivity equations are linear in the sensitivities. Therefore, if one is interested in the steady state sensitivities, instead of (3.3) one may directly treat its stationary version

$$(5.2) \qquad \frac{\partial F_1'}{\partial x} + \frac{\partial F_2'}{\partial y} = 0.$$

Since (5.2) is linear in the components of Q', one does not need to consider marching algorithms in order to compute a steady sensitivity. One merely discretizes (5.2) and solves the resultant linear system, which has the form

$$(5.3) \qquad \mathcal{A}(Q_h)Q_h' = \mathcal{B}(Q_h),$$

where Q_h' denotes the discrete approximation to the steady sensitivity. The matrix $\mathcal{A}$ and vector $\mathcal{B}$ differ from the A and B of (5.1) because we have discretized different differential equations and boundary conditions. Note that $\mathcal{A}$ and $\mathcal{B}$ in (5.3) depend only on the steady state Q_h and thus (5.3) is a *linear system of algebraic equations* for the discrete sensitivity Q_h'.

The cost of finding a solution of (5.3) is similar to that for finding the solution of (5.1) for a single value of n, i.e. for a single time step. The differences in the assembly of the coefficient matrices and right-hand-sides of (5.1) and (5.3) are minor. Thus, in theory at least, *one can obtain a steady sensitivity in the same computer time it takes to perform one time step in a state calculation.* If one wants to obtain all the sensitivities, e.g., $K + 1$ in our example, one can do so at a cost similar to , e.g., K+1 time steps of the state calculation. This is very cheap compared to the multiple state calculations necessary in order to compute sensitivities through the use of difference quotients.

Although (5.3) is in theory no more complex than one time step in (5.1), we can solve (5.2) by using the same iterative (or another) scheme. The simplest approach (but certainly not the optimal approach) is to use the PARC code to solve (5.2) by time marching. In particular, assume that $Q_h^{(n)}$ is a solution to (5.1), then the system

$$(5.4) \qquad \left[I + \Delta t A'(Q_h^{(n)}) \right] (Q')_h^{(n+1)} = \left[(Q')_h^{(n)} + \Delta t B'(Q_h^{(n)}) \right]$$

can be used to find $(Q')_h^{(n+1)}$ given $(Q')_h^{(n)}$. Thus, one makes an initial guess for $Q_h^{(0)}$ and $(Q_h')^{(0)}$ and then iterates (5.1) and (5.4) simultaneously. Also, the same scheme can be used to compute any $Q_k = \frac{\partial Q}{\partial P_k}$, i.e.,

$$(5.5) \qquad \left[I + \Delta t A'(Q_h^{(n)}) \right] (Q_k)_h^{(n+1)} = \left[(Q_k)_h^{(n)} + \Delta t B'(Q_h^{(n)}) \right].$$

In practice, these "optimal" estimates of speed up are rarely achieved. Moreover, as noted above, it is important to note that finite difference (FD) and sensitivity equation (SE) methods do not necessarily produce the same results. Since the ultimate goal is to find useful and cheap gradients for optimization, the most important issue is whether or not the SE method combined with an optimization algorithm produces a convergent optimal design as fast as possible. We have tested this scheme on the forebody design problem and the next section contains a summary of these results.

6. An optimal design example. In order to illustrate the SE method and to test its use in an optimization problem, we used the PARC code as described above to compute sensitivities and the used these sensitivities in a BFGS/Trust Region scheme to find an optimal shortened forebody simulator. As shown in Figure 2.2, data was generated by solving the Euler equations over the long forebody at a Mach number of 2.0. The objective is to find a forebody simulator with length one half of the long forebody and such that the resulting flow matches the data as well as possible, i.e. minimizes $\mathcal{J}$ along the outflow boundary.

The shortened forebody was parameterized by a Bezier curve using two parameters. Thus, there are three design parameters $p = \left(M_0^2, P_1, P_2 \right)$. The algorithm used in this numerical experiment was based on using the PARC code to simultaneously march to the steady state solutions of the flow and sensitivity equations. We made no attempt to optimize the algorithm since the main goal was to test for convergence.

The design algorithm proceeds as follows. First, an initial guess for the optimal design is made, i.e., we select a $p^0 = \left(\left(M_0^2\right)^0, P_1^0, P_2^0\right)$. A good selection of initial parameters can be made knowing the operating conditions of the aircraft and some rough guess of the shape from the aircraft forebody. In our example, we chose M_0^2 as the inlet Mach number from the computation which generated our data. The initial guess for the parameters were those used to generate the long forebody (although corresponding to different x-locations). These parameters, p^0, are used to generate a grid, the inflow and forebody boundary conditions for both the flow (2.1) and sensitivity equations ((3.3) and (4.4)) and an initial guess for both $Q_h^{(0)}$ and $\left(\frac{\partial}{\partial p}Q\right)_h^{(0)}$. In our example, a rough guess for the flow field $Q_h^{(0)}$ uses the constant inflow boundary condition throughout the flow domain. Likewise, the initial guess for $(Q')_h^{(0)}$ is taken as the inflow boundary conditions (given in equation (3.10)) throughout the flow domain. The initial guess for $(Q_k)_h^{(0)}$ is initially taken as zero (except on the forebody). The systems (5.1), (5.4) and (5.5) are then solved simultaneously (in our case the left hand side matrix is the same for (5.1) as for the sensitivity equations (5.4) and (5.5), i.e. $A = A'$) for the updated $Q_h^{(1)}$, $(Q')_h^{(1)}$ $\left(\frac{\partial Q}{\partial P_1}\right)_h^{(1)}$ and $\left(\frac{\partial Q}{\partial P_2}\right)_h^{(1)}$. The updated $Q_h^{(n)}$ is then used to formulate (5.1), (5.4) and (5.5) and solve for $(Q_h)^{(n+1)}$ and $\left(\frac{\partial}{\partial p}Q\right)_h^{(n+1)}$. Then one iterates until the desired convergence is achieved. In our example, the residuals, $\Delta Q_h = \left[Q_h^{(n+1)} - Q_h^{(n)}\right]$ were converged to approximately 10^{-15} (in 800 time steps). The outflow data Q_h and $\left(\frac{\partial}{\partial p}Q\right)_h$ are then used to compute $\mathcal{J}(p^0)$ and $\nabla\mathcal{J}(p^0)$.

The optimization algorithm consisted of a BFGS secant method coupled with a "hook" step model trust region method [5]. The initial Hessian was obtained by finite differences on $\nabla\mathcal{J}(\tilde{p})$. The function and gradient information needed by the optimization algorithm is obtained by calling the modified PARC code with $p = \tilde{p}$.

This algorithm was tested for the case where the forebody simulator was allowed to have the full length of the body generating the data. In this case the optimization algorithm produced exact data fits, i.e. $\mathcal{J}(p^*) = 0$ and it recovered the parameters used to generate the data. However, the more realistic test (constraining the length of the forebody simulator) also produced a convergent design and reduced the cost functional significantly.

Figure 6.1 shows the flow field over the long forebody. Observe, that there is a shock in the flow. As noted in [2], shocks can cause difficulties if one is not careful in the selection of an appropriate numerical scheme. High order schemes can produce (numerically generated) local minimum that can cause the optimization loop to fail. This problem is avoided here because the numerical viscosity in PARC (required for stability) is sufficient to "smooth" the cost functional (see [2] for details).

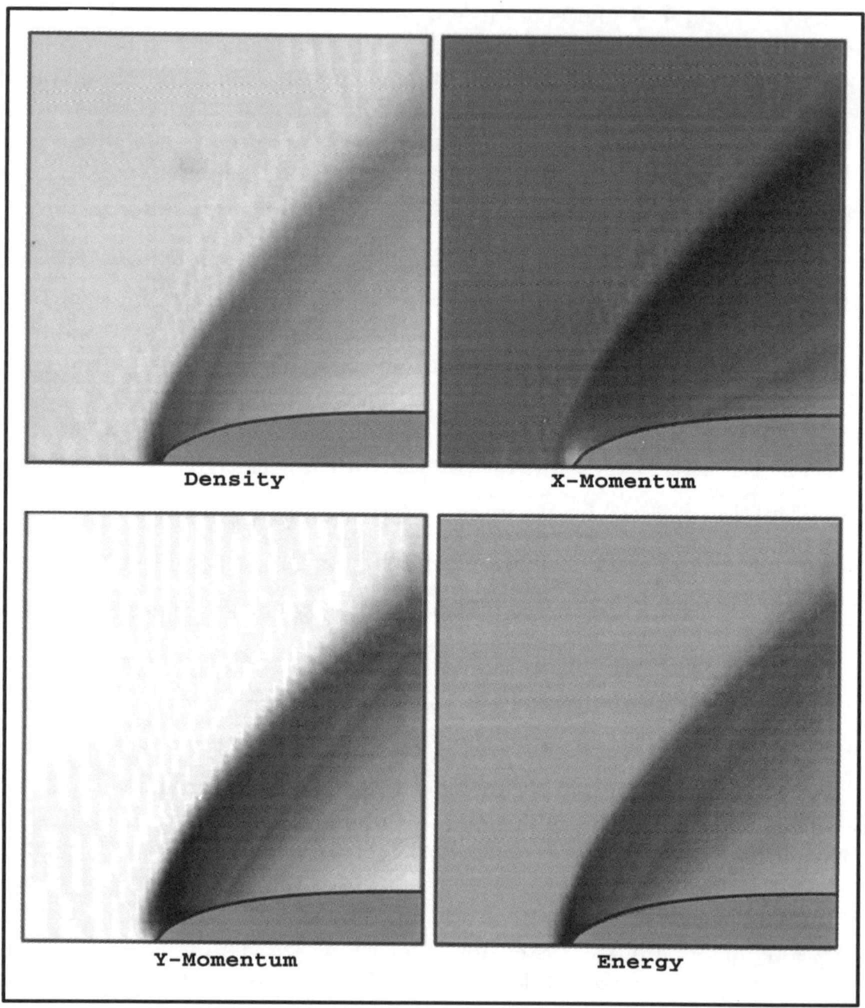

FIGURE 6.1: *Long Forebody (Outflow Data to be Matched)*

Figure 6.2 shows the shape and flow field of the optimal shortened fore body. This design was obtained after 12 iterations of the optimization loop. Figures 6.3–6.6 show the 1st, 2nd, 3rd, 5th and 12th iterations for each of the flow variables.

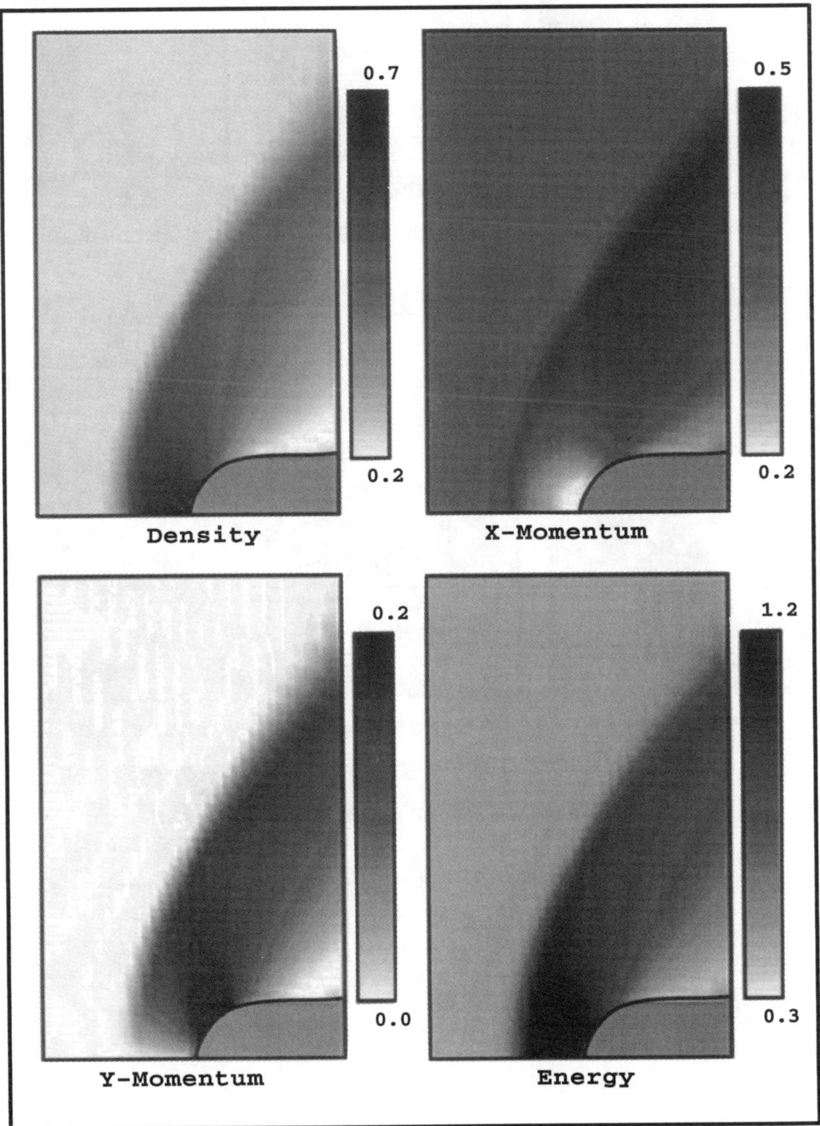

FIGURE 6.2: *Optimal Shortened Forebody Flow*

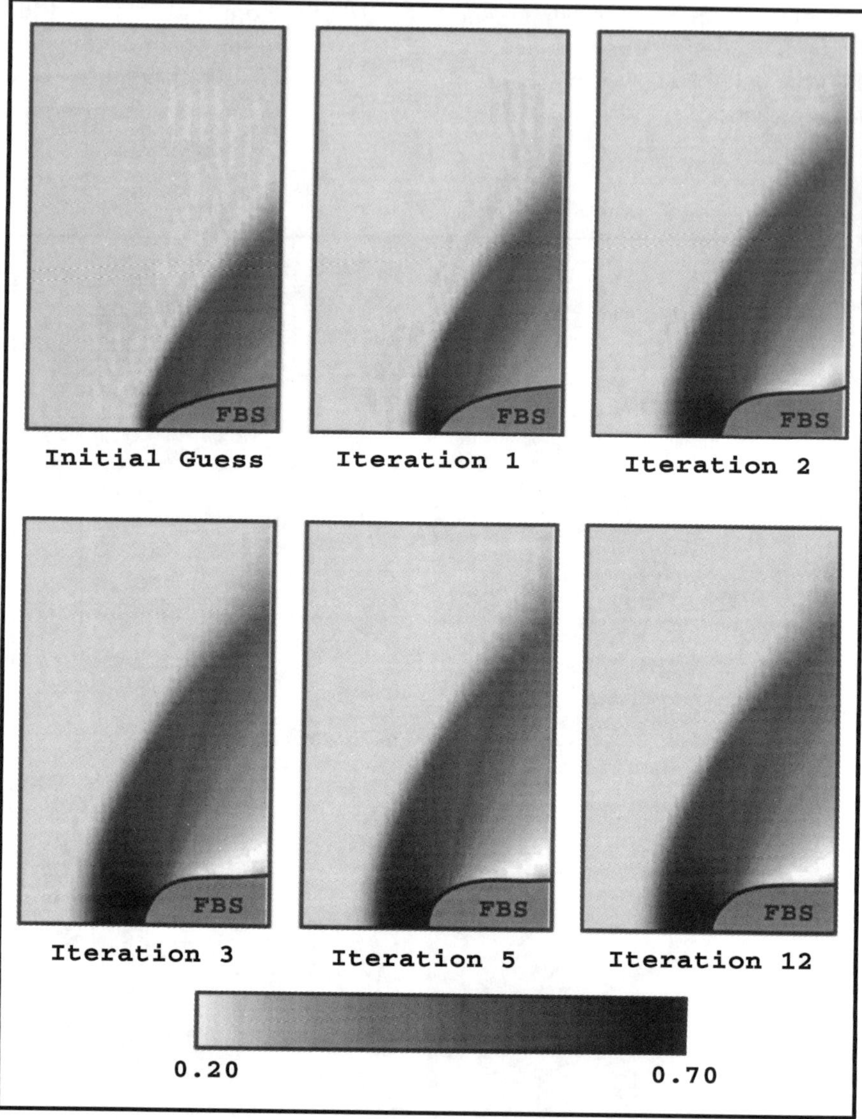

FIGURE 6.3: *Iteration to Optimal Forebody Design: Density*

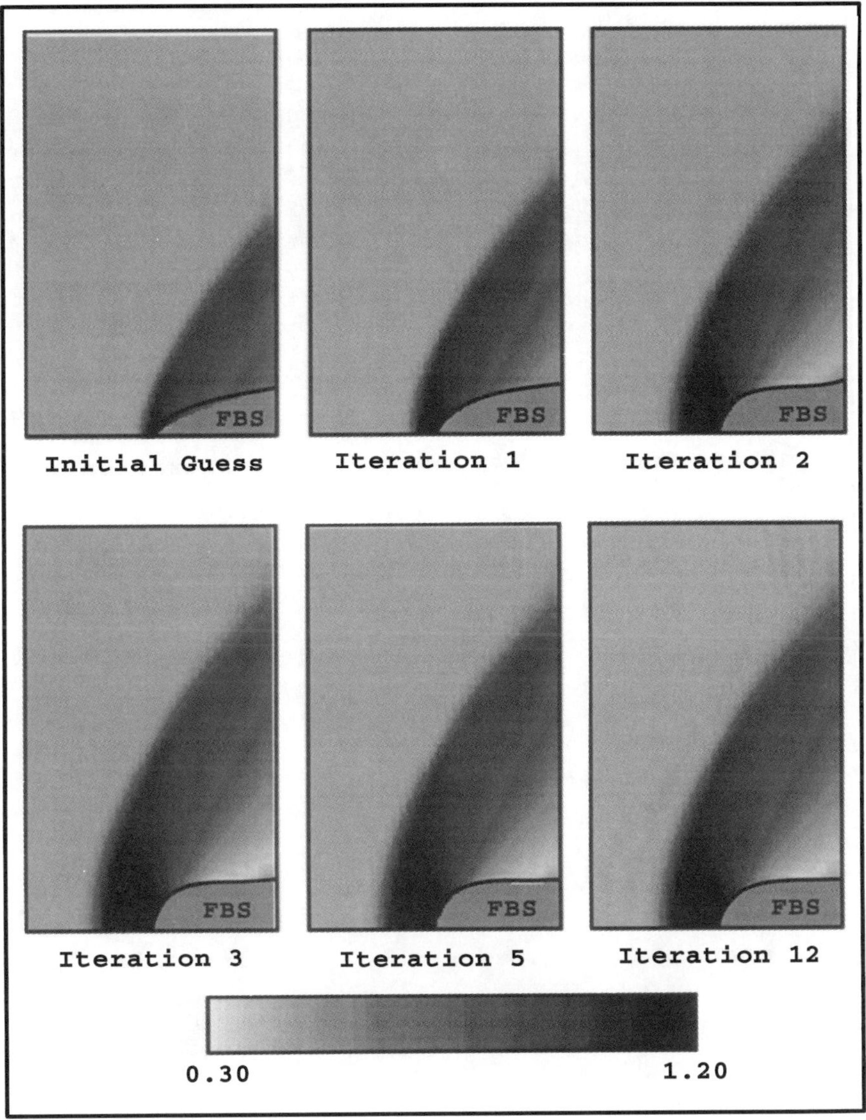

FIGURE 6.4: *Iteration to Optimal Forebody Design Energy*

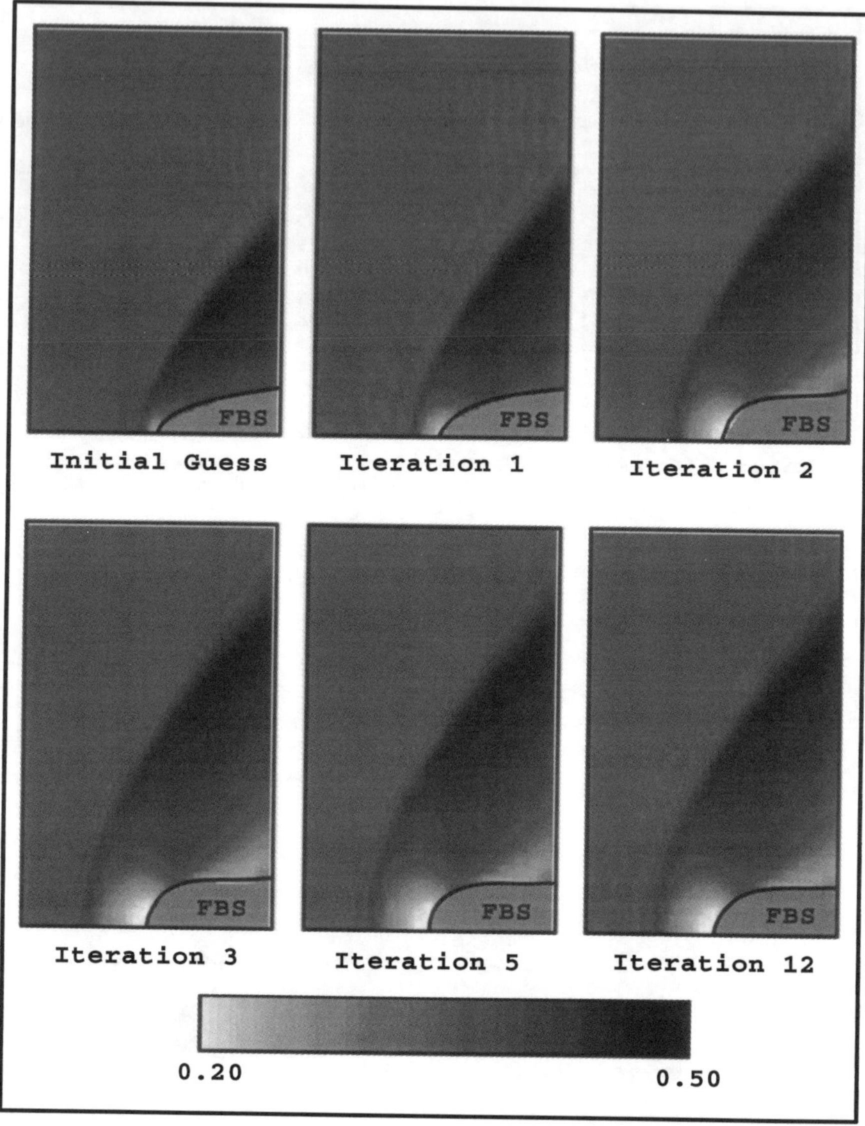

FIGURE 6.5: *Iteration to Optimal Forebody Design: X-Component of Momentum*

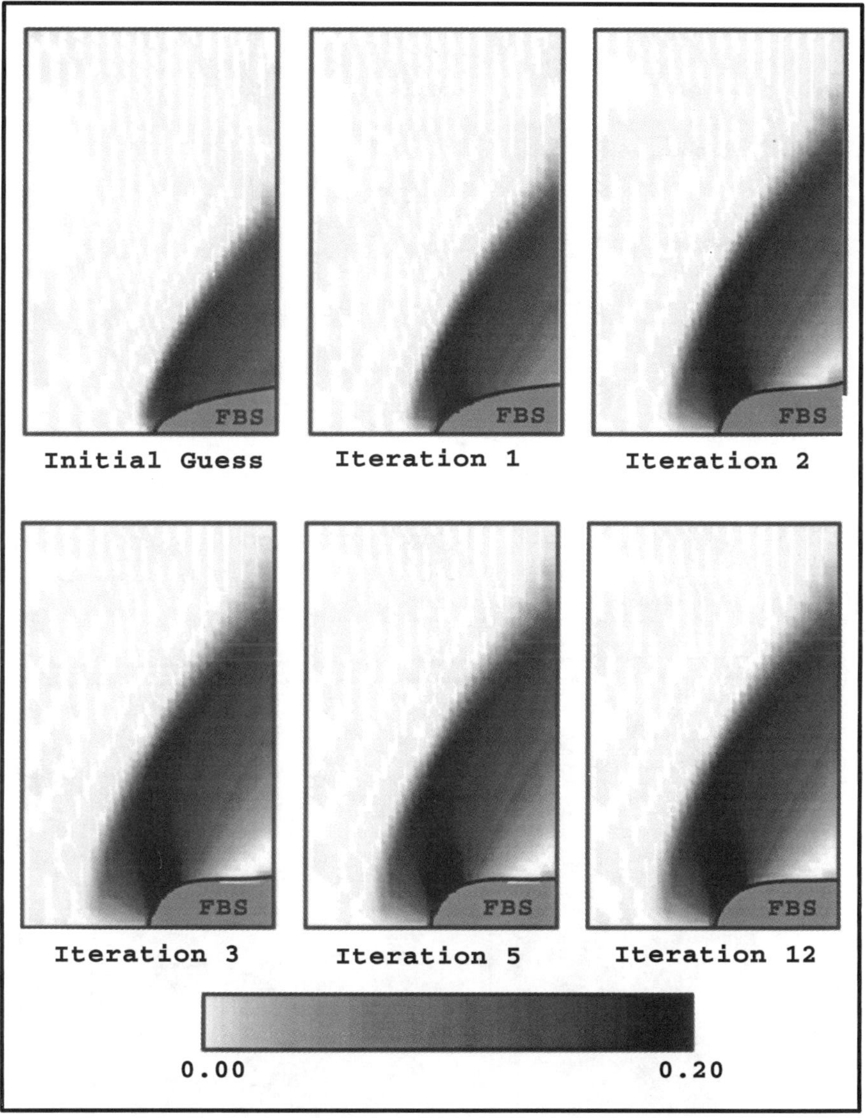

FIGURE 6.6: *Iteration to Optimal Forebody Design: Y-Component of Momentum*

The initial guess for the parameters were

$$p^0 = \left(\left(M_0^2 \right)^0, P_1, P_2 \right) = (2.0, 0.10, 0.15)$$

and

$$\mathcal{J}(p^0) = 3.2339.$$

The "converged" optimal parameters are

$$p^* = p^{12} = (2.020, 0.294, 0.156)$$

with

$$\mathcal{J}(p^*) = 0.2229.$$

Observe that the cost function was decreased by more than 93%. Figures 6.7–6.10 show a comparison of the flow fields for the optimal shortened forebody simulator and the data. The optimization loops converged rapidly. For example, $\mathcal{J}(p^3) = 0.2334$ and $\mathcal{J}(p^5) = 0.2289$. This is due to the fact that the shock location was found quickly.

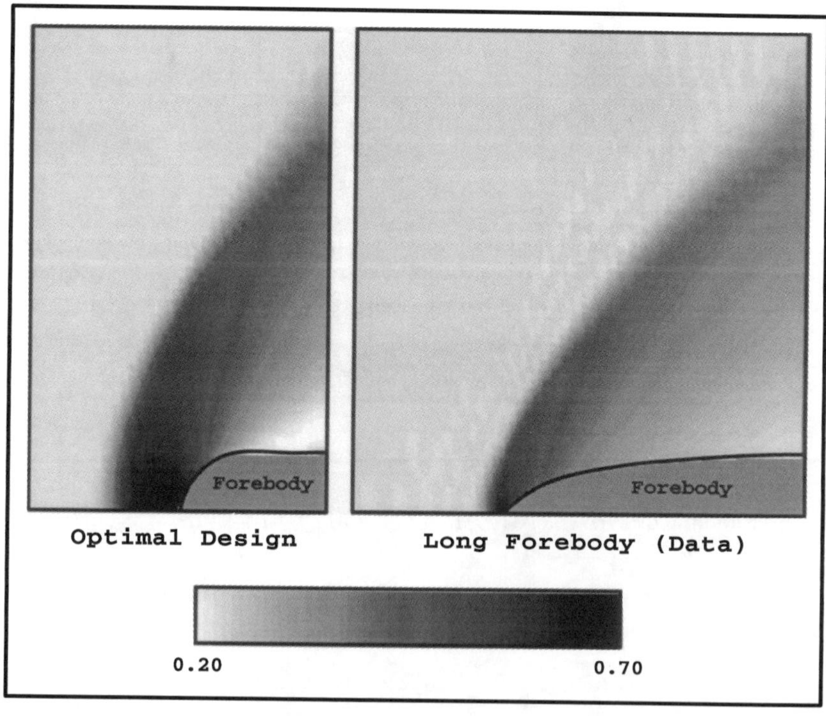

FIGURE 6.7: *Comparison of Optimal Shortened and Long Forebody: Density*

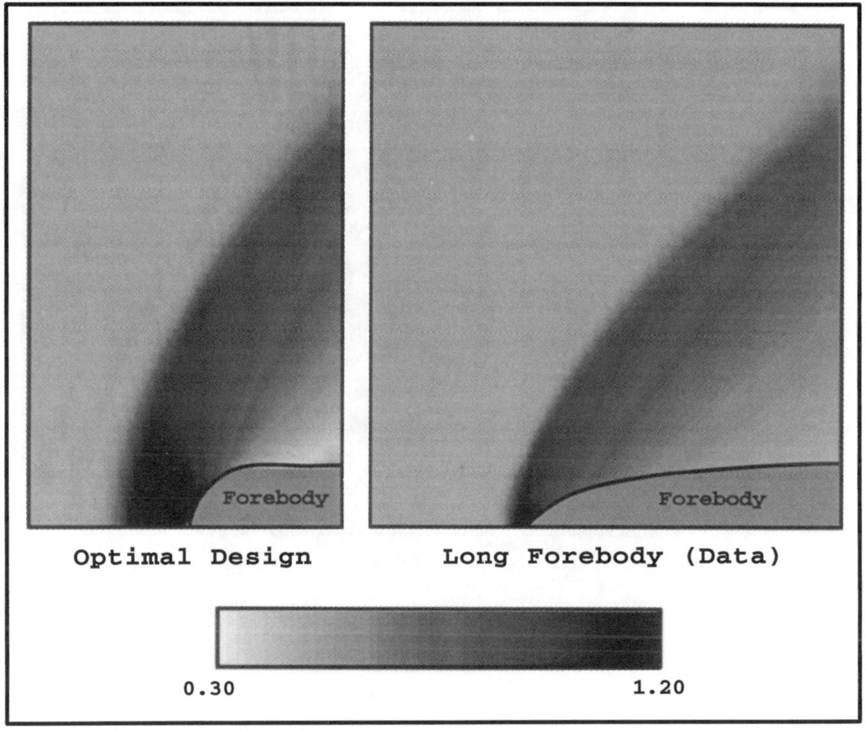

FIGURE 6.8: *Comparison of Optimal Shortened and Long Forebody: Energy*

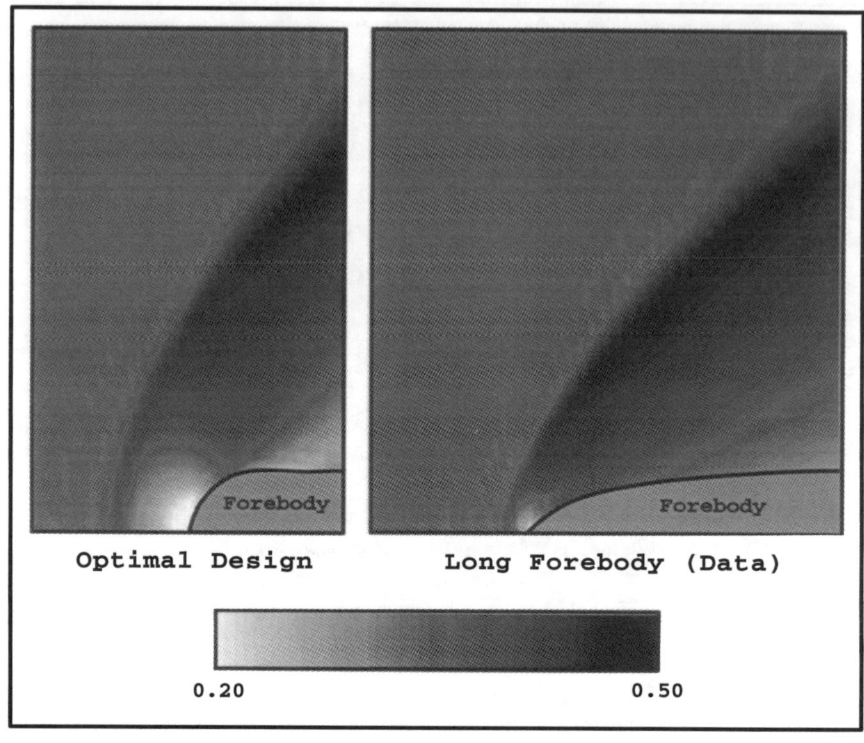

FIGURE 6.9:   *Comparison of Optimal Shortened and Long Forebody: X-Component of Momentum*

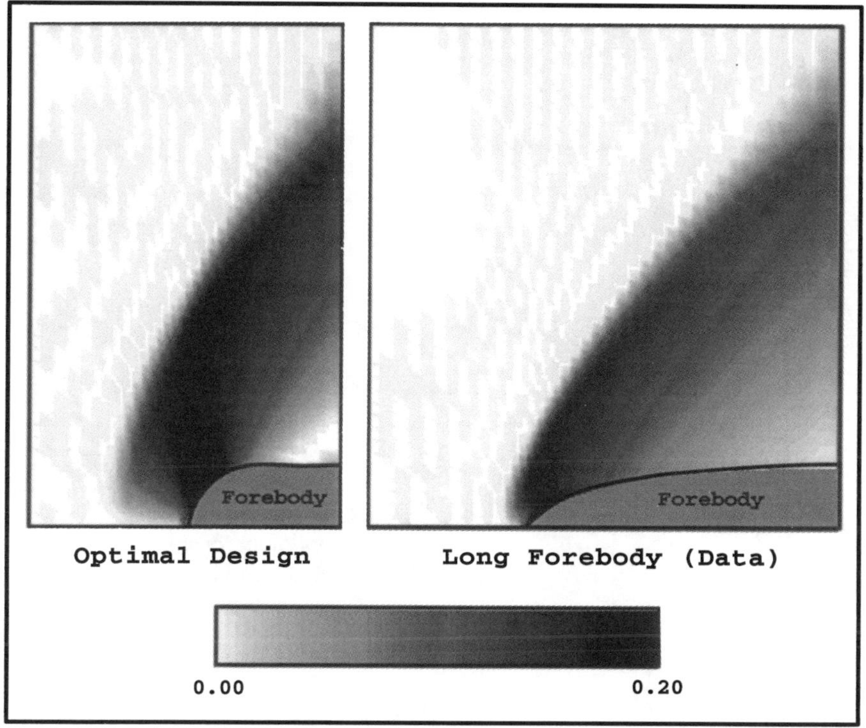

FIGURE 6.10:   *Comparison of Optimal Shortened and Long Forebody: Y-Component of Momentum*

Note that although the flows are close, there is a significant error near the forebody. This can also be seen in the plots in Figures 6.11–6.14. It is worthwhile to note that the match is good considering the fact the shortened forebody is constrained to be one half the length of the "real" forebody and only two Bezier parameters are used to model $\Gamma(\cdot)$. It is also important to note that the shock is captured by the optimal design. In particular, observe in Figures 6.3–6.6 how the optimization algorithm "shapes" the shortened forebody so that the optimal shape has a blunt nose. This is necessary in order to generate the correct shock location at the outflow.

J. T. BORGGAARD AND J. A. BURNS

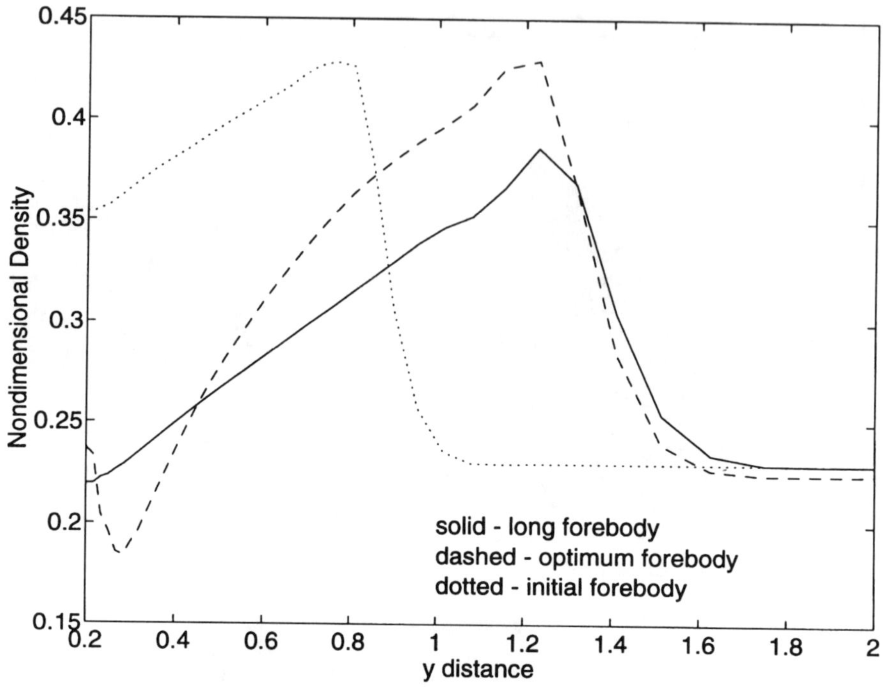

FIGURE 6.11: *Comparison of Outflow Data: Density*

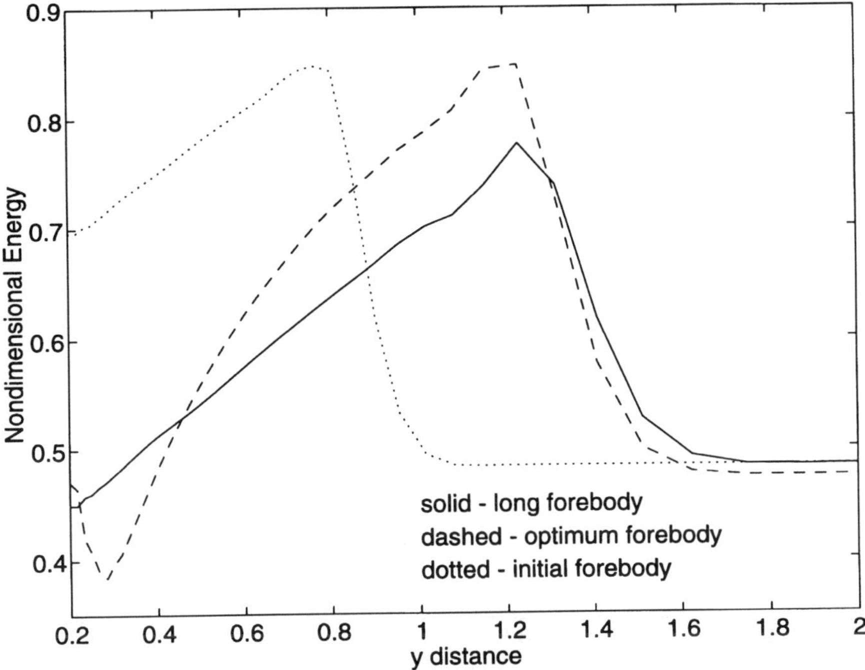

FIGURE 6.12: *Comparison of Outflow Data: Energy*

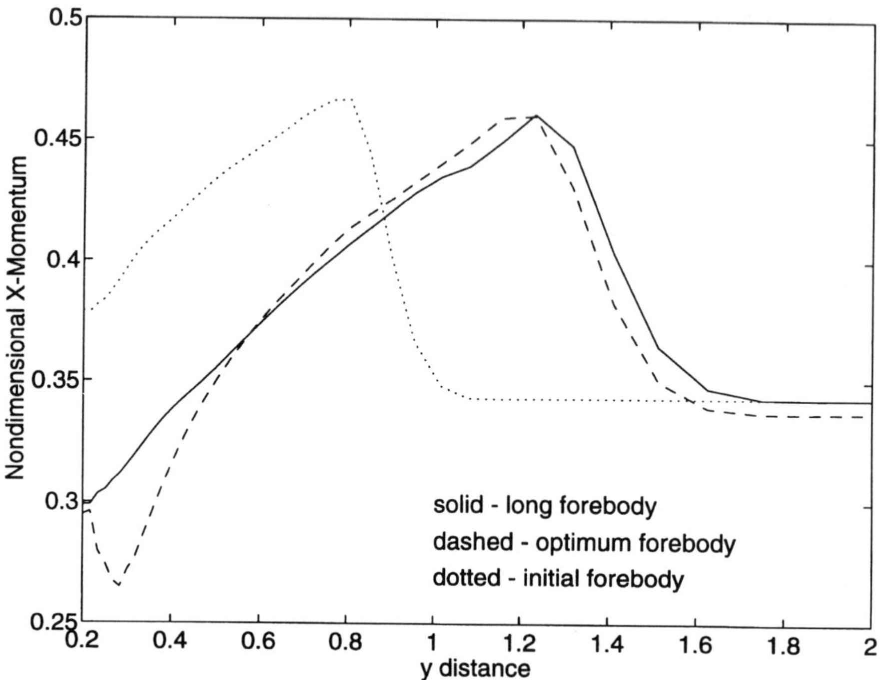

FIGURE 6.13: *Comparison of Outflow Data: X-Component of Momentum*

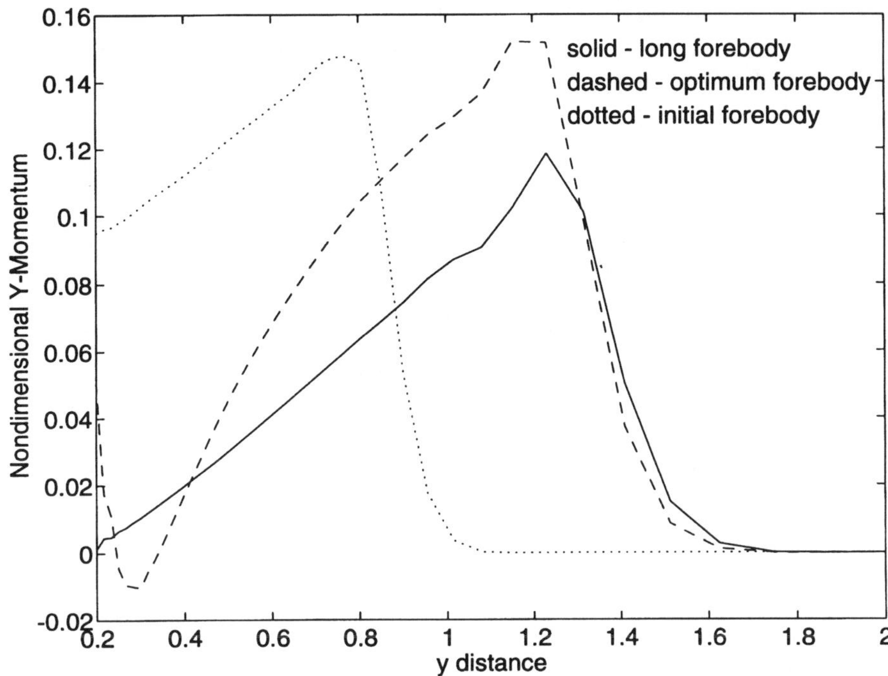

FIGURE 6.14: *Comparison of Outflow Data: Y-Component of Momentum*

7. Conclusions. The numerical experiment above illustrates that the SE method can produce sensitivities suitable for optimization based design. There are a number of interesting theoretical issues that need to be addressed in order to analyze the convergence of this approach. Moreover, one should investigate "fast solvers" for the sensitivity equations (multi-grid, etc.) as well as develop numerical schemes that are not only fast, but produces consistent derivatives when possible.

Finally, we note that we have conducted a number of timing tests which compute sensitivities to compare the SE method with the finite difference method. In particular, we observed that for the problem above (with three design parameters), the SE method needed only 58% of the CPU time required by finite differencing. When twenty design parameters were used, the SE method produced these sensitivities in about 38% of the time required by finite differencing. These early numerical results indicate that considerable computational savings may be possible if one extends and refines the basic SE method presented here.

REFERENCES

[1] Beale, D. and Collier, M., *Validation of a Free-jet Technique for Evaluating Inlet-Engine Compatibility*, AIAA Paper 89-2325, AIAA/ASME/SAE/ASEE 25th Joint Propulsion Conference, Monterey, CA, July 1989.

[2] Borggaard, J. T., *On the Presence of Shocks on Domain Optimization of Euler Flows*, in this volume.

[3] Borggaard, J., Burns, J., Cliff, E. and Gunzburger, M., *Sensitivity Calculations for a 2D, Inviscid, Supersonic Forebody Problem*, in Identification and Control of Distributed Parameter Systems, H.T. Banks, R. Fabiano and K. Ito, Eds., SIAM Publications, Philadelphia, PA, 1993, pp. 14-24.

[4] Cooper, G. and Phares, W., *CFD Applications in an Aerospace Engine Test Facility*, AIAA Paper 90-2003, AIAA/ASME/SAE/ASEE 26th Joint Propulsion Conference, Orlando, FL, July 1990.

[5] Dennis Jr., J. E. and Schnabel, R. B., *Numerical Methods for Unconstrained Optimization and Nonlinear Equations*, Prentice Hall, Inc., Englewood Cliffs, New Jersey, 1983.

[6] Farin, G., *Curves and Surfaces for Computer Aided Geometric Design: A Practical Guide*, Academic Press, Inc., San Diego, CA, 1988.

[7] Frank, P. and Shubin G., *A Comparison of Optimization-Based Approaches for a Model of Computational Aerodynamics Design Problem*, Journal of Computational Physics 98, (1992), pp.74-89.

[8] Huddleston, D., *Aerodynamic Design Optimization Using Computational Fluid Dynamics*, Ph.D. Dissertation, University of Tennessee, Knoxville, TN, December, 1989.

[9] Huddleston, D., *Development of a Free-Jet Forebody Simulator Design Optimization Method*, AEDC-TR-90-22, Arnold Engineering Development Center, Arnold AFB, TN, December, 1990.

[10] Taylor III, A. C., Hou, G. W. and Korivi, V. M., *A Methodology for Determining Aerodynamic Sensitivity Derivatives With Respect to Variation of Geometric Shape*, Proceedings of the AIAA/ASME/ASCE/AHS/ASC 32nd Structures, Structural Dynamics, and Materials Conference, April 8-10, Baltimore, MD, AIAA Paper 91-1101.

[11] Taylor III, A. C., Hou, G. W. and Korivi, V. M., *Sensitivity Analysis, Approximate Analysis and Design Optimization for Internal and External Viscous Flows*, AIAA Paper 91-3083, AIAA Aircraft Design Systems and Operations Meeting, Baltimore, MD, September 1991.

QUASI-ANALYTICAL SHAPE MODIFICATION FOR NEIGHBORING STEADY-STATE EULER SOLUTIONS

J.S. BROCK* AND W.F. NG*

Abstract. Aerodynamic inverse design methods which are governing equation consistent are generally limited to the Full Potential equations. Consistent design methods use identical governing equations for all fluid dynamic segments of the algorithm, including shape modification. This ensures that all relevant physical information is included within each design estimate, and therefore, a minimum number of analysis/design iterations are required. This report presents a new, and consistent, shape modification method for future use within a direct-iterative inverse design algorithm. The method is simple, being developed from a truncated quasi-analytical Taylor's series expansion of the global governing equations. The method is general, since it may use either the Euler or Navier-Stokes equations, any combination of numerical techniques, and any number of spatial dimensions. The proposed method also includes a unique iterative algorithm, and new geometry/grid constraints, to solve the over-determined design problem. An upwind, cell-centered, finite-volume formulation of the two-dimensional Euler equations is used within the present effort. The method is evaluated within a symmetric channel where the design variable is a mid-channel ramp angle which is nominally $\theta = 5^\circ$. Tests were conducted for three target ramp angle perturbations, $\Delta\theta = 2\%$, 10%, and 40%, and three inlet Mach numbers, $M = 0.30, 0.85$, and 2.00. For a single design estimate, using design-like test conditions, the new method is demonstrated to accurately predict geometry shape changes. This includes the transonic test case with an extreme 40% design variable perturbation where the target geometry was predicted with 95% accuracy.

1. Introduction. The ability of Computational Fluid Dynamics (CFD) methods to solve direct, or analysis, problems has progressed rapidly in the last two decades. Direct solutions for complex two or three dimensional (2-D or 3-D) configurations using the Euler or Navier-Stokes (N-S) equations are common. The direct problem is characterized by the specification of the geometry and boundary conditions (BC's), followed by a solution of the field equations governing continuity, momentum, and energy exchange.

Of equal importance to the CFD community is the aerodynamic design problem. The design problem determines the geometry required to support a given set of BC's, physical constraints, and target design goals. The target design goals may be a surface function, such as surface pressure, or a global parameter, such as a shock free flow field. Development of more efficient and effective aerodynamic design technologies continues to receive great emphasis, and is the focus of the present research.

There are many ways in which aerodynamic design methods can be catergorized [1-3]. This report considers two general categories: optimization design and inverse design. Inverse design methods may be subdivided as classical [4-7], shock-free [8,9], direct-iterative [10-12], and stream-tube methods [13-16]. All of these methods possess unique strengths and specializations, and each will continue to serve the design community.

---

* Mechanical Engineering Department, Virginia Polytechnic Institute and State University, Blacksburg, VA 24061-0238.

Optimization design methods are generally considered the more advanced of the two aerodynamic design categories. One reason may be the capability to perform design tasks using the Euler or N-S equations for all fluid dynamic portions of the method. This includes both the analysis and sensitivity derivative codes which are coupled within a design optimization algorithm.

Inverse design methods use various sets of governing equations and implementation algorithm's. Some methods use the Euler of N-S equations as part of the algorithm, while others use the Full Potential (FP) equations exclusively. Some methods couple a boundary layer (BL) model with the FP equations for the initial and intermediate direct solutions. However, the relationship within each inverse design algorithm which actually predicts shapes, the design methodologies, are in general limited to the FP equations.

Direct-iterative inverse design methods are conceptually simple, relatively easy to implement, and so the most commonly used inverse design method. These are also considered to be the more advanced inverse design method since they may use any existing CFD analysis code, and therefore governing equations, as a portion of the method. Direct-iterative methods require an initial geometry, BC's, an initial solution, and a target surface pressure profile.

Direct-iterative inverse design methods use two distinct code portions. A shape modification code is coupled with an analysis code, and the design geometry is determined iteratively. The shape modification code contains the relationship between the difference in the initial or current surface function and the target function, to the change in geometry necessary to obtain the target. These relationships are termed Body Shape Rule's (BSR) [12].

Separation of the direct and design portions of the direct-iterative method provides benefits and disadvantages. The most advanced analysis code, using the most descriptive set of governing equations, may be used for initial and intermediate direct solutions. The algorithm separation also allows relatively simple BSR's to be distinct from, but equally valid for, any set of analysis governing equations or CFD methods. However, current BSR's are based on Mach number dependent potential theory. This disadvantage requires a different BSR to be used within each flow regime; subsonic, transonic, and supersonic.

Another disadvantage of these popular inverse design methods is that BSR's are only local, or surface, applications of potential theory. This is in contrast to the global Euler or N-S equations. Governing equation compatibility, that is analysis and shape modification with the same governing equations, at either the Euler or N-S level is therefore not presently possible. Compatible, or consistent, direct-iterative shape modification, and so inverse design, is then limited to the FP equations.

The limitation of consistent inverse design at the FP level does have exceptions. A unique stream-tube method has been successfully demonstrated for inverse design using the potential equations and includes rotational effects [16]. Another type of stream-tube method uses the quasi $1 - D$ Euler equations within a $2 - D$ coordinate system [15]. These methods provide both direct and inverse design solutions in $2 - D$ which satisfy their respective equations. However, consistent extension of these design methods to include viscous effects, or a $3 - D$ extension, is not possible.

A general inverse design method would be equally valid for any set of governing equations, CFD technologies, and number of spatial dimensions. Also, governing equation consistency would ensure the inclusion of all relevant physical information within each shape modification estimate. This may reduce the number of shape modifications required to satisfy the design targets, and would be inherently Mach number independent. The goal of the present research is to develop and test a shape modification method with these qualities which my be incorporated within a direct-iterative inverse design algorithm.

Following this introduction, the fundamental development of the proposed new shape modification method will be presented. A truncated Taylor's series expansion of the discrete, global governing equations is the basis of development. The truncated series relates solution and geometry changes with quasi-analytical flux Jacobian matrices. This simple and general concept has previously been demonstrated to provide consistent neighboring steady-state solution predictions and sensitivity derivatives [17-22]. The attributes of the truncated series satisfy the goals of the present research and it provides a general and consistent means for shape modification.

The present research is the first attempt to use the truncated series within the area of shape modification, and therefore inverse design. A number of unique challenges exist in this effort which were not of concern in the previous studies. Implementation considerations for the $2 - D$ Euler equations and a unique solution algorithm will be presented. A simple channel geometry, and design-like test parameters will then be defined. Results of the new method will be presented and discussed for tests including subsonic, transonic, and supersonic inlet Mach numbers, and a summary of the research will conclude the document.

2. Theory. In this section the basic theory and fundamental equations for the new shape modification method are presented. This initial development is general in nature. In a following section the particular set of governing equations used in this research will be presented, along with the specific details of the implementation.

The non-linear, time-dependent, coupled partial differential equations for either the Euler or N-S equations can be expressed as

$$(2.1) \qquad \frac{\partial Q}{\partial t} + R(Q) = 0$$

where Q is the vector of conversed variables. The vector Q contains combinations of density, component velocities, and energy terms. The size of Q and the residual vector, $R(Q)$, depends on the number of spatial dimensions chosen. The residual represents the steady-state form of the governing equations and is an explicit function of the conserved variables. At steady-state conditions the residual is exactly zero and the governing equations, together with proper BC's, are satisfied.

The numerical description or discrete version of the residual can take many forms. Two choices herein are either finite-difference or finite-volume spatial discretization, and either upwind or central difference flux evaluation. The culmination of these and other decisions determine the set of CFD technologies used, which in sum determine the CFD method. Irrespective of the CFD method chosen to describe the residual, the governing partial differential equations must be discretized over the domain in question. The semi-discrete form is then expressed as

$$(2.2) \qquad \frac{dQ_{j,k}}{dt} + R_{j,k}(Q) = 0$$

where the (j,k) indices are used here only as an example for (x,y) coordinates in $2-D$. This expression represents one equation within a system of non-linear, coupled, ordinary differential equations. The system of equations may be integrated in time for unsteady solutions given a proper set of BC's and initial conditions (IC).

A common practice in determining steady state solutions to the governing equations is to integrate the coupled system in pseudo-time from a reasonably selected set of IC's. This is performed in either an explicit or an implicit manner iteratively, where implicit integration is preferred. the Euler implicit, or backward Euler, time integration method is commonly used.

$$(2.3) \qquad \left[\frac{I}{\Delta t} + \frac{\partial R^n(Q)}{\partial Q} \right] \{^n \Delta Q\} = -\{R^n(Q)\} + 0(\Delta t)$$

Here $\{^n \Delta Q\}$ is the finite difference for the vector of discrete conserved variables between the $(n+1)$ st and the (n)th time level.

$$(2.4) \qquad \{^n \Delta Q\} = \{Q^{n+1}\} - \{Q^n\}$$

The explicit dependence of the governing equations on the conserved variables, Q has been emphasized above. What is understood, but not explicitly shown, is the dependence on the discretized domain, the grid, which on the boundaries includes the body geometry. This geometry/grid dependence is generally not expressed since direct solutions generally use a fixed grid with only the discrete values of the conserved variables being of interest.

To this point, only common and well understood analysis concepts have been presented for subsequent comparison. However, the geometry/grid dependence of the residual becomes equally important when computational design methods are considered. The discrete residual vector is then explicitly defined to be a function of both the discrete solution vector, Q, and discrete geometry/grid vector, $\overline{X}$.

$$(2.5) \qquad \{R\} = \{R(Q, \overline{X})\}$$

The vector $\overline{X}$ represents the physical (x, y) coordinates of the discretized geometry and domain.

$$(2.6) \qquad \overline{X} = (x, y)^T$$

The new shape modification method begins with the discrete residual system of equations, Equation 2.5. Consider two non-linear, steady-state solutions to the governing equations, Q_1 and Q_2, which were obtained on two similar geometries/grids, $\overline{X}_1$ and $\overline{X}_2$. A relationship between the solutions and geometries/grids can be obtained with a Taylor series expansion, in both Q and $\overline{X}$, from the first to the second solution and geometry/grid.

$$(2.7) \qquad \{R_2(Q, \overline{X})\} = \{R_1(Q, \overline{X})\} \;+\; \left[\frac{\partial R_1(Q, \overline{X})}{\partial Q}\right] \{\Delta Q\}$$
$$+\; \left[\frac{\partial R_1(Q, \overline{X})}{\partial \overline{X}}\right] \{\Delta \overline{X}\}$$
$$+\; 0[(\Delta Q)^2, \Delta Q \Delta \overline{X}, (\Delta \overline{X})^2]$$

The vectors ΔQ and $\Delta \overline{X}$ are defined to be the finite change in conserved variables and geometries/grids between the solutions.

$$(2.8) \qquad \{\Delta Q\} = \{Q_2\} - \{Q_1\}$$
$$(2.9) \qquad \{\Delta \overline{X}\} = \{\overline{X}_2\} - \{\overline{X}_1\}$$

If the two solutions and geometries/grids are closely related then the higher-order terms of the series can be truncated for a formally first-order accurate equation. Also, since both solutions are at steady-state conditions, both residual vectors are exactly zero and can be dropped to obtain Equation 2.10.

$$(2.10) \qquad \left[\frac{\partial R_1(Q, \overline{X})}{\partial Q}\right] \{\Delta Q\} + \left[\frac{\partial R_1(Q, \overline{X})}{\partial \overline{X}}\right] \{\Delta \overline{X}\} = 0$$

This expression is termed the standard prediction/design equation and was first developed by Taylor, et. al. in 1991 [17]. The functional dependence of the discrete residual on both the solution vector and the geometry/grid coordinates is not a new concept. However, exploitation of this property in conjunction with the truncated Taylor's series expansion is a simple yet powerful tool which has not been fully explored.

The relationship between any finite change in solution, ΔQ, to the corresponding finite change in geometry/grid, $\Delta \overline{X}$, is quite evident in Equation 2.10. The vectors ΔQ and $\Delta \overline{X}$ are related by two flux Jacobian matrices, $\partial R/\partial Q$ and $\partial R/\partial \overline{X}$, which are derived from the global, discrete form of the governing equations. Therefore, both Jacobian matrices are quasi-analytical expressions and are explicitly derived from the flux evaluation method of choice.

The sensitivity of the governing equations to the solution variables, $\partial R/\partial Q$, is a standard matrix used within implicit analysis algorithms, such as Equation 2.3. The shortened phrase flux Jacobian is the commonly used term for this matrix. The sensitivity of the governing equations to the geometry/grid, $\partial R/\partial \overline{X}$, is the focus of the present research and represents a new BSR for direct-iterative inverse design methods. This is a relatively new matrix within the CFD community and requires special distinction. The short and simple phase metric Jacobian, while strictly a misnomer, is suggested and used throughout this report.

The standard prediction/design equation can be used in many ways. If a finite change in the geometry/grid is specified, $\Delta \overline{X}$, then the geometric forcing function, F_x, is known and a change in solution can be predicted, ΔQ. In this format the expression is termed the standard prediction equation and the results are referred to as geometric solution predictions.

$$(2.11) \qquad \left[\frac{\partial R(Q, \overline{X})}{\partial Q} \right] \{\Delta Q\} = - \left[\frac{\partial R(Q, \overline{X})}{\partial X} \right] \{\Delta \overline{X}\} = -\{F_x\}$$

Note that the subscripts have been dropped here for expedience, and will be shown subsequently only when necessary for clarity.

Geometric solution predictions have been demonstrated for both the $2-D$ Euler [17] and Thin Layer N-S (TLNS) [21] equations. Non-geometric solution predictions have also been demonstrated with a modified version of Equation 2.11 [18]. These solution predictions are driven by variation of non-geometric design variables such as inlet Mach number, angle of attack, and exit pressure. The non-geometric form of the standard prediction equation is developed in the same manner as above after the discrete residual is expressed as an explicit function of the non-geometric design variables.

The major use of the standard prediction equation has been in efficiently obtaining sensitivity derivatives for use within optimization design algorithms [19-22]. This is also accomplished by functional dependence modification of the discrete residual to include specific design variables, together with repeated use of the chain rule. These sensitivity derivatives have been obtained with the Euler and TLNS equations for both finite-difference and finite-volume discretization methods, central difference and upwind flux evaluations, and in both 2-D and 3D.

The present research builds on the success and utility of the standard prediction/design equation in the area of inverse design through shape modification. If a finite solution change is specified, ΔQ, then the solution forcing function, F_q, is known and a shape modification prediction can be found, $\Delta \overline{X}$. In this format the expression is termed the standard design equation and the results are referred to as design predictions.

$$(2.12) \qquad \left[\frac{\partial R(Q, \overline{X})}{\partial \overline{X}} \right] \{\Delta \overline{X}\} = - \left[\frac{\partial R(Q, \overline{X})}{\partial Q} \right] \{\Delta Q\} = -\{F_q\}$$

The standard design equation is the fundamental equation for the proposed new shape modification method. This equation enables shape modification to be performed with the same governing equations, and successful CFD techniques, as originally selected for the analysis solution. This is the essence of consistent design and is guaranteed since the method begins with the discrete version of the governing equations. Also, since the governing equations apply for all flow regimes, the consistent shape modification method is similarly Mach number independent.

Since the higher-order terms were truncated in the Taylor series expansion, each shape modification is strictly first-order accurate. Use of all domain and BC equations also produces a global system of equations which is more costly to solve than current local methods. However, consistent design ensures that all relevant near and far-field physical information is included in the metric Jacobian BSR. Incorporating all the relevant physics within each shape modification estimate may reduce the total number of analysis/design iterations.

In summary, the dual functional dependence of the discrete residual on Q and $\overline{X}$, in conjunction with the truncated Taylor series expansion, provides a new shape modification method which is simple in concept and straightforward in application. As no restrictions were placed in the development, the new method has general applicability. That is, it is not restricted to any set of governing equations, CFD technologies, or spatial dimensions. In the following sections specific details of implementation will be discussed for the 2-D Euler equations.

3. Euler Equations. The governing equations used in the present research are the 2-D Euler equations. After a transformation from cartesian (x, y) to generalized (ξ, η) coordinates these equations may be written as

$$(3.1) \qquad \frac{1}{J} \frac{\partial Q}{\partial t} + \hat{R}(Q) = 0$$

where J is the determinant of the Jacobian matrix of the coordinate transformation. The conserved variables and residual are

$$(3.2) \qquad Q = \{\rho, \rho u, \rho v, \rho e_0\}^T$$

$$(3.3) \qquad \hat{R}(Q) = \frac{\partial \hat{F}(Q)}{\partial \xi} + \frac{\partial \hat{G}(Q)}{\partial \eta}$$

where ρ is the density, u and v are the velocity components in cartesian coordinates, and e_0 is the total energy

$$(3.4) \qquad e_0 = e + \frac{u^2 + v^2}{2}$$

and e is the specific internal energy. Also, the $2-D$ primitive variables are defined below and include the pressure, P.

$$(3.5) \qquad q = \{\rho, u, v, P\}^T$$

The generalized coordinate flux vectors are given as

$$(3.6) \qquad \begin{aligned} \hat{F}(Q) &= \tfrac{\xi_x}{J} F(Q) + \tfrac{\xi_y}{J} G(Q) \\ \hat{G}(Q) &= \tfrac{\eta_x}{J} F(Q) + \tfrac{\eta_y}{J} G(Q) \end{aligned}$$

with the cartesian counterparts given in Equation 3.7.

$$(3.7) \qquad \begin{aligned} F(Q) &= [\rho u, \rho u^2 + P, \rho uv, \rho u h_0]^T \\ G(Q) &= [\rho v, \rho uv, \rho v^2 + P, \rho v h_0]^T \end{aligned}$$

The stagnation enthalpy, h_0, is defined in Equation 3.8

$$(3.8) \qquad h_0 = e_0 + \frac{P}{\rho}$$

and the calorically perfect ideal gas law is used to evaluate the pressure, P, with the ratio of specific heats is equal to $\gamma = 1.4$

$$(3.9) \qquad P = (\gamma - 1)\rho \left(e_0 - \frac{u^2 + v^2}{2} \right)$$

These governing equations are computationally described in an integral, conservation law form, using an upwind, cell-centered finite-volume formulation [23]. This formulation is identical for both the analysis and shape modification solutions, and is intended for structured $H, 0$, or C type grids within the present research.

To ensure that additional errors are not added to the original series expansion, Van Leer's continuously differentiable flux vector splitting method is used [24]. Second-order upwind and third-order upwind biased primitive variable extrapolation is performed in the stream-wise and normal directions respectively. Also, both flux and metric Jacobian matrices require proper linearization of all boundary equations.

4. Incremental Normal Equations. The global nature of quasi-analytical shape modification requires the inverse of the metric Jacobian matrix. Since the present research is the first assessment of the standard design equations potential for shape modification, it is also the first attempt of this inversion. This section discusses two inversion difficulties and presents the Incremental Normal Equations (INE) as a simple solution algorithm which circumvents these. However, first consider the vector ΔQ in Equation 2.12 as known. Together with the flux Jacobian matrix this vector defines the solution forcing function, F_q, for the standard design equation.

The first difficulty encountered in solving the standard design equation is the evaluation of each Jacobian matrix, and the inversion of the metric Jacobian matrix. Both the flux and metric Jacobian matrices must be evaluated exactly to get the proper solution for $\Delta \overline{X}$. This requires the proper linearization and inclusion of the BC equations within each matrix. The proper solution for $\Delta \overline{X}$ also requires a non-iterative inversion of the metric Jacobian which is costly in terms of storage and computational effort.

The second difficulty encountered in solving the standard design equation for $\Delta \overline{X}$ requires an examination of the size of the metric Jacobian matrix. Recall that the governing equations were transformed from a cartesian, (x, y), to a generalized, (ξ, η), coordinate system and the domain is discretized. In the present effort, (j, k) corresponds to the (ξ, η) directions respectively, with JDIM and KDIM defined as the maximum (j, k) dimensions.

The total number of domain governing and boundary BC equations, and the total number of physical (x, y) coordinate unknowns within the domain are given by m and n respectively. Both the number of equations and the coordinate unknowns are functions of JDIM and KDIM as defined in Equation 4.1.

$$(4.1) \qquad \begin{aligned} m &= (JDIM + 1)(KDIM + 1)(4) \\ n &= (JDIM)(KDIM)(2) \end{aligned}$$

These values define the size of the flux and metric Jacobian matrices, and the ΔQ and $\Delta \overline{X}$ vectors as shown in Equation 4.2.

$$(4.2) \qquad \left[\frac{\partial \hat{R}(Q, \overline{X})}{\partial \overline{X}} \right]_{m \times n} \{\Delta \overline{X}\}_{n \times 1} = - \left[\frac{\partial \hat{R}(Q, \overline{X})}{\partial Q} \right]_{m \times m} \{\Delta Q\}_{m \times 1}$$
$$= -\{F_q\}_{m \times 1}$$

There are approximately twice the number of equations, m, as there are unknowns, n for the $2-D$ equations considered. (In practice, BC equations are not solved at the domain corners and therefore m should be reduced by sixteen but this will not be further noted.)

Consistent aerodynamic inverse design with the Euler or $N-S$ equations is a naturally over-determined problem. The original $2-D$ fluid dynamic partial differential equation has four equations while only two unknowns for shape modification, (x, y). Governing and BC equations are also solved at more positions within the discretized domain than there are physical coordinate positions. Together, these factors produce a massively over-determined system of linear equations and so an exact solution for $\Delta \overline{X}$ is not possible.

In summary, two difficulties exist for the quasi-analytical shape modification solution. The first is the requirement of exact matrix inversion which is costly. This is feasible for $2 - D$ cases, but it is prohibitive for $3 - D$. The second problem is the over-determined nature of the system. Neither of these difficulties is unique to the $2 - D$ Euler equations. The first problem is universal to governing equations, CFD techniques, and spatial dimensions. The second problem, the over-determined nature of the system, is also universal. The ratio of the number of equations to the number of unknowns will change slightly for $3 - D$, but a massively over-determined system of equations will remain.

Two simple techniques are now applied to the standard design equation to overcome both difficulties and to obtain the best solution possible. First, it is beneficial to define new terms and re-cast Equation 4.2.

$$(4.3) \qquad A_{m \times n} = \left[\frac{\partial \hat{R}(Q, \overline{X})}{\partial \overline{X}} \right]_{m \times n} \quad ; \quad z_{n \times 1} = \{\Delta \overline{X}\}_{n \times 1}$$

$$(4.4) \qquad b_{m \times 1} = \left[\frac{\partial \hat{R}(Q, \overline{X})}{\partial Q} \right]_{m \times m} \{\Delta Q\}_{m \times 1} = \{F_q\}_{m \times 1}$$

The standard design equation is now simply defined in Equation 4.5.

$$(4.5) \qquad A_{m \times n} z_{n \times 1} + b_{m \times 1} = 0$$

The first technique is to apply Newton's iterative method to the linear system in Equation 4.5 as shown in Equation 4.6.

$$(4.6) \qquad A_{m \times n} \Delta z_{n \times 1} = -(A_{m \times n} z_{n \times 1} + b_{m \times 1})_{m \times 1}$$

This technique is typically applied to root finding methods for systems of non-linear equations. Recently however, this method was successfully demonstrated to reduce the storage and computational cost of the solution for a linear system [25].

The standard design equation is now in incremental, or iterative, form. As the iterations converge, Δz approaches zero such that any approximations to the left-hand-side (LHS) metric Jacobian matrix will not affect the solution of the right-hand-side (RHS) equations. The LHS matrix may be partitioned such that large $3 - D$ problems can be efficiently solved. Approximations may also stabilize its inversion and reduce the cost of its evaluation. These approximations include inconsistent LHS/RHS numerical evaluation and the addition of a diagonal term. Also, any common iterative algorithm can be used to solve these equations.

The over-determined nature of the standard design equation remains as the second inversion difficulty. The best solution possible for $\Delta \overline{X}$ would occur if each equation in the system is satisfied in a least-squared sense. Singular Value Decomposition (SVD) is one popular least-squared method [26]. However, the SVD method is storage intensive and involves arbitrary tolerance filtering of the singular values. Both of these properties eliminate SVD from consideration for fluid dynamic problems which are typically large and include complex physical phenomena.

The normal equations method of solution also obtains a least-squared solution for an over-determined linear system [26]. Simple pre-multiplication of both sides of Equation 4.6 by the transpose of the metric Jacobian matrix, A^T, defines the normal equation format as shown in Equation 4.7.

$$(4.7) \quad [A^T_{n \times m} A_{m \times n}]_{n \times n} \Delta z_{n \times 1} = -A^T_{n \times m}(A_{m \times n} Z_{n \times 1} + b_{m \times 1})_{m \times 1}$$

The linear system is now determined in Δz, and $\Delta \overline{X}$ is found in a least-squared sense. This total method, Newton's root finding method cast in incremental formulation, defines the INE's. This algorithm is the key development within this research, and therefore is an integral part of the proposed method. The INE's provide an algorithm which is not limited to any set of governing equations, CFD technologies, or number of spatial dimensions.

Recall that the INE's provide the flexibility to approximate the LHS matrix and still obtain the least-squared solution for $\Delta \overline{X}$. Therefore, both metric Jacobian matrices within the normal matrix, $A^T A$, may be different in a numerical sense than either of the two which appear on the RHS. The original normal matrix may then be replaced by an approximate one as shown in Equation 4.8.

$$(4.8) \qquad [A^T_{n \times m} A_{m \times n}]_{n \times n} \rightarrow [\tilde{A}^T_{n \times m} \tilde{A}_{m \times n}]_{n \times n}$$

An efficient method of approximating the normal matrix would use identical metric Jacobians for its evaluation, and this also produces a symmetric normal matrix. All Jacobians on the RHS must be evaluated in a numerically consistent manner to obtain the least-squared solution. Therefore, both the LHS and RHS sides of the INE's are separately consistent in a numerical sense, while each side may be distinct from the other.

While the INE's provide the best possible solution for the over-determined linear system, a least-squared solution, additional errors are introduced. This minimized error solution does not satisfy each equation within the system exactly and $\Delta z = \Delta(\Delta \overline{X})$ is found only in a least-squared sense, Δz_{LS}. The least-squared error of the solution, ϵ_{LS}, is then additional to the second-order terms truncated in the original Taylor's series expansion. This error is the square of the l_2 norm of the standard design equation with the converged vector z_{LS} as shown in Equation 4.9.

$$(4.9) \qquad \|(A_{m \times n} z_{LS,n \times 1} + b_{m \times 1})_{m \times 1}\|_2 = \sqrt{\epsilon_{LS}}$$

Further details of the metric Jacobian and normal matrices structure and elements are given in Appendix A. However, general properties of the INE's are of special interest. The normal matrix is sparse and contains nine diagonals, A through I, each element of which is a block 2×2 matrix. One equation within the INE system is given in Equation 4.10 and clearly shows the structure of the system.

$$
\begin{aligned}
& A_{j,k}\Delta z_{j,k-1} + B_{j,k}\Delta z_{j,k} + C_{j,k}\Delta z_{j,k+1} + \\
(4.10) \quad & D_{j,k}\Delta z_{j-1,k-1} + E_{j,k}\Delta z_{j-1,k} + F_{j,k}\Delta z_{j-1,k+1} + \\
& G_{j,k}\Delta z_{j+1,k-1} + H_{j,k}\Delta z_{j+1,k} + I_{j,k}\Delta z X_{j+1,k+1} = -F_{Nj,k}
\end{aligned}
$$

The INE forcing function, F_N, shown in Equation 4.11, is a product of the metric Jacobian and the standard design equation.

$$
(4.11) \qquad F_{N\ n \times 1} = A^T_{n \times m}(A_{m \times n}z_{n \times 1} + b_{m \times 1})_{m \times 1}
$$

The number of diagonals in the normal matrix is independent of the solution variable extrapolation order used in either coordinate direction. This is in contrast to the flux Jacobian matrix where the extrapolation order determines the number of diagonals. The bandwidth of the normal matrix is approximately one half of the flux Jacobians bandwidth which requires less storage and inversion costs. Also, if the normal matrix is symmetric the cost of storage and solution are further reduced.

5. Geometry and Grid Constraints. Unlike direct analysis solutions of the fluid dynamic equations, physical constraints for aerodynamic design solutions must be included [27, 28]. These constraints are both geometric and aerodynamic in nature, and each must be satisfied within a design algorithm. Geometric constraints refer to the restriction of shape modification to certain conditions or limits. Examples are the fixed length of a diffuser, the maximum diameter of an inlet, and closed leading and trailing edges for a blade or an airfoil.

Optimization design methods maximize or minimize aerodynamic quantities such as lift and drag. Aerodynamic constraints for these methods would then define bounds for a design region. However, inverse design methods attempt to satisfy target functions. These methods require some means of determining whether the target function, together with geometric constraints and BC's, are physically possible. If a target function is not physically possible, then aerodynamic constraints alter the initial target such that a valid solution of the flow equations is possible. The resulting geometry then satisfies the adjusted design targets.

Since the present research is focused on shape modification, and not inverse design, these aerodynamic constraints are not included at this time. However, the new shape modification method presents unique geometric constraint requirements which current design methods are not required to consider.

Geometric constraints within current design methods are applied locally at the body, which occupies only a small portion of the discretized boundary. Governing equation consistency is the major advantage of the proposed new method and is inherently global in nature. A global shape modification method requires consideration of global geometric constraints, including the interior of the grid and all free or solid surface boundaries.

These new global geometric constraints, like current local geometric and aerodynamic constraints, are problem dependent and based on physical considerations of the body and domain. Examples are the outer boundary of an airfoil grid, the inlet and exit planes of an impeller, and the centerline of a combustion chamber. These surfaces should be held fixed in their original positions since they are typically not considered design variables.

Therefore, both current local and new global, or grid, geometric constraints must be addressed within the quasi-analytical shape modification method. These combined geometric/grid constraints are expressed as additional equations, and should be satisfied simultaneously with the standard design equation. In practice, these additional equations are of the form $\Delta \overline{X}_{j,k} = 0$, since the initial geometry/grid naturally satisfy the constraints.

However, the geometry/grid constraint equations cannot simply be added to the design system of equations, Equation 4.5. All equations, including the constraints, are only satisfied within the INE's in a least-squared sense. Alternately, if these equations are added to the design system, and proper adjustments made to the metric Jacobian and solution forcing function, they would be exactly satisfied. However, both of these options require dynamic storage definitions in the basic code structure that would be problem and constraint dependent. This would greatly increase the complexity of the code and is not recommended at this time.

A third alternative, which maintains simple coding and general storage requirements, includes adjusting the system to reflect the constraint, and then to replace one equation within the standard design system for each constraint [29]. This method is applied to the standard design equation, Equation 4.5, and exactly satisfies the constraints which appear within the solution vector. This process does, however, violate consistency to some degree. While no attempt to quantify the additional error is presently attempted, this method is considered the best alternative. Details of the implementation are given in Appendix B.

6. INE Singularity. The solution of an over-determined system of equations with the normal equation technique is subject to singularity problems. A non-singular normal matrix for the INE's would be ensured if each equation within the standard design system of equations is linearly independent. Both aerodynamic and geometric singularity issues for the INE's are discussed in this section. A simple method of guaranteeing a non-singular normal matrix is proposed and is an integral part of the present shape modification method.

The unknowns in the global shape modification method are the physical (x, y) coordinates of each grid point within the domain. If two grid points within the initial, or an intermediate, grid occupy the same point in space, then linear independence is violated for two equations within the system. A singular normal matrix may then be produced if this occurred often within the grid.

Viscous and highly non-linear inviscid problems require some level of grid refinement. For these conditions grid points may lie close together and the normal matrix may be nearly singular. However, due to the inherently independent discretization methods employed in CFD, both of these geometric singularity problems should not occur. However, another possible singularity problem does exist for the quasi-analytical shape modification method which is aerodynamic in nature.

Consider a solution prediction with Equation 2.11 where a change in geometries/grids, $\Delta \overline{X} \neq 0.0$, is used to predict the change in solution variables, ΔQ. Specialize this case to one with localized regions of uniform flow. If the grid is simply shifted over this region, the proper result is that no change in solution variables be predicted, $\Delta Q = 0.0$. This result is guaranteed since the local metric Jacobians associated with uniform flow are zero. This may however be detrimental for quasi-analytical shape modification.

Each equation within the standard design system which is associated with uniform flow conditions contains metric Jacobian matrix elements which are zero. If the uniform flow area is large enough then the normal matrix, may be singular or nearly singular. The solution forcing function would also be zero since $\Delta Q = 0.0$. A number of options are available to avoid this problem and still solve the standard design system with the INE's.

One alternative is to test the metric Jacobian matrix and identify those row elements or equations which may cause the singularity. In practice this is the evaluation of solution variable gradients by some arbitrary tolerance level. These equations may then be replaced with some benign identity statement since they do not contribute to the system of equations. This would avoid dynamic storage adjustment and is analogous to the SVD's method of filtering singular values. However, to maintain a general and simple method another alternative is recommended.

Recall that one benefit of an incremental formulation was the flexibility to approximate the LHS matrix to ensure stability while not affecting the final results. In this case the normal matrix would be altered to ensure a stable inversion. However, simply altering this matrix to begin the iterations would be of no benefit since the original singularity problem would be encountered again at convergence. Therefore, a combination of techniques is recommended and tested for all cases within the present effort.

The first technique is to add the identity matrix to the normal matrix as shown in Equation 6.1.

$$(6.1) \qquad [\tilde{A}^T_{n\times m}\tilde{A}_{m\times n}]_{n\times n} = [A^T_{n\times m}A_{m\times n}]_{n\times n} + [I]_{n\times n}$$

This LHS approximation ensures a stable inversion. The second technique is to restrict convergence, measured by the l_2 norm of the $\Delta z = \Delta(\Delta \overline{X})$ vector, to engineering accuracy. Combined, these techniques solve the important equations within the system to an acceptable level of accuracy. At the same time these techniques provide a general, simple, and robust solution of the INE's.

Incomplete convergence of the INE's however does not satisfy the least-squared solution of the normal equations and is therefore an additional error. The total error within the quasi-analytic shape modification method includes a least-squared error, a geometry/grid constraint error, and the incomplete convergence error. If theses are equivalent to the second order terms truncated in the original Taylor's series expansion, then they contribute no additional error to the method. At this time no attempt is made to estimate these errors, however a general assessment of the method's strengths is made with results that follow.

7. Design-Like Test. The forcing function for the standard design equation is the flux Jacobian matrix, $\partial R/\partial Q$, post-multiplied by the ΔQ vector. The flux Jacobian matrix is evaluated with the initial, or current solution and grid, and therefore is known at all times. The vector ΔQ is the finite change in solution variables from the current to the desired solution. This vector is defined at all discrete points within the domain, and contains four or five conserved variables for either a $2-D$ or $3-D$ problem. Unlike the flux Jacobian matrix, the vector ΔQ is only partially known for each shape modification estimate within an iterative design algorithm.

Recall that direct-iterative inverse design methods specify a target surface function as the design goal. This target function is typically only defined over a portion of the surface. On the remaining portion of the boundary, and within the interior, ΔQ is unknown. The target function is also generally only one of the four or five primitive variables defined at each discrete point in the design region. Therefore, the known portion of the vector ΔQ is very small. Current BSR's are applied locally and this limited, or partial, ΔQ is sufficient. However, the quasi-analytical method is a global method, and therefore requires a full ΔQ vector.

The focus of the present research is an initial assessment of the standard design equations and the INE's ability to accurately predict shape modification. A design-like test is then defined as one in which the full ΔQ vector is specified. This provides a simple means of implementing the test, focuses on the methods ability, and is the best-case scenario for evaluation. The assumption implied herein is that if this new method performs well within a design-like environment, then further development effort is warranted. It may then be modified at a later time to perform within a

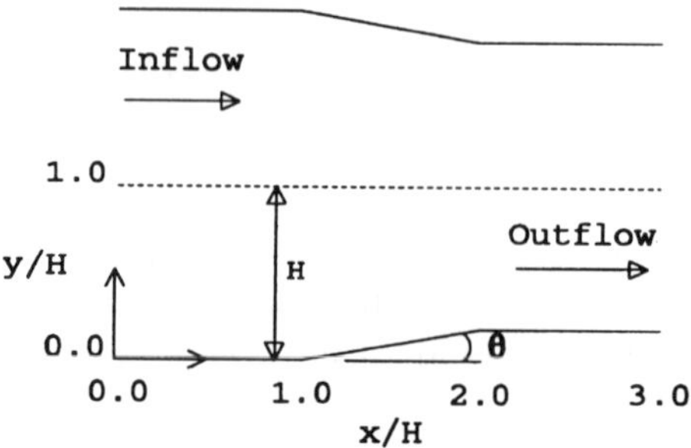

FIG. 8.1. *Channel Geometry*

direct-iterative inverse design algorithm.

The design-like test begins by defining a baseline geometry and identifying one or more design variables. A similar geometry, with perturbed design variables, defines the target geometry. Grids, and non-linear solutions for both the baseline and target geometries are then obtained, and the difference between the solutions defines ΔQ. The baseline grid and solution, and the target solution, but not the target grid, are then used for one shape modification estimate. The goal is to predict the target geometry and grid. Success of the method will be measured by the comparison of the estimated geometry/grid to the known target.

8. Geometry and Test Parameters. The geometry used in the present design-like test is the symmetrical channel shown in Figure 8.1. The channel contains three equal length sections, with a ramp in the middle section. The channel inlet half height, H, is the reference length and the total channel length is three times this value. The ramp angle, θ, is the design variable. Target geometries are those with perturbed ramp angles, while each sections length, and the inlet height remain constant. Also, to test the Mach number independence of the method, subsonic, transonic, and supersonic inlet Mach numbers test cases were used; $M = 0.30, 0.85$, and 2.0.

Since the test geometry is symmetrical about the channel centerline, only the lower portion was computationally modeled. Initially, grids containing ramp angles of $\theta = 5.0^o$, 5.1^o, 5.5^o, and 7.0^o were generated. Each grid contained 31 and 21 lines in the x and y-direction respectively, and were evenly distributed. The baseline grid, with $\theta = 5.0^o$, is shown in Figure 8.2. The three target grids represent design variable changes of $\Delta\theta = 2\%$, 10%, and 40%.

A tangency BC was applied along the lower wall and symmetry was enforced along the channel centerline for all cases. For the subsonic and

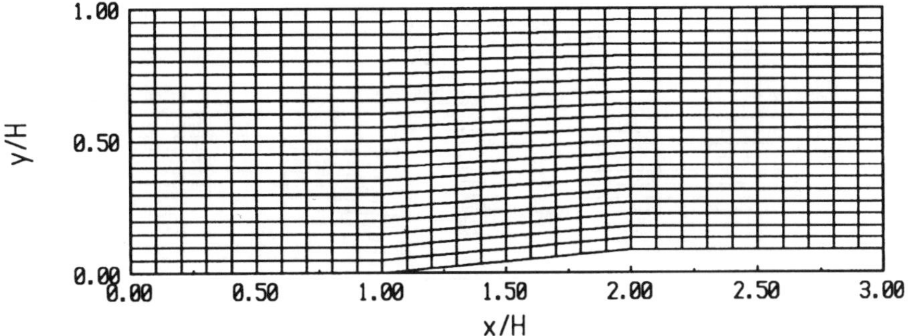

FIG. 8.2. *Baseline Channel Grid* ($\theta = 5.0^o$)

transonic test cases, the inlet BC held the stagnation enthalpy and entropy fixed at free-stream values. Also, the vertical component of velocity was zero and the pressure was extrapolated from the interior. The outlet BC for these cases extrapolated the density and both components of velocity from the interior, and set the back pressure ratio $P_b/P_\infty = 1.0$. The inlet BC for the supersonic test case set the ratios $\rho/\rho_\infty = 1.0$, $u/u_\infty = 1.0$, $v/u_\infty = 0.0$, and $P/P_\infty = 1.0$. The supersonic outlet BC extrapolated all primitive variables from the interior.

Non-linear Euler solutions were then obtained with all four grids at each Mach number. All INE solutions used a banded matrix direct solver [30] and completed 500 iterations of the INE's which converged the l_2 norm of the Δz vector by at least three orders of magnitude. The goal for these tests is to predict the $\theta = 5.1^o$, 5.5^o, and 7.0^o geometries and grids using the INE's.

9. Results. Results for two sets of design-like test cases are presented within this section. The first set is for an unconstrained shape modification test where the geometry/grid constraints are not included. These tests were completed for all nine target geometry and inlet Mach number cases, and are shown for comparison purposes only. The more important, and physically meaningful, set of nine constrained shape modification estimates are also presented.

The first unconstrained test case is the target ramp angle of $\theta = 5.1^o$ at the subsonic inlet Mach number of 0.30. The predicted geometry and grid are shown in Figure 9.1. The results are excellent in a global, qualitative sense, with the predicted grid being similar to the baseline grid. However, more revealing results are predictions of local geometry changes, $\Delta \overline{X}$, from which the actual predicted geometry is easily inferred.

Figure 9.2 illustrates the local variation of geometry, Δy, used in this report to assess the success of each shape modification. All subsequent

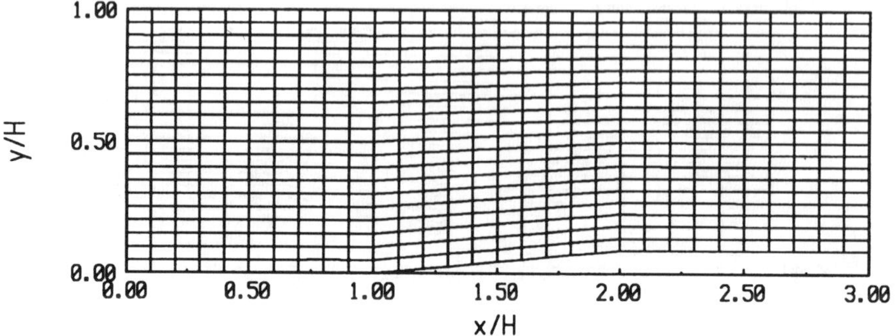

FIG. 9.1. *Unconstrained Predicted Channel Grid* $(M = 0.30, \theta = 5.1^{\circ})$

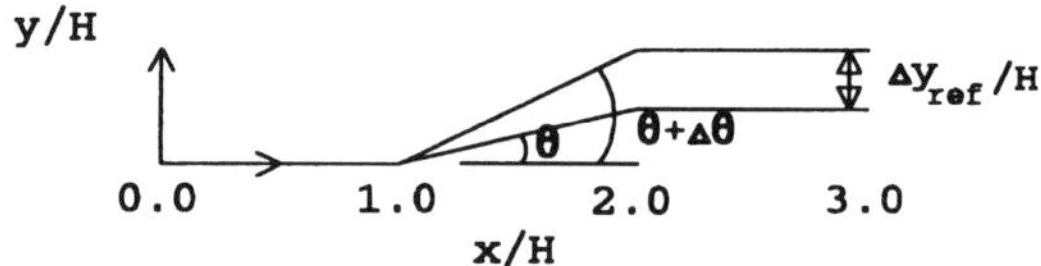

FIG. 9.2. *Channel Geometry Modifications*

geometry shape changes will compare the known target function of Δy to the predicted function. Both the target and the predicted shape changes are normalized by the known change in height along the channel exit plate, Δy_{ref}, which varies for each target geometry. The local predictions for the $M = 0.30$ and $\theta = 5.1^{\circ}$ test case are presented in Figure 9.3.

The predicted Δy changes in channel inlet, ramp, and exit plates are straight lines that are everywhere parallel to the target geometry but are shifted slightly downward. The Δy changes for the channel centerline exhibited these same characteristics with a smaller vertical shift, but are not shown. Also, minor horizontal translations, Δx, for both the centerline and lower wall were noted in the results but are not shown. These translations did not alter the total channel length, or effectively the end points of the ramp, which can be inferred from Figure 9.3.

The results in Figure 9.3 imply that the target $\theta = 5.1^{\circ}$ ramp geometry is obtained since the predicted lines are parallel to the targets, and the

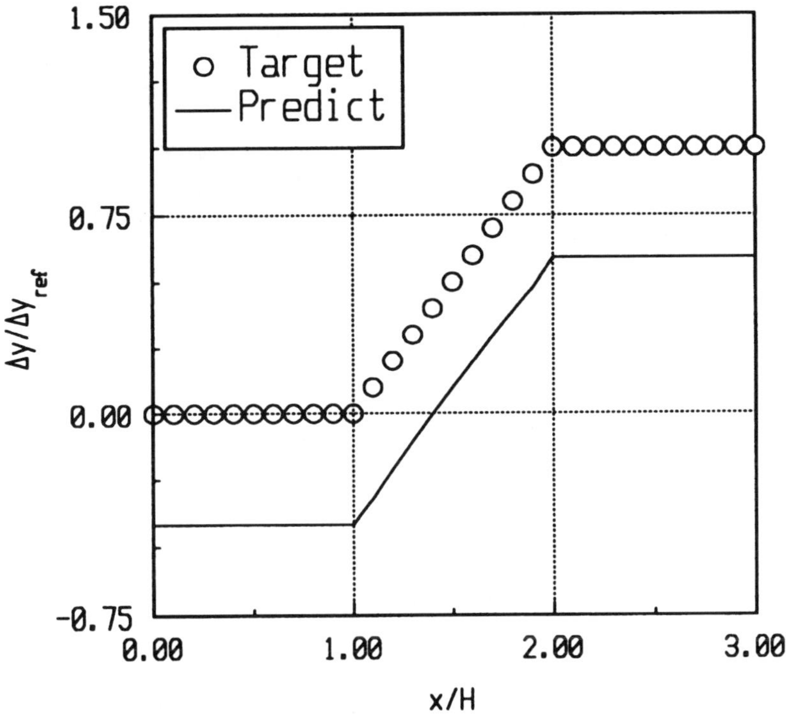

FIG. 9.3. *Unconstrained Predicted Channel Geometry* $(M = 0.30, \theta = 5.1^\circ)$

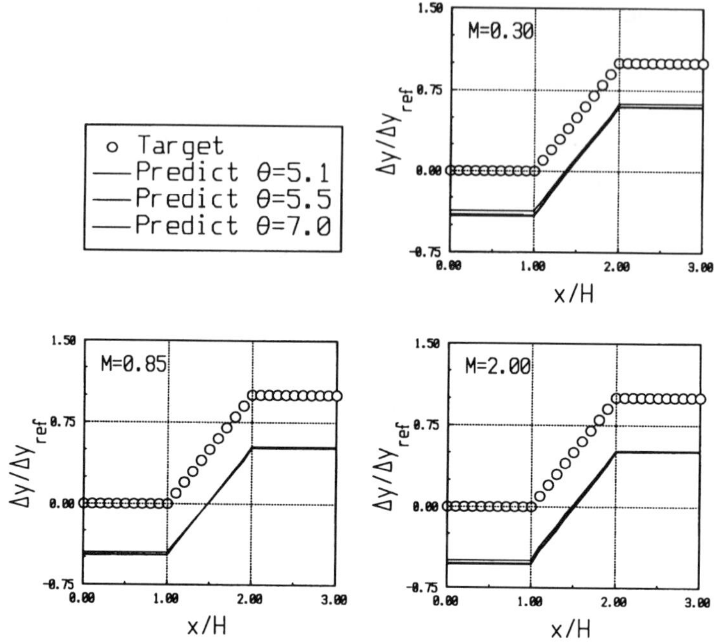

FIG. 9.4. *Unconstrained Predicted Channel Geometries*

lengths and horizontal positions are unchanged. However, the entire domain has been shifted slightly down and expanded in height. This vertical
shifting and stretching, and the excellent global results in Figure 9.1, are
common traits for all of the unconstrained results. Lower wall results for
all unconstrained test cases are shown in Figure 9.4. Each graph represents
a different inlet Mach number test case, $M = 0.30, 0.85$, and 2.0, and each
contains results for all target geometries, $\theta = 5.1^o$, 5.5^o, and 7.0^o. For
each inlet Mach number and target ramp angle, the desired geometry is
obtained but again the domain is vertically shifted.

Collectively, the unconstrained results are very encouraging. Each grid
point within the domain was free to move and yet the target geometry was
obtained with only minor vertical shifting and stretching of the domain.
The next set of results are a constrained version of those just presented,
and therefore represent a more practical application of the INE's.

The geometry/grid constraints for the symmetrical channel fix the
(x, y) positions of the channel inlet plane, the centerline, and the entrance
plate. All x coordinates within the domain are also fixed at the baseline positions. All y coordinates within the domain, on the exit plane, on
the ramp, and on the exit plate are free to move. These constraints effectively fix the channel inlet height and the total length while allowing
geometry changes of the ramp, the channel exit plate, and the exit plane.

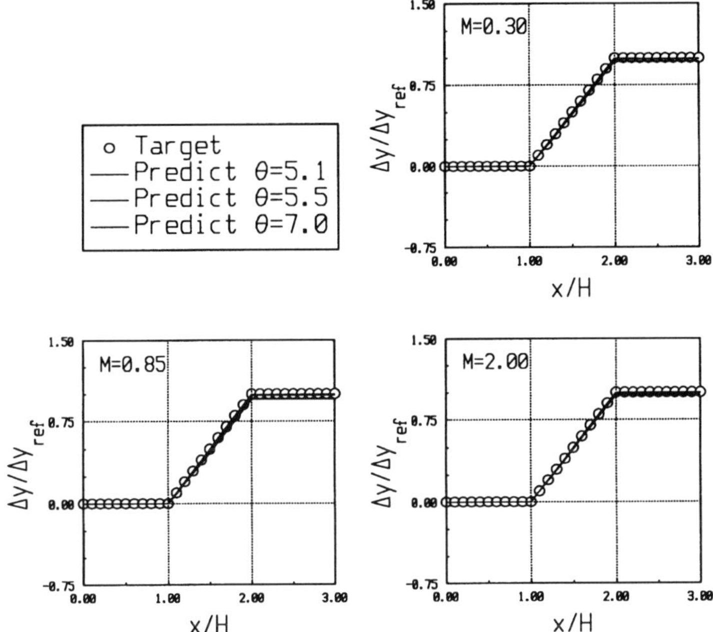

FIG. 9.5. *Constrained Predicted Channel Geometries*

The equations for these constrained test cases however remain massively over-determined.

The predicted grids for each of the nine constrained test cases, while not shown, were again excellent in a global sense. The lower wall results for each of these tests are shown in Figure 9.5. Each graph is again for a different inlet Mach number, $M = 0.30, 0.85$, and 2.0, and each contains results for all target geometries, $\theta = 5.1^o$, 5.5^o, and 7.0^o.

For each inlet Mach number test case, the $\theta = 5.1^o$ and 5.5^o target geometries are obtained. These two cases, with a maximum design variable change of $\Delta\theta = 10\%$, are considered within a normal design range. However, even the extreme ramp angle change of $\Delta\theta = 40\%$ was predicted to within 5% of its target value for the transonic test case. Therefore, the quasi-analytical method, with proper geometric/grid constraints and design-like tests, does provide accurate and physically meaningful shape modification estimates.

The l_2 norm of the standard design equation with z_{LS}, and the l_2 norm of the target grid minus the predicted grid, $(\overline{X}_t - \overline{X}_p)$, are also measures of success for this shape modification method. However, any discussion of these vector norms, or any other error term evaluation, using design-like tests results would be misleading. The most relevant evaluation of this method must be made within a true design environment in which only a small portion of the ΔQ vector is known. Therefore, a through assessment of error terms and equation norms is not included within the present report.

10. Summary. A new method for shape modification was proposed for future inclusion within a direct-iterative aerodynamic inverse design algorithm. The method is based on a truncated quasi-analytical Taylor's series expansion of the global governing equations. The method is general and provides consistent governing equation shape modification for either the Euler or $N - S$ equations, any combination of CFD techniques, and any number of spatial dimensions.

An iterative solution algorithm, the Incremental Normal Equations (INE), was developed to provide a least-squared solution for the inherently global and over-determined consistent shape modification problem. Global geometry/grid constraints were also included to provide practical shape modification estimates. An upwind, cell-centered, finite-volume formulation of the Euler equations in $2 - D$ was used within the present effort for both the initial direct solutions and the shape modification estimates.

The method was evaluated with a symmetric channel which contained a mid-channel ramp. The baseline geometry defined the design variable ramp angle at $\theta = 5^o$. A total of nine tests cases were defined which included combinations of three target ramp angle perturbations $\Delta\theta = 2\%$, 10%, and 40%, and three inlet Mach numbers $M = 0.30, 0.85,$ and 2.00. The global finite change in solution variables from the baseline to the target solutions was provided for testing within a design-like environment.

The quasi-analytical shape modification method was demonstrated to accurately predict target geometries for both an unconstrained and constrained set of design-like tests. This includes the transonic test with an extreme 40% change in the design variable. The constrained version of the method provides more physically meaningful results since geometry changes were effectively restricted to the channel geometry. All test case results were obtained with a single design estimate and clearly reflects the power of consistent shape modification. These results also demonstrate that the method is Mach number independent.

Acknowledgement

The first author is supported by the Graduate Student Researchers Program through the NASA Lewis Research Center (LeRC), Dr. Francis J. Montegani, Program Director. The authors would like to thank Drs. Lonnie Reid, Louis A. Povinelli, and D.R. Reddy of the Internal Fluid Mechanics Division of the NASA LeRC for their support in this effort. Also, a special thanks to Drs. A.C. Taylor and P.L. Andrew for their assistance in both the research and review of this manuscript.

REFERENCES

[1] Slooff, J.W., *Computational Methods for Subsonic and Transonic Aerodynamic Design*, AGARD-R-712, May 1983, pp. 3.1–40.

[2] Dulikravich, G.S., *Aerodynamic Shape Design*, AGARD-R-780, May 1990, pp. 1.1–10.

[3] Volpe, G., *Transonic Shock Free Wing Design*, AGARD-R-780, May 1990, pp. 5.1–16.

[4] Tranen, T.L., *A Rapid Computer Aided Transonic Airfoil Design Method*, AIAA Paper 74-0501, 1974.

[5] Shankar, V., *A Full Potential Inverse Procedure for Wing Design Based on a Density Linearization Scheme*, NASA CR-165991, October 1982.

[6] Volpe, G. and Melnik, R.E., *Method for Designing Closed Airfoils for Arbitrary Supercritical Speed Distributions*, Journal of Aircraft, Vol. 23, No. 10, 1986, pp. 775–782.

[7] Gally, T.A. and Carlson, L.A., *Inviscid Transonic Wing Design Using Inverse Methods in Curvilinear Coordinates,* AIAA Paper 87-2551, 1987.

[8] Bauer, F., Garabedian, P., and Korn, D., *Supercritical Wing Sections III*, Springer-Verlag, New York, 1977.

[9] Sobieczky, H., Yu, N.J., Fung, K.Y., and Seebass, A.R., *New Method for Designing Shock-Free Transonic Configurations*, AIAA Journal, Vol. 17, No. 7, July 1979, pp. 722–729.

[10] Davis, W.H., *Technique for Developing Design Tools from the Analysis Methods of Computational Aerodynamics*, AIAA Journal, Vol. 18, No. 9, 1980, pp. 1080–1087.

[11] Campbell, R.L. and Smith, L.A., *A Hybrid Algorithm for Transonic Airfoil and Wing Design*, AIAA Paper 87-2552, 1987.

[12] Lee, J. and Mason, W.H., *Development of an Efficient Inverse Method for Supersonic and Hypersonic Body Design*, AIAA Paper 91-0395, 1991.

[13] Meauze, G., *An Inverse Time Marching Method for the Definition of Cascade Geometry*, ASME Journal of Engineering for Power, Vol. 104, July 1982, pp. 650–656.

[14] Schmidt, E. and Berger, P., *Inverse Design of Supercritical Nozzles and Cascades*, International Journal For Numerical Methods in Engineering, Vol. 22, 1986, pp. 417–432.

[15] Giles, M.B. and Drela, M., *Two-Dimensional Transonic Aerodynamic Design Method*, AIAA Journal, Vol. 25, No. 9, 1987, pp. 1199–1206.

[16] Dedoussis, V., Chaviaropoulos, P., and Papailiou, K.D., *Rotational Compressible Inverse Design Method for Two-Dimensional, Internal Flow Configurations*, AIAA Journal, Vol. 31, No. 3, 1993, pp. 551–558.

[17] Taylor, A.C., Korivi, V.M. and Hou, G.W., *Sensitivity Analysis Applied to the Euler Equations: A Feasibility Study with Emphasis on Variation of Geometric Shape*, AIAA Paper 91-0173, 1991.

[18] Taylor, A.C., Hou, G.W., and Korivi, V.M., *An Efficient Method for Estimating Neighboring Steady-State Numerical Solutions to the Euler Equations*, AIAA Paper 91-1680, 1991.

[19] Taylor, A.C., Hou, G.W., and Korivi, V.M., *A Methodology for Determining Aerodynamic Sensitivity Derivatives with Respect to Variation of Geometric Shape*, AIAA Paper 91-1101, 1991.

[20] Baysal, O. and Eleshaky, M.E., *Aerodynamic Design Optimization Using Sensitivity Analysis and Computational Fluid Dynamics*, AIAA Paper 91-0471, 1991.

[21] Taylor, A.C. Korivi, V.M. and Hou, G.W., *Approximate Analysis and Sensitivity Analysis Methods for Viscous Flow Involving Variation of Geometric Shape*, AIAA Paper 91-1569, 1991.

[22] Korivi, V.M., Taylor, A.C., Hou, G.W., Newman, P.A., and Jones, H.E., *Sensitivity Derivatives for a 3D Supersonic Euler Code Using the Incremental Iterative Strategy*, AIAA 11th Computational Fluid Dynamics Conference, Orlando, Florida, July 1993.

[23] Thomas, J.L. and Walters, R.W., *Upwind Relaxation Algorithms for the Navier-Stokes Equations*, AIAA Journal, Vol. 25, No. 4, 1987, pp. 527-534.

[24] Van Leer, B., *Flux-Vector Splitting for the Euler Equations*, ICASE Report 82-30, September 1982.

[25] Korivi, V.M., Taylor, A.C., Newman, P.A., Hou, G.W., and Jones, H.E., *An Incremental Strategy for Calculating Consistent Discrete CFD Sensitivity Derivatives*, NASA TM-104207, February 1992.

[26] Golub, G.H. and Van Loan, C.F., *Matrix Computations*, 2nd ed., Johns Hopkins University Press, Baltimore, 1989, pp. 193-259.

[27] Volpe, G. and Melnik, R.E., *The Role of Constraints in the Inverse Design Problem for Transonic Airfoils*, AIAA Paper 81-1233, 1981.

[28] Volpe, G., *Geometric and Surface Pressure Restrictions in Airfoil Design*, AGARD-R-780, May 1990, pp. 4.1-14.

[29] Reddy, J.N., *An Introduction to the Finite Element Method*, 1st ed., McGraw-Hill, New York, 1984, pp. 173-174.

[30] Riggins, D.W. and Walters, R.W., *The Use of Direct Solvers for Compressible Flow Computations*, AIAA Paper 88-0229, 1988.

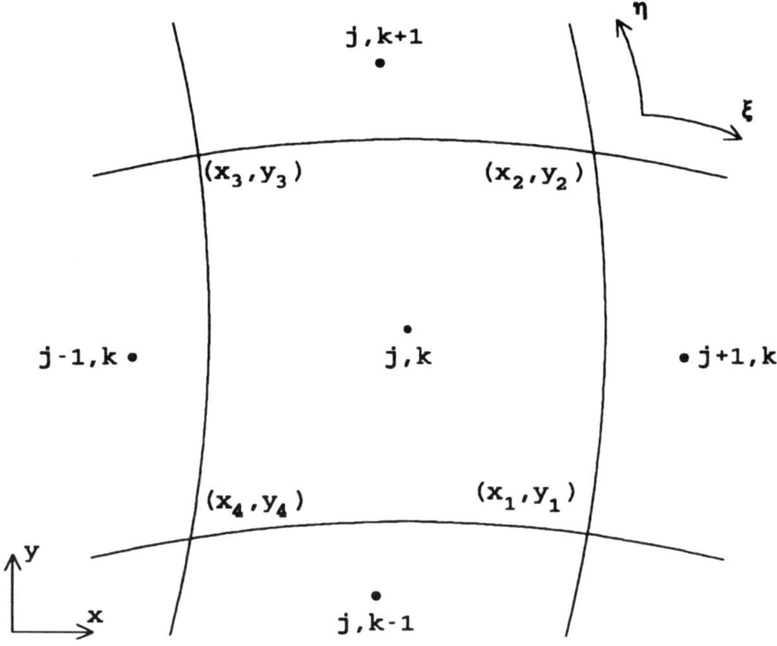

FIG. A.1. *(j,k)th Computational Cell*

A. Appendix. The purpose of this appendix is to complete the presentation of the metric Jacobian and normal matrices, and their associated system of equations. The sparse and systematic structure of each Jacobian matrix is a function of the domain discretization used in the present research. An illustration of the typical (j, k)th computational cell is shown in Figure A.1. The (x, y) grid points surrounding each domain cell and boundary cell-face are labeled one through four, but are only for local designation purposes.

Each equation within the standard design system of equations, Equation 4.2, is a function of four local metric Jacobians, $W1 \rightarrow W4$, four local physical coordinate vectors, $\overline{X}_1 \rightarrow \overline{X}_4$, and one solution forcing function vector, F_q. One equation within the standard design system is given in Equation A.1.

$$\text{(A.1)} \quad \begin{aligned} W1_{j,k}\Delta\overline{X}_{1j,k} + W2_{j,k}\Delta\overline{X}_{2j,k}+ \\ W3_{j,k}\Delta\overline{X}_{3j,k} + W4_{jk}\Delta\overline{X}_{4j,k} = -F_{qj,k} \end{aligned}$$

The local metric Jacobian matrices, physical coordinate vectors, and solution forcing function vectors are of size $4X2$, $2X1$, and $4X1$ respectively. Details of the local metric Jacobian matrices, and the evaluation of these for Van Leer's flux vector splitting method, are given in reference [17].

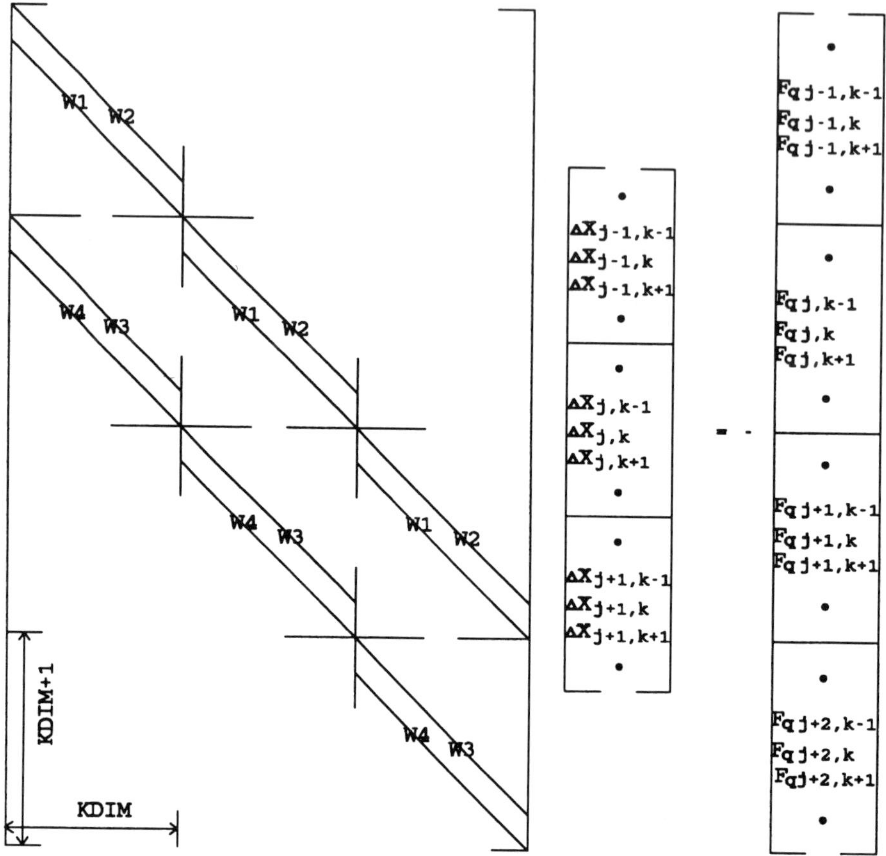

FIG. A.2. *Standard Design Equations*

The structure of the global metric Jacobian matrix and the standard design system of equations are presented in Figure A.2. The size of the global metric Jacobian matrix is $m \times n$. The number of equations within the system, $m = (JDIM+1)(KDIM+1)(4)$, is equal to the number of domain cell-center governing and boundary cell-face BC equations. The number of columns of the metric Jacobian matrix, $n = (JDIM)(KDIM)(2)$, is equal to the number of (x, y) unknowns within the domain. The system in Figure A.2 is shown such that KDIM controls the structure of the metric Jacobian matrix.

The metric Jacobian matrix has JDIM column sections, each of length (KDIM)(2). There are also JDIM+1 row sections in the matrix, each of length (KDIM+1)(4). For simplicity, Figure A.2 is drawn for JDIM=3. Also recall, the number of equations defined throughout the domain is sixteen less than m. However, these extra equations are included in Figure

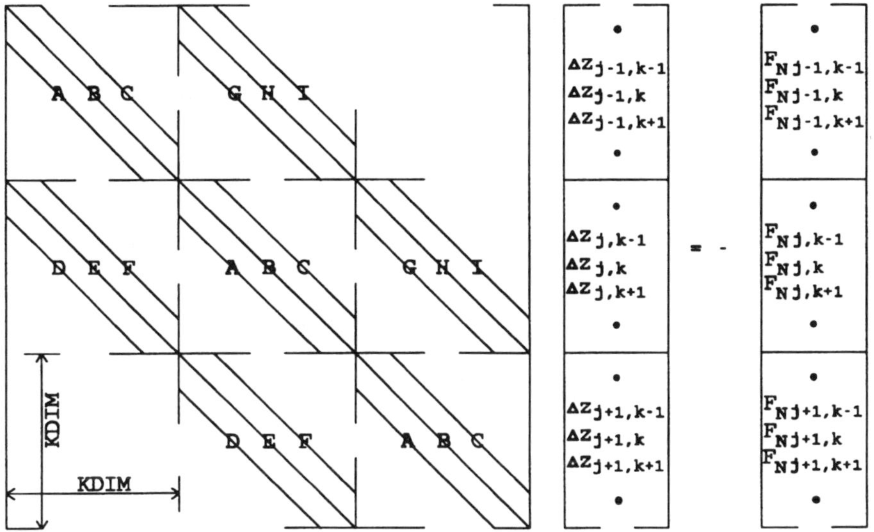

FIG. A.3. *Incremental Normal Equations*

A.2 to maintain a systematic structure.

The over-determined standard design system of equations is solved with the INE's, Equation 4.7, which requires the normal matrix, $A^T A$. One equation within this system was given in Equation 4.10. The sparse diagonal structure of the normal matrix, and the INE system of equations are shown in Figure A.3. Again, the example in this figure is for JDIM=3 and KDIM controls the normal matrix structure. The molecule which represents each equation within the system and the grid correspondence of the normal matrix diagonals are shown in Figure A.4.

The normal matrix is square and has JDIM row and column sections, each of length (KDIM)(2). The half bandwidth of the normal matrix is (KDIM)(2) and is approximately one half the size of the square flux Jacobian matrix. Each element of the normal matrices nine diagonals is a linear combinations of local metric Jacobian multiplications and are given

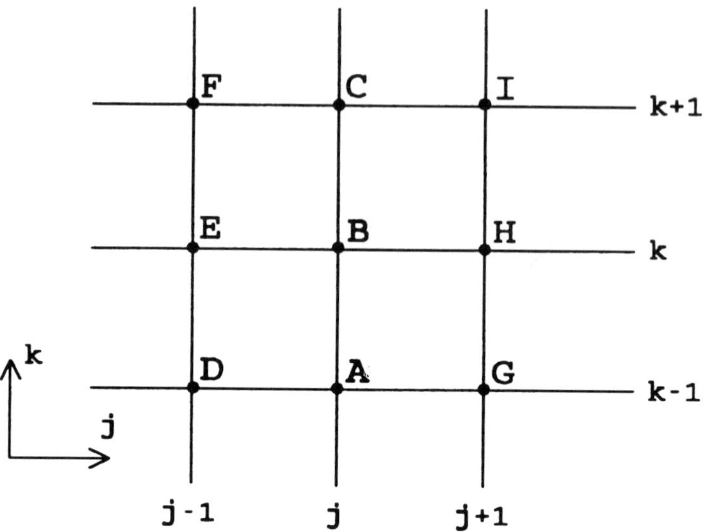

FIG. A.4. *INE Molecule*

in Equation A.2.

$$
\begin{aligned}
A_{j,k} &= W2_{j,k}^{T} W1_{j,k} + W3_{j+1,k}^{T} W4_{j+1,k} \\
B_{j,k} &= W1_{j,k+1}^{T} W1_{j,k+1} + W2_{j,k}^{T} W2_{j,k} + \\
        &\quad W3_{j+1,k}^{T} W3_{j+1,k} + W4_{j+1,k+1}^{T} W4_{j+1,k+1} \\
C_{j,k} &= W1_{j,k+1}^{T} W2_{j,k+1} + W4_{j+1,k+1}^{T} W3_{j+1,k+1} \\
D_{j,k} &= W2_{j,k}^{T} W4_{j,k} \\
E_{j,k} &= W1_{j,k+1}^{T} W4_{j,k+1} + W2_{j,k}^{T} W3_{j,k} \\
F_{j,k} &= W1_{j,k+1}^{T} W3_{j,k+1} \\
G_{j,k} &= W3_{j+1,k}^{T} W1_{j+1,k} \\
H_{j,k} &= W4_{j+1,k+1}^{T} W1_{j+1,k+1} + W3_{j+1,k}^{T} W2_{j+1,k} \\
I_{j,k} &= W4_{j+1,k+1}^{T} W2_{j+1,k+1}
\end{aligned}
\tag{A.2}
$$

Recall that the normal matrix coefficients are defined at cell-corners and the metric Jacobians are defined at cell-centers and boundary cell-faces. Two sets of indices are then used; one set for the cell-centers, and another set for the cell-corners. A graphical representation of Equation A.2, and the dual set of cell-center and cell-corner indices, are presented in Figure A.5.

The normal matrix is symmetric if both metric Jacobian matrices used in its evaluation are identical. The symmetrical correspondence between the lower the upper diagonal terms is evident in Equation A.2. Symmetry of the main diagonal B, and the lower diagonal symmetry correspondence

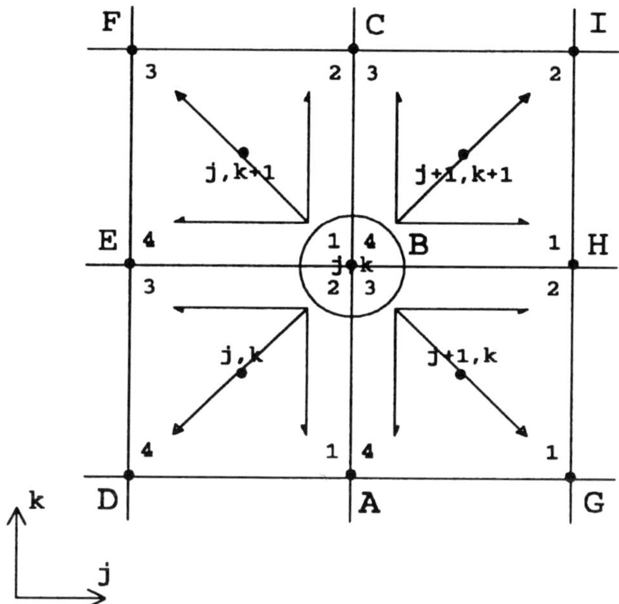

FIG. A.5. *INE Normal Matrix Diagonals*

to upper diagonals are emphasized in Equation A.3.

$$
\begin{aligned}
B_{j,k} &= B_{j,k}^{T} \\
A_{j,k} &= C_{j,k-1}^{T} \\
F_{j,k} &= G_{j-1,k+1}^{T} \\
E_{j,k} &= H_{j-1,k}^{T} \\
D_{j,k} &= I_{j-1,k-1}^{T}
\end{aligned}
$$

(A.3)

Also, the INE normal forcing function, F_N, is defined in Equation A.4, and a graphical representation is shown in Figure A.6.

$$
F_{Nj,k} = W1_{j,k+1}^{T} F_{qj,k+1} + W2_{j,k}^{T} F_{qj,k} \\
+ W3_{j+1,k}^{T} F_{qj+1,k} + W4_{j+1,k+1}^{T} F_{qj+1,k+1}
$$

(A.4)

B. Appendix. The purpose of this appendix is to present the details of implementation for the geometry/grid constraints which are necessary for physically meaningful quasi-analytical shape modification. The method used in the present research includes adjusting the standard design system of equations, Equation 4.5 , to account for each constraint. This is followed by replacing one equation within the system with each constraint. This is a simple method which guarantees that the constraint is exactly satisfied and naturally appears within the solution vector.

In the present code the metric Jacobian matrix is stored such that JDIM controls the structure. This is the opposite storage sequence of that

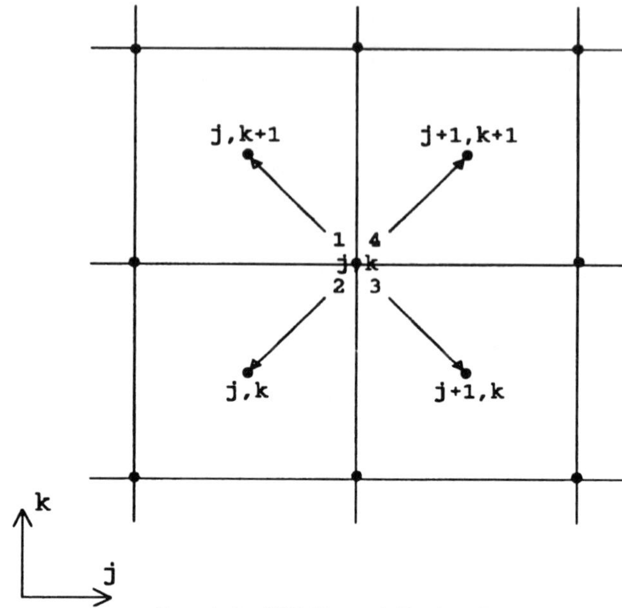

FIG. A.6. *INE Normal Forcing Function*

represented in Figure A.2, but results in a similar matrix structure. (In practice only the local metric Jacobian vectors, $W1 \rightarrow W4$, are stored, but for this discussion consider the entire metric Jacobian matrix as stored).

Each of the (JDIM)(KDIM) column pairs within the metric Jacobian matrix are associated with an x and y variable in that order. Each of the (JDIM+1)(KDIM+1) row sections of the standard design system of equations are associated with a set of four cell-center governing or cell-face BC equations. These equations are generally ordered in a continuity, ξ-momentum, η-momentum, and energy equation sequence. The metric Jacobian matrix and solution forcing function are represented by $A_{m \times n}$ and $F_{q\ m \times 1}$ respectively. Each element within the matrix and the forcing function are denoted by $A_{i,j}$ and $F_{q\ i}$, where $i = 1, m$ and $j = 1, n$.

Each geometry/grid constraint is expressed as $\Delta s_{me,ne} = \epsilon$ where Δs is a generic variable for either the change in x or y. The (j)th column of the matrix, and so the (j)th x or y coordinate, determines the value of both me and ne. To adjust the standard design system of equations for each constraint the following sequence of operations is performed.

$$
\begin{aligned}
F_{qi} &\rightarrow F_{qi} - A_{i,ne} * \epsilon \quad ; \quad i = 1, m \\
A_{i,ne} &\rightarrow 0 \quad ; \quad i = 1, m
\end{aligned}
$$

(B.1)

The (me)th equation is also selected to remove the system, and is replaced by the constraint equation. This involves the following sequence of operations for each constraint.

$$(\text{B.2}) \qquad \begin{aligned} A_{me,j} &\to 0 \quad ; \quad j = 1, n \\ A_{me,ne} &\to 1 \quad and \quad F_{q\ me} \to \epsilon \end{aligned}$$

This method of implementation adjusts all equations within the system for each geometric constraint. However, only the first n, of the total m, equations are considered for replacement. Given the present storage sequence of the metric Jacobian matrix, this method effectively concentrates the constraint enforcement to the lower half of the domain. An equation in the lower grid section may be replaced by a constraint which was written for an upper section grid point. This is considered inappropriate for a general design algorithm, but is used here only for the initial assessment of the quasi-analytical shape modification method.

CONTROL OF STEADY INCOMPRESSIBLE 2D CHANNEL FLOW

JOHN BURKARDT[*] AND JANET PETERSON[†]

Abstract. We consider steady incompressible flows in a 2D channel with flow quantities measured along some fixed, transverse sampling line. From a set of allowable flows it is desired to produce a flow that matches a given set of measurements as closely as possible. Allowable flows are completely specified by a set of control parameters which determine the shape of the inflow at the boundary and the shape of an internal bump which partially obstructs the flow. Difficulties concerning the transformation of this problem into a standard optimization problem are discussed, including the correct choice of functional and algorithm, and the existence of local minima.

1. Introduction. If a log falls into a stream, it disrupts the flow, creating a pattern of ripples and whirls. If the log lies hidden under a bridge, a wise observer standing on the bridge and staring downstream could nonetheless detect the change in the flow, and make a guess as to the size and position of the obstacle. But as the flow rushes on, it rapidly destroys this information, and just a few yards downstream there will be no discernible record of the intrusion into the flow.

In aeronautical design, a similar problem occurs. Instead of a stream, a wind tunnel is used, through which a steady flow of air is driven. It is not a log, but a mockup of an aircraft wing, or fuselage, or forebody, which is deliberately inserted into the flow. Instead of an observer on a bridge, a string of measuring devices are used to record the velocity and pressure of the flow at a fixed position downstream from the obstacle.

For a given orientation and position of the object within the wind tunnel, and for a given pattern of inflowing air, the measured values of velocity and pressure can be regarded as the "signature" of the obstacle. Generally, if two objects differ in shape, their signatures will differ. However, it is possible for one shape to "forge" or approximate the signature of another. This fact can be very useful for some kinds of wind tunnel tests. Certain parts of a plane come "after" other parts; that is, they are further downstream in the airflow. Thus the flow field that strikes the downstream part has already been changed by its interactions with the upstream part and so although it's possible to test a propeller, say, by itself in a wind tunnel, to test a tail assembly requires a mockup of the entire forebody of the airplane as well.

---

[*] Department of Mathematics, Interdisciplinary Center for Applied Mathematics, Virginia Polytechnic Institute and State University, Blacksburg, Virginia, 24061. Supported by the Air Force Office of Scientific Research under grant AFOSR 93-I-0061.

[†] Department of Mathematics, Interdisciplinary Center for Applied Mathematics, Virginia Polytechnic Institute and State University, Blacksburg, Virginia, 24061. Supported by the Office of Naval Research under grant N00014-91-J-1493.

Sometimes a full model cannot be tested because it is too large for the wind tunnel. But it may be possible to find a smaller shape which will fit in the wind tunnel ahead of the object to be tested and which will have the same "signature" as the true forebody. Such a forebody "simulator" should have the property that the velocity and pressure of the air flow, along some transverse downstream plane, will closely match the values associated with the true forebody. A complete description of the forebody simulator problem is available in Huddleston [1] and in Borggaard, Burns, Cliff and Gunzburger [2].

The problem described above is the motivation for the preliminary study given here. This is an ongoing project whose goal is the development of algorithms to select, from an allowable family, a set of flow parameters, and a shape which can be inserted into that flow, which will most closely match a given set of downstream measurements. In Section 2 we give the equations which model steady, viscous, incompressible flow in a channel. We choose to use finite elements to discretize these equations, so in Section 3 we discuss the choice of approximating spaces and give the set of nonlinear equations which must be solved. In Section 4 we discuss the optimization problem using flow sensitivities. The next three sections describe problems in which we allow one or more parameters to vary in order to obtain a flow which matches a given velocity profile. The first problem is simple channel flow with no obstacle in the flow field; here we allow the inflow to vary in order to match a given flow. For the second problem we allow an obstacle to be placed in the flow field but we require that it be modeled by a single parameter. In this case the inflow is fixed and the shape of the bump is allowed to vary. For the third problem we combine the first two and also allow the bump and inflow to be described by more than one parameter. We conclude the report by discussing future work in Section 8.

2. Mathematical model . The equations governing steady, viscous, incompressible flows are the Navier-Stokes equations which can be written in terms of the velocity $u = (u, v)$ and the pressure p as

$$(2.1) \qquad -\nu \Delta u + u \cdot grad\, u + grad\, p \;=\; f \quad \text{in } \Omega$$

$$(2.2) \qquad div\, u \;=\; 0 \quad \text{in } \Omega$$

plus appropriate boundary conditions. Here ν is the constant inverse Reynolds number, f the given forcing function and Ω the domain in $\mathbb{R}^2$ modeling the wind tunnel. We make the assumption that the problem can be restricted to two dimensions; that is, we assume that the behavior of the wind tunnel, the flow field, and the shape are all constant along the z-direction. For the case of simple channel flow with no obstacle, Ω is formed by two parallel horizontal walls. The boundary conditions chosen describe a flow entering the region from the left and passing out of the region at the right. At the inflow we set $u = q(y)$, $v = 0$; at the top and bottom of the channel we set both components of the velocity to zero; and at the outflow we set the usual conditions $v = 0$ and $\dfrac{\partial u}{\partial x} = 0$.

The weak formulation Equations (2.1)–(2.2) which we consider follows [3]. We seek $u \in H^1(\Omega)$ and $p \in L_0^2(\Omega)$ such that

$$\nu \int_\Omega grad\, u : grad\, w \; d\Omega + \int_\Omega u \cdot grad\, u \cdot w \; d\Omega - \int_\Omega p\, div\, w \; d\Omega$$

(2.3)
$$= \int_\Omega fw \; d\Omega$$

(2.4)
$$\int_\Omega \phi\, div\, u \; d\Omega = 0 \; ,$$

and such that $u = v = 0$ on the top and bottom walls, $u = q(y)$, $v = 0$ at the inflow, and $v = 0$ at the outflow. Here $H^1(\Omega)$ represents the space of vector-valued functions each of whose components is in $H^1(\Omega)$, the standard Sobolev space of real-valued functions with square integrable derivatives of order up to one. $L_0^2(\Omega)$ is defined by all functions in $L^2(\Omega)$ with zero mean over Ω. Again see [3] for details.

3. Finite element approximations. In order to approximate the flow we must choose a particular discretization. Our choice here is to use the finite element method, although clearly other discretization methods could be employed. Using the standard techniques of finite elements we discretize our flow region into a finite number of subregions called *elements*, inside each of which we will assume that the flow has a simple structure. For our problems, we choose triangles to create this mesh. For our first problem, which has no internal obstacle, the flow region is rectangular and so it is a simple matter to divide the region up into rectangles, each of which can be split to form two triangles. However, for the problem with an internal obstacle, the flow region is thought of as being a mild distortion of a rectangle and so we are forced to use elements with curvilinear sides. This will require the use of isoparametric elements.

Having represented the region by a mesh of finite elements, we now approximate the continuously varying physical quantities u and p by functions which can be determined from a finite set of data associated with each finite element. Typically these functions will be represented over the entire region by continuous, piecewise polynomials. An examination of the error estimates for the velocity and pressure indicate that one should usually choose one degree higher polynomial for the velocity than for the pressure. For our computations, the velocities are represented by quadratic polynomials and the pressure by linear polynomials. The finite data which represents the velocity, for instance, is then simply the value of the velocity at six particular *nodes* in the element, which are the vertices of the triangle and the midpoints of its sides. Similarly, the pressure is specified by its value at just the three vertices.

We now define a problem which will yield approximate solutions of the weak formulation given in Equations (2.3)–(2.4). Let V^h be the space of

114   J.V. BURKARDT AND J.S. PETERSON

vector-valued functions whose components are continuous piecewise quadratic polynomials over the triangles, let S^h be the space of piecewise linear polynomials over the triangles, and let S_0^h denote the functions in S^h which are constrained to have zero mean. Then we seek a $u^h \in V^h$ and $p^h \in S_0^h$ satisfying

$$\nu \int_\Omega grad\, u^h : grad\, w^h \, d\Omega \;+\; \int_\Omega u^h \cdot grad\, u^h \cdot w^h \, d\Omega$$

$$-\; \int_\Omega p^h \, div\, w^h \, d\Omega = \int_\Omega f w^h \, d\Omega \quad \forall w^h \in V^h$$

$$\int_\Omega \phi^h \, div\, u^h \, d\Omega \;=\; 0 \quad \forall \phi^h \in S^h .$$

The essential boundary conditions are enforced in the usual manner. In particular, at the inflow, $u^h = q^h(y)$, where $q^h(y)$ is the piecewise quadratic interpolant of $q(y)$.

Using standard techniques from the theory of finite elements, we can write these equations as a set of algebraic equations. Each pair of unknown velocities is uniquely associated with a velocity node at which we have two scalar equations. Similarly, each unknown pressure corresponds to a pressure node and a pressure equation. Thus, we should be able to solve the system and compute the values of the flow quantities at each node.

The finite element equations are *nonlinear*, and so they must be solved via an iterative method. The iterative method we employ is Newton's method. See [4] for the formulation and convergence results for the Navier-Stokes equations.

4. The optimization problem. Our goal in this study is to specify the values of some of the flow quantities along a line in the region Ω and then to deduce from that a flow over the whole region, whose values of u and p match (or come as close as possible to) the original given values along some sampling line. We will assume that we have some family of possible flows from which to select. In fact, we will assume that there are one or more *parameters* which characterize this family, so that specifying the value of the parameters completely specifies a flow. In such a case, we may regard the flow quantities as *functions* of the parameters. We will use the letters λ and α for typical parameters.

In order to solve the matching problem, we must first specify a mathematical measure of how well an arbitrary flow matches the given data. It would be desirable to produce a "score", that is, a single number which represents the closeness of the fit, and which is minimized for a perfect fit. One possible choice of a functional to minimize is the integral of the square of the differences between the data and the computed horizontal velocity variables; i.e.,

$$(4.1) \qquad\qquad f_1(u,p) = \int_S (u(x_s,y) - u_s(y))^2 \, dy ,$$

where S denotes the sampling line, x_s denotes the x-coordinate of the sampling line, and $u_s(y)$ represents the sampled velocity data we are attempting to match. Other choices of functionals to be minimized will be discussed in Section 7.

Once we have chosen a particular functional, we can formulate a *minimization problem*, which is to find a flow (u, p) which minimizes the given functional. If we have a single free parameter λ, we can phrase this problem as follows:

> Given a functional $f(u, p)$, where u, p are functions of a parameter λ, find the value of λ that minimizes f.

Clearly there are many different approaches to solving this one-dimensional minimization problem. Rather than seeking to minimize the functional $f(u, p)$ itself, we choose to seek a zero of the derivative of the functional with respect to λ. It was not considered feasible to compute the derivative of the functional with respect to the parameter directly, since the effect of the parameter on the functional is expressed only indirectly, through the flow field. Instead, equations for the *flow sensitivities* are used to approximate the required derivative.

Suppose we can represent a flow field that satisfies a set of flow equations involving a single parameter λ as

$$G(u, v, p, \lambda) = 0\,.$$

Then the corresponding flow sensitivities

$$(\frac{du}{d\lambda}, \frac{dv}{d\lambda}, \frac{dp}{d\lambda})$$

are defined by the linear equations

$$\frac{\partial G}{\partial u}\frac{du}{d\lambda} + \frac{\partial G}{\partial v}\frac{dv}{d\lambda} + \frac{\partial G}{\partial p}\frac{dp}{d\lambda} = -\frac{\partial G}{\partial \lambda}\,.$$

If the original nonlinear flow equations have just been solved, the corresponding flow sensitivities are inexpensive to compute; this is because Newton's method, which is used to solve the nonlinear system, uses an iteration matrix which converges to the sensitivity matrix as the iterates converge to the correct solution. Thus, if the iteration has been deemed to converge, the current, factored iteration matrix may then be used to immediately solve for the sensitivities at very low computational cost.

Now we can reframe the problem of finding a minimum of the optimization functional $f(u, p) = f(\lambda)$ in terms of finding a zero of the derivative of f with respect to λ given by

$$\frac{df}{d\lambda} = \frac{\partial f}{\partial u}\frac{du}{d\lambda} + \frac{\partial f}{\partial v}\frac{dv}{d\lambda} + \frac{\partial f}{\partial p}\frac{dp}{d\lambda}\,.$$

Once we have a method of computing $df/d\lambda$ we can pose the problem of finding a zero of this derivative, and hope that such a zero corresponds to a minimum of the original optimization function, that is, a best match to the flow data.

There are numerous choices for finding the zeros of a function. Since calculating the derivative of the function $df/d\lambda$ requires calculating the second derivative of f, the scalar secant method is an obvious choice when we have only one parameter since it uses only function evaluations. Of course, such a method is not guaranteed to find a minimum; a zero value of the derivative is just as likely to represent a maximum or inflection point. This problem can be forestalled by beginning the optimization with starting points close enough to the correct solution so that convergence to a minimum was very likely. Thus the usefulness of this approach is limited to testing one's code and some very simple problems. When we report on the solution of multiparameter problems in Section 7, we discuss the choice of a suitable optimization package.

5. Simple channel flow. Our first example is the simple case of channel flow with no obstacles in the flow field. These computations were made to test the underlying flow solver and to begin to get some experience with the optimization techniques necessary to solve a general problem.

The channel is modeled by two parallel horizontal walls separated by 3 units and extending from $0 \leq x \leq 10$ units. The boundary conditions chosen for this problem describe a simple, parallel flow entering the region from the left and passing out of the region at the right. The inflow profile is required to be parabolic, but the actual strength of the inflow is allowed to vary, according to the value of a parameter λ. In particular, we set

$$u(0, y) = \lambda y(3 - y) \,.$$

As usual, the pressure must be required to satisfy an additional condition such as having zero mean or fixing its value at some point.

To simulate the experimental process of making measurements and then trying to produce a flow configuration that matched them, a "target" value of λ was chosen. The flow was determined for this value and the flow profile at the sampling line was recorded. It was this "experimental data" that we attempted to match. Because the target flow was actually generated by a particular value of λ, we knew that the minimum value of the functional was zero. This made it easy to determine when the search should halt or when the search was not converging.

A simple test case was set up where the correct solution was $\lambda = 1.0$ and the code was started with the two nearby estimates $\lambda_1 = 0.1$, $\lambda_2 = 0.5$. The secant method was used to find the zero of $df/d\lambda$ where f given by Equation (4.1). For the family of solutions controlled by λ, the functional was actually a quadratic function of λ. Hence its derivative was linear, and the secant method converged to the solution *in one step*. The channel flow results were only useful in that they gave us confidence that the flow field was being solved correctly and that the sensitivities were correctly being used to evaluate the derivative function. In the next problem we allow an obstacle to lie in the flow field, but still require the obstacle to be characterized by just one parameter.

6. Flow over a bump using one parameter. For the second problem, we use a single parameter as a means of selecting a geometric shape which lies in the flow field as an obstacle. The inflow does not vary for this problem, but rather has a fixed strength and parabolic shape.

A "bump" is placed at a fixed location on the bottom of the channel which is again modeled by two parallel walls of length $0 \leq x \leq 10$ units separated by 3 units. The bump is required to be parabolic in shape and extend horizontally from $1 \leq x \leq 3$, but the height of the bump is allowed to vary, being characterized by a parameter α.

The boundary conditions are similar to those for the channel flow, with two exceptions. First, since the inflow does not vary, the inflow equations simplify to

$$u(0, y) = y(3 - y) \, .$$

Secondly, because the *height* of the lower boundary between $1 \leq x \leq 3$ now varies, the boundary conditions for that portion of the lower wall are rewritten as

$$\begin{aligned} u(x, y(x, \alpha)) &= 0 \\ v(x, y(x, \alpha)) &= 0 \\ y(x, \alpha) &= \alpha(x - 1)(3 - x) \, , \end{aligned}$$

for $1 \leq x \leq 3$.

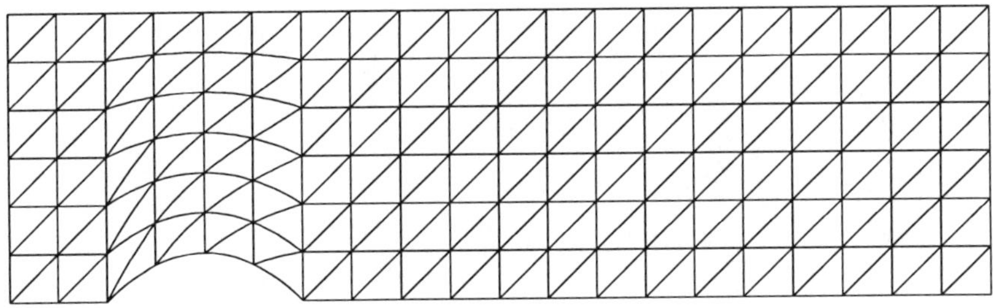

FIGURE 1: A typical region with a bump showing elements.

Because of the curvature of the bump, the computational mesh of the region must also be curved, at least in the vicinity of the bump. Instead of triangular elements with straight sides, *isoparametric* elements were used so that the curved edges of the bump could be modeled. A typical triangulation of the channel with a bump is shown in Figure 1.

Another complication in this problem resulted from the fact that at each step of the optimization, a new value of α was produced for which the corresponding flow had to be computed. Because each α changed the shape of the region, all of the mesh calculations had to be redone at the beginning of every optimization. Thus, the geometry of the region changed at each step, with effects that were harder to predict than those that were caused by simply varying the inflow as in the first example of simple channel flow.

For our computations we set the Reynolds number to one and chose the secant method to find the zero of the functional given by Equation (4.1). This is analogous to the first problem described in Section 5.

The choice of the location of the profile sampling line considerably affected the results. If the profile sampling line was set near the outflow, say at $x_s = 9$, then we often encountered problems. For starting parameters that were quite close to the target value, we were able to get convergence, but often what seemed only slightly greater perturbations of the starting point would cause the program to take many more steps, or in some cases even to fail to converge. We concluded that the difficulty rested in the combined problem of the location of the sampling line relative to the bump and the low Reynolds number of the flow. We can use the sensitivities to see the problem. In Figure 2 the plot shows a bump of height 0.5 and the velocity sensitivity field. Each vector represents the effect that a unit increase in the height parameter would have on the local velocity. As is obvious from the graph, the influence is extremely strong above the bump, but drops off dramatically within a few units downstream. This illustrates the fact that low Reynolds numbers are problematic for flow optimization; i.e., for such cases large changes in the control parameters produce only small changes in the flow.

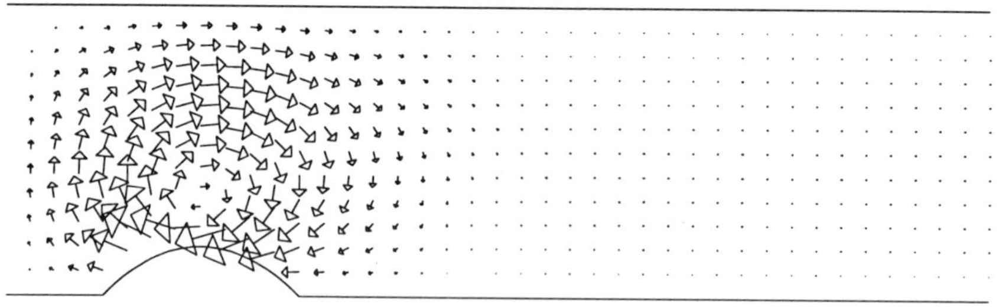

FIGURE 2: The velocity sensitivity field for a bump solution.

For this reason, the profile line was moved to $x_s = 3$, immediately behind the bump. This vastly improved the responsiveness of the functional to changes in the bump.

7. Multiparameter flow past an obstacle. This problem is a generalization, as well as a combination, of the first two problems. In this example we parameterize both the inflow and the shape of the obstacle and allow the number of parameters for each to be greater than one. Due to the more complicated nature of this problem, a major change in the optimization method was needed, since the secant method was not suitable for further use.

A search was made for a more suitable optimization package to use with the code. Robustness and flexibility were key considerations. We chose ACM TOMS algorithm 611 [5], which uses a model/trust region approach for choosing the step, and a BFGS procedure for updating an approximate Hessian matrix. Some of the advantages of this code included: access to the source code, good documentation, good portability with machine dependent quantities handled through calls to a machine dependent function, a *reverse communication* formulation which made it easy to integrate the package into the existing program, three versions of the code in case Hessian or gradients are not available, the fact that it will not accept an iterate if its functional value is higher than that of the current approximate solution, and the fact that it handles an arbitrary number of dimensions.

Another consideration was how to handle more complicated shapes. It was assumed that this would be done by adding more parameters, but it was not clear how those parameters should be used to determine the shapes. In the first two problems discussed above, a single parameter controlled the height of a simple, parabolic shape. In order to model more complicated shapes, we had to choose a reasonable set of shapes and a finite set of parameters to catalog them; we chose to use cubic splines [6] to represent the bumps. Such a representation requires the value of the shape at a specified sequence of nodes. Then a shape is produced which is a piecewise cubic polynomial between the nodes, and which is continuous, with continuous derivatives, at the nodes. In order to complete the system, typical spline representations also require that the slope of the shape be given at the end nodes.

For our shapes, whether they represent an inflow, or an obstacle, we set the value at the first and last nodes to zero. We did not specify the slopes at the end nodes, but rather, used the "not-a-knot" option. This permitted us to define a shape by specifying only the values. The penalty for this simplification was that the shape was required to have one greater degree of continuity at the first and last interior nodes.

Once any shape could be specified uniquely by giving its values at a sequence of nodes, it was natural to consider those values to be the parameters that would be varied in the optimization. One set of nodes would be placed along the inflow boundary. The value of the inflow at those nodes could be used to specify an inflow function defined along the entire line. A second set of nodes would be placed on the bottom of the channel, for $1 \leq x \leq 3$, where the internal obstacle would be placed. The height of the obstacle at each node would be enough to define the whole shape of the obstacle. The number of nodes placed at either location was arbitrary, and hence we could study different inflows or complicated obstacles or both.

The inflow parameters are given in a vector $\boldsymbol{\lambda}$ and the bump parameters in a vector $\boldsymbol{\alpha}$. The parameters enter the flow problem through the boundary conditions. In particular, the value of the inflow velocity at any inlet point $(0, y)$ is given by a function of y and the inflow parameter vector:

$$
\begin{aligned}
u(0, y) &= inflow(y, \boldsymbol{\lambda}) \\
v(0, y) &= 0
\end{aligned}
$$

and the height of the lower channel for $1 \leq x \leq 3$ is determined as a function of x and the bump parameter vector:

$$
\begin{aligned}
u(x, y(x, \boldsymbol{\alpha})) &= 0 \\
v(x, y(x, \boldsymbol{\alpha})) &= 0 \\
y(x, \boldsymbol{\alpha}) &= height(x, \boldsymbol{\alpha}).
\end{aligned}
$$

Secondly, the vector of parameters requires that the optimization search be conducted in $I\!R^M$ rather than $I\!R^1$. The sensitivities are now defined as *partial* derivatives of the flow quantities with respect to the several parameters.

Various problems became apparent as we attempted to solve these multiparameter problems. The first difficulty we encountered was when the optimization code seemed to "get stuck" on an incorrect minimizer. The optimization code at first produced a rapid decrease in the functional value and a correspondingly better approximation to the known solution. However, after a few steps the convergence ground to a halt. The optimization code took progressively smaller steps, and "converged" to a point that was still a significant distance from the target solution. A study of the data showed that the problem was rooted in a discrepancy between the functional and the approximate derivative data we were supplying. We had been computing the sensitivities of the functional with respect to the parameters. These quantities are easily computed from the same linear system used during the Newton-type iteration that produces the flow field itself. They are only approximations to the derivatives of the functional with respect to the parameters, and their accuracy depends on the fineness of the grid of the region. When the optimization code had gotten fairly close to the correct solution, greater accuracy in the derivatives was required than the sensitivities could deliver. In fact, at the false convergence point, the dot product of the sensitivities with the direction vector pointing towards the true solution was positive, suggesting incorrectly that the functional would increase in that direction, when in fact it was monotonically decreasing. This problem disappeared if the mesh was refined, which improved the sensitivities enough to permit convergence. On the other hand, if we wanted to do calculations on a coarse mesh, an alternative was to use the derivative-free version of the ACM code, in which derivatives with respect to the parameters are approximated internally by finite differences and the accuracy could be improved. This increased the cost of computation on a coarse mesh greatly, but also allowed the optimization method to reach the correct target solution.

A second problem that arose was in the choice of the cost functional given in Equation (4.1). To analyze the difficulty, consider the following problem which has one inflow and three bump parameters. The starting point was given as

$$
\begin{aligned}
\lambda &= (0) \\
\alpha &= (0, 0, 0)
\end{aligned}
$$

and the "target" profile was generated at

$$
\begin{aligned}
\lambda &= \left(\frac{1}{2} \right) \\
\alpha &= \left(\frac{3}{8}, \frac{1}{2}, \frac{3}{8} \right),
\end{aligned}
$$

which corresponds to an inflow with parabolic shape and strength $\frac{1}{2}$ and a bump which "happened" to be a parabola of height $\frac{1}{2}$ although it lies in a space of more complicated shapes. The computation proceeded satisfactorily at first, but after four or five steps, the solution ceased to approach the target solution. Instead, the second and fourth components of the parameter vector became negative! In fact, after about twenty iterations, the optimization code returned with the message that the iteration had "converged", though the computed solution was not our intended target solution. The computed solution is shown in Figure 3.

A graph of the shape corresponding to the converged values shows that the resulting bump had roughly the same height as the target bump, but with a "gutter" before and after it. To the eye, at least, the resulting horizontal flow at the sampling line looks "close" to the target values.

The cost functional we used, f_1, seemed to have a local minimum which the optimizer had found. We tested this belief by marching along the line between the computed solution and the target solution. The corresponding functional values are shown in Figure 4 and clearly display a "double dip" curve.

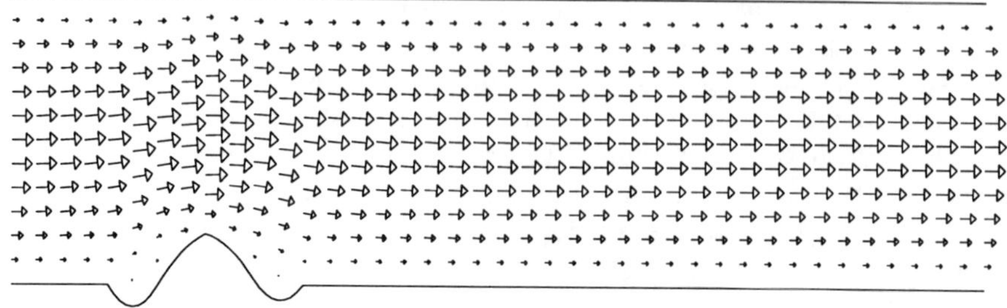

FIGURE 3: A "local minimum" solution that was not the "target".

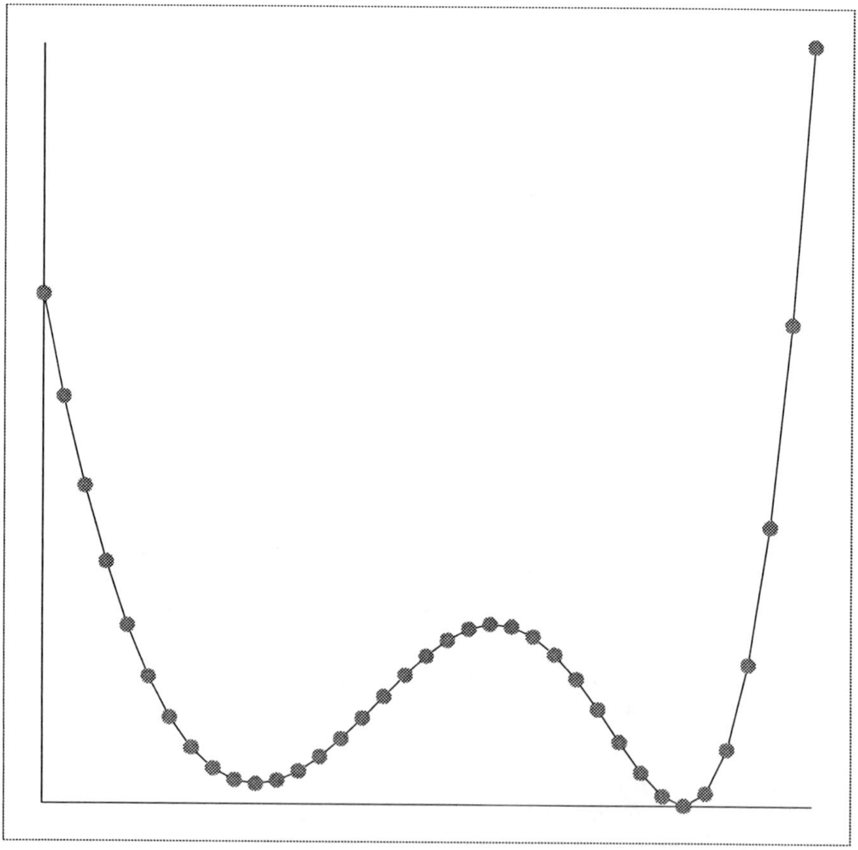

FIGURE 4: The functional values from local minimum to target.

The question then arose as to whether this was actually a local minimum or a *spurious* numerical solution. As other local minima solutions were found, they all tended to share the property of being oscillatory. That is, the shape would either be a crest sandwiched between two valleys, or a valley between two crests. When the mesh was refined for a particular case, the "valleys" doubled in depth, while the crest remained roughly where it had been. This suggested that the program was not approximating a real local minimum, which would have a fixed, finite shape. Rather, some instability in the program or in the problem formulation was generating numerical behavior that did not correspond to a physical solution.

One obvious modification to the problem formulation which might alleviate the problem was to increase the value of the Reynolds number. The fact that the functional seemed to be so insensitive to large changes in geometry was possibly due to the very low Reynolds of the problem. In such a setting, the viscous effects could be expected to dominate the flow, and quickly overwhelm disturbances that the functional would be trying to measure.

Some tests of the program seemed indirectly to bear out this statement. In one test we controlled the inflow with three parameters and the target profile was generated by specifying a parabolic inflow. The program produced as a solution an inflow with a "double hump", having a deep drop in the middle. Nonetheless, this contorted inflow assumed essentially a parabolic shape within two mesh units. In another problem we set up a simple channel flow with no bump. There were 11 equally spaced nodes along the left hand boundary at which an inflow velocity was specified with only the first node having a nonzero velocity. Nonetheless, as demonstrated in Figures 5 and 6, the flow very quickly took on a parabolic profile.

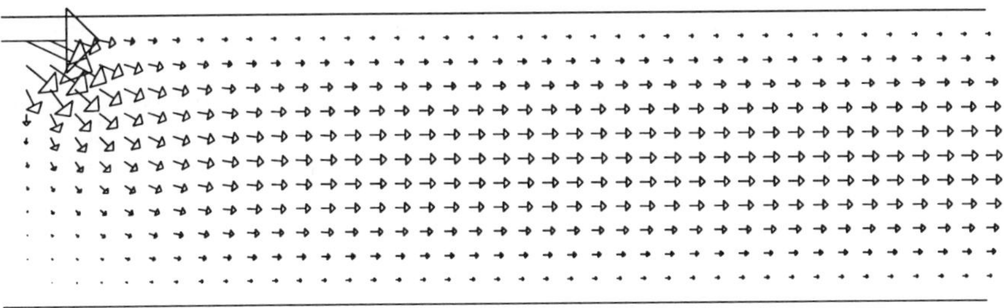

FIGURE 5: Velocity vectors for the single inflow node test.

FIGURE 6: Streamlines for the single inflow node test.

Therefore, it would be natural to expect that only at higher Reynolds number (faster inflow or lower viscosity), would the program be able to distinguish between the convex target bump, and the oscillatory solution that had been found. Calculations using higher Reynolds numbers can be done with our code by incorporating a continuation method. We plan to do this in future work. Another modification to the problem formulation is a change in the cost function f_1 given by Equation (4.1). One possible change was to include the discrepancies in the vertical velocity and pressure along the sampling line as well. This gave the functional

$$f_2(u, v, p) = \left(\int_0^3 (u(x_s, y) - u_s(y))^2 + (v(x_s, y) - v_s(y))^2 + (p(x_s, y) - p_s(y))^2 \, dy \right)^{\frac{1}{2}}.$$

We used this cost functional f_2 for the same problem that the functional f_1 had displayed a local minimum. The new functional also produced a local minimum, differing from the previous one only in that the bump did not actually have "gutters", but rather "low shoulders".

Another choice of the cost functional was one that included the cost of the control. An obvious choice was to estimate the cost of the bump control by approximating the L^2 norm of the height of the bump about the channel bottom:

$$g_1(\alpha) = \left(\int_1^3 (height(x, \alpha))^2 dx \right)^{\frac{1}{2}}.$$

However, this still allowed oscillatory solutions. We only began to get smoother solutions when we took the L^2 norm of the *slope* of the height:

$$g_2(\alpha) = \left(\int_1^3 (height_x(x, \alpha))^2 dx \right)^{\frac{1}{2}}.$$

The smoothing was so effective that we immediately added the corresponding cost function for the inflow control:

$$h_2(\lambda) = \left(\int_0^3 (inflow_y(y, \lambda))^2 dy \right)^{\frac{1}{2}}.$$

Combining these, we have the cost function

$$Cost(\alpha, \lambda) = f_2(\alpha, \lambda) + g_2(\alpha) + h_2(\lambda).$$

This produced smooth solutions, but not correct ones. The effect of the two added cost integrals depends partly on their scale relative to the first integral. If g_2 and h_2 are relatively small, then the optimization code will work for most of the time on minimizing f_2, and only towards the end of the optimization will the control costs perturb the solution slightly. But since a flow with a flat bump and a zero inflow will minimize the control integrals, it's clear that a mistake in scale would cause the optimizer to spend most of its effort smoothing a poor solution.

To avoid this problem, we modified the cost functional to include weights and then allowed these weights to be changed at any time during the run. If the weights were modified, however, the optimizer had to be restarted, since the functional was changed. The new cost function had the form

$$Cost(\alpha, \lambda) = w_1 f_2(\alpha, \lambda) + w_2 g_2(\alpha) + w_3 h_2(\lambda).$$

With these modifications, a typical run of the code would involve several steps. A first run of the code would start from a zero solution, and find a minimizer of the cost function with weights (1, 0.001, 0.001). The solution would be used as the starting point for a second optimization of the cost function with weights (1, 0.00001, 0.00001). Finally, this solution would be used as a starting point for an optimization of the cost function with weights (1, 0, 0). Thus, the control costs kept the shapes from wiggling too much while a "crude" solution was being sought. After a few such procedures, with decreasing weight, the "crude" solution was close enough to the true solution that minimization of the original functional would produce the true solution. Using the control costs allowed the program to avoid the undesirable local minima that had trapped the original program.

8. Future work. The original problem that motivated this investigation sought to find a flow obstacle from a given test set which had a downstream profile similar to one generated by a particular shape which was *not* a member of the test set. We have not handled such cases yet, although they require no change to the program. The optimizer does not require that the functional to be minimized achieve a value of zero.

The spurious solutions that were encountered with the multiparameter problem corresponded to a bump that had a very sharply varying profile. Because the simple gridding routine that was used determines the grid of the region by the shape of the boundary, sharp variations in the bump caused distortions of the grid. These distortions, if severe enough, could cause the calculations to become unreliable. Thus, a better gridding method is desirable, whose accuracy will be less dependent on the smoothness of the boundary.

We chose to represent the space of allowable shapes by cubic splines. We imposed no convexity or positivity requirements on the shapes generated by the splines. We found that the program often generated unacceptable shapes, sometimes only as trial solutions for an iteration, but on occasion as the final solutions of the overall optimization. It is possible that we could avoid this problem by choosing a more restrictive set of allowable shapes.

Finally, we note that we need to be able to handle higher Reynolds number flows. This is so for several reasons. The wind tunnel flow we are ultimately interested in has a very high Reynolds number. Also, our method of trying to match a downstream profile is hampered when the viscosity effects dampen out the perturbations caused by the obstacle.

9. Acknowledgements. The authors would like to thank the IMA for providing a forum for the discussion of flow control issues and they would like to thank Max Gunzburger for many helpful suggestions.

REFERENCES

[1] Huddleston, *Development of a Free-Jet Forebody Simulator Design Optimization Method*, AEDC-TR-90-22, Arnold Engineering Development Center, Arnold AFB, TN, December 1990.
[2] Borggaard, Burns, Cliff and Gunzburger *Sensitivity Calculations for a 2D, Inviscid Supersonic Forebody Problem*, in Identification and Control of Distributed Parameter Systems, to appear.
[3] M. D. Gunzburger & J. S. Peterson, *On Conforming Finite Element Methods for the Inhomogeneous Stationary Navier-Stokes Equations*, Numer. Math. 42, 173–194.
[4] Ohannes A Karakashian, *On a Galerkin-Lagrange Multiplier Method for the Stationary Navier-Stokes Equations* SIAM Journal of Numerical Analysis, Volume 19, Number 5, October 1982 pages 909-923
[5] David Gay, *Algorithm 611, Subroutines for Unconstrained Minimization Using a Model/Trust Region Approach* ACM Transactions on Mathematical Software, Volume 9, Number 4, December 1983 pages 503-524.
[6] Carl DeBoor *A Practical Guide to Splines* Springer Verlag, New York, 1978.

OPTIMALITY CONDITIONS FOR SOME CONTROL PROBLEMS OF TURBULENT FLOWS*

EDUARDO CASAS[†]

Abstract. In this article, we are concerned with the control of the turbulence of viscous, incompressible flows. The control are the body forces or the heat flux through the boundary of the domain occupied by the fluid. The state is the velocity of the fluid and the turbulence is measured by some integral involving the vorticity within the flow. We consider steady and time-dependent three-dimensional flows described by the Navier-Stokes equations, sometimes coupled with the heat equation. We prove existence of optimal controls and derive some first order optimality conditions.

Key words. optimal control problems, Navier-Stokes equations, Boussinesq equations, optimality conditions

AMS(MOS) subject classifications. 49J20, 49K20, 35Q30, 35Q35, 76D05

1. Introduction. We consider the problem of controlling the turbulence behaviour of viscous, incompressible three-dimensional flows. The control variables are the body forces or the heat flux through the boundary of the domain occupied by the fluid. The state is the velocity of the fluid and the cost functional involves the norm of the vorticity of the fluid. This norm gives a good measure of the turbulence within the flow. The relation between the control and the state, that is, the state equation, is described by the Navier-Stokes equations, coupled with the heat equation when the control is the heat flux. The first paper dealing with this problem was published by Abergel and Temam [2]. They considered two-dimensional flows described by evolution equations, the three-dimensional case being more difficult because of the lack of an existence and uniqueness theorem of solution for the evolution Navier-Stokes equations. Other papers dealing with the optimal control of these equations are: Casas [4], Fattorini and Sritharan [7], [6], [8], Sritharan [19], [20]. In [7], [6] and [20] existence of an optimal control was investigated. In [7] and [19], a Pontryagin maximum principle was proved by using the semigroup theory to deal with the state equations. In [4], exitence of an optimal control and optimality conditions of first order were studied, by using variational methods in the study of the state equations.

When steady flows are considered, the nonuniqueness of solution of the state equations occurs in dimensions two and three. To simplify the exposition, we will only consider three-dimensional flows in this paper, but the results and methods are readily extended to the two-dimensional case.

---

* The research of this author was partially supported by Dirección General de Investigación Científica y Técnica (Madrid).

† Dpto. de Matemática Aplicada y Ciencias de la Computación, E.T.S.I. de Caminos, Universidad de Cantabria, 39071 Santander, Spain. E-mail: casasccucvx.unican.es.

Two different approaches have been considered to derive the optimality
conditions in this case. The first one, followed by Gunzburger et al. [11],
[12], uses an abstract theorem for optimality due to Ioffe and Tikhomirov
[13]. The second one, due to Abergel and Casas [1], consists in introducing
a family of approximate control problems, obtained by linearization of the
state equation with the help of an additional control and setting a penalty
term in the cost functional; these approximate problems are well possed and
it is easy to derive the optimality conditions for them; finally, it is passed
to the limit in these optimality systems. This last approach provides a
numerical method to deal with these ill-possed state equations and solve
the control problems.

The plan of this paper is as follows. In §2 and §3, we study the stationary case corresponding to the distributed control of the Navier-Stokes
equation and the boundary control of the system coupled with the heat
equation, respectively. In these sections we describe the method used in [1]
to derive the optimality conditions. In §4 and §5 the corresponding time-
dependent cases are considered. To derive the conditions for optimality
in the evolution case, we must make a suitable formulation of the control
problem, different of that one of [2]. This formulation requires every feasible state of the control problem to be a strong solution of the Navier-Stokes
equations. Here we follow the idea developed in Casas [4].

Let us give some notation, which we will follow in this paper. First, we
assume that the fluid occupies a physical domain $\Omega \subset R^3$, which is bounded
and has a Lipschitz boundary Γ. In Ω we consider the usual Sobolev spaces
$W^{m,p}(\Omega)$ and $W_0^{m,p}(\Omega)$; see, for instance, Adams [3] or Neças [17]. When
$p = 2$ we write $H^m(\Omega)$ and $H_0^m(\Omega)$ instead of $W^{m,2}(\Omega)$ and $W_0^{m,2}(\Omega)$,
respectively. We also put

$$(1.1) \qquad Y = \{\vec{y} \in H^1(\Omega)^3 : \operatorname{div} \vec{y} = 0\} \quad \text{and} \quad Y_0 = Y \cap H_0^1(\Omega)^3,$$

where div denotes the divergence operator. It is easy to check that Y
and Y_0 are separable Hilbert spaces when they are endowed with the inner
products

$$(\vec{y}, \vec{z})_Y = (\vec{y}, \vec{z})_{L^2(\Omega)^3} + a(\vec{y}, \vec{z})$$

and

$$(\vec{y}, \vec{z})_{Y_0} = a(\vec{y}, \vec{z}),$$

respectively, where

$$(1.2) \qquad a(\vec{y}, \vec{z}) = \sum_{j=1}^{3} \int_\Omega \nabla y_j(x) \nabla z_j(x) dx \quad \forall \vec{y}, \vec{z} \in H^1(\Omega)^3.$$

Now, given $T > 0$, we denote $\Omega_T = \Omega \times (0, T)$ and $\Sigma_T = \Gamma \times (0, T)$.

Following Lions and Magenes [16, Vol. 1] we write

$$H^{2,1}(\Omega_T) = \left\{ y \in L^2(\Omega_T) : \frac{\partial y}{\partial x_i}, \frac{\partial^2 y}{\partial x_i x_j}, \frac{\partial y}{\partial t} \in L^2(\Omega_T), \ 1 \le i, j \le 3 \right\}$$

and

$$\|y\|_{H^{2,1}(\Omega_T)} = \left\{ \int_{\Omega_T} \left(|y|^2 + \left| \frac{\partial y}{\partial t} \right|^2 \right) dx\,dt \right.$$

$$\left. + \sum_{i=1}^{3} \int_{\Omega_T} \left| \frac{\partial y}{\partial x_i} \right|^2 dx\,dt + \sum_{i,j=1}^{3} \int_{\Omega_T} \left| \frac{\partial^2 y}{\partial x_i x_j} \right|^2 dx\,dt \right\}^{1/2}.$$

In [16, Vol. 1] it is proved that every element of $H^{2,1}(\Omega_T)$, after a modification over a zero measure set, is a continuous function from $[0, T] \longrightarrow H^1(\Omega)$, so we can consider $H^{2,1}(\Omega_T) \subset C([0, T], H^1(\Omega))$, moreover the inclusion is continuous.

2. Steady flow: distributed control. Let us consider a stationary viscous incompressible flow in Ω; the equations of motion are

$$(2.1) \qquad \begin{aligned} -\nu \Delta \vec{y} + (\vec{y} \cdot \nabla)\vec{y} + \nabla p &= \vec{f} + \mathcal{C}u \text{ in } \Omega, \\ \operatorname{div} \vec{y} &= 0 \text{ in } \Omega, \ \vec{y} = \vec{\phi}_\Gamma \text{ on } \Gamma, \end{aligned}$$

where $\nu > 0$, $\vec{f} \in H^{-1}(\Omega)^3$, $\mathcal{C} \in \mathcal{L}(U, H^{-1}(\Omega)^3)$, $u \in U$, U being a Hilbert space, and $\vec{\phi}_\Gamma \in H^{1/2}(\Gamma)^3$. We assume that

$$(2.2) \qquad \exists \vec{\chi} \in H^2(\Omega)^3 \text{ such that } \ \vec{\phi}_\Gamma = (\nabla \times \vec{\chi})|_\Gamma,$$

where

$$(2.3) \qquad \nabla \times \vec{\chi} = (\partial_{x_2}\chi_3 - \partial_{x_3}\chi_2, \partial_{x_3}\chi_1 - \partial_{x_1}\chi_3, \partial_{x_1}\chi_2 - \partial_{x_2}\chi_1).$$

Under this hypothesis, it is obvious that the usual compatibility condition holds:

$$(2.4) \qquad \int_\Gamma \vec{\phi}_\Gamma(x) \cdot \vec{n}(x) d\sigma(x) = 0,$$

$\vec{n}(x)$ denoting the outward unit normal vector to Γ at the point x; see Nečas [17]. Assumptions on $\vec{\phi}_\Gamma$ allowing to prove the existence of $\vec{\chi}$ satisfying (2.2) are given in Lions [15] and Temam [21].

In (2.1), $\vec{y}$ denotes the velocity, p the pressure, $\vec{f}$ the body forces and u is the control that can act over all domain Ω, or only over a part of Ω, or even only in a given direction of the space. All these possibilities can be treated by choosing a suitable space U and the corresponding linear mapping $\mathcal{C}$.

It is well known that (2.1) has at least one solution $(\vec{y}, p) \in H_0^1(\Omega)^3 \times L^2(\Omega)$; see, for example, [15] or [21]. However there is not general uniqueness results. Moreover, we have the following estimate for the solutions of (2.1):

PROPOSITION 2.1. *Let* $\vec{y} \in H^1(\Omega)^3$ *satisfy* (2.1) *for some* $p \in L^2(\Omega)$. *Then there exist constants* $M_i > 0$, $i = 1, 2$, *independent of* $u \in U$ *and* $\vec{y}$ *such that*

$$(2.5) \qquad \|\vec{y}\|_{H^1(\Omega)^3} \leq M_1 \left(\|\vec{f}\|_{H^{-1}(\Omega)^3} + \|\mathcal{C}\|\|u\|_U \right) + M_2.$$

Moreover $M_2 = 0$ *whenever* $\vec{\phi}_\Gamma = \vec{0}$.

Proof. Let $d(x, \Gamma)$ denote the distance from x to Γ. Then there exists a function of class C^2 in R^3 such that

$$\gamma_\epsilon = 1 \text{ in some neighbourhood of } \Gamma,$$

$$(2.6) \qquad \gamma_\epsilon = 0 \text{ if } \rho(x) \geq 2\delta(\epsilon), \ \delta(\epsilon) = \exp\left(-1/\epsilon\right),$$

$$|\partial_{x_j}\gamma_\epsilon(x)| \leq \frac{\epsilon}{\rho(x)} \text{ if } \rho(x) \leq 2\delta(\epsilon), \ 1 \leq j \leq n.$$

Thus, given $\mu > 0$ arbitrary, we can take $\epsilon > 0$ small enough and $\vec{\phi}_\epsilon = \nabla \times (\gamma_\epsilon \vec{\chi})$, note (2.2), in such a way that $\vec{\phi}_\epsilon|_\Gamma = \vec{\phi}_\Gamma$ and

$$(2.7) \qquad \sum_{i,j=1}^{3} \int_\Omega |y_i \partial_{x_i} z_j \phi_{\epsilon j}|\, dx \leq \mu\|\vec{y}\|_{H_0^1(\Omega)^3}\|\vec{z}\|_{H_0^1(\Omega)^3} \quad \forall \vec{y}, \vec{z} \in Y_0;$$

see [15] or [21].

Let us take $\epsilon > 0$ such that (2.7) holds with $\mu = \nu/2$ and let us set $\vec{z} = \vec{y} - \vec{\phi}_\epsilon \in Y_0$. Then multiplying (2.1) by $\vec{z}$ we obtain

$$(2.8) \qquad \nu a(\vec{z} + \vec{\phi}_\epsilon, \vec{z}) + b(\vec{z} + \vec{\phi}_\epsilon, \vec{z} + \vec{\phi}_\epsilon, \vec{z}) = \langle \vec{f} + \mathcal{C}u, \vec{z}\rangle,$$

where

$$(2.9) \qquad b(\vec{z}^1, \vec{z}^2, \vec{z}^3) = \sum_{i,j=1}^{3} \int_\Omega z_i^1 \partial_{x_i} z_j^2 z_3^j\, dx.$$

Recalling that

$$b(\vec{z} + \vec{\phi}_\epsilon, \vec{z}, \vec{z}) = 0$$

and

$$b(\vec{z}, \vec{\phi}_\epsilon, \vec{z}) = -b(\vec{z}, \vec{z}, \vec{\phi}_\epsilon),$$

we deduce from (2.8) that

$$\nu\|\vec{z}\|_{Y_0}^2 \leq C_1\left(\|\vec{f}\|_{H^{-1}(\Omega)^3} + \|\mathcal{C}\|\|u\|_U\right)\|\vec{z}\|_{Y_0} + |b(\vec{z},\vec{z},\vec{\phi}_\epsilon)| + |b(\vec{\phi}_\epsilon,\vec{\phi}_\epsilon,\vec{z})|.$$

Finally, due to the inclusion $H^1(\Omega) \subset L^4(\Omega)$, we obtain from (2.7), the above inequality and the well known relation

$$|b(\vec{z}^1,\vec{z}^2,\vec{z}^3)| \leq \|\vec{z}^1\|_{L^4(\Omega)^3}\|\vec{z}^2\|_{H^1(\Omega)^3}\|\vec{z}^3\|_{L^4(\Omega)^3}$$

that

$$\nu\|\vec{z}\|_{Y_0}^2 \leq C_1\left(\|\vec{f}\|_{H^{-1}(\Omega)^3} + \|\mathcal{C}\|\|u\|_U\right)\|\vec{z}\|_{Y_0} + \frac{\nu}{2}\|\vec{z}\|_{Y_0}^2 + C_2\|\vec{\phi}_\epsilon\|_{H^1(\Omega)^3}^2\|\vec{z}\|_{Y_0},$$

which, together with the inequality

$$\|\vec{y}\|_{H^1(\Omega)^3} \leq \|\vec{z}\|_{H^1(\Omega)^3} + \|\vec{\phi}_\epsilon\|_{H^1(\Omega)^3} \leq C_3\|\vec{z}\|_{Y_0} + \|\vec{\phi}_\epsilon\|_{H^1(\Omega)^3},$$

leads to (2.5). $\qquad\qquad\qquad\qquad\qquad\qquad\qquad\qquad\qquad\qquad\qquad\qquad\square$

Now we define the functional $J : H^1(\Omega)^3 \times U \longrightarrow R$ by

$$J(u,\vec{y}) = \frac{1}{2}\int_\Omega |\nabla \times \vec{y}|^2 dx + \frac{N}{2}\|u\|_U^2,$$

with $N \geq 0$ and $\nabla \times \vec{y}$ denoting the vorticity of the flow, defined as in (2.3). The physically relevant term in J is of course

$$\frac{1}{2}\int_\Omega |\nabla \times \vec{y}|^2 dx,$$

which provides an estimate of the level of turbulence within the flow; the other term is put there for technical reasons and it is not necessary if the set of admissible controls is bounded.

Given a nonempty convex closed subset K of U, we formulate the optimal control problem as follows:

$$(P1)\left\{\begin{array}{l} \text{Minimize } J(u,\vec{y}) \\ (u,\vec{y}) \in K \times H^1(\Omega)^3 \text{ satisfying (2.1) for some } p \in L^2(\Omega). \end{array}\right.$$

The first question to study is the existence of a solution for (P1).

THEOREM 2.1. *Assumed that $N > 0$ or K is bounded in U, then (P1) has at least one solution.*

To prove this result, it is enough to consider, as usual, a minimizing sequence, which is bounded in $U \times H^1(\Omega)^3$, due to the assumptions of the theorem, and to take into account the convexity and continuity of J; see [1] for the details.

Our next goal is to derive the optimality conditions for Problem (P1), which is made in the following theorem.

THEOREM 2.2. *If $(u_0, \vec{y}_0) \in U \times H^1(\Omega)^3$ is a solution of* (P1), *then there exist a number $\alpha \geq 0$ and some elements $\vec{\varphi}_0 \in H^1(\Omega)^3$ and $p_0, \pi_0 \in L^2(\Omega)$ verifying*

$$(2.10) \qquad \alpha + \|\vec{\varphi}_0\|_{H^1(\Omega)^3} > 0;$$

$$(2.11) \qquad \begin{aligned} &-\nu\triangle\vec{y}_0 + (\vec{y}_0 \cdot \nabla)\vec{y}_0 + \nabla p_0 = \vec{f} + \mathcal{C}u_0 \ \text{in} \ \Omega, \\ &\operatorname{div} \vec{y}_0 = 0 \ \text{in} \ \Omega, \ \vec{y}_0 = 0 \ \text{on} \ \Gamma; \end{aligned}$$

$$(2.12) \qquad \begin{aligned} &-\nu\triangle\vec{\varphi}_0 - (\vec{y}_0 \cdot \nabla)\vec{\varphi}_0 + (\nabla\vec{y}_0)^T \vec{\varphi}_0 + \nabla\pi_0 = \alpha\nabla \times (\nabla \times \vec{y}_0) \ \text{in} \ \Omega, \\ &\operatorname{div} \vec{\varphi}_0 = 0 \ \text{in} \ \Omega, \ \vec{\varphi}_0 = 0 \ \text{on} \ \Gamma; \end{aligned}$$

$$(2.13) \qquad (\mathcal{C}^\star\vec{\varphi}_0 + \alpha N u_0, u - u_0)_U \geq 0 \quad \forall u \in K.$$

Let us remark that sometimes it is possible to get (2.11)–(2.13) with $\alpha = 1$. Indeed, following Gunzburger et al. [12] we say that the control set K has property C at $(u_0, \vec{y}_0)$ if for any nonzero solution $(\vec{\varphi}, \pi) \in H^1(\Omega)^3 \times L^2(\Omega)$ of the system

$$(2.14) \qquad \begin{aligned} &-\nu\triangle\vec{\varphi} - (\vec{y}_0 \cdot \nabla)\vec{\varphi} + (\nabla\vec{y}_0)^T \vec{\varphi} + \nabla\pi = 0 \ \text{in} \ \Omega, \\ &\operatorname{div} \vec{\varphi} = 0 \ \text{in} \ \Omega, \ \vec{\varphi} = 0 \ \text{on} \ \Gamma, \end{aligned}$$

we can find $u \in K$ such that

$$(2.15) \qquad (\mathcal{C}^\star\vec{\varphi}, u - u_0) < 0.$$

Convention will have it that property C is to hold vacuously if there are no nonzero solutions of (2.14).

COROLLARY 2.1. *If K has property C at $(u_0, \vec{y}_0)$, then there exist $\vec{\varphi}_0 \in H^1(\Omega)^3$ and $p_0, \pi_0 \in L^2(\Omega)$ verifying* (2.11)–(2.13) *with $\alpha = 1$.*

Proof. It is enough to remark that (2.14) and (2.15) implies that $\alpha \neq 0$ in (2.11)–(2.13). Then we can replace $\vec{\varphi}_0$ by $\vec{\varphi}_0/\alpha$ and so deduce the desired result. $\qquad\square$

Remark 1. It is obvious that if $U = K = L^2(\Omega)^3$ and $\mathcal{C} =$ inclusion operator from $L^2(\Omega)^3$ into $H^{-1}(\Omega)^3$, then K has property C at $(u_0, \vec{y}_0)$.

For a detailed demonstration of this theorem, the reader is referred to Abergel and Casas [1]. In the §2.1 and §2.2 we sketch the proof.

2.1. The problems (P1_ϵ). The main difficulty to prove Theorem 2.2 is the multivalued character of the relation between the control and the state. To prove this theorem, we introduce a family of problems (P1_ϵ), with well-possed state equations, whose solutions converge toward the solution $(u_0, \vec{y}_0)$, then we derive the optimality conditions for these problems and finally we pass to the limit in these conditions.

For every $\epsilon > 0$ we define the functional $J_\epsilon : U \times Y \longrightarrow R$ by

$$J_\epsilon(u, \vec{w}) \;=\; J(u, \vec{y}(u, \vec{w})) + \frac{1}{2\epsilon} \sum_{j=1}^{3} \int_\Omega |\nabla y_j(u, \vec{w}) - \nabla w_j|^2 dx$$

$$+ \;\; \frac{1}{2} \sum_{j=1}^{3} \int_\Omega |y_j - y_{0j}|^2 dx + \frac{1}{2}\|u - u_0\|_U^2,$$

where $\vec{y}(u, \vec{w})$ is the unique element of $H^1(\Omega)^3$ that satisfies, together with some $p \in L^2(\Omega)$, the system

$$(2.16) \qquad \begin{aligned} -\nu\triangle\vec{y} + (\vec{w} \cdot \nabla)\vec{y} + \nabla p &= \vec{f} + \mathcal{C}u \text{ in } \Omega, \\ \operatorname{div} \vec{y} = 0 \text{ in } \Omega, \; \vec{y} &= \vec{\phi}_\Gamma \text{ on } \Gamma, \end{aligned}$$

To check that this system has a unique solution, it is enough to consider its variational formulation

$$\begin{aligned} &\text{Find } \vec{y} \in Y \text{ such that} \\ &\nu a(\vec{y}, \vec{z}) + b(\vec{w}, \vec{y}, \vec{z}) = \langle \vec{f} + \mathcal{C}u, \vec{z}\rangle \;\; \forall \vec{z} \in Y_0, \end{aligned}$$

and to apply the Lax-Milgram theorem, noting that the $H_0^1(\Omega)^3$-coercivity of the bilinear form is an immediate consequence of the following orthogonality property of b:

$$b(\vec{w}, \vec{z}, \vec{z}) = 0 \quad \forall \vec{w} \in Y \;\; \text{and} \;\; \vec{z} \in H_0^1(\Omega)^3.$$

Note (2.9) for the definition of b.

Now we formulate the problems (P1_ϵ) in the following way

$$(\text{P1}_\epsilon) \begin{cases} \text{Minimize } J_\epsilon(u, \vec{w}), \\ (u, \vec{w}) \in K \times Y \text{ and } \vec{w} = \vec{\phi}_\Gamma \text{ on } \Gamma. \end{cases}$$

The next proposition states that the problems (P1_ϵ) constitute an approximating family of (P1).

PROPOSITION 2.2. *For every $\epsilon > 0$ there exists at least one solution $(u_\epsilon, \vec{w}_\epsilon)$ of (P1_ϵ). Moreover if we denote by $\vec{y}_\epsilon$ the solution of (2.16) corresponding to $(u_\epsilon, \vec{w}_\epsilon)$, then we have*

$$(2.17) \qquad \lim_{\epsilon \to 0} \|u_\epsilon - u_0\|_U = \lim_{\epsilon \to 0} \frac{1}{2\epsilon} \sum_{j=1}^{3} \int_\Omega |\nabla y_{\epsilon j} - \nabla w_{\epsilon j}|^2 dx = 0,$$

$$(2.18) \qquad \vec{w}_\epsilon \to \vec{y}_0 \;\; and \;\; \vec{y}_\epsilon \to \vec{y}_0 \;\; weakly \text{ in } Y,$$

$$(2.19) \qquad \lim_{\epsilon \to 0} J_\epsilon(u_\epsilon, \vec{w}_\epsilon) = J(u_0, \vec{y}_0).$$

Taking into account that System (2.16) has a unique solution $\vec{y} \in H^1(\Omega)^3$ for every pair of controls $(u, \vec{w}) \in U \times Y$, the relation $(u, \vec{w}) \to \vec{y}$ being C^1, and

$$K \times \{\vec{w} \in Y : \vec{w}|_\Gamma = \vec{\phi}_\Gamma\}$$

is convex, it is enough some computations to prove the following result

PROPOSITION 2.3. *Let us suppose that* $(u_\epsilon, \vec{w}_\epsilon)$ *is a solution of* $(\mathrm{P1}_\epsilon)$, *then there exist elements* $\vec{y}_\epsilon, \vec{\varphi}_\epsilon \in H^1(\Omega)^3$ *and* $p_\epsilon, \pi_\epsilon \in L^2(\Omega)$ *such that the following system holds*

$$(2.20) \qquad \begin{aligned} &-\nu\triangle\vec{y}_\epsilon + (\vec{w}_\epsilon \cdot \nabla)\vec{y}_\epsilon + \nabla p_\epsilon = \vec{f} + \mathcal{C}u_\epsilon \ in \ \Omega, \\ &\mathrm{div}\,\vec{y}_\epsilon = 0 \ in \ \Omega, \ \vec{y}_\epsilon = \vec{\phi}_\Gamma \ on \ \Gamma; \end{aligned}$$

$$(2.21) \qquad \begin{aligned} &-\nu\triangle\vec{\varphi}_\epsilon - (\vec{w}_\epsilon \cdot \nabla)\vec{\varphi}_\epsilon + (\nabla\vec{y}_\epsilon)^T\vec{\varphi}_\epsilon + \nabla\pi_\epsilon \\ &= \nabla \times (\nabla \times \vec{y}_\epsilon) + \vec{y}_\epsilon - \vec{y}_0 \ in \ \Omega, \\ &\mathrm{div}\,\vec{\varphi}_\epsilon = 0 \ in \ \Omega, \ \vec{\varphi}_\epsilon = 0 \ on \ \Gamma; \end{aligned}$$

$$(2.22) \qquad (\mathcal{C}^\star\vec{\varphi}_\epsilon + Nu_\epsilon + u_\epsilon - u_0, u - u_\epsilon)_U \geq 0 \quad \forall u \in K.$$

Remark 2. The method described in this section provides an efficient numerical scheme to solve Problem (P1); obviously, the functional J_ϵ should be modified by removing the last two terms. Then Proposition 2.2 may fail to be true, but, under the assumptions of Theorem 2.1, it is still possible to prove that $\{u_\epsilon\}_{\epsilon>0}$ is a bounded sequence in U and every weak limit point, when $\epsilon \to 0$, is a solution of (P1). In fact these subsequences converge strongly in U if $N > 0$. Furthermore $\vec{y}_\epsilon \to \vec{y}_0$ weakly in $H^1(\Omega)^3$ and $\inf(\mathrm{P1}_\epsilon) \to \inf(\mathrm{P1})$.

2.2. Sketch of proof of Theorem 2.3. We are going to pass to the limit in the system (2.20)–(2.22) with the help of Proposition 2.2. In this process the essential point is the boundedness of $\{\vec{\varphi}_\epsilon\}_{\epsilon>0}$ in $H^1(\Omega)^3$. First, let us assume that $\{\vec{\varphi}_\epsilon\}_{\epsilon>0}$ is bounded in $L^2(\Omega)^3$. Multiplying (2.21) by $\vec{\varphi}_\epsilon$ and using the orthogonality property of b, defined in (2.9), we get

$$(2.23) \qquad \|\vec{\varphi}_\epsilon\|^2_{H^1(\Omega)^3} \leq C_1\|\vec{\varphi}_\epsilon\|^2_{L^4(\Omega)^3} + C_2\|\vec{\varphi}_\epsilon\|_{H^1(\Omega)^3}.$$

From the inequality (Temam [21, page 296])

$$(2.24) \qquad \|\vec{\varphi}_\epsilon\|_{L^4(\Omega)^3} \leq \sqrt{2}\|\vec{\varphi}_\epsilon\|^{1/4}_{L^2(\Omega)^3}\|\vec{\varphi}_\epsilon\|^{3/4}_{H^1(\Omega)^3}$$

and (2.23), we obtain

$$\|\vec{\varphi}_\epsilon\|^2_{H^1(\Omega)^3} \leq C_3\|\vec{\varphi}_\epsilon\|^{3/2}_{H^1(\Omega)^3} + C_2\|\vec{\varphi}_\epsilon\|_{H^1(\Omega)^3},$$

which proves the boundedness of $\{\vec{\varphi}_\epsilon\}_{\epsilon>0}$ in $H^1(\Omega)^3$. Then we can extract a subsequence, denoted in the same way, and an element $\vec{\varphi}_0 \in H^1(\Omega)^3$ such that $\vec{\varphi}_\epsilon \rightharpoonup \vec{\varphi}_0$ weakly in $H^1(\Omega)^3$. Now it is easy to pass to the limit in (2.20)–(2.22) and to deduce (2.11)–(2.13) with $\alpha = 1$.

If $\{\vec{\varphi}_\epsilon\}_{\epsilon>0}$ is not bounded in $L^2(\Omega)^3$ we take

$$\alpha_\epsilon = \frac{1}{\|\vec{\varphi}_\epsilon\|_{L^2(\Omega)^3}} \to 0 \quad \text{when } \epsilon \to \infty$$

and again we denote $\alpha_\epsilon\vec{\varphi}_\epsilon$ by $\vec{\varphi}_\epsilon$. Now repeating the previous argument, we derive (2.10)–(2.13) with $\alpha = 0$. It remains to prove (2.10) or equivalently that $\vec{\varphi}_0 \neq 0$. From the weak convergence $\vec{\varphi}_\epsilon \rightharpoonup \vec{\varphi}_0$ in $H^1(\Omega)^3$ and Rellich's theorem, follows the strong convergence of $\{\vec{\varphi}_\epsilon\}_{\epsilon>0}$ to $\vec{\varphi}_0$ in $L^2(\Omega)^3$, which proves that

$$\|\vec{\varphi}_0\|_{L^2(\Omega)^3} = \lim_{\epsilon \to 0} \|\vec{\varphi}_\epsilon\|_{L^2(\Omega)^3} = 1.$$

3. Steady flow: boundary control. Now we consider the applicability of the method introduced in the previous section to a more realistic problem. In this section, the state equations are the equations of thermohydraulics in the Boussinesq approximation:

$$\text{(3.1)} \quad \begin{array}{l} -\nu\triangle\vec{y} + (\vec{y}\cdot\nabla)\vec{y} + \nabla p = \vec{f} + \vec{\beta}\tau \text{ in } \Omega, \\ -\kappa\triangle\tau + \vec{y}\cdot\nabla\tau = g \text{ in } \Omega, \\ \operatorname{div}\vec{y} = 0 \text{ in } \Omega, \ \vec{y} = \vec{\phi}_\Gamma \text{ on } \Gamma, \\ \tau = h \text{ on } \Gamma_0, \ \partial_n\tau = u \text{ on } \Gamma_1, \end{array}$$

where $\nu, \kappa > 0$, $\vec{f} \in H^{-1}(\Omega)^3$, $\vec{\beta} \in L^\infty(\Omega)^3$, $\vec{\phi}_\Gamma \in H^{1/2}(\Gamma)^3$, $g \in L^{6/5}(\Omega)$, $h \in H^{1/2}(\Gamma_0)$, $u \in L^2(\Gamma_1)$, $\Gamma = \Gamma_0 \cup \Gamma_1$, $\Gamma_0 \cap \Gamma_1 = \emptyset$ and $\sigma(\Gamma_0), \sigma(\Gamma_1) > 0$. We still assume that (2.2) holdsand moreover

$$\text{(3.2)} \qquad \vec{\phi}_\Gamma(x) \cdot \vec{n}(x) = 0 \ \ a.e. \ x \in \Gamma_1.$$

Here $\vec{y}$, p and $\vec{f}$ are the same things as in (2.1), τ is the temperature inside the fluid and u is the heat flux through the boundary. Let us remark that the hypothesis $g \in L^{6/5}(\Omega)$ is made to give a sense to the Neumann boundary condition of (3.1). Thanks to this assumption the term $\partial_n\tau$ is well defined and the usual variational formulation of this problem is equivalent to (3.1); see, for instance, Casas and Fernández [5].

The control problem is formulated in the following way

$$\text{(P2)} \quad \begin{cases} \text{Minimize } J(u, \vec{y}), \\ (u, \vec{y}) \in K \times H^1(\Omega)^3 \text{ and satisfies (3.1) for some } (p, \tau), \end{cases}$$

with $J : H^1(\Omega)^3 \times L^2(\Gamma_1) \longrightarrow R$ defined by

$$J(u, \vec{y}) = \frac{1}{2}\int_\Omega |\nabla \times \vec{y}|^2 dx + \frac{N}{2}\int_{\Gamma_1} |u|^2 d\sigma,$$

$N \geq 0$ and $K \subset L^2(\Gamma_1)$ nonempty, convex and closed. In this problem the role of the control is to cool suitably the fluid from a part of the boundary in order to minimize the turbulence inside the flow. The reader is referred to [2] for an application of Problem (P2) to the case of a fluid in a driven cavity.

To prove the existence of a solution of (3.1), it is usual to make some assumption on the size of the viscosity ν and the diffusion coefficient κ; see, for example, Gaultier and Lezaun [10]. However we show in the next theorem that it is not necessary. Let us note that, in general, there is no uniqueness of solution (see Rabinowitz [18]), therefore we are again dealing with a multistate equation.

THEOREM 3.1. *Under the above conditions, System (3.1) has at least one solution* $(\vec{y}, p, \tau) \in H^1(\Omega)^3 \times L^2(\Omega) \times H^1(\Omega)$. *Furthermore there exist constants* $M_3, M_4 > 0$ *such that*

$$(3.3) \quad \|\tau\|_{H^1(\Omega)} + \|\vec{y}\|_{H^1(\Omega)^3}$$

$$\leq M_3 \left(\|\vec{f}\|_{H^{-1}(\Omega)^3} + \|g\|_{L^{6/5}(\Omega)} + \|h\|_{H^{1/2}(\Gamma_0)} + \|u\|_{L^2(\Gamma_1)} \right) + M_4,$$

where M_4 *depends on* $\vec{\phi}_\Gamma$, *being zero when this function is zero.*

The proof of this theorem uses the following lemma:

LEMMA 3.1. *Let us assume that* $\vec{\phi} \in H^1(\Omega)^3$, *with* $\operatorname{div}\vec{\phi} = 0$ *and* $\vec{\phi} \cdot \vec{n} = 0$ *on* Γ_1, *and* $B \in \mathcal{L}(Y, Y_0')$, *with*

$$(3.4) \qquad |\langle B(\vec{z}), \vec{z} \rangle| \leq \frac{\nu}{2} \|\vec{z}\|_{Y_0}^2, \quad \forall \vec{z} \in Y_0.$$

Then there exists at least one solution $(\vec{z}, p, \tau) \in H^1(\Omega)^3 \times L^2(\Omega) \times H^1(\Omega)$ *of the system*

$$(3.5) \quad \begin{aligned} &-\nu\Delta\vec{z} + (\vec{z}\cdot\nabla)\vec{z} + B(\vec{z}) + \nabla p = \vec{f} + \vec{\beta}\tau \ \text{in } \Omega, \\ &-\kappa\Delta\tau + (\vec{z}+\vec{\phi})\cdot\nabla\tau = g \ \text{in } \Omega, \\ &\operatorname{div}\vec{z} = 0 \ \text{in } \Omega, \ \vec{z} = \vec{0} \ \text{on } \Gamma, \\ &\tau = h \ \text{on } \Gamma_0, \ \partial_n\tau = u \ \text{on } \Gamma_1. \end{aligned}$$

Furthermore, there exists $M_5 > 0$ *such that*

$$(3.6) \quad \|\tau\|_{H^1(\Omega)} + \|\vec{z}\|_{H^1(\Omega)^3}$$

$$\leq M_5 \left(\|\vec{f}\|_{H^{-1}(\Omega)^3} + \|g\|_{L^{6/5}(\Omega)} + \|h\|_{H^{1/2}(\Gamma_0)} + \|u\|_{L^2(\Gamma_1)} \right).$$

Proof. Let us take

$$X = \{\vec{w} \in L^4(\Omega)^3 : \operatorname{div}\vec{w} = 0 \ \text{ and } \ \vec{w}\cdot\vec{n} = 0 \ \text{on } \Gamma_1\}.$$

Let us note that $\vec{w}\cdot\vec{n}$ is defined in a trace sense on Γ, with $\vec{w}\cdot\vec{n} \in W^{-1/4,4}(\Gamma)$ and

$$\int_\Omega \vec{w}\cdot\nabla\psi \, dx = \langle \vec{w}\cdot\vec{n}, \psi \rangle \quad \forall\psi \in W^{1,4/3}(\Omega);$$

see Casas and Fernández [5]. Space X, endowed with the norm of $L^4(\Omega)^3$, is a Banach space.

Now we define a mapping $F : X \longrightarrow X$, with $F(\vec{w}) = \vec{z}$ being the solution, together with some $\tau \in H^1(\Omega)$ and $p \in L^2(\Omega)$, of the system

$$(3.7) \quad \begin{aligned} &-\nu\Delta\vec{z} + (\vec{w} \cdot \nabla)\vec{z} + B(\vec{z}) + \nabla p = \vec{f} + \vec{\beta}\tau \text{ in } \Omega, \\ &-\kappa\Delta\tau + (\vec{w} + \vec{\phi}) \cdot \nabla\tau = g \text{ in } \Omega, \\ &\operatorname{div}\vec{z} = 0 \text{ in } \Omega, \ \vec{z} = \vec{0} \text{ on } \Gamma, \\ &\tau = h \text{ on } \Gamma_0, \ \partial_n\tau = u \text{ on } \Gamma_1. \end{aligned}$$

Indeed, this system is uncoupled, then we can solve first the problem in τ, which is immediate from Lax-Milgram theorem, and then obtain $\vec{z}$. The existence and uniqueness of $\vec{z}$ can be proved in the same way than for Stokes problem, it is enough to note that

$$(3.8) \quad \begin{aligned} &\nu a(\vec{z}, \vec{z}) + b(\vec{w}, \vec{z}, \vec{z}) + \langle B(\vec{z}), \vec{z}\rangle \\ = \ &\nu a(\vec{z}, \vec{z}) + \langle B(\vec{z}), \vec{z}\rangle \\ \geq \ &\nu\|\vec{z}\|_{Y_0}^2 - \frac{\nu}{2}\|\vec{z}\|_{Y_0}^2 = \frac{\nu}{2}\|\vec{z}\|_{Y_0}^2. \end{aligned}$$

Let us estimate $\|\vec{z}\|_{Y_0}$, To do this, we first take $\psi \in H^1(\Omega)$ such that $\psi = h$ on Γ_0 and $\partial_n\psi = 0$ on Γ_1, for example ψ can be the solution of

$$\begin{aligned} -\Delta\psi &= 0 \quad \text{in } \Omega, \\ \psi &= h, \quad \text{on } \Gamma_0 \\ \partial_n\psi &= 0 \quad \text{on } \Gamma_1. \end{aligned}$$

Let now $\rho_\epsilon \in D(R^3)$ verifying

$$\rho_\epsilon(x) = \left\{ \begin{array}{ll} 1 & \text{if } d(x, \Gamma) \leq \epsilon/2, \\ 0 & \text{if } d(x, \Gamma) \geq \epsilon. \end{array} \right.$$

Given $\delta > 0$ arbitrary, redefining ψ as $\rho_\epsilon\psi$ and taken ϵ small enough, we can suppose that

$$(3.9) \quad \|\psi\|_{L^4(\Omega)} \leq \delta.$$

We set $\zeta = \tau - \psi$, then we have

$$\begin{aligned} \int_\Omega [(\vec{w} + \vec{\phi}) \cdot \nabla\zeta]\zeta\,dx &= \langle(\vec{w} + \vec{\phi}) \cdot \vec{n}, \zeta^2\rangle - \int_\Omega \operatorname{div}(\vec{w} + \vec{\phi})\zeta^2\,dx \\ &\quad - \int_\Omega [(\vec{w} + \vec{\phi}) \cdot \nabla\zeta]\zeta\,dx \\ &= \langle(\vec{w} + \vec{\phi}) \cdot \vec{n}, \zeta^2\rangle_{\Gamma_1} - \int_\Omega [(\vec{w} + \vec{\phi}) \cdot \nabla\zeta]\zeta\,dx \\ &= -\int_\Omega [(\vec{w} + \vec{\phi}) \cdot \nabla\zeta]\zeta\,dx, \end{aligned}$$

hence

$$\int_{\Omega} [(\vec{w} + \vec{\phi}) \cdot \nabla \zeta]\zeta \, dx = 0.$$

Analogously we can prove

$$\int_{\Omega} [(\vec{w} + \vec{\phi}) \cdot \nabla \zeta]\psi \, dx = -\int_{\Omega} [(\vec{w} + \vec{\phi}) \cdot \nabla \psi]\zeta \, dx.$$

Now multiplying the second equation of (3.7) by ζ, integrating by parts and using the two above identities follows

$$\kappa \int_{\Omega} |\nabla \zeta|^2 dx = \int_{\Omega} g\zeta \, dx + \int_{\Gamma_1} u\zeta \, dx - \kappa \int_{\Omega} \nabla \psi \nabla \zeta \, dx + \int_{\Omega} (\vec{w} + \vec{\phi}) \cdot \nabla \zeta]\psi \, dx.$$

From this equality and (3.9) we obtain that

$$\kappa\|\zeta\|_{H^1(\Omega)} \leq C_1 \left(\|g\|_{L^{6/5}(\Omega)} + \|u\|_{L^2(\Gamma_1)} + \kappa\|\psi\|_{H^1(\Omega)} + \|\vec{w} + \vec{\phi}\|_{L^4(\Omega)^3}\delta \right).$$

Estimating ψ in terms of h we get

(3.10)
$$\|\tau\|_{H^1(\Omega)} \leq \|\psi\|_{H^1(\Omega)} + \|\zeta\|_{H^1(\Omega)}$$
$$\leq C_2 \left(\|g\|_{L^{6/5}(\Omega)} + \|h\|_{H^{1/2}(\Gamma_0)} + \|u\|_{L^2(\Gamma_1)} + \|\vec{\phi}\|_{L^4(\Omega)^3} \right) + \delta C_3 \|\vec{w}\|_{L^4(\Omega)^3},$$

with C_3 independent of δ.

On the other hand, multiplying the first equation of (3.7) by $\vec{z}$, we derive with (3.8) and (3.11) that

(3.11)
$$\frac{\nu}{2}\|\vec{z}\|_{Y_0} \leq C_4 \left(\|\vec{f}\|_{H^{-1}(\Omega)^3} + \|g\|_{L^{6/5}(\Omega)} + \|h\|_{H^{1/2}(\Gamma_0)} \right.$$
$$\left. + \|u\|_{L^2(\Gamma_1)} + \|\vec{\phi}\|_{L^4(\Omega)^3} \right) + \delta C_5 \|\vec{w}\|_{L^4(\Omega)^3},$$

with C_5 independent of δ. This inequality implies that

$$\|\vec{z}\|_{Y_0} \leq C_6 + \delta C_7 \|\vec{w}\|_{L^4(\Omega)^3}.$$

Thus choosing δ in such a way that $C_7\delta \leq 1/2$ and setting $r = 2C_6$, we deduce that F applies the ball $\bar{B}_r(0)$ de X into itself. Finally the existence of a fixed point of F follows from the compactness of the inclusion $Y_0 \subset X$ and Schauder's theorem. Estimate (3.6) follows from (3.11) and (3.11). $\square$

Proof of Theorem 3.1. Let us take $\vec{\chi}$ and γ_ϵ satisfying (2.2) and (2.6), respectively. We take $\epsilon > 0$ small enough, so that the following inequality holds

(3.12)
$$\left(\sum_{i,j=1}^{3} \|\psi_i \phi_{\epsilon j}\|_{L^2(\Omega)}^2 \right)^{1/2} \leq \frac{\nu}{2}\|\vec{\psi}\|_{H_0^1(\Omega)^3} \quad \forall \vec{\psi} \in H_0^1(\Omega)^3;$$

see Temam [21, page 177]. Then we define

$$B(\vec{z}) = (\vec{\phi}_\epsilon \cdot \nabla)\vec{z} + (\vec{z} \cdot \nabla)\vec{\phi}_\epsilon.$$

From (3.12) and the orthogonality property of b we have

$$|\langle B(\vec{z}), \vec{z}\rangle| \le |b(\vec{\phi}_\epsilon, \vec{z}, \vec{z})| + |b(\vec{z}, \vec{\phi}_\epsilon, \vec{z})| = |b(\vec{z}, \vec{z}, \vec{\phi}_\epsilon)|$$

$$\le \sum_{i,j=1}^{3} \int_\Omega |z_i \partial_{x_i} z_j \phi_{\epsilon j}| dx \le \sum_{i,j=1}^{3} \|\partial_{x_i} z_j\|_{L^2(\Omega)} \|z_i \phi_{\epsilon j}\|_{L^2(\Omega)} \le \frac{\nu}{2}\|\vec{z}\|^2_{H_0^1(\Omega)^3}$$

Now we take a solution $(\vec{z}, \tau, p)$ of (3.5), changing the right hand side of the first equation for $\vec{f} + \vec{\beta}\tau - (\vec{\phi}_\epsilon \cdot \nabla)\vec{\phi}_\epsilon$. It is enough to set $\vec{y} = \vec{z} + \vec{\phi}_\epsilon$ to conclude the proof of the theorem.   □

A different proof of Theorem 3.1 was given by Abergel and Casas [1], where Brouwer's theorem, after a discretization of the state equations (3.1), was used instead of Schauder's theorem.

Once we have proved the existence of solutions of the state equations, we can establish a theorem of existence of a solution for Problem (P2) analogous to Theorem 2.1.

THEOREM 3.2. *If $N > 0$ or K is bounded in $L^2(\Gamma_1)$, then (P2) has at least one solution.*

Also we have the following conditions for optimality

THEOREM 3.3. *Let $(u_0, \vec{y}_0)$ be a solution of (P2), then there exist a constant $\alpha \ge 0$ and elements $\vec{\varphi}_0 \in H^1(\Omega)^3$, $\tau_0, \psi_0 \in H^1(\Omega)$ and $p_0, \pi_0 \in L^2(\Omega)$ such that*

$$(3.13) \qquad\qquad \alpha + \|\psi_0\|_{H^1(\Omega)} > 0;$$

$$(3.14) \qquad \begin{aligned} &-\nu\Delta\vec{y}_0 + (\vec{y}_0 \cdot \nabla)\vec{y}_0 + \nabla p_0 = \vec{f} + \vec{\beta}\tau_0 \ in \ \Omega, \\ &-\kappa\Delta\tau_0 + \vec{y}_0 \cdot \nabla\tau_0 = g \ in \ \Omega, \\ &\operatorname{div}\vec{y}_0 = 0 \ in \ \Omega, \ \vec{y}_0 = \vec{\phi}_\Gamma \ on \ \Gamma, \\ &\tau_0 = h \ on \ \Gamma_0, \ \partial_n\tau_0 = u_0 \ on \ \Gamma_1; \end{aligned}$$

$$(3.15) \qquad \begin{aligned} &-\nu\Delta\vec{\varphi}_0 - (\vec{y}_0 \cdot \nabla)\vec{\varphi}_0 + (\nabla\vec{y}_0)^T \vec{\varphi}_0 + \nabla\pi_0 \\ &\quad = \tau_0\nabla\psi_0 + \alpha\nabla \times (\nabla \times \vec{y}_0) \ in \ \Omega, \\ &-\kappa\Delta\psi_0 - \vec{y}_0 \cdot \nabla\psi_0 = \vec{\beta}\vec{\varphi}_0 \ in \ \Omega, \\ &\operatorname{div}\vec{\varphi}_0 = 0 \ in \ \Omega, \ \vec{\varphi}_0 = 0 \ on \ \Gamma, \\ &\psi_0 = 0 \ on \ \Gamma_0, \ \partial_n\psi_0 = 0 \ on \ \Gamma_1; \end{aligned}$$

$$(3.16) \qquad \int_{\Gamma_1} (\psi_0 + \alpha N u_0)(u - u_0)d\sigma \ge 0 \quad \forall u \in K.$$

The proof of this theorem follows the same steps as that of Theorem 2.2; see [1] for the details. Similarly to Theorem 2.2 here we could formulate a statement analogous to that of Corollary 2.1, which would allow to conclude (3.14)–(3.16) with $\alpha = 1$ if K had property C at $(\vec{y}_0, u_0)$.

4. Time-dependent flow: distributed control. This section and the following one are devoted to the control of the turbulence in time-dependent flows. Here we study the case of a distributed control and the state equations are the Navier-Stokes equations:

$$(4.1) \qquad \frac{\partial \vec{y}}{\partial t} - \nu \triangle_x \vec{y} + (\vec{y} \cdot \nabla_x)\vec{y} + \nabla_x p = \vec{f} + \mathcal{C}u \quad \text{in } \Omega_T,$$

$$\operatorname{div}_x \vec{y} = 0 \quad \text{in } \Omega_T, \ \vec{y}(0) = \vec{\phi}_0 \quad \text{in } \Omega, \ \vec{y} = 0 \quad \text{on } \Sigma_T,$$

where $\nu > 0$ is a constant, $\vec{f} \in L^2([0,T], L^2(\Omega)^3)$, $\mathcal{C} \in \mathcal{L}(U, L^2([0,T], L^2(\Omega)^3))$, $u \in U$, U being a Hilbert space, and $\vec{\phi}_0 \in Y_0$ is the initial velocity. We will henceforth assume that Γ is of class C^2.

The existence of a weak solution of (4.1) is well known; see, for instance, Ladyzhenskaya [14], Lions [15], Temam [21], etc. However its uniqueness is an open question so far. We recall that an element $\vec{y} \in L^2([0,T], Y_0) \cap C_w([0,T], L^2(\Omega)^3)$ is said a weak solution of (4.1) if it satisfies

Find $\vec{y} \in L^2([0,T], Y_0)$ such that

$$(4.2) \qquad \frac{d}{dt}(\vec{y}(t), \vec{\psi})_{L^2(\Omega)^3} + \nu a(\vec{y}(t), \vec{\psi})_{Y_0} + b(\vec{y}(t), \vec{y}(t), \vec{\psi})$$

$$= (\vec{f}(t) + \mathcal{C}u(t), \vec{\psi})_{L^2(\Omega)^3} \ \ \forall \vec{\psi} \in Y_0, \ a.e. \ t \in (0,T),$$

$$\vec{y}(0) = \vec{\phi}_0,$$

and the energy inequality

$$(4.3) \qquad \|\vec{y}(t)\|_{L^2(\Omega)^3}^2 + 2\nu \int_0^t \|\vec{y}(s)\|_{Y_0}^2 ds$$

$$\leq \ \|\vec{\phi}_0\|_{L^2(\Omega)^3}^2 + 2\int_0^t (\vec{f}(s) + \mathcal{C}u(s), \vec{y}(s))_{L^2(\Omega)^3} ds \ \forall t \in [0,T].$$

With $C_w([0,T], L^2(\Omega)^3)$ we denote the space of functions $\vec{y} : [0,T] \longrightarrow L^2(\Omega)^3$ weakly continuous; that is, $\vec{y}$ is continuous when $L^2(\Omega)^3$ is endowed with the weak topology. Thus the initial conditions $\vec{y}(0) = \vec{\phi}_0$ makes sense. Once a solution $\vec{y}$ of (4.2) has been found, the existence of the pressure $p \in D'(\Omega_T)$ can be proved, in such a way that $(\vec{y}, p)$ is a solution of (4.1), satisfying the partial differential equations in the distribution sense, the boundary condition in the trace sense and the initial condition weakly in $L^2(\Omega)^3$. The pressure is unique up to the addition of a real distribution in $(0,T)$.

Therefore, if we formulate an optimal control problem letting the weak solutions of (4.1) to be feasible states, we find the same type of difficulty than in problems (P1) and (P2) to derive the optimality conditions: the relation control $\rightarrow$ state is not well defined. It seems natural to try the same technique than in the previous sections to derive these conditions for optimality. Thus we can introduce a new control $\vec{w} \in L^2([0,T], Y_0)$ and consider the new system

$$\frac{\partial \vec{y}}{\partial t} - \nu \triangle_x \vec{y} + (\vec{w} \cdot \nabla_x)\vec{y} + \nabla_x p = \vec{f} + \mathcal{C}u \ \text{ in } \Omega_T,$$

$$\text{div}_x \vec{y} = 0 \ \text{ in } \Omega_T, \ \vec{y}(0) = \vec{\phi}_0 \ \text{ in } \Omega, \ \vec{y} = 0 \ \text{ on } \Sigma_T.$$

Unfortunately, we can not prove uniqueness of a weak solution for this problem; one falls essentially on the same difficulties than in the study of the uniqueness of (4.1). Consequently, the method used in the previous problems does not work for time-dependent flows.

To overcome this difficulty motivated for the lack of uniqueness of (4.1), we can consider a more restrictive class of solutions, namely, strong solutions. We say that $\vec{y}$ is a strong solution of (4.1) if it is a weak solution and $\vec{y} \in L^8([0,T], L^4(\Omega)^3)$. It is well known that (4.1) does not have more than one strong solution. Strong solutions satisfy the energy equality instead of the inequality (4.3). So they seem to be physically more significant than weak solutions. Unfortunately, there is no existence result of strong solutions. However, we can formulate the optimal control problem in such a way that the only feasible states are strong solutions. This means that we will work with a subset of controls providing strong solutions of (4.1). Moreover, the relation between the control and the state becomes differentiable when the controls are taken in this set. To attain this goal, instead of taking the cost functional as in [2], we put

$$J(u, \vec{y}) = \frac{1}{6} \int_0^T \left(\int_\Omega |\nabla_x \times \vec{y}|^2 dx \right)^3 dt + \frac{N}{2} \|u\|_U^2.$$

Then the optimal control problem is formulated in the following way

$$(\text{P3}) \left\{ \begin{array}{l} \text{Minimize } J(u, \vec{y}), \\ (u, \vec{y}) \in K \times H^{2,1}(\Omega_T)^3 \text{ satisfying (4.1) for some } p \in L^2([0,T], H^1(\Omega)). \end{array} \right.$$

The fact of taking $H^{2,1}(\Omega_T)^3$ as state space is motivated for the following result, whose proof can be found in Casas [4]:

THEOREM 4.1. *Let us assume that $(\vec{y}, p)$ is a strong solution of System* (4.1), *then $\vec{y} \in H^{2,1}(\Omega_T)^3 \cap C([0,T], Y_0)$ and $p \in L^2([0,T], H^1(\Omega))$. Moreover*

$$\|\vec{y}\|_{H^{2,1}(\Omega_T)^3} \leq \eta \left(\|\vec{\phi}_0\|_{Y_0} + \|\vec{f}\|_{L^2([0,T], L^2(\Omega)^3)} + \|\vec{y}\|_{L^8([0,T], L^4(\Omega)^3)} \right),$$

(4.4)
where $\eta : [0, +\infty) \longrightarrow [0, +\infty)$ is an increasing function depending only on Ω and ν.

The first term of the cost functional gives a measure of the turbulence in the flow through the norm of the vorticity in the space $L^6([0, T], L^2(\Omega)^3)$. The reason of the choice of this norm is that any weak solution of (4.1) verifying $J(u, \vec{y}) < +\infty$ is a strong solution, which reduces the feasible states of (P3) to strong solutions of (4.1). The following proposition proves this statement.

PROPOSITION 4.1. *Let $\vec{y}$ be a weak solution of (4.1) verifying $J(u, \vec{y}) < +\infty$, then $\vec{y}$ is a strong solution. Moreover*

$$(4.5) \qquad \|\vec{y}\|_{L^8([0,T],L^4(\Omega)^3)} \leq M_6,$$

for some constant M_6 depending on $J(u, \vec{y})$ and $\|u\|_U$.

Proof. Let us begin noting that there exists a constant $C_1 > 0$ such that

$$(4.6) \qquad \|\vec{z}\|_{Y_0} \leq C_1 \|\nabla_x \times \vec{z}\|_{L^2(\Omega)^3} \quad \forall \vec{z} \in Y_0;$$

see, for instance, Temam [21, Lemma 1.6; page 465]. Then the inequality $J(u, \vec{y}) < +\infty$ implies that $\vec{y} \in L^6([0, T], Y_0)$. On the other hand, since $\vec{y}$ is a weak solution, we have that $\vec{y} \in L^\infty([0, T], L^2(\Omega)^3)$. Therefore, from (2.23) we obtain that

$$\|\vec{y}(t)\|_{L^4(\Omega)^3} \leq \sqrt{2}\|\vec{y}\|_{L^\infty([0,T],L^2(\Omega)^3)}^{1/2}\|\vec{y}(t)\|_{Y_0}^{3/4} = C_2\|\vec{y}(t)\|_{Y_0}^{3/4},$$

which implies that

$$\|\vec{y}\|_{L^8([0,T],L^4(\Omega)^3)}^8 \leq C_2^8\|\vec{y}\|_{L^6([0,T],Y_0)}^6 < +\infty.$$

Finally, (4.5) follows from this inequality and (4.6). $\square$

To prove the existence of a solution for (P3), in addition to the standard hypothesis assumed in theorems 2.1 and 3.2, we must suppose the existence of a feasible pair $(u, \vec{y})$ for (P3). To check the existence of these pairs, we can take an element $\vec{y} \in H^{2,1}(\Omega_T)^3 \cap C([0, T], Y_0)$, with $\vec{y}(0) = \vec{\phi}_0$, and obtain u from the partial differential equations. If u is an element of K, then the assumption is satisfied. For instance, this is the case if $K = U = L^2([0, T], L^2(\Omega)^3)$ and $\mathcal{C}$ is the identity. The precise result is formulated as follows

THEOREM 4.2. *Let us assume that the following two hypotheses hold:*
   1. There exists a feasible pair $(u, \vec{y}) \in K \times H^{2,1}(\Omega_T)^3$ satisfying (4.1).
   2. Either $N > 0$ or K is bounded in U.
Then there exists at least one optimal solution $(u_0, \vec{y}_0)$ of (P3).

The next step is to derive the conditions for optimality satisfied by these optimal solutions. The crucial point is that the set of controls of U having associated a strong solution form an open set; moreover the relations between the control and the state is differentiable on this set.

THEOREM 4.3. *If System (4.1) has a strong solution for some element u of U and some $\vec{\phi}_0 \in Y_0$, then there exists an open neighbourhood $\mathcal{U}$ of* *u in U such that the Navier-Stokes equations with body forces $\vec{g} = \vec{f} + Cv$, $v \in \mathcal{U}$, and initial condition equal to ϕ_0 have a strong solution $\vec{y}_v$. Moreover the mapping $G : \mathcal{U} \longrightarrow H^{2,1}(\Omega_T)^3 \cap C([0,T], Y_0)$, defined by $G(u) = \vec{y}_u$, is of class C^∞. Finally, if $\vec{z} = DG(u) \cdot v$, for some $u \in \mathcal{U}$ and some $v \in U$, then $\vec{z}$ is the unique strong solution of the problem*

$$(4.7)\qquad \frac{\partial \vec{z}}{\partial t} - \nu \triangle_x \vec{z} + (\vec{y}_u \cdot \nabla_x)\vec{z} + (\vec{z} \cdot \nabla_x)\vec{y}_u + \nabla_x p = Cv \quad in \ \Omega_T,$$

$$\operatorname{div}_x \vec{z} = 0 \quad in \ \Omega_T, \ \ \vec{z}(0) = 0 \quad in \ \Omega, \ \ \vec{z} = 0 \quad on \ \Sigma_T,$$

for some $p \in L^2([0,T], H^1(\Omega))$, which is unique up to the addition of a function of $L^2(0,T)$.

By using this result, it is not difficult to prove the following theorem

THEOREM 4.4. *Let us assume that $(u_0, \vec{y}_0)$ is a solution of (P3) and p_0 the pressure corresponding to the velocity $\vec{y}_0$. Then there exist a unique element $\vec{\varphi}_0 \in H^{2,1}(\Omega_T)^3 \cap C([0,T], Y_0)$ and a function $\pi_0 \in L^2([0,T], H^1(\Omega))$, unique up to the addition of a function of $L^2(0,T)$, such that the following system is satisfied*

$$(4.8)\qquad \frac{\partial \vec{y}_0}{\partial t} - \nu \triangle_x \vec{y}_0 + (\vec{y}_0 \cdot \nabla_x)\vec{y}_0 + \nabla_x p_0 = \vec{f} + Cu_0 \quad in \ \Omega_T,$$

$$\operatorname{div}_x \vec{y}_0 = 0 \quad in \ \Omega_T, \ \ \vec{y}_0(0) = \vec{\phi}_0 \quad in \ \Omega, \ \ \vec{y}_0 = 0 \quad on \ \Sigma_T;$$

$$-\frac{\partial \vec{\varphi}_0}{\partial t} - \nu \triangle_x \vec{\varphi}_0 - (\vec{y}_0 \cdot \nabla_x)\vec{\varphi}_0 + (\nabla_x \vec{y}_0)^T \vec{\varphi}_0 + \nabla_x \pi_0$$

$$(4.9)\qquad = \|\nabla_x \times \vec{y}_0\|_{L^2(\Omega)^3}^4 [\nabla_x \times (\nabla_x \times \vec{y}_0)] \quad in \ \Omega_T,$$

$$\operatorname{div}_x \vec{\varphi}_0 = 0 \quad in \ \Omega_T, \ \ \vec{\varphi}_0(T) = 0 \quad in \ \Omega, \ \ \vec{\varphi}_0 = 0 \quad on \ \Sigma_T;$$

$$(4.10)\qquad (C^\star \vec{\varphi}_0 + Nu_0, u - u_0)_U \geq 0 \quad \forall u \in K.$$

The detailed proofs of the two previous theorems can be found in [4].

5. Time-dependent flow: boundary control. As in the previous section, we assume Γ to be of class C^2, $\nu > 0$, $\vec{f} \in L^2([0,T], L^2(\Omega)^3)$ and $\vec{\phi}_0 \in Y_0$. Moreover we take $\vec{\beta} \in L^\infty(\Omega_T)$, $g \in L^2([0,T], L^2(\Omega))$, $\theta_0 \in L^3(\Omega)$, $\Sigma_T^0 = \Gamma_0 \times (0,T)$, $\Sigma_T^1 = \Gamma_1 \times (0,T)$, with Γ_0 and Γ_1 as in §3, and

$u \in L^2(\Sigma_T^1)$. Then the state equations are

$$\frac{\partial \vec{y}}{\partial t} - \nu \triangle_x \vec{y} + (\vec{y} \cdot \nabla_x)\vec{y} + \nabla_x p = \vec{f} + \vec{\beta}\tau \ \text{ in } \Omega_T,$$

$$\frac{\partial \tau}{\partial t} - \kappa \triangle_x \tau + \vec{y} \cdot \nabla_x \tau = g \ \text{ in } \Omega_T,$$

(5.1)

$$\text{div}_x \vec{y} = 0 \ \text{ in } \Omega_T, \ \vec{y}(0) = \vec{\phi}_0 \ \text{ in } \Omega, \ \vec{y} = 0 \ \text{ on } \Sigma_T,$$

$$\tau(0) = \theta_0 \ \text{ in } \Omega, \ \tau = 0 \ \text{ on } \Sigma_T^0, \ \partial_n \tau = u \ \text{ on } \Sigma_T^1.$$

The physical interpretation of τ and u is as in §3.

This system has similar properties to (4.1). So the existence of a weak solution $(\vec{y}, \tau, p)$, with $\vec{y} \in C_w([0,T], L^2(\Omega)^3) \cap L^2([0,T], Y_0)$, $\tau \in C([0,T], L^2(\Omega)) \cap L^2([0,T], H^1(\Omega))$ and $p \in D'(\Omega_T)$, satisfying an energy inequality can be proved by using the methods of [14], [15] or [21]; see also Foias et al. [9]. Again the uniqueness is an open question. We follow the method of §4 to control the turbulence of the fluid described by (5.1). Therefore we consider the cost functional

$$J(u, \vec{y}) = \frac{1}{6} \int_0^T \left(\int_\Omega |\nabla_x \times \vec{y}|^2 dx \right)^3 dt + \frac{N}{2} \int_0^T \int_{\Gamma_1} |u|^2 d\sigma dt.$$

Then the optimal control problem is formulated in the following way

$$\text{(P4)} \left\{ \begin{array}{l} \text{Minimize } J(u, \vec{y}), \\ (u, \vec{y}) \in K \times H^{2,1}(\Omega_T)^3 \text{ satisfying (5.1) together with some } (\tau, p), \end{array} \right.$$

with $(\tau, p) \in C([0,T], L^2(\Omega)) \cap L^2([0,T], H^1(\Omega)) \times L^2([0,T], H^1(\Omega))$ and $K \subset L^2(\Sigma_T^1)$ nonempty, convex and closed.

We say that $(\vec{y}, \tau, p)$ is a strong solution of (5.1) if it is a weak solution and moreover $\vec{y} \in L^8([0,T], L^4(\Omega)^3)$. The following result is an immediate consequence of Theorem 4.1.

THEOREM 5.1. *Let us assume that $(\vec{y}, \tau, p)$ is a strong solution of System (5.1), then $\vec{y} \in H^{2,1}(\Omega_T)^3 \cap C([0,T], Y_0)$ and $p \in L^2([0,T], H^1(\Omega))$. Moreover*

$$(5.2) \qquad \|\vec{y}\|_{H^{2,1}(\Omega_T)^3} + \|\tau\|_{L^2([0,T], H^1(\Omega))} + \|\tau\|_{L^\infty([0,T], L^2(\Omega))}$$

$$\leq \eta \Big(\|\vec{\phi}_0\|_{Y_0} + \|\vec{f}\|_{L^2([0,T], L^2(\Omega)^3)} + \|\vec{y}\|_{L^8([0,T], L^4(\Omega)^3)} + \|u\|_{L^2(\Sigma_T^1)} + \|\theta_0\|_{L^3(\Omega)} \Big),$$

where $\eta : [0, +\infty) \longrightarrow [0, +\infty)$ is an increasing function depending only on Ω, κ and ν.

Proposition 4.1 remains true and an existence theorem analogous to 4.2 can be stated for Problem (P4). We have also the following result about differentiability of mapping $u \to \vec{y}_u$

THEOREM 5.2. *If System* (5.1) *has a strong solution for some element* u *of* $L^2(\Sigma_T^1)$ *and some* $\vec{\phi}_0 \in Y_0$ *and* $\theta_0 \in L^3(\Omega)$, *then there exists an open neighbourhood* $\mathcal{U}$ *of* u *in* $L^2(\Sigma_T^1)$ *such that System* (5.1), *with* v *instead of* u *as Neumann condition, has a strong solution* $\vec{y}_v$. *Moreover the mapping* $G : \mathcal{U} \longrightarrow H^{2,1}(\Omega_T)^3 \cap C([0,T], Y_0)$, *defined by* $G(u) = \vec{y}_u$, *is of class* C^∞. *Finally, if* $\vec{z} = DG(u) \cdot v$, *for some* $u \in \mathcal{U}$ *and some* $v \in L^2(\Sigma_T^1)$, *then* $\vec{z}$ *is the unique strong solution of the problem*

$$
(5.3) \quad
\begin{aligned}
&\frac{\partial \vec{z}}{\partial t} - \nu \Delta_x \vec{z} + (\vec{y}_u \cdot \nabla_x)\vec{z} + (\vec{z} \cdot \nabla_x)\vec{y}_u + \nabla_x p = \vec{\beta}\zeta \quad in \ \Omega_T, \\[2mm]
&\frac{\partial \zeta}{\partial t} - \kappa \Delta_x \zeta + \vec{z} \cdot \nabla_x \tau_u + \vec{y}_u \cdot \nabla_x \zeta = 0 \quad in \ \Omega_T, \\[2mm]
&\operatorname{div}_x \vec{z} = 0 \quad in \ \Omega_T, \ \vec{z}(0) = 0 \quad in \ \Omega, \ \vec{z} = 0 \quad on \ \Sigma_T, \\[2mm]
&\zeta(0) = 0 \quad in \ \Omega, \ \zeta = 0 \quad on \ \Sigma_T^0, \ \partial_n \zeta = v \quad on \ \Sigma_T^1.
\end{aligned}
$$

for some $p \in L^2([0,T], H^1(\Omega))$, *which is unique up to the addition of a function of* $L^2(0,T)$.

The proof of this theorem follows from the implicit function theorem and regularity $H^{2,1}(\Omega)^3$ for strong solutions; see Casas [4]. Finally, by using this theorem, it is immediate to derive the following conditions for optimality

THEOREM 5.3. *Let us assume that* $(u_0, \vec{y}_0)$ *is a solution of* (P4) *and* τ_0 *and* p_0 *are the temperature and the pressure, respectively, corresponding to the velocity* $\vec{y}_0$. *Then there exist two unique elements* $\vec{\varphi}_0 \in H^{2,1}(\Omega_T)^3 \cap C([0,T], Y_0)$ *and* $\psi_0 \in C([0,T], L^2(\Omega)) \cap L^2([0,T], H^1(\Omega))$ *and a function* $\pi_0 \in L^2([0,T], H^1(\Omega))$, *unique up to the addition of a function of* $L^2(0,T)$, *such that the following system is satisfied*

$$
(5.4) \quad
\begin{aligned}
&\frac{\partial \vec{y}_0}{\partial t} - \nu \Delta_x \vec{y}_0 + (\vec{y}_0 \cdot \nabla_x)\vec{y}_0 + \nabla_x p_0 = \vec{f} + \beta \tau_0 \quad in \ \Omega_T, \\[2mm]
&\frac{\partial \tau_0}{\partial t} - \kappa \Delta_x \tau_0 + \vec{y}_0 \cdot \nabla_x \tau_0 = g \quad in \ \Omega_T, \\[2mm]
&\operatorname{div}_x \vec{y}_0 = 0 \quad in \ \Omega_T, \ \vec{y}_0(0) = \vec{\phi}_0 \quad in \ \Omega, \ \vec{y}_0 = 0 \quad on \ \Sigma_T, \\[2mm]
&\tau_0(0) = \theta_0 \quad in \ \Omega, \ \tau_0 = 0 \quad on \ \Sigma_T^0, \ \partial_n \tau_0 = u_0 \quad on \ \Sigma_T^1;
\end{aligned}
$$

$$-\frac{\partial \vec{\varphi}_0}{\partial t} - \nu \Delta_x \vec{\varphi}_0 - (\vec{y}_0 \cdot \nabla_x)\vec{\varphi}_0 + (\nabla_x \vec{y}_0)^T \vec{\varphi}_0 + \nabla_x \pi_0$$

$$= \tau_0 \nabla_x \psi_0 + \|\nabla_x \times \vec{y}_0\|_{L^2(\Omega)^3}^4 [\nabla_x \times (\nabla_x \times \vec{y}_0)] \quad in \ \Omega_T,$$

$$(5.5) \qquad -\frac{\partial \psi_0}{\partial t} - \kappa \Delta_x \psi_0 - \vec{y}_0 \cdot \nabla_x \psi_0 = \vec{\beta}\vec{\varphi}_0 \quad in \ \ \Omega_T,$$

$$\mathrm{div}_x \vec{\varphi}_0 = 0 \ \ in \ \Omega_T, \ \ \vec{\varphi}_0(T) = 0 \ \ in \ \Omega, \ \ \vec{\varphi}_0 = 0 \ \ on \ \Sigma_T,$$

$$\psi_0(T) = 0 \ \ in \ \Omega, \ \ \psi_0 = 0 \ \ on \ \Sigma_T^0, \ \ \partial_n \psi_0 = 0 \ \ on \ \Sigma_T^1;$$

$$(5.6) \qquad \int_0^T \int_{\Gamma_1} (\psi_0 + N u_0)(u - u_0) d\sigma dt \geq 0 \quad \forall u \in K.$$

The detailed proofs of the theorems of this section will be given in a forthcoming paper.

REFERENCES

[1] F. ABERGEL AND E. CASAS, *Some optimal control problems of multistate equations appearing in fluid mechanichs*, M^{2}AN, 27 (1993), pp. 223–247

[2] F. ABERGEL AND R. TEMAM, *On some control problems in fluid mechanics*, Theoret. Comput. Fluid Dynamics, 1 (1990), pp. 303–325.

[3] R. ADAMS, *Sobolev Spaces*, Academic Press, New York, 1975.

[4] E. CASAS, *An optimal control problem governed by the evolution Navier-Stokes equations*, to appear in Optimal Control of Viscous Flows, S. Sritharan, ed., Philadelphia, 1993, Frontiers in Applied Mathematics, SIAM.

[5] E. CASAS AND L. FERNÁNDEZ, *A Green's formula for quasilinear elliptic operators*, J. Math. Anal. Appl., 142 (1989), pp. 62–72.

[6] H. FATTORINI AND S. SRITHAHRAN, *Existence of optimal controls for viscous flow problems*, Proceedings Royal Soc. London, 39 (1992), pp. 81–102.

[7] ———, *Necessary and sufficient conditions for optimal controls in viscous flow problems*, Proc. Royal Soc. of Edinburgh, (To appear)

[8] ———, *Optimal chattering controls for viscous flow*, (To appear).

[9] C. FOIAS, O. MANLEY, AND R. TEMAM, *Attractors for the Bénard problem: existence and physical bounds on their fractal dimensions*, Nonlinear Anal. TMA., 11 (1987), pp. 939–967.

[10] M. GAULTIER AND M. LEZAUN, *Equations de Navier–Stokes couplées à des équations de la chaleur: résolution par une méthode de point fixe en dimension infinie*, Ann. Sc. Math. Québec, 13 (1989), pp. 1–17.

[11] M. GUNZBURGER, L. HOU, AND T. SVOBODNY, *Analysis and finite element approximations of optimal control problems for the stationary Navier-Stokes equations with Dirichlet conditions*, M^{2}AN, 25 (1991), pp. 711–748.

[12] ———, *Boundary velocity control of incompressible flow with an application to viscous drag reduction*, SIAM J. Control Optim., 30 (1992), pp. 167–181.

[13] A. IOFFE AND V. TIKHOMOROV, *Extremal Problems*, North-Holland, Amsterdam, 1979.

[14] O. LADYZHENSKAYA, *The Mathematical Theory of Viscous Incompressible Flow*, Gordon and Breach, New York, second edition ed., 1969. English translation.

[15] J. LIONS, *Quelques Méthodes de Résolution des Problèmes aux Limites non Linéaires*, Dunod, Paris, 1969.

[16] J. LIONS AND E. MAGENES, *Problèmes aux Limites non Homogènes*, Dunod, Paris, 1968.

[17] J. NEČAS, *Les Méthodes Directes en Théorie des Equations Elliptiques*, Editeurs Academia, Prague, 1967.

[18] P. RABINOWITZ, *Existence and nonuniqueness of rectangular solutions of the Bénard problem*, Arch. Rational Mech. Anal., 29 (1968), pp. 32–57.

[19] S. SRITHARAN, *Dynamic programming of the Navier-Stokes equations*, Systems & Control Letters, 16 (1991), pp. 299–307.

[20] ———, *An optimal control problem in exterior hydrodynamics*, Proc. Royal Soc. of Edinburgh, 121A (1992), pp. 5–32.

[21] R. TEMAM, *Navier-Stokes Equations*, North-Holland, Amsterdam, 1979.

ON CONTROLLABILITY OF CERTAIN SYSTEMS SIMULATING A FLUID FLOW

ANDREI V. FURSIKOV* AND OLEG YU. IMANUVILOV[†]

Abstract. Approximate controllability of the Stokes system is established by a constructive method when control is a right-hand-side concentrated in subdomain i.e. in the case of local distributed control. Approximate uncontrollability of the Burgers equation is shown in the cases of boundary and local distributed controls. A local theorem of exact controllability for the Burgers equation with boundary control is proved. With its help it is shown that the controlled trajectory going out an arbitrary initial point can achieve the attractor of the Burgers equation during a finite time and after that belongs to attractor. The sets possessing such property we call an absorbing set of reachability. For the boundary and local distributed controls the description of absorbing points of reachability for the Burgers equation is given.

Key words. approximate controllability, exact controllability, absorbing set of reachability.

AMS(MOS) subject classifications. 93B05

Introduction. This paper is devoted to the investigation of controllability of certain distributed systems which simulate a fluid flow. In Section 1 we study the approximate controllability of the nonstationary Stokes system defined in a domain Ω. It is considered the case when a control is a density of external forces concentrated in an arbitrary fixed subdomain ω of the domain Ω. We call such kind of control as local distributed one. The approximate controllability of the Stokes system with such control has been proved by A.V. Fursikov and O. Yu. Imanuvilov (see [1], [2]). Here we discuss a method of construction of the controls concentrated in ω which generate the solutions of the Stokes system approximating a given solenoidal vector field. This method is based on application of an extremal problem depending on parameter. The solution of this extremal problem determines the control sought for. For analyzing of the constructed control the boundary value problem is applied which is the optimality system of the extremal problem. Note that this method was applied earlier by A.V. Fursikov in [3], [4] for the investigation of analogous problems in the case of the Cauchy problem for an elliptic operator of the second order.

The rest of this paper is devoted to the investigation of nonlinear models. First of all the problem of approximate controllability in nonlinear case arises. In the papers of C. Fabre, J.-P. Puel, E. Zuazua [5], [6] approximate controllability has been proved in the case of semilinear heat equation with a local distributed control as well as with a local Dirichlet boundary

---

* Department of Mechanics and Mathematics, Moscow State University, Lenin Hills 119899, Moscow, RUSSIA.

† Department of Applied Mathematics, Moscow Forest-technical Institute, 141000 Mytischi-1, Moscow Region, Russia.

control. In that papers the assumption that the nonlinear term satisfies the global Lipschitz condition is essential. The situation changes cardinally when this condition is broken. Below, in Section 2 (see also A.V. Fursikov, O. Yu. Imanuvilov [2]) an estimate of solution of the Burgers equation has been obtained which shows that this equation is not approximately controllable with respect to boundary control as well as with respect to local distributed control. Similar negative results have been obtained in the case of semilinear equations with a power nonlinearity (see A. Bamberger in [7], J.I. Diaz [8]). Note that the conjecture on approximate controllability of the Navier-Stokes system formulated by J.L. Lions (see [9], [10]) remains open until now.

In this situation it is natural to look for new formulations of controllability problem for nonlinear models. It would be possible to formulate a problem of investigation of reachable sets i.e. such sets in the phase space which can be achieved by the controlled trajectory going out an arbitrary initial point when controls from a given set are applied. But we think that it is interesting to study more narrow class of sets. The point is that, usually, in applications one has to achieve some set of controlled trajectory and not only to achieve but to hold it on this set or in its small neighbourhood. A subset of the phase space will be called an absorbing set of reachability if it can be achieved by the controlled trajectory going out an arbitrary initial point and this trajectory can be held on this set during the rest of time by means of controls from a given class.

Below, in Section 3 we study absorbing points of reachability in the case of the simple model of the Burgers equation with zero right-hand-side and the boundary control. The complete description of all absorbing points of reachability is given. The analogous result has been obtained also in the case of local distributed control.

The situation when a dynamical system has the attractor with a complicate structure is much more difficult. Sections 4,5 are devoted to the proof of a fundamental theorem which can be applied in this situation. (It is applied in Section 3 also). This theorem is as follows: Let $\hat{y}(t, x)$ be a solution of the Burgers equation with a fixed right-hand-side $g(x)$. Then for an arbitrary initial function $y_0(x)$ from a sufficiently small neighbourhood of $\hat{y}(0, \cdot)$ there exists such boundary control $v(t)$ that the solution $y(t, x)$ of the mixed boundary problem for the Burgers equation satisfies relation $y(T, x) \equiv \hat{y}(x)$. This method consists in deduction the nonlinear problem to the exact controllability problem for a linear parabolic equation with variable coefficients by means of the Schouder fixed point theorem. The exact controllability of a linear equation is proved with help of some Carleman estimate. Similar estimate was applied earlier by O.Yu.Imanuvilov in [11],[12], and [13] for the case of semilinear parabolic equations with a sublinear nonlinearity. Note that besides the proof of existence of a solution we have to choose the solution depending compactly on coefficients of the linear equation. Such choice of a solution is realized with help of a certain extremal problem. The linear problem of exact controllability is studied in Section 4. In Section 5 the local theorem of exact controllability is proved. And, besides, we give some corollaries of this theorem. In particular, it is shown that the attractor of the dynamical system defined by the Burgers equation with zero boundary conditions is an absorbing set of reachability by means of boundary controls. After that we show that singular points of the same dynamical system (and, in particular, hyperbolic singular points) can be made stable if one would apply a boundary control. The last fact does more clear some points of problem formulated by J.L. Lions: "Are there connections between turbulence and controllability?"

1. A constructive proof of approximate controllability of the Stokes system.

1.1. Preliminaries. We consider the Stokes system which describes a viscous uncompressible fluid flow in a bounded domain $\Omega \subset R^d$:

$$(1.1) \qquad \partial_t y(t, x) - \Delta y(t, x) - \nabla q(t, x) = u(t, x), \quad \mathrm{div}\, y(t, x) = 0$$

where $x = (x_1, \ldots, x_d) \in \Omega$, $t \in [0, T]$, $y = (y_1, \ldots, y_d)$ is a velocity vector field, $\nabla q(t, x)$ is a pressure gradient, $\partial_t y = \partial y / \partial t$, $u(t, x) = (u_1, \ldots, u_d)$ is a density of external forces which will be a control in this section. It is assumed that $u(t, x)$ is concentrated in a given subdomain of the domain:

$$(1.2) \qquad \forall t \in [0, T] \quad \mathrm{supp}\, u(t, \cdot) \subset \omega, \omega \subset \Omega$$

We suppose that the boundary condition

$$(1.3) \qquad y|_{\partial\Omega} = 0$$

and the initial condition

$$(1.4) \qquad y(t, x)|_{t=0} = 0$$

hold.

Let U, Y, Q, H be Banach spaces and for every $u \in U$ the unique solution $(y, q) \in Y \times Q$ of problem (1.1), (1.3), (1.4) exists. We denote by γ_T the operator of restriction of a function $y(t, x)$ at $t = T$: $\gamma_T y = y(T, \cdot)$ and suppose that the operator $\gamma_T : Y \to H$, is continuous. We remind

DEFINITION 1.1. *Problem (1.1), (1.3), (1.4) is called H-approximate controllable with respect to a control space U if for arbitrary $\hat{y} \in H$ and any $\varepsilon > 0$ there exists such control $u \in U$ that for the solution (y, q) of problem (1.1), (1.3), (1.4) the inequality*

$$\|\gamma_T y - \hat{y}\|_Y < \varepsilon$$

holds.

Let us introduce concrete spaces to study the approximate controllability of the Stokes system. For a domain $G \subset \Omega$ we set

$$V(G) = \{v(x) \in (C_0^\infty(\Omega))^d : \operatorname{supp} v \subset G, \quad \operatorname{div} v = 0\}$$

$$
\begin{aligned}
(1.5) \qquad H^0(G) &= \text{ the closure of } V(G) \text{ in } (L_2(\Omega))^d \\
(1.6) \qquad H^1(G) &= \text{ the closure of } V(G) \text{ in } (W_2^1(\Omega))^d \\
(1.7) \qquad H^2(G) &= (W_2^2(\Omega))^d \cap H^1(G)
\end{aligned}
$$

where $W_2^k(\Omega)$ is the Sobolev space of functions defined on Ω having the finite norm

$$(1.8) \qquad \|u\|_{W_2^k(G)}^2 = \sum_{|a| \le k} \int_\Omega |D^\alpha u(x)|^2 dx$$

Here $a = (a_1, \ldots, a_d)$ is a multiindex, $|a| = a_1 + \ldots + a_d$, $D^a u = \partial^{|a|} u / \partial x_1^{a_1} \ldots \partial x_d^{a_d}$.

We shall consider the space

$$U = L_2(0, T; H^0(\omega))$$

as a space of controls. Then

$$
\begin{aligned}
Y &= \{y(t, \cdot) \in L_2(0, T; H^2(\Omega)) : \partial_t y \in L_2(0, T; H^0(\Omega))\}, \\
(1.9) \\
Q &= \{q(t, x) \in D([0, T] \times \Omega) : \nabla q \in (L_2((0, T) \times \Omega))^d\}
\end{aligned}
$$

where $D([0, T] \times \Omega)$ is the space of distributions on $[0, T] \times \Omega$. It is known (J.L. Lions, E. Magenes [14]) that $\gamma_T Y = H^1(\Omega) \subset H^0(\Omega)$ and, hence, for U, Y, Q indicated above it is possible to take $H^i(\Omega), i = 0, 1$ as a space H. We shall consider the case

$$H = H^0(\Omega).$$

In papers by A.V. Fursikov and O.Yu. Imanuvilov [1], [2] the $H^0(\Omega)$-approximate controllability of the Stokes system with respect to the control space $L^2(0, T; H^0(\omega))$, has been proved. Below, we will give an independent constructive proof of the same theorem applying the theory of extremal problems.

1.2. An extremal problem and its system of optimality. Let $G \subset R^d$ be a domain. We denote by π_G the orthoprojector of the space $(L_2(G))^d$ onto $H^0(G)$ and set $\pi_\Omega = \pi$ when $G = \Omega$. Applying the operator

π to the both sides of (1.1) and taking into account that $u \in L_2(0, T; H^0(\omega))$ and $y \in Y$, where Y is space (1.9), we obtain the equation

$$(1.10) \qquad \partial_t y(t, x) - \pi \Delta y(t, x) = u(t, x).$$

Let us consider the extremal problem

$$(1.11) \qquad J_\varepsilon(y, u) = \frac{1}{2}\|\gamma_T y - \hat{y}\|^2_{H^0(\Omega)} + \frac{\varepsilon}{2} \int_0^T \|u(\tau)\|^2_{H^0(\omega)} d\tau \to \inf$$

which is defined on the space of couples $(y, u) \in Y \times L_2(0, T; H^0(\omega))$ satisfying equation (1.10), (1.4).

PROPOSITION 1.1.   *For an arbitrary $\varepsilon > 0$ there exists the unique solution $(y_\varepsilon, u_\varepsilon) \in Y \times L_2(0, t; H^0(\omega))$ of problem (1.11), (1.10), (1.4).*

This proposition can be proved by means of well known methods. (see, for example, A.V. Fursikov [4], [15]).

To prove the $H^0(\Omega)$-approximate controllability of the Stokes system with respect to $L_2(0, T; H^0(\omega))$ it is sufficient to show that

$$(1.12) \qquad \|\gamma_T y_\varepsilon - \hat{y}\|_{H^0(\Omega)} \to 0 \quad \text{as } \varepsilon \to 0$$

We will prove (1.12) by means of the optimality system of problems (1.11), (1.10), (1.4).

PROPOSITION 1.2.   *A couple $(y, u) \equiv (y_\varepsilon, u_\varepsilon) \in Y \times L_2(0, T; H^0(\omega))$ is a solution of problem (1.11), (1.10), (1.4) if and only if it satisfies (1.10), (1.4) and there exists such $p \in Y$ that*

$$(1.13) \qquad -\partial_t p(t, x) - \pi \Delta p \;=\; 0, \quad p|_{\partial\Omega} = 0$$
$$(1.14) \qquad \qquad\qquad p(T, \cdot) \;=\; \hat{y} - y(T, \cdot)$$
$$(1.15) \qquad \qquad\qquad \varepsilon u(t, \cdot) \;=\; \hat{\pi}_\omega p(t, \cdot)$$

where the operator $\hat{\pi}_\omega$ is a superposition of three operators: the restriction operator $|_\omega$ onto ω, the operator π_ω and the operator L_ω of extending of functions by zero outside ω:

$$(1.16) \qquad \hat{\pi}_\omega p = L_\omega \pi_\omega (p|_\omega)$$

Proof. We apply the Lagrange principle for smooth problems (see V.M. Alekseev, V.M. Tikhomirov, S.V. Fomin [16], A.V. Fursikov [4]): Let Z, W be Banach spaces and $\hat{z}$ be a solution of the extremal problem

$$(1.17) \qquad g(z) \longrightarrow \inf, \quad Gz = 0$$

where $g : Z \longrightarrow R$ is a continuously differentiable strictly convex functional, $G : Z \longrightarrow W$ is a linear continuous operator such that $ImG = W$. Then there exists such linear continuous functional w^* on W that the Lagrange function $L(z, w^*) = g(z) + \langle Gz, w^* \rangle_W$ satisfies the equality

$$(1.18) \quad \langle L'_z(\hat{z}, w^*), h \rangle_Z = \langle g'(\hat{z}), h \rangle_Z + \langle Gh, w^* \rangle_W = 0 \qquad \forall h \in Z$$

where $\langle \cdot, \cdot \rangle_V$ is duality between a Banach space V and its conjugated V^*. Besides, if $z \in Z$ satisfies (1.18), (1.17$_2$) then z is the solution of problem (1.17).

We take $z = (y, u)$, $Gz = \partial_t y - \pi \Delta y - u$, $g(z) = J_\varepsilon(y, u)$ (see (1.11)), $W = L_2(0, T; H^0(\Omega))$, $Z = Y_0 \times L_2(0, T; H^0(\omega))$ where $Y_0 = \{y(t, \cdot) \in Y : y(0, \cdot) = 0\}$. Then the condition $ImG = W$ follows from the theorem of the unique solvability of the boundary value problem for the Stokes system (see O.A. Ladyzhenskaya [17], R. Temam [18]). The Lagrange function is as follows:

$$L(y, u, p) = J_\varepsilon(y, u) + (\partial_t y - \pi \Delta y - u, p)_{L_2(0,T;H^0(\Omega))}$$

and (1.18) can be written as two equations:

$$(1.19) \quad (\gamma_T y - \hat{y}, \gamma_T h)_{H^0(\Omega)} + (\partial_t h - \pi \Delta h, p)_{L_2(0,T;H^0(\Omega))} = 0 \quad \forall h \in Y_0$$

$$(1.20) \quad \varepsilon \int_0^T (u(t), v(t))_{H^0(\omega)} dt - (v, p)_{L_2(0,T;H^0(\Omega))} = 0$$

$$\forall v \in L_2(0, T; H^0(\omega)).$$

Equalities (1.13), (1.14) and the inclusion $p \in Y$ are derived from (1.19) as in A.V. Fursikov [4]. Equation (1.20) together with the equality

$$(v, p)_{L_2(0,T;H^0(\Omega))} = (v, p)_{L_2(0,T;H^0(\omega))} \qquad \forall v \in L_2(0, T; H^0(\omega))$$

imply (1.15). $\square$

Substituting (1.15) into (1.10) and taking into account (1.4) we obtain the equalities

$$(1.21) \qquad \partial_t y(t, x) - \pi \Delta y = \frac{1}{\varepsilon}(\hat{\pi}_\omega p)(t, x), \quad y|_{t=0} = 0$$

Let us solve problem (1.13), (1.14), (1.21). As in M.I. Vishik, A.V. Fursikov [19] we deduce from (1.13), (1.14) that

$$(1.22) \qquad p(t, \cdot) = e^{\pi \Delta (T-t)}(\hat{y} - y(T, \cdot)),$$

and from (1.21) that

$$(1.23) \qquad y(t, \cdot) = \frac{1}{\varepsilon} \int_0^t e^{\pi \Delta(t-\tau)} \hat{\pi}_\omega p(\tau, \cdot) d\tau$$

Substituting (1.22) into (1.23) and taking $t = T$ we obtain equality

$$(1.24) \qquad y(T, \cdot) = \frac{1}{\varepsilon} \int_0^T e^{\pi\Delta(T-\tau)} \hat{\pi}_\omega (e^{\pi\Delta(T-\tau)}(\hat{y} - y(T))) d\tau$$

Denote

$$(1.25) \qquad Rz = \int_0^T e^{\pi\Delta(T-\tau)} \hat{\pi}_\omega (e^{\pi\Delta(T-\tau)} z) d\tau$$

Then it is possible to rewrite (1.24) in the form

$$(1.26) \qquad (I + \varepsilon^{-1} R) y(T, \cdot) = \varepsilon^{-1} R \hat{y}$$

To solve (1.26) we must study R.

1.3. Properties of the operator R.

LEMMA 1.1. *Operator*

$$(1.27) \qquad R : H^0(\Omega) \longrightarrow H^0(\Omega)$$

defined by (1.25) is a compact self-adjoint and nonnegative one.

Proof. The self-adjointness and negative definiteness of $\pi\Delta$ imply the self-adjointness in $H^0(\Omega)$ of the operator $e^{\pi\Delta(T-t)}$. Therefore, taking (1.16) into account we obtain the equalities

$$(Rz_1, z_2)_{H^0(\Omega)} = \int_0^T (e^{\pi\Delta(T-\tau)}(\hat{\pi}_\omega e^{\pi\Delta(T-\tau)} z_1), z_2)_{H^0(\Omega)} d\tau$$

$$(1.28) \quad = \int_0^T \left(\left(\pi_\omega \left(e^{\pi\Delta(T-\tau)} z_1 \right) |_\omega \right), \left(e^{\pi\Delta(T-\tau)} z_2 \right) |_\omega \right)_{H^0(\omega)} d\tau =$$

$$= \int_0^T \left(z_1, e^{\pi\Delta(T-\tau)} \hat{\pi}_\omega \left(e^{\pi\Delta(T-\tau)} z_2 \right) \right)_{H^0(\Omega)} d\tau = (z_1, Rz_2)_{H^0(\Omega)}$$

Inequality

$$(1.29) \qquad (Rz, z)_{H^0(\Omega)} = \int_0^T \|\hat{\pi}_\omega e^{\pi\Delta(T-\tau)} z\|_{H^0(\Omega)}^2 d\tau \geq 0$$

can be proved as in (1.28). It was shown in M.I. Vishik, A.V. Fursikov [19, p. 27] that the mapping $z \longrightarrow e^{\pi\Delta(T-\tau)}z$ acts continuously from $H^0(\Omega)$ into $L_2(0, T; H^2(\Omega))$, therefore the operator $z \longrightarrow \hat{\pi}_\omega(e^{\pi\Delta(T-t)}z)$ is continuous from $H^0(\Omega)$ into $L_2(0, T; H^0(\Omega))$ and, hence, again by means of M.I. Vishik, A.V. Fursikov [19, p. 27] operator (1.25) acts continuously from $H^0(\Omega)$ into $H^1(\Omega)$. Since the embedding $H^1(\Omega) \Subset H^0(\Omega)$ is compact then operator (1.27) is compact also. $\square$

The following property of operator R will be essential in a future.

LEMMA 1.2. *The equality* $\operatorname{Ker} R = 0$ *holds.*

Proof. [1] Suppose that for a certain $z_0 \in H^0(\Omega)$ the equality $Rz_0 = 0$ holds. By (1.29) we have:

$$(1.30) \qquad (Rz_0, z_0)_{H^0(\Omega)} = \int_0^T \|\pi_\omega(p(\tau, \cdot)|_\omega)\|^2_{H^0(\omega)} d\tau = 0$$

where

$$(1.31) \qquad p(\tau, \cdot) = e^{\pi\Delta(T-\tau)}z_0.$$

Let G be a bounded domain with a boundary ∂G of class C^1. We note that for an arbitrary vector field $w \in (L_2(G))^d$ the Weyl decomposition

$$(1.32) \qquad w = \pi_G w + \nabla\varphi$$

holds where $\pi_G w \in H^0(G), \varphi \in W_2^1(G)$. Applying to the both parts of (1.32) the operator div we obtain that φ is a solution of the Neumann problem

$$(1.33) \qquad \Delta\varphi = \operatorname{div} w, \quad \partial\varphi/\partial n|_{\partial G} = (w, n)|_{\partial G}$$

where n is the vector field of external normals to ∂G. Let Q be the operator which transforms a right-hand-side div w and a boundary value $(w, u)|_{\partial G}$ to the solution φ of problem (1.33) which satisfies the condition $\int_G \varphi dx = 0$:

$$(1.34) \qquad Q(\operatorname{div} w, (w, n)|_{\partial G}) = \varphi$$

Relations (1.32), (1.34) involve the formula defining π_G:

$$(1.35) \qquad \pi_G w = w - Q(\operatorname{div} w, (w, n)|_{\partial G})$$

The function $p(t, x)$ is a solution of the following Stokes problem with the inverse time:

$$(1.36) \qquad -\partial_t p(t, x) - \pi\Delta p(t, x) = 0, \quad p|_{\partial\Omega} = 0, \quad p|_{t=T} = z_0.$$

---

[1] Actually, lemma has been proved in A.V. Fursikov and O. Yu. Imanuvilov papers [1], [2]. We give the proof here only for the completeness of an account.

In virtue of the theorem of the smoothness of solutions of this problem (see O.A. Ladyzhenskaya [17]) we have that $p(t,x) \in C^\infty((0,T) \times \Omega) \cap L_2(0,T;H^2(\Omega))$. By (1.30) the equality

$$p(t,x)|_\omega = \nabla q, \quad q \in W_2^1(\omega)$$

holds for an arbitrary $t \in (0,T)$. It follows from this relation and (1.35) that

$$(1.37) \qquad p(t,\cdot)|_\omega = \nabla Q(\operatorname{div} p, (p,n)|_{\partial\omega}) = \nabla Q(0, (p,n)|_{\partial\omega})$$

because $\operatorname{div} p = 0$. Substituting (1.35) with $G = \Omega$ and $w = \Delta p$ into (1.36) we obtain that $p(t,x)$ satisfies the equation

$$(1.38) \qquad -\partial_t p(t,x) - \Delta p + \nabla Q(0, (\Delta p, n)|_{\partial\Omega}) = 0$$

We denote $w = \Delta p$. Applying the operator Δ to the both parts of (1.38) we obtain that

$$(1.39) \qquad -\partial_t w(t,x) - \Delta w(t,x) = 0, \quad t \in (0,T), x \in \Omega$$

Applying the operator Δ to the both parts of (1.37) we obtain the equality

$$(1.40) \qquad w(t,x) = 0 \quad t \in (0,T), \quad x \in \omega$$

The function $w(t,x)$ is a solution of inverse heat equation (1.39) and hence it is analytic with respect to x. Therefore by (1.40) we have the equality $w(t,x) = 0$, $t \in (0,T)$, $x \in \Omega$ which implies together with (1.36_2), the relations

$$\Delta p(t,x) = 0, \quad p|_{\partial\Omega} = 0$$

and, hence, the relation $p(t,x) \equiv 0$, $t \in (0,T)$, $x \in \Omega$ holds. Thus, using (1.36_3) we obtain that $z_0 = 0$. $\square$

1.4. Proof of the main results. Lemmas 1.1, 1.2 and the Gilbert-Schmidt theorem involve that the operator R has a denumerable system of eigenfunctions $\{e_j\}$ with eigenvalues $\lambda_1 \geq \lambda_2 \cdots \lambda_j \longrightarrow 0$ as $j \longrightarrow \infty$. Moreover, $\{e_j\}$ forms an orthonormal basis in $H^0(\Omega)$ and for an arbitrary $z \in H^0(\Omega)$ we have:

$$(1.41) \qquad \text{if } z = \sum_{j=1}^\infty z_j e_j \quad \text{then } Rz = \sum_{j=1}^\infty \lambda_j z_j e_j$$

It follows from (1.26), (1.41) that if $\hat{y} = \sum_{j=1}^\infty \hat{y}_j e_j$ then

$$(1.42) \qquad \hat{y} - y(T,\cdot) = \sum_{j=1}^\infty \frac{\varepsilon \hat{y}_j e_j}{(\varepsilon + \lambda_j)}$$

THEOREM 1.1. *For an arbitrary $\varepsilon > 0$ problem (1.13), (1.14), (1.21) has the unique solution $(p_\varepsilon(t), y_\varepsilon(t))$ and*

$$(1.43) \qquad p_\varepsilon(t) = e^{\pi\Delta(T-t)}\varepsilon(\varepsilon I + R)^{-1}\hat{y},$$

where R is operator (1.25) and $y_\varepsilon(t)$ is determined by (1.23). Besides, relation (1.12) holds.

Proof. Equality (1.43) follows from (1.22), (1.42). Relations (1.43), (1.23) involve the existence and uniqueness of a solution of problem (1.13), (1.14), (1.21). Let us prove (1.12). It is evident that

$$(1.44) \quad \|\hat{y} - y(T)\|^2_{H^0(\Omega)} = \sum_{j=1}^{\infty} \frac{\varepsilon^2|\hat{y}_j|^2}{(\varepsilon + \lambda_j)^2} \leq \sum_{j=1}^{N} \frac{\varepsilon^2|\hat{y}_j|^2}{(\varepsilon + \lambda_j)^2} + \\ + \sum_{j=N+1}^{\infty} |\hat{y}_j|^2$$

For any $\delta > 0$ there exists such N that the second term in the right-hand-side of inequality (1.44) is less than δ. For this N and for sufficiently small ε the first term will be less than δ also. $\square$

THEOREM 1.2. *Problem (1.1), (1.3), (1.4) is $H^0(\Omega)$-approximate controllable with respect to the control space $L_2(0,T; H^0(\omega))$. Besides, if $(y_\varepsilon(t,\cdot), u_\varepsilon(t,\cdot))$ is the solution of problem (1.1), (1.3), (1.4) with control*

$$(1.45) \qquad u_\varepsilon(t) = \hat{\pi}_\omega \left(e^{\pi\Delta(T-t)}(\varepsilon I + R)^{-1}\hat{y} \right)$$

then y_ε satisfies (1.12).

Theorem 1.2 follows from theorem 1.1 immediately.

REMARK 1.1. *In A.V. Fursikov, O. Yu. Imanuvilov [2] approximate controllability of the Stokes system has been proved unconstructively with respect to the following classes of controls (besides the case was considered above). 1. Densities of external forces having the form $\delta(t - t_0)v(x)$ where $\delta(t-t_0)$ is the Dirac measure and supp $v \subset \omega \subset \Omega$. 2. Densities of external forces concentrated on hypersurface $S \subset \Omega$. 3. Initial value concentrated in a subdomain $\omega \subset \Omega$. 4. Dirichlet boundary values concentrated in subdomains of the boundary $\partial\Omega$. The methods of this section can be applied to all cases mentioned above.*

2. On approximate uncontrollability of the Burgers equation.

In this section we show that the Burgers equation is not approximately controllable on an arbitrary finite time interval.

2.1. The main estimate. Let us consider the Burgers equation

$$(2.1) \quad \partial_t y(t,x) - \partial_{xx}^2 y(t,x) + y\partial_x y = u(t,x),\, x \in (0,a),\, t \in (0,T),$$

where $a > 0$, $T > 0$ are arbitrary fixed numbers. We suppose that a solution $y(t,x)$ satisfies zero boundary and initial conditions

$$(2.2) \quad\quad\quad\quad y(t,0) = y(t,a) = 0, \quad y(0,x) = 0.$$

Assume that $u(t,x) \in L_2([0,T] \times [0,a])$ and for any $t \in (0,T)$ the inclusion

$$(2.3) \quad\quad\quad\quad \text{supp } u(t,x) \subset (b,c), \quad 0 < b < c < a.$$

holds.

It is well-known that for an arbitrary $u \in L_2([0,T] \times [0,a])$ there exists the unique solution $y(t,x) \in L_2(0,T; W_2^2(0,a))$ of problem (2.1), (2.2). It is possible to see, expressing $\partial y/\partial t$ from (2.1) that $\partial y(t,x)/\partial t \in L_2((0,T) \times (0,a))$. We deduce one estimate for a solution $y(t,x)$ of problem (2.1), (2.2) which simply involves the uncontrollability of this problem.

LEMMA 2.1. *Let $u \in L_2([0,T] \times [0,a])$ satisfy condition (2.3), and $y(t,x)$ be a solution of problem (2.1), (2.2). Denote $y_+(t,x) = \max(y(t,x),0)$. Then for arbitrary $N > 5$ the estimate*

$$(2.4) \quad\quad\quad\quad \frac{\partial}{\partial t} \int_0^b (b-x)^N y_+^4(t,x)dx < \alpha(N)b^{N-5}$$

holds where b is the constant from (2.3) and $\alpha(N) > 0$ is a constant, depending on N only.

Proof. We multiply both sides of (2.1) by $(b-x)^N y_+^3(t,x)$ and integrate them with respect to x from 0 upto b. Integrating by parts in the second term of the left-hand-side of the obtained identity we shall have

$$\int_0^b (b-x)^N (\partial_t y) y_+^3\, dx \;+\; \int_0^b (b-x)^N 3y_+^2 (\partial_x y_+)(\partial_x y)dx \;-$$

$$(2.5)$$

$$-\int_0^b N(b-x)^{N-1} y_+^3 \partial_x y dx \;+\; \int_0^b (b-x)^N y_+^4 \partial_x y dx = 0.$$

It follows from the theorem on smoothness of solution of Burgers equation that $y(t,x) \in C^0((0,t) \times (0,a))$. Denote $y_- = \min(y,0)$. Then

$$y_+^3 \partial_t y = y_+^3 (\partial_t y_+ + \partial_t y_-) = y_+^3 \partial_t y_+ = \frac{1}{4}\partial_t y_+^4.$$

160   ANDREI V. FURSIKOV AND OLEG YU. IMANUVILOV

The following identities are proved similarly:

$$y_+^2 \frac{\partial y_+}{\partial x} \frac{\partial y}{\partial x} = y_+^2 \left(\frac{\partial y_+}{\partial x}\right)^2 , y^k \frac{\partial y_+}{\partial x} = \frac{1}{k+1} \frac{\partial y_+^{k+1}}{\partial x}$$

Using this equalities and integrating by parts in the last two terms of equation (2.5) we obtain

$$\int_0^b (b-x)^N \frac{1}{4} \partial_t y_+^4 \, dx \;\; + \;\; \int_0^b (b-x)^N 3 y_+^2 (\partial_x y_+)^2 \, dx \; -$$

$$(2.6)$$

$$- \int_0^b \frac{N}{4}(N-1)(b-x)^{N-2} y_+^4 \, dx \;\; + \;\; \int_0^b \frac{N}{5}(b-x)^{N-1} y_+^5 \, dx = 0.$$

By the Hölder inequality

$$\int_0^b (b-x)^{N-2} y_+^4 \, dx \;\; \leq \;\; \left(\int_0^b (b-x)^{N-6} dx\right)^{1/5} \left(\int_0^b (b-x)^{N-1} y_+^5 \, dx\right)^{4/5} =$$

$$(2.7)$$

$$= \frac{b^{(N-5)/5}}{(N-5)^{1/5}} \left(\int_0^b (b-x)^{N-1} y_+^5 \, dx\right)^{4/5}$$

Using the Young inequality we shall have

$$(2.8) \quad \frac{N}{5} \int_0^b (b-x)^{N-1} y_+^5 \, dx \; -$$

$$-\frac{N(N-1)}{4(N-5)^{1/5}} b^{(N-5)/5} \left(\int_0^b (b-x)^{N-1} y_+^5 \, dx\right)^{4/5} \geq -\alpha(N) b^{N-5}.$$

where $\alpha(N)$ is a positive constant, depending on $N > 5$ only. Substituting (2.7), (2.8) into (2.6) we obtain (2.4). $\square$

2.2. The results on approximate uncontrollability.

THEOREM 2.1. *Let $T > 0$ be an arbitrary finite number. Then problem (2.1), (2.2) is not $L_2(0,a)$-approximately controllable with respect to set of controls $u \in L_2((0,T) \times (0,a))$ satisfying (2.3).*

Proof. Let $\hat{y}(x) \in L_2(0, a), \hat{y}(x) \geq 0$, y be a solution of problem (2.1), (2.2) and $T > 0$. Then

$$\left(\int_0^a |\hat{y}(x) - y(T, x)|^2 dx \right)^{1/2} > \left(\int_0^{b/2} |\hat{y}(x) - y_+(T, x)|^2 dx \right)^{1/2} \geq$$

(2.9)

$$\geq \|\hat{y}\|_{L_2(0, b/2)} - \|y_+(T, \cdot)\|_{L_2(0, b/2)}$$

By the Cauchy-Bunyakovskii inequality we have:

$$\|y_+(T, \cdot)\|_{L_2(0, b/2)} \leq$$

(2.10)

$$\left(\int_0^{b/2} (b - x)^{-N} dx \right)^{1/2} \left(\int_0^{b/2} (b - x)^N |y_+(T, x)|^4 dx \right)^{1/2} \leq$$

$$\leq \left(\frac{b^{1-N}(2^{N-1} - 1)}{N - 1} \right)^{1/2} \left(\int_0^{b/2} (b - x)^N |y_+(T, x)|^4 dx \right)^{1/2}.$$

In virtue of (2.4) for any $T > 0$ inequality

(2.11)

$$\int_0^b (b - x)^N |y_+(T, x)|^4 dx \leq T\alpha(N) b^{N-5}$$

holds. Let $T > 0$ be fixed and $\hat{y}(x) \in L_2(0, a)$ satisfies condition

(2.12) $$\|\hat{y}\|_{L_2(0, b/2)} > \left(\frac{b^{1-N}(2^{N-1} - 1)}{N - 1} T\alpha(N) b^{N-5} \right)^{1/2} + 1.$$

Then it follows from (2.9)–(2.12) that for any control $u \in L_2((0, T) \times (0, a))$ satisfying (2.3) the solution y of problem (2.1), (2.2) satisfies inequality

$$\|\hat{y} - y(T, \cdot)\|_{L_2(0, a)} > 1.$$

This inequality ascertains the approximate uncontrollability of problem (2.1), (2.2). $\square$

Now we consider the Burgers equation with boundary control u:

(2.13) $$\partial_t y(t, x) - \partial_{xx}^2 y + y \partial_x y = 0, x \in (0, a), t \in (0, T)$$

(2.14) $$y(t, 0) = 0, y(t, a) = u(t), y|_{t=0} = 0, u \in L_2(0, T).$$

THEOREM 2.2. *Problem (2.13), (2.14) is not $L_2(0, a)$-approximately controllable with respect to the control space $L_2(0, T)$ for arbitrary $T > 0$.*

Proof. Estimate (2.4) holds for a solution y of problem (2.13), (2.14) and its proof does not differ from the proof of Lemma 2.1. We obtain the assertion of theorem by means of this estimate after repeating the proof of Theorem 2.1. $\square$

REMARK 2.1. *Actually, estimate (2.4) is based on the following property of solutions of the Hopf equation (i.e. the Burgers equation without the term $\partial_{xx}^2 y$): a positive wave moves to the right and a negative one moves to the left. Therefore it seems real to generalize estimate (2.4) on the case of the equation*

$$(2.15) \qquad \partial_t y + \partial_x f(y) - \partial_{xx}^2 y = 0$$

with some $f(y)$. It is interesting to consider the case when (2.15) is a system of equations as well as other one-dimensional parabolic quasilinear systems of equation. It is possible to begin to study many-dimensional case from the many-dimensional Burgers equation

$$(2.16) \qquad \partial_t v + \sum_{j=1}^{3} v_j \partial_{x_j} v = \Delta v, \operatorname{rot} v = 0, v|_{t=0} = -\nabla \xi(x)$$

where $v(t, x) = (v_1, v_2, v_3)$ is unknown vector-field and $\xi(x)$ is a given scalar function.

3. Absorbing points of reachability for the Burgers equation. In previous section the approximate uncontrollability of the Burgers equation has been proved. The analogous situation takes place for a number semilinear equations (see [7], [8]). Therefore it seems to be expedient to consider some new formulations of the controllability problem for nonlinear partial differential equations with nonlinearities of a power growth.

3.1. Absorbing points of reachability in the case of the boundary control. We consider the Burgers equation

$$(3.1) \qquad \partial_t y(t, x) - \partial_{xx}^2 y(t, x) + \partial_x y^2(t, x) = 0, \quad x \in [0, a], t > 0$$

with a boundary control

$$(3.2) \qquad y(t, 0) = u_0(t), \quad y(t, a) = u_1(t)$$

and with an initial condition

$$(3.3) \qquad y(t, x)|_{t=0} = y_0(x)$$

where $y_0(x) \in L_2(0, a)$ is a given function.

DEFINITION 3.1. *A function $\hat{y}(x) \in L_2(0, a)$ is called an absorbing point of reachability for the Burgers equation with a boundary control if for* *an arbitrary initial function $y_0(x) \in L_2(0, a)$ there exist $T = T(y_0) > 0$ and such controls $u_j(t) \in L_2^{\mathrm{loc}}(R_+)$ $j = 0, 1$, that the solution $y(t, x)$ of problem (3.1)–(3.3) with the indicated data satisfies condition*

$$\|y(t, \cdot) - \hat{y}\|_{L_2(0,a)} \equiv 0 \quad \forall t > T(y_0).$$

We will need also the other notion which looks more weak from the formal point of view.

DEFINITION 3.2. *A function $\hat{y}(x) \in L_2(0, a)$ is called approximately absorbing point of reachability for the Burgers equation with a boundary control if for arbitrary initial function $y_0(x) \in L_2(0, a)$ there exist such controls $u_j(t) \in L_2^{\mathrm{loc}}(R_+)$, $j = 0, 1$, that the solution of problem (3.1)– (3.3) satisfies conditions*

$$(3.4) \qquad \|y(t, \cdot) - \hat{y}\|_{L_2(0,a)} \longrightarrow 0 \quad as \quad t \longrightarrow \infty$$

$$(3.5) \qquad \forall \varphi \in C_0^\infty(0, a) \int_0^a \partial_t y(t, x)\varphi(x)dx \longrightarrow 0 \quad as \quad t \longrightarrow \infty.$$

Firstly, we describe the set of all approximately absorbing points of reachability of the Burgers equation with a boundary control.

Suppose, that $\hat{y}(x) \in L_2(0, a)$ is an approximately absorbing point of reachability, $y(t, x)$ is the solution of the Burgers equation satisfying conditions (3.4) and (3.5) and

$$(3.6) \qquad w(t, x) = y(t, x) - \hat{y}(x).$$

Let us substitute $w(t, x) + \hat{y}(x)$ into (3.1) and scale in $L_2(0, a)$ the obtaining equality on $\varphi \in C_0^\infty(0, a)$. Then we obtain the identity

$$(3.7) \qquad \int_0^a (-\partial_{xx}^2 \hat{y}(x) + \partial_x \hat{y}^2(x))\varphi(x)/dx = \int_0^a (-\partial_t w \cdot \varphi + w\partial_{xx}^2\varphi +$$

$$+(2\hat{y}w + w^2)\partial_x\varphi) \, dx$$

It is easy to pass to limit in the right-hand-side of equality (3.7) with an arbitrary function $\varphi \in C_0^\infty(0, a)$ if we will use (3.4), (3.5), (3.6). As a result we obtain relation

$$(3.8) \qquad -\partial_{xx}^2 \hat{y}(x) + \partial_x \hat{y}^2(x) = 0.$$

Thus, we have proved

LEMMA 3.1. *If $\hat{y}(x)$ is an approximately stable point of reachability for the Burgers equation with a boundary control then $\hat{y}(x)$ satisfies equation (3.8).*

We set

$$(3.9) \qquad \hat{y}(0) = \alpha_1, \hat{y}(a) = \alpha_2$$

and show that an arbitrary solution of problem (3.8), (3.9) with finite α_1, α_2 is the approximately absorbing point of reachability. For this we, firstly, solve problem (3.8), (3.9).

LEMMA 3.2. *For arbitrary finite $\alpha_1 \leq \alpha_2$ there exists the unique solution $\hat{y}(x)$ of problem (3.8), (3.9). Moreover*

$$(3.10) \qquad \begin{cases} if\ \alpha_2 - \alpha_1 > a\alpha_1\alpha_2\ then\ \hat{y}(x) = \sqrt{c}\,ctg(\sqrt{c}(x+d)) \\ if\ \alpha_2 - \alpha_1 = a\alpha_1\alpha_2\ then\ \hat{y}(x) = -1/(x+d) \\ if\ \alpha_2 - \alpha_1 < a\alpha_1\alpha_2\ then\ \hat{y}(x) = -\sqrt{c}\,cth(\sqrt{c}(x+d)). \end{cases}$$

For $\alpha_1 \geq \alpha_2$ problem (3.8), (3.9) has the solution

$$(3.11) \qquad \hat{y}(x) \ \equiv\ \alpha_1, \quad if\ \alpha_1 = \alpha_2$$
$$(3.12) \qquad \hat{y}(x) \ =\ -\sqrt{c}\,th(\sqrt{c}(x+d)), \ if\ \alpha_1 > \alpha_2$$

and the constants c, d in (3.10)–(3.12) are determined uniquely by α_1, α_2.

Proof. Integrating (3.8) one time we obtain

$$(3.13) \qquad \partial_x \hat{y} = \hat{y}^2 + c.$$

If $c > 0$ then integrating (3.13) we obtain the equality

$$(3.14) \qquad \frac{1}{\sqrt{c}} \arctan \frac{\hat{y}}{\sqrt{c}} = x + d$$

which implies (3.10). We show that the constants $c > 0, d$ in this equality is determined uniquely by α_1, α_2. It follows from (3.14), (3.9) that

$$a\sqrt{c} = \arctan \frac{\alpha_2}{\sqrt{c}} - \arctan \frac{\alpha_1}{\sqrt{c}}$$

Applying to the both parts of this equality the operation tg we obtain that

$$\mathrm{tg}\,(a\sqrt{c}) = \sqrt{c}(\alpha_2 - \alpha_1)/(c + \alpha_1\alpha_2).$$

Solving this equation by the method of graphics we obtain that if α_1, α_2 satisfy condition (3.10_1) then the unique positive solution c of this equation exists.

If $c = 0$ then we obtain (3.10_2) after integrating (3.13). Equation (3.13) with $c < 0$ implies the equality

$$(3.15) \qquad \left| \frac{\hat{y} - \sqrt{c_1}}{\hat{y} + \sqrt{c_1}} \right| = e^{2\sqrt{c_1}(x+d)}$$

where $c_1 = -c$. It follows from (3.15), (3.9) that

$$e^{2\gamma a} = \left| \frac{(\alpha_2 - \gamma)(\alpha_1 + \gamma)}{(\alpha_2 + \gamma)(\alpha_1 - \gamma)} \right|$$

where $\gamma = \sqrt{c_1}$. Solving this equation by method of graphics, it is easy to show that this equation has the unique positive solution if α_1, α_2 satisfy condition (3.12), (3.10$_3$). The case (3.11) is evident. Thus, we have obtained the complete substantiation of (3.10)–(3.12). $\square$

THEOREM 3.1. *Let* α_1, $\alpha_2 \in R$ *satisfy condition* $\alpha_2 \geq \alpha_1$ *and* $\hat{y}(x)$ *is a solution of problem (3.8), (3.9). Then* $\hat{y}(x)$ *is approximately absorbing point of reachability which can be approached by solution* $y(t, x)$ *of problem (3.1)–(3.3) with control* $u_0(t) \equiv \alpha_1, u_1(t) \equiv \alpha_2$. *Moreover*

$$\|w(t, \cdot)\|_{L_2(0,a)}^2 \leq e^{-\lambda t}\|y_0 - \hat{y}\|_{L_2(0,a)}^2,$$

(3.16)

$$\int_0^\infty \|\partial_x w(t, \cdot)\|_{L_2(0,a)}^2 dt \leq \|y_0 - \hat{y}\|_{L_2(0,a)}^2$$

where w *is function (3.6),* $\lambda > 0$.

Proof. Let $y(t, x)$ be the solution of (3.1)–(3.3) with $u_0(t) = \alpha_1, u_1(t) = \alpha_2$ and w be function (3.6). In virtue of (3.1)–(3.3), (3.8), $w(t, x)$ is a solution of the problem

$$(3.17) \qquad \partial_t w - \partial_{xx}^2 w + 2\partial_x(w\hat{y}) + \partial_x w^2 = 0$$

$$(3.18) \qquad w(t, 0) = w(t, a) = 0, \quad w(0, x) = y_0(x) - \hat{y}(x)$$

Scaling in $L_2(0, a)$ both parts of (3.17) on $w(t, x)$ and taking into account (3.18) we obtain after simple transformations, that

$$(3.19) \quad \frac{1}{2}\partial_t\|w(t, \cdot)\|_{L_2}^2 + \|\partial_x w(t, \cdot)\|_{L_2}^2 + \int_0^a (\partial_x\hat{y})w^2(t, x)dx = 0.$$

Let λ_1 be the minimal eigen-value of the spectral problem

$$-\partial_{xx}^2 v(x) + (\partial_x\hat{y}(x))v(x) = \lambda v(x), \quad v(0) = v(a) = 0$$

Since by Lemma 3.2 the inequality $\partial_x\hat{y}(x) \geq 0$ holds, then $\lambda_1 > 0$. It follows from (3.19) that

$$\frac{1}{2}\partial_t\|w(t, \cdot)\|_{L_2(0,a)}^2 + \lambda_1\|w(t, \cdot)\|_{L_2(0,a)}^2 \leq 0.$$

This inequality and (3.19) imply (3.16). Relation (3.5) is deduced easily from (3.16), (3.17). $\square$

Let $\hat{y}(t,x) \in W_2^{1,2}((0,T) \times (0,a)) = \{y \in L_2(0,T;W_2^2(0,a)) : \partial_t y \in L_2(Q)\}$ be a solution of equation (3.1) and $B_r(y_0) = \{z(x) \in W_2^1(0,a) : \|z - y_0\|_{W_2^1} < r\}$ be the ball of radius r with the center $y_0 \in W_2^1(0,a)$.

THEOREM 3.2. *For sufficiently small r and for an arbitrary $z_0(x) \in B_r(\hat{y}(0,x))$ there exists the solution $z(t,x) \in W_2^{1,2}((0,T) \times (0,a))$ of equation (3.1) which satisfies conditions*

$$z(0,x) = z_0(x), \quad z(T,x) = \hat{y}(T,x)$$

We denote $z(t,x)|_{x=0} = u_0(t), z(t,x)|_{x=a} = u_1(t)$. By means of Theorem 3.2 boundary controls $u_0(t)$, $u_1(t)$ transform the solution $z(t,x)$ of (3.1), (3.2), (3.3) with $y_0 = z_0$ to the given solution $\hat{y}$ at moment T : $z(T,x) = \hat{y}(T,x)$. The proof of one more general assertion than Theorem 3.2 will be given below at subsection 5.1 of Section 5 (See Theorem 5.1).

THEOREM 3.3. *Let α_1, $\alpha_2 \in R^1$ satisfy condition $\alpha_2 \geq \alpha_1$ and $\hat{y}(x)$ be a solution of problem (3.8), (3.9). Then $\hat{y}(x)$ is an absorbing point of reachability for the Burgers equation with a boundary control.*

Proof. Let $y(t,x)$ be the solution of problem (3.1)–(3.3) with $u_0(t) = \alpha_1$, $u_1(t) = \alpha_2$. We apply Theorem 3.1. By virtue of (3.16) for r as small as we want there exists such t_0 that

$$\|w(t_0,x)\|_{W_2^1(0,a)} = \|y(t_0,x) - \hat{y}(x)\|_{W_2^1(0,a)} < r$$

Now we apply Theorem 3.2 with $\hat{y}(t,x) = \hat{y}(x)$, $z_0(x) = y(t_0,x)$. In virtue of this theorem by means of correct choice of boundary controls it is possible to do that the corresponding solution $z(t,x)$ will coincide with $\hat{y}(x)$ when $t = T$. $\square$

The solution $\hat{y}(x)$ of (3.8), (3.9) is an absorbing point of reachability in the case when $\alpha_1 > \alpha_2$. This assertion will be proved at the end of section 5.

3.2. Absorbing points of reachability in the case of local distributed control. We consider the Burgers equation with distributed control

$$(3.20) \qquad \partial_t y(t,x) - \partial_{xx}^2 y(t,x) + \partial_x y^2 = u(t,x)$$

with periodic boundary conditions

$$(3.21) \qquad y(t, x + 2\pi) = y(t,x), u(t, x + 2\pi) = u(t,x)$$

and with initial condition (3.3). We assume that the support of control $u(t, x)$ is concentrated on subinterval

$$(3.22) \qquad \operatorname{supp} u(t, \cdot) \subset \bigcup_{k \in \mathbb{Z}} ((a, 2\pi) + 2\pi k) \quad \forall t > 0.$$

Note that by (3.21) we can suppose that equation (3.20) is defined on the circumference $S = \{x \in (0, 2\pi);$ the points 0 and 2π are identified$\}$.

As in Definition 3.1 a function $\hat{y} \in L_2(S)$ will be called an absorbing point of reachability for problem (3.20)–(3.22), (3.3) if for an arbitrary initial function $y_0 \in L_2(S)$ there exists such control $u(t, x) \in L_2^{\mathrm{loc}}(\mathbb{R}_+ \times S)$ satisfying (3.22) that for the solution $y(t, x)$ of problem (3.20), (3.21), (3.3) the relation

$$(3.23) \qquad \|y(t, \cdot) - \hat{y}\|_{L_2(S)} = 0 \text{ when } t > t_1$$

holds where $t_1 = t_1(y_0, \hat{y})$ is a sufficiently large number.

We show that an arbitrary solution $\hat{y}(x)$ of equation

$$(3.24) \qquad -\partial_{xx}^2 \hat{y}(x) + \partial_x \hat{y}^2(x) = f(x),$$
$$\text{where } f(x) \in L_2(S), \quad \operatorname{supp} f \in (a, 2\pi)$$

is the stable point of reachability.

THEOREM 3.4. *Let $\hat{y}(x)$, $x \in S$ satisfy equation (3.24). Then $\hat{y}$ is the absorbing point of reachability for problem (3.20)–(3.22), (3.3).*

Proof. We denote

$$\alpha_1 = \hat{y}(0), \quad \alpha_2 = \hat{y}(a).$$

Then $\hat{y}(x)$ is a solution of problem (3.8), (3.9) on interval $(0, a)$. Therefore in virtue of Theorem 3.3 $y(t, x) - \hat{y}(x) = 0$ if $t > t_1$ for the solution $y(t, x)$ of problem (3.1)–(3.3) with the controls $u_0(t)$, $u_1(t)$ chosen correctly.

Let $\varphi_j(x) \in C^\infty(a, 2\pi)$, $j = 1, 2$, and

$$\varphi_1(x) = \begin{cases} 1, & x \in \left(a, a + \frac{2\pi - a}{3}\right) \\ 0, & x \in \left(a + \frac{2}{3}(2\pi - a), 2\pi\right) \end{cases}, \varphi_2(x) = 1 - \varphi_1(x) \ .$$

Let $w(t, x)$ be function (3.6) defined for $x \in (0, a)$. We extend the function $w(t, x)$ from $x \in (0, a)$ up to $x \in S$ by formula

$$(3.25) \quad w_1(t, x) \ = \ \begin{cases} w(t, x), x \in (0, a) \\ \varphi_1(x) \left(4w\left(t, \frac{3a}{2} - \frac{x}{2}\right) - 3w(t, 2a - x)\right) + \\ \quad + \ \varphi_2(x) \left(4w\left(t, \pi - \frac{x}{2}\right) - 3w(t, 2\pi - x)\right), x \in (a, 2\pi) \end{cases}$$

It follows immediately from (3.25) that

$$
\begin{aligned}
w_1(t, a+0) &= w_1(t, a-0), \partial_x w_1(t, x)|_{x=a+0} = \partial_x w_1(t, x)|_{x=a-0} \\
w_1(t, 2\pi) &= w_1(t, 0), \partial_x w_1(t, x)|_{x=2\pi} = \partial_x w_1(t, x)|_{x=0}
\end{aligned}
$$

By means of this formulas and (3.25) it is easy to deduce estimates

$$
\begin{aligned}
(3.26) && \|\partial_t w_1(t, \cdot)\|_{L_2(S)} &\leq c\|\partial_t w(t, \cdot)\|_{L_2(0,a)} \\
(3.27) && \|w_1(t, \cdot)\|_{W_2^2(S)} &\leq c\|w(t, \cdot)\|_{W_2^2(0,a)}
\end{aligned}
$$

with a constant c which does not depend on t. It follows from (3.25) and from the method of construction of $w(t, x)$ that

$$
(3.28) \qquad w_1(t, x) \equiv 0 \quad \text{when } t > t_1
$$

We denote

$$
(3.29) \qquad y_1(t, x) = w_1(t, x) + \hat{y}(x)
$$

and define the function $u(t, x)$ by the equality

$$
(3.30) \qquad u(t, x) = \partial_t y_1 - \partial_{xx}^2 y_1 + \partial_x y_1^2
$$

It follows from estimates (3.26), (3.27) that $u(t, x) \in L_2^{\mathrm{loc}}(R_+ \times S)$. Besides, equalities (3.29), (3.25) imply that $y_1(t, x) \equiv y(t, x)$ for $x \in (0, a)$ and therefore by (3.30) the inclusion $\operatorname{supp} u(t, x) \subset R_+ \times [a, 2\pi]$ holds. Note that it follows from (3.28), (3.29) that $y_1(t, x) \equiv \hat{y}(x)$, $u(t, x) \equiv f(x)$ when $t > t_1$. Thus, the control $u(t, x)$ defined in (3.30) transforms the initial function $y_0(x)$ by trajectory $y_1(t, x)$ to $\hat{y}(x)$ during a finite time. $\square$

REMARK 3.1. *Apparently, some generalizations of the section's 3 theory can be done on the case of semilinear one-dimensional parabolic equations as well as on the case of equations (2.15). It is possible to try to construct the theory of section 3 in the case of the many-dimensional Burgers equation (2.16), taking into account that this equation can be reduced to the heat-equation as in one-dimensional case.*

4. Exact controllability of a linear parabolic equation. To prove the local theorem of exact controllability for the Burgers equation we establish in this section one theorem on exact controllability of parabolic equations with variable coefficients in the Sobolev space $W_2^{1,2}(Q)$. Note that analogous results on the exact controllability of linear parabolic equations in the Sobolev space $W^{\frac{1}{4},\frac{1}{2}}(Q)$ was obtained by O. Yu. Imanuvilov ([11]– [13]). Here we use such important tools of these works as Carleman estimates.

4.1. Formulation of the problem. Reduction to the homogeneous boundary conditions. In the domain $Q = (0,T) \times \Omega$, where $\Omega = (-2,2)$ we consider the linearized Burgers equation

$$(4.1) \quad Ly = \partial_t y(t,x) - \partial_{xx}^2 y(t,x) + \partial_x(z(t,x)y(t,x)) = 0, \quad (t,x) \in Q$$

where $z(t,x) \in W_2^{1,2}(Q)$ is a datum and $y(t,x) \in W_2^{1,2}(Q)$ is an unknown function. It is assumed that $y(t,x)$ satisfies conditions:

$$(4.2) \qquad\qquad y(t,x)|_{t=0} \;=\; y_0(x)$$
$$(4.3) \qquad\qquad y(t,x)|_{t=T} \;=\; 0,$$

where $y_0(x) \in W_2^1(\Omega)$ is a datum.

We use the following functional spaces: the Sobolev space $W_2^k(\Omega)$ of functions defined in Ω and possessing finite norm (1.8), the space $(\overset{\circ}{W}{}_2^k(\Omega)$, that is the closure of $C_0^\infty(\Omega)$ in norm (1.8) and, at last, the space

$$W_2^{k,2k} = \left\{ y(t,x) \in L_2(0,T;W_2^{2k}(\Omega)) : \right.$$
$$\left. : \partial_t^j y(t,\cdot) \in L_2(0,T;W_2^{2(k-j)}(\Omega)), \quad j = 1,\cdots k \right\},$$

$$\|y\|_{W_2^{k,2k}}^2 = \sum_{j=0}^{k} \int_0^T \|\partial_t^j y(t,\cdot)\|_{W_2^{2(k-j)}(\Omega)}^2 \, dt.$$

The problem of exact controllability of equation (4.1) is as follows: one must find such boundary controls $v_j(t) \in L_2(0,T), j = 0,1$, i.e.

$$(4.4) \qquad y(t,x)|_{x=-2} = v_-(t), y(t,x)|_{x=2} = v_+(t)$$

that the solution of mixed boundary problem (4.1), (4.2), (4.4) would satisfy condition (4.3).

We reduce the problem of exact controllability to a similar one having initial function $y_0(x)$ from (4.2) which equals zero. Let $\chi(t,x)$ be the solution of problem (4.1), (4.2), (4.4) with the boundary conditions $v_\pm(t) \equiv 0$. The solution $\chi(t,x)$ belongs to $W_2^{1,2}(Q)$.

Let $\varphi(t) \in C^\infty(0,T)$, $\varphi(t) = 1$ when $t \in (0,\frac{T}{3})$, $\varphi(t) \equiv 0$ when $t \in (\frac{2}{3}T, T)$. Denote

$$(4.5) \qquad \hat{y}(t,x) = \chi(t,x)\varphi(t), \quad L\hat{y} = -f_0(t,x)$$

It is evident, that inequalities

$$(4.6) \qquad \|f_0\|_{L_2(Q)} \le c_1 \|\hat{y}\|_{W_2^{1,2}(Q)} \le c_2 \|y_0\|_{W_2^1(\Omega)}$$

hold where the constants c_1, c_2 depend continuously on $\|z\|_{W_2^{1,2}(Q)}$.

We set

$$(4.7) \qquad\qquad y(t,x) = w(t,x) + \hat{y}(t,x)$$

where $\hat{y}(t, x)$ is function (4.5). It is easy to show that the following assertion holds.

PROPOSITION 4.1. *A function $y(t, x) \in L_2(Q)$ is a solution of exact controllability problem (4.1)–(4.3) if and only if the function $w(t, x)$ defined in (4.7) satisfies equalities*

$$(4.8) \qquad Lw \;=\; f_0,$$

$$(4.9) \qquad w(t, x)|_{t=0} \;=\; 0, \quad w(t, x)|_{t=T} = 0$$

where f_0 is the function defined in (4.5).

4.2. Boundary value problem. Thus, our problem has been reduced to construction of a function $w(t, x) \in L_2(Q)$ satisfying (4.8), (4.9). We consider the following extremal problem: To minimize the functional

$$(4.10) \qquad J(w) = \frac{1}{2} \int_Q w^2(t, x)\,dx\,dt \longrightarrow \inf$$

on the set of functions w satisfying (4.8), (4.9)

LEMMA 4.1. *If $w(t, x) \in L_2(Q)$ is a solution of problem (4.10), (4.8), (4.9) then there exists the function $p(t, x)$ satisfying the relations*

$$(4.11) \quad L^*p = \partial_t p(t, x) + \partial^2_{xx} p(t, x) + z(t, x)\partial_x p(t, x) = w(t, x), (t, x) \in Q$$
$$(4.12) \quad p(t, x)|_{x=\pm 2} = 0, \quad \partial_x p(t, x)|_{x=\pm 2} = 0.$$

The proof of this lemma can be realized, for instance, as in O. Yu. Imanuvilov [20] in spite of it's complication. But we do not need to have this proof, as well as a proof of the existence theorem for extremal problem (4.8)–(4.10). The point is that to prove the solvability of problem (4.8), (4.9) it is sufficient to prove the solvability of problem (4.8), (4.9), (4.11), (4.12). Problem (4.8)–(4.10) and Lemma 4.1 are useful only to understand how boundary value problem (4.8), (4.9), (4.11), (4.12) was obtained.

To exclude from (4.8), (4.9), (4.11), (4.12) unknown function $w(t, x)$ we apply to both parts of (4.11) the operator L. By means of (4.8) we obtain equation

$$(4.13) \qquad LL^*p = f_0 \quad (t, x) \in Q$$

Besides boundary conditions (4.12), equation (4.13) satisfies the boundary conditions

$$(4.14) \qquad L^*p|_{t=0} = 0, \quad L^*p|_{t=T} = 0$$

These conditions arise from (4.9), (4.11).

4.3. An apriori estimate of Carleman type.

THEOREM 4.1. *Let $p(t, x)$ satisfies relations (4.11), (4.12) where $w(t, x) \in L_2(Q)$. Then for an arbitrary $\tau \in (0, T)$ and sufficiently large k the estimate*

$$(4.15) \quad \int_Q (|\partial_t p|^2 + |\partial_{xx}^2 p|^2 + |\partial_x p|^2 + p^2) e^{-\frac{k}{(T-t)^2}} \, dx\, dt +$$

$$+ \int_\Omega (|\partial_x p(\tau, x)|^2 + p^2(\tau, x)) e^{-\frac{k}{(T-\tau)^2}} \, dx$$

$$\leq c \int_Q w^2(t, x) dx\, dt$$

holds where the constant c depends continuously on $\|z\|_{W_2^{1,2}(Q)}$.

Proof. We consider the function

$$(4.16) \qquad \varphi(t, x) = \ln(x + 4)((T - t)^{-2} + t^{-2}).$$

For an arbitrary $s > 0$ we define the operator

$$Mu = e^{-s\varphi} L^*(e^{s\varphi} u)$$

where L^* is operator (4.11). The operator M can be written in the form

$$Mu = L^* u + s(\partial_t \varphi)u + 2s(\partial_x \varphi)\partial_x u + (s^2(\partial_x \varphi)^2 + s\partial_{xx}^2 \varphi)u + zs(\partial_x \varphi)u.$$

We introduce also the operators

$$(4.17) \qquad M_1 = \partial_{xx}^2 + s^2(\partial_x \varphi)^2, \quad M_2 = \partial_t + 2s(\partial_x \varphi)\partial_x$$

Denote

$$(4.18) \quad u = e^{-s\varphi} p, \quad w_1 = e^{-s\varphi} w, \quad f_s = -z\partial_x u - us(\partial_t \varphi + \partial_{xx}^2 \varphi + z\partial_x \varphi)$$

It is easy to see that equation (4.11) is equivalent to equality

$$(4.19) \qquad (M_1 + M_2)u = w_1 + f_s,$$

where $M_1, M_2, u, w_1 f_s$ are defined in (4.17), (4.18). It follows from (4.19) that

$$(4.20) \quad \|f_s + w_1\|_{L_2(Q)}^2 = \|M_1 u\|_{L_2(Q)}^2 + \|M_2 u\|_{L_2(Q)}^2 + 2(M_1 u, M_2 u)_{L_2(Q)}$$

We transform the last term in the right-hand-side of (4.20). Taking into account that by (4.12), (4.16), (4.18) the relations

$$(4.21) \qquad u|_{\partial\Omega} = 0, \quad \partial_x u|_{\partial\Omega} = 0, \quad u|_{\substack{t=0 \\ t=T}} = 0,$$

hold we obtain

$$(4.22) \qquad 2\left(\partial_{xx}^2 u + s^2(\partial_x\varphi)^2 u, \quad \partial_t u + 2s(\partial_x\varphi)\partial_x u\right) =$$

$$= 2\int_Q \left[-\frac{1}{2}\partial_t(\partial_x u)^2 + \frac{s^2(\partial_x\varphi)^2}{2}\partial_t u^2 + \right.$$

$$\left. + s(\partial_x\varphi)\partial_x(\partial_x u)^2 + s^3(\partial_x\varphi)^3\partial_x u^2 \right] dx\,dt =$$

$$= -\int_Q s^2\partial_t(\partial_x\varphi)^2 u^2\,dx\,dt +$$

$$+ \int_Q (-\partial_{xx}^2\varphi)(2s(\partial_x u)^2 + 6s^3(\partial_x\varphi)^2 u^2)\,dx\,dt$$

It follows from (4.18) that

$$(4.23)\ \|f_s + w_1\|_{L_2(Q)}^2 \leq c_0(\|w_1\|_{L_2(Q)}^2 + \|z\|_{W_2^{1,2}(Q)}^2\|\partial_x u\|_{L_2(Q)}^2 +$$

$$+ s^2(\|(\partial_t\varphi)u\|_{L_2(Q)}^2 + \|z\|_{W_2^{1,2}(Q)}^2\|(\partial_x\varphi)u\|_{L_2(Q)}^2 + \|(\partial_{xx}\varphi)u\|_{L_2(Q)}^2))$$

Substituting (4.22), (4.23) into (4.20) we obtain that

$$(4.24)\ 2\int_Q (-\partial_{xx}^2\varphi)[3s^3 u^2(\partial_x\varphi)^2 + s(\partial_x u)^2]\,dx\,dt +$$

$$+ \|M_1 u\|_{L_2(Q)}^2 + \|M_2 u\|_{L_2(Q)}^2 \leq$$

$$\leq c_1(\|w_1\|_{L_2(Q)}^2 + \|z\|_{W_2^{1,2}(Q)}^2\|\partial_x u\|_{L_2(Q)}^2 + s^2(\|(\partial_t\varphi)u\|_{L_2(Q)}^2 +$$

$$+ \|z\|_{W_2^{1,2}(Q)}^2\|(\partial_x\varphi)u\|_{L_2(Q)}^2 + \|u\partial_{xx}^2\varphi\|_{L_2(Q)}^2 +$$

$$+ \int_Q 2|\partial_t(\partial_x\varphi)^2|u^2\,dx\,dt))$$

Taking into account that by (4.16) $-\partial_{xx}^2\varphi > c > 0, \partial_x\varphi > c > 0$ and

$$|\partial_t\varphi|^2 \leq c_2(-\partial_{xx}^2\varphi)(\partial_x\varphi)^2, |\partial_t(\partial_x\varphi)^2| \leq c_2(-\partial_{xx}^2\varphi)(\partial_x\varphi)^2$$

we obtain that for sufficiently large s inequality (4.24) involves the estimate

$$(4.25) \qquad \int_Q (-\partial_{xx}^2\varphi)[3s^3 u^2(\partial_x\varphi)^2 + s(\partial_x u)^2]\,dx\,dt + \|M_1 u\|_{L_2(Q)}^2 +$$

$$+ \|M_2 u\|_{L_2(Q)}^2 \leq c_3\|w_1\|_{L_2(Q)}^2$$

where c_3 is a constant which depends continuously on $\|z\|_{W_2^{1,2}(Q)}^2$.

Note that the following relations hold

$$(4.26) \quad \frac{1}{s}\int_Q \frac{1}{\partial_x\varphi}|\partial_t u|^2 dxdt = \frac{1}{s}\int_Q \frac{1}{\partial_x\varphi}(M_2 u - 2s(\partial_x\varphi)\partial_x u)^2 dxdt \leq$$

$$\int_Q \left(\frac{2}{s\partial_x\varphi}|M_2 u|^2 + 8s\partial_x\varphi|\partial_x u|^2\right) dxdt \leq$$

$$\leq c_4 \int_Q (|M_2 u|^2 + s\partial_x\varphi|\partial_x u|^2)dxdt$$

$$(4.27) \quad \int_Q \frac{1}{s\partial_x\varphi}(\partial_{xx}^2 u)^2 dxdt = \int_Q \frac{1}{s\partial x\varphi}(M_1 u - s^2(\partial_x\varphi)^2 u)^2 dxdt \leq$$

$$\leq c_5 \int_Q (|M_1 u|^2 + s^3(\partial_x\varphi)^3 |u|^2)dxdt$$

It follows from (4.25)–(4.27) that

$$(4.28) \quad \int_Q \left(\frac{1}{s\partial_x\varphi}(|\partial_t u|^2 + |\partial_{xx}^2 u|^2) + s\partial_x\varphi|\partial_x u|^2 + \right.$$

$$\left. + (s^3(\partial_x\varphi)^3 u^2)\right) dxdt \leq c\|w_1\|_{L_2(Q)}^2$$

It is easily deduced from (4.28) that for an arbitrary $t \in (0, T)$

$$(4.29) \quad \int_\Omega \left(\frac{1}{\partial_x\varphi(t, x)}|\partial_x u(t, x)|^2 + (\partial_x\varphi(t, x))u^2(t, x)\right) dx \leq$$

$$\leq c\|w_1\|_{L_2(Q)}^2.$$

Returning in (4.28), (4.29) from the variables u, w_1 to the variables p, w we obtain the inequality

$$(4.30) \quad \int_Q \left[\frac{1}{s\partial_x\varphi}((\partial_t p)^2 + (\partial_{xx}^2 p)^2) + s(\partial_x\varphi)(\partial_x p)^2 + \right.$$

$$\left. + s^3(\partial_x\varphi)^3 p^2\right] e^{-2s\varphi} dxdt + \int_\Omega \left(\frac{|\partial_x p(t, x)|}{\partial_x\varphi(t, x)} + \right.$$

$$\left. + \partial_x\varphi(t, x)p^2(t, x)\right) dx \leq c \int_Q e^{-2s\varphi} w^2 dxdt.$$

Since p satisfies inverse parabolic equation (4.11) and boundary conditions (4.12) then for an arbitrary $t \in (0, T)$ the following estimate holds:

$$(4.31) \quad \|p\|_{W_2^{1,2}((0,t)\times\Omega))}^2 \leq c\left(\|p(t, \cdot)\|_{W_2^1(\Omega)}^2 + \|w\|_{L_2(Q)}^2\right)$$

Inequality (4.15) follows from (4.30), (4.31). □

4.4. Unique solvability of the boundary problem. We define the functional space Φ of functions defined on cylinder Q for which the norm

$$(4.32) \qquad \|p\|_{\Phi}^2 = \|L^*p\|_{L_2(Q)}^2 + \int_Q e^{-\frac{k}{(T-t)^2}} \left((\partial_t p)^2 + (\partial_{xx}^2 p)^2 + \right.$$
$$\left. +(\partial_x p)^2 + p^2\right) dx dt$$

is finite and boundary conditions (4.12) hold.

DEFINITION 4.1. *A function $p \in \Phi$ is called generalized solution of problem (4.13), (4.14) if it satisfies the equation*

$$(4.33) \qquad (L^*p, L^*g)_{L_2(Q)} = -(f_0, g)_{L_2(Q)} \quad \forall g \in \Phi$$

THEOREM 4.2. *There exists unique generalized solution $p \in \Phi$ of problem (4.13), (4.14). For an arbitrary subdomain $\theta \subset \Omega$ the function p belongs to $W_2^{2,4}(\theta)$ and satisfies equation (4.13) as well as boundary conditions (4.14) which are understood as equalities in the space $W_2^{-1}(\Omega)$.*

Proof. By virtue of (4.5) supp $f_0 \in \left[\frac{T}{3}, \frac{2T}{3}\right] \times \Omega$ and, hence, applying Theorem 4.1 we obtain the inequality

$$(4.34) \qquad |(f_0, g)_{L_2(Q)}| \le \|f\|_{L_2(Q)} \left(\int_{T/3}^{2T/3} g^2 dx dt\right)^{1/2} \le$$
$$\le c\|f\|_{L_2(Q)}\|g\|_{\Phi}$$

This estimate shows that the functional $g \to (f_0, g)_{L_2(Q)}$ is a continuous one on Φ. It follows from Theorem 4.1 that the norm generated by the scalar product $(L^*p, L^*q)_{L_2(Q)}$ is equivalent to the norm $\|\cdot\|_{\Phi}$. Therefore the existence of the unique function $p \in \Phi$ satisfying (4.33) follows from the Riesz theorem on the representation of a functional on a Hilbert space.

Equality (4.33) with $g \in C_0^\infty(Q)$ implies equation (4.13) understood in the sense of distributions theory. Since the operator LL^* is hypoelliptic (see L. Hörmander [21]) then $p \in W^{2,4}(\theta)$ where $\theta \subset Q$ is an arbitrary subdomain of Q. Using the denotion $w = L^*p$ we obtain by (4.32) that $w \in L_2(Q)$ and by (4.13) that

$$(4.35) \qquad Lw = f_0 \in L_2(Q)$$

where L is operator (4.1). Since $w \in L_2(Q)$ then $\partial_{xx}^2 w \in L_2(0, T; W_2^{-2}(\Omega))$ and expressing $\partial_t w$ from (4.35) we obtain that $\partial_t w \in L_2(0, T; W_2^{-2}(\Omega))$.

Therefore, using the theorem on restrictions (J.-L. Lions, E. Magenes [14]) it is easy to show that the restrictions $w(0, \cdot)$¡ $w(T, \cdot)$ of w is defined in the space $W_2^{-1}(\Omega)$. Integrating by parts in (4.33) with $g \in C^\infty(\overline{Q})$ satisfying (4.12) we deduce that $w(T, \cdot) = w(0, \cdot) = 0$. This proves (4.14). $\square$

4.5. Compact dependence on coefficient.

LEMMA 4.2. *Let p_n be the solution of problem (4.13), (4.14) with coefficient $z = z_n$ (see (4.1), (4.11)) and $w_n = L^* p_n$. Suppose that $z_n \rightharpoonup z_0$ weakly in $W_2^{1,2}(Q)$ as $n \to \infty$. Then*

$$(4.36) \qquad w_n \to w_0 \text{ strongly in } L_2(Q) \text{ as } n \to \infty.$$

Proof. It follows from definition (4.5) of f_0 that

$$(4.37) \qquad f_0(z_n) \to f_0(z_0) \qquad \text{strongly in } L_2(Q) \text{ as } n \to \infty$$

$$(4.38) \qquad f_0(z_n)(t, x) \equiv 0 \qquad \text{when } t \in \left(\frac{2T}{3}, T\right) \forall n$$

We prove (4.37). Let $z_n \rightharpoonup z_0$ weakly in $W_2^{1,2}(Q)$ and χ_n be the solution of problem (4.1), (4.2), (4.4) where $v_\pm \equiv 0$ and coefficient $z = z_n$ in (4.1). Denote $v_n = \chi_0 - \chi_n$. Then

$$(4.39) \qquad \partial_t v_n - \partial_{xx}^2 v_n + \partial_x(z_0 v_n) = \partial_x((z_n - z_0)\chi_n),$$

$$v_n|_{t=0} = 0, \quad v_n|_{x=\pm 2} = 0$$

In virtue of compactness of the embeddings $W_2^{1,2}(Q) \Subset C(\overline{Q})$, $W_2^{1,2}(Q) \Subset L_2(0, T; W_2^1(\Omega))$ and boundedness of $\|\chi_n\|_{W_2^{1,2}(Q)}$ we have that

$$\partial_x((z_n - z_0)\chi_n) = (\partial_x(z_n - z_0))\chi_n + (z_n - z_0)\partial_x\chi_n \to 0 \text{ in } L_2(Q)$$

$$\text{as } z_n \rightharpoonup z_0 \text{ weakly in } W_2^{1,2}(Q).$$

Therefore, taking into account a well known estimate for solutions of problem (4.39) we obtain that

$$\|v_n\|_{W_2^{1,2}(Q)} \to 0 \text{ as } z_n \rightharpoonup z_0 \text{ weakly in } W_2^{1,2}(Q).$$

Relation (4.37) follows from this one. Let p_n be the solution of equation (4.33) with the coefficient $z = z_n$. We substitute $z = z_n$, $p = g = p_n$ into (4.33). Then taking into account (4.37), (4.38) and estimate (4.15) we obtain as in (4.34) that

$$\|w_n\|_{L_2(Q)}^2 \le c\|f_0\|_{L_2(Q)}\|w_n\|_{L_2(Q)} \text{ where } w_n = L^* p_n$$

and, hence,

$$\|w_n\|_{L_2(Q)} \le c\|f_0\|_{L_2(Q)}$$

where c does not depend on n. It follows from this estimate and from uniqueness of the solution of problem (4.33) that

$$(4.40) \qquad w_n \rightharpoonup w_0 \text{ weakly in } L_2(Q).$$

Applying (4.15) we establish analogously that

$$(4.41) \qquad p_n \rightharpoonup p_0 \text{ weakly in } L_2\left(\left(0, \frac{2T}{3}\right) \times \Omega\right)$$

Substituting into (4.33) $p = g = p_n$ and taking into account (4.41), (4.37), (4.38) we can pass to the limit in (4.33) as $n \to \infty$. As result we can obtain that

$$(4.42) \quad \begin{aligned} \|w_n\|^2_{L_2(Q)} &= (f_0(z_n), p_n)_{L_2(Q)} \to (f_0(z_0), p_0)_{L_2(Q)} = \\ &= \|w_0\|^2_{L_2(Q)} \end{aligned}$$

as $n \to \infty$.

Relation (4.36) follows from (4.40), (4.42). $\square$

We prove also one lemma which will let to establish the compact dependence the function w on the coefficient z in the space $W_2^{1,2}(Q)$.

LEMMA 4.3. *Let $w(t, x) \in L_2(Q)$ satisfy the relations*

$$(4.43) \qquad Lw(t, x) = f(t, x), (t, x) \in Q, w_{|t=0} = 0$$

where L is operator (4.1), $f \in L_2(Q)$. Denote $\rho(x) = (4 - x^2)$. Then for the function w the following estimates hold:

$$(4.44) \quad \sup_{t \in [0,T]} \int_\Omega w^2 \rho^2 dx + \int_Q (\partial_x w)^2 \rho^2 dx dt \le$$

$$\le \gamma \left(1 + \|z\|_{W_2^{1,2}(Q)}\right) \left(\|w\|^2_{L_2(Q)} + \int_Q f^2 \rho^4 dx dt\right)$$

$$(4.45) \quad \int_Q \left((\partial_t w)^2 + (\partial^2_{xx} w)^2\right) \rho^4 dx dt \le c_1 \left(\|w\|^2_{L_2(Q)} + \int_Q f^2 \rho^4 dx dt\right) \times$$

$$\times \gamma_1 \left(1 + \|z\|_{w_2^{1,2}(Q)}\right)$$

where $\gamma(\lambda) > 0$, $\gamma_1(\lambda) > 0$ are continuous functions.

Proof. We scale in $L_2(Q)$ both parts of (4.43) on $w\rho^2$. After simple transformations we obtain the equality

$$\frac{1}{2}\partial_t \int_\Omega w^2 \rho^2\, dx \;+\; \int_\Omega (\partial_x w)^2 \rho^2\, dx + \frac{1}{2}\int_\Omega \partial_x w^2 \partial_x \rho^2\, dx -$$
$$- \int_\Omega zw\left((\partial_x w)\rho^2 + w\partial_x \rho^2\right) dx = \int_\Omega fw\rho^2\, dx$$

After integrating both parts of this equality with respect to t we obtain the estimate

$$\int_\Omega w^2(t,x)\rho^2(x)dx + \int_0^t \int_\Omega (\partial_x w)^2 \rho^2\, dx dt \le c \int_0^t \int_\Omega w^2 dx dt\; \cdot$$
$$\cdot\left(1 + \|z\|^2_{c(\overline{Q})} + \|z\|_{c(\overline{Q})}\right) + \frac{1}{2}\int_0^t \int_\Omega (\partial_x w)^2 \rho^2\, dx dt + \int_0^t \int_\Omega f^2 \rho^4 dx dt$$

Carrying the term with $(\partial_x w)^2$ from the right side to the left one, we obtain (4.44).

Multiplying (4.43) on ρ^2 and doing simple transformations we obtain the equality

$$(4.46) \qquad \begin{aligned} \partial_t(w\rho^2) - \partial^2_{xx}(w\rho^2) + \partial_x(zw\rho^2) = \\ f\rho^2 - 2(\partial_x w)\partial_x \rho^2 - w\partial^2_{xx}\rho^2 + zw\partial_x \rho^2 \end{aligned}$$

Function $w\rho^2$ satisfies equation (4.46) as well as the following initial and boundary conditions:

$$(4.47) \qquad\qquad w\rho^2|_{t=0} = 0, \quad w\rho^2|_{x=\pm 2} = 0$$

Applying to the solution $w\rho^2$ of mixed boundary problem (4.46), (4.47) well-known inequality

$$\|w\rho^2\|_{W_2^{1,2}(Q)} \le c\|f\rho^2 - 2(\partial_x w)\partial_x \rho^2 - w\partial^2_{xx}\rho^2\|_{L_2(Q)},$$

where c depends continuously on $\|z\|_{W_2^{1,2}(Q)}$ and estimating the right-hand-side of this inequality by means of (4.44) we obtain (4.45). $\square$

4.6. Termination of solution of the exact controllability problem. Thus, we study the following problem of exact controllability: In the domain $Q_1 = (0,T) \times \Omega_1$ where $\Omega_1 = (-1,1)$ the equation

$$(4.48) \qquad \partial_t y(t,x) - \partial^2_{xx} y(t,x) + \partial_x(z(t,x)y(t,x)) = 0$$

is considered with initial condition

$$(4.49) \qquad y(t, x)|_{t=0} = y_1(x)$$

One has to find a solution of equation (4.48) which, besides (4.49), satisfied the condition

$$(4.50) \qquad y(t, x)|_{t=T} = 0.$$

THEOREM 4.3. *Let functions $z_1(t, x) \in W_2^{1,2}(Q_1)$, $y_1(x) \in W_2^1(\Omega_1)$ be data. Then there exists a solution $y(t, x) \in W_2^{1,2}(Q_1)$ of problem (4.48)–(4.50). Besides, it is possible to define the map, transforming coefficient $z(t, x)$ to a solution $y(t, x)$ of problem (4.48)–(4.50), which act compactly from $W_2^{1,2}(Q_1)$ to $W_2^{1,2}(Q_1)$.*

Proof. Let $R : W_2^{1,2}(Q_1) \to W_2^{1,2}(Q)$ be the linear continuous operator, which extends $f(t, x)$, $(t, x) \in Q_1$ up to a function $Rf(t, x)$, $(t, x) \in Q = (0, T) \times (-2, 2)$ and $R_1 : W_2^1(\Omega_1) \to W_2^1(\Omega)$ be a linear continuous operator of extension of a function g from Ω_1 up to $\Omega = (-2, 2)$ such that $R_1 g|_{\partial\Omega} = 0$. Denote $z = Rz_1$, $y_0 = R_1 y_1$. Then

$$(4.51) \quad \|z\|_{W_2^{1,2}(Q)} \le c\|z_1\|_{W_2^{1,2}(Q_1)}, \quad \|y_0\|_{W_2^1(Q)} \le c_1\|y_1\|_{W_2^1(\Omega_1)}$$

where constants c, c_1 don't depend on z_1, y_1 correspondingly. We consider instead of (4.48)–(4.50) problem (4.1)–(4.3) defined in the wider domain $Q = (0, T) \times (-2, 2)$. To solve it we pass to problem (4.8), (4.9) by transformation (4.7). To solve (4.8), (4.9) we use boundary value problem (4.13), (4.14) which has the unique solution $p \in \Phi$ in virtue of Theorem 4.2. As it was shown in the proof of this theorem, the function $w = L^* p$ where L^* is operator (4.11), satisfies relations (4.8), (4.9) and the inclusion $w \in L_2(Q)$. By means of Lemma 4.3 the restriction onto Q_1 of this function belongs to the space $W_2^{1,2}(Q_1)$. Denote by Ψ the map that transforms coefficient $z \in W_2^{1,2}(Q)$ from (4.1) to the function w and let χ be the operator of restriction of a function from Q onto Q_1. We show that the operator $\chi\Psi$ acts compactly from $W_2^{1,2}(Q)$ to $W_2^{1,2}(Q_1)$. In Lemma 4.2 it was established that the operator

$$(4.52) \qquad \Psi : W_2^{1,2}(Q) \to L_2(Q)$$

is compact. To mark the dependence of operator L from (4.1) on z we denote it by $L(z)$. Let

$$(4.53) \qquad L(z)w = f_0, \quad L(z + \delta)w_1 = f_0$$

and w, w_1 are equal zero when $t = 0$, $t = T$.

We denote

$$(4.54) \qquad\qquad v = w - w_1$$

Substituting (4.53_2) from (4.53_1) we obtain the equality

$$(4.55) \qquad\qquad \partial_t v - \partial_{xx}^2 v + \partial_x(zv) = \partial_x(\delta w_1).$$

Applying estimate (4.45) to equation (4.55) we will have:

$$(4.56) \qquad \int_Q \left[((\partial_t v)^2 + (\partial_{xx}^2 v)^2)\rho^4 + (\partial_x v)^2 \rho^2 \right] dx\,dt \le$$

$$\le c \left(1 + \|z\|_{W_2^{1,2}(Q)} \right) \left(\|v\|_{L_2(Q)}^2 + \right.$$

$$+ \|\delta\|_{c(\overline{Q})}^2 \int_Q (\partial_x w_1)^2 \rho^2 \, dx\,dt +$$

$$\left. + \|\delta\|_{L_2(0,T;W_2^1(\Omega))}^2 \int_Q w_1^2 \rho^2 \, dx\,dt \right)$$

Applying inequality (4.44) to the equation $L(z+\delta)w_1 = f_0$ we can see that for $\|\delta\|_{W_2^{1,2}(Q)} \le$ const the inequality

$$\int_Q w_1^2 \rho^2 \, dx\,dt + \int_Q (\partial_x w_1)^2 \rho^2 \, dx\,dt \le c$$

holds where c does not depend on δ. Therefore by the compactness of operator (4.52) and by the compactness of embeddings $C(\overline{Q}) \ni W_2^{1,2}(Q)$¡ $L_2(0,T;W_2^1(\Omega)) \ni W_2^{1,2}(Q)$ it follows from (4.56) that

$$(4.57) \qquad \int_Q \left[((\partial_t v)^2 + (\partial_{xx}^2 v)^2)\rho^4 + (\partial_x v)^2 \rho^2 \right] dx\,dt \to 0$$

as $\delta \rightharpoonup 0$ weakly in $W_2^{1,2}(Q)$. Relation (4.57) implies the compactness of operator

$$(4.58) \qquad\qquad \chi\Psi : W_2^{1,2}(Q) \to W_2^{1,2}(Q_1)$$

Applying estimates (4.51), the continuity of the operator $\chi\Psi$ and the compact dependence of the function $\hat{y}$ from (4.5) on z proved in Lemma 4.2 we obtain assertion of the theorem on the compact dependence of the solution y on the coefficient z_1. $\square$

5. The local theorem of exact controllability of the Burgers equation.

5.1. The main theorem. Let in the domain $Q_1 = (0, T) \times \Omega_1$ where $\Omega_1 = (-1, 1)$ the Burgers equation

$$(5.1) \quad G(y) = \partial_t y(t, x) - \partial_{xx}^2 y(t, x) + \partial_x y^2(t, x) = g(x), (t, x) \in Q_1$$

be defined where $g(x) \in L_2(\Omega_1)$ is a fixed function. We consider a function $\hat{y}(t, x) \in W_2^{1,2}(Q_1)$ satisfied equation (5.1). Introduce the denotions

$$(5.2) \qquad \hat{y}(0, x) = \hat{y}_0(x), \quad \hat{y}(T, x) = \hat{y}_T(x).$$

Obviously, $\hat{y}_0(x) \in W_2^1(\Omega_1)$, $\hat{y}_T(x) \in W_2^1(\Omega_1)$. Let

$$(5.3) \qquad B_r(z) = \left\{ y \in W_2^1(\Omega_1) : \|y - z\|_{W_2^1(\Omega)} < r \right\}$$

be the ball of radius r in $W_2^1(\Omega_1)$ with the center at the point z. The problem of local exact controllability of the Burgers equation is as follows: To find for an arbitrary initial function $y_0(x) \in B_r(\hat{y}_0)$ where r is sufficiently small number, such boundary control $(v_-(t), v_+(t))$ that the solution $y(t, x) \in W_2^{1,2}(Q_1)$ of equation (5.1) with the boundary conditions

$$(5.4) \qquad y(t, -1) = v_-(t), \quad y(t, 1) = v_+(t)$$

and with the initial condition

$$(5.5) \qquad y(t, x)|_{t=0} = y_0(x)$$

satisfies the equality

$$(5.6) \qquad y(t, x)|_{t=T} = \hat{y}_T(x)$$

where $\hat{y}_T(x)$ is defined in (5.2).

THEOREM 5.1. *If r is sufficiently small then for an arbitrary $y_0(x) \in B_r(\hat{y}_0)$ there exists such control $(v_-(t), v_+(t)) \in (C(0, T))^2$ that the solution $y(t, x)$ of problem (5.1), (5.4), (5.5) belongs to $W_2^{1,2}(Q)$ and satisfies condition (5.6).*

Proof. It is sufficient to prove the existence of such ball $B_r(0) \subset W_2^1(\Omega_1)$ that for arbitrary $\xi_0(x) \in B_r(0)$ a solution $\xi(t, x) \in W_2^{1,2}(Q_1)$ of problem

$$(5.7) \qquad \partial_t \xi(t, x) - \partial_{xx}^2 \xi + 2\partial_x(\hat{y}(t, x)\xi(t, x)) + \partial_x \xi^2 = 0 \quad (t, x) \in Q$$

$$(5.8) \qquad \xi(t, x)|_{t=0} = \xi_0(x)$$

$$(5.9) \qquad \xi(t, x)|_{t=T} = 0$$

exists. Indeed, if we possess such a function $\xi(t, x)$, then the function

$$y(t, x) = \xi(t, x) + \hat{y}(t, x)$$

satisfies all assertions of the theorem, because it is the solution of problem (5.1), (5.4), (5.5) with $y_0(x) = \hat{y}_0(x) + \xi_0(x)$ and with the boundary control $(v_-(t), v_+(t)) \subset (C(0, T))^2$ which is the restriction of the function $y(t, x)$ at $x = \pm 1$.

For an arbitrary function $\delta \in W_2^{1,2}(Q_1)$ we consider the operator

$$(5.10) \qquad L(\delta)\xi = \partial_t \xi - \partial_{xx}^2 \xi + 2\partial_x(\hat{y}\xi) + \partial_x(\delta\xi) = 0$$

and look for a function $\xi \in W_2^{1,2}(Q_1)$ satisfying (5.8)–(5.10). We denote by θ the operator

$$(5.11) \qquad \theta : W_2^{1,2}(Q_1) \to W_2^{1,2}(Q_1)$$

which transforms a function $(2\hat{y} + \delta)$ to the solution ξ of (5.8)–(5.10) which has been built in Theorem 4.3. Operator (5.11) is compact as it was shown in Theorem 4.3. Besides, relations (4.6), (4.8), (4.9), (4.44), and (4.45) imply the inequality

$$(5.12) \qquad \|\xi\|_{W_2^{1,2}(Q_1)} \leq \|\theta(2y + \delta)\|_{W_2^{1,2}(Q_1)} \leq$$

$$\leq \gamma\left(\|2y + \delta\|_{W_2^{1,2}(Q_1)}\right) \|\xi_0\|_{W_2^1(\Omega)}$$

where $\gamma(\lambda) > 0$ is a certain continuous function with respect to $\lambda > 0$. It follows from (5.12) that for any $\xi_0 \in B_r(0)$ where radius r is sufficiently small the operator $\delta \to \theta(y + \delta)$ transforms the ball $B_1 = \left\{\|\delta\|_{W_2^{1,2}(Q_1)} < 1\right\}$ into itself. Hence, by the Shauder fixed point theorem (see, for example, L. Nirenberg [22]) there exists $\xi \in B_1$ such that $\xi = \theta(2\hat{y} + \xi)$. This function ξ is a solution of problem (5.7)–(5.9). $\square$

5.2. Some corollaries of the main theorem.

DEFINITION 5.1. *A set $R \subset W_2^1(\Omega)$ is called an absorbing set of reachability for the Burgers equation if for an arbitrary function $y_0 \in W_2^1(\Omega)$ there exist such time moment T and a control $(v_-(t), v_+(t))$ $t \geq 0$ that the solution $y(t, x)$ of the boundary problem*

$$Gy = g(x), \quad y|_{t=0} = y_0, \quad y(t, -1) = v_-(t), \quad y(t, 1) = v_+(t)$$

belongs to the set R for any $t > T$.

We consider the Burgers equation (5.1) with zero boundary conditions:

$$(5.13) \qquad y(t, x)|_{x=\pm 1} = 0$$

It is known that the dynamical systems generated by (5.1), (5.13) possesses the $(\overset{\circ}{W}{}^{1}_{2}(\Omega), (\overset{\circ}{W}{}^{1}_{2}(\Omega))$-attractor (see A.V. Babin, M.I. Vishik [23]).

THEOREM 5.2. *The attractor of dynamical system (5.1), (5.2) is an absorbing set of reachability of the Burgers equation.*

Proof. It is known (A.V. Babin, M.I. Vishik [23]) that the attractor A of dynamical system (5.1), (5.13) is a bounded, closed set in $W^1_2(\Omega_1)$. By virtue of the definition of an attractor for an arbitrary trajectory $y(t, \cdot)$ of dynamical system (5.1), (5.13) and for any $\delta > 0$ there exists such time moment T_0 that

$$\underset{\overset{\circ}{W}{}^{1}_{2}(\Omega)}{\operatorname{dist}} (A, y(t, \cdot)) \le \delta \quad \forall t \ge T_0.$$

Choosing δ sufficiently small we transfer the trajectory y of the dynamical system onto attractor by means of boundary control $(v_-(t), v_+(t))$ which existence has been proved in theorem 5.1. After that take $v_\pm(t) \equiv 0$. By the invariantness of the set A the trajectory will remain on the attractor during all posterior time. $\square$

Let for a function $g(x) \in L_2(\Omega)$ the boundary value problem

$$-\partial^2_{xx} y(x) + \partial_x y^2(x) = g(x), \quad y(x)|_{x=\pm 1} = 0$$

has several solutions $y_1(x), \ldots, y_N(x)$. (Surely, they are singular points of dynamical system (5.1), (5.13)).

THEOREM 5.3. *Let $\{y_i\} \subset W^2_2(\Omega)$ are singular points of system (5.1), (5.13). Then for any $j = 1, \ldots, N$ there exists a number $r_i = r_i(y_i) > 0$ such that for an arbitrary $y_0 \in B_{r_i}(y_i)$ the solution $y(t, x) \in W^{1,2}_2(Q_1)$ of the exact controllability problem*

$$Gy = g(x), \quad y|_{t=0} = y_0, \quad y|_{t=T} = y_i$$

exists.

This theorem is an easy corollary of Theorem 5.1. It means that an arbitrary singular point of dynamical system (5.1), (5.13) (for instance, the hyperbolic one) is the stable if we can use a boundary control for stabilization.

The knowledge of attractor's properties can be applied to the investigation of absorbing sets of reachability. As an example we give the following

PROPOSITION 5.1. *Let a singular point $y_i(x)$ of system (5.1), (5.13) possesses the property: An arbitrary trajectory of system (5.1), (5.13) belonging to the attractor A intersects a sufficiently small neighbourhood of y_i. Then y_i is the absorbing point of reachability.*

Proof. We consider an arbitrary initial condition $y_0(x) \in \overset{\circ}{W}{}^1_2(\Omega_1)$. Letting the trajectory of system (5.1), (5.13) to go out y_0 and applying theorem 5.2 we will be found after some time on a trajectory belonging to the attractor A of system (5.1), (5.13). By the assumption of Proposition 5.1 after an other period of time a point moving along our trajectory will be close to y_i enough for application of Theorem 5.3. At this moment we apply Theorem 5.3. $\square$

PROPOSITION 5.2.   *Let $\hat{y}(x)$ be a solution of (3.8), (3.9) with the boundary conditions α_1, α_2 satisfying inequality $\alpha_1 > \alpha_2$. Then $\hat{y}(x)$ is an absorbing point of reachability for the Burgers equation with the boundary control.*

Proof. Let $\alpha_1 > \alpha_2$ but $\alpha_1 - \alpha_2$ is thus far small that the minimal eigenvalue λ_1 of the spectral problem written below (3.19) is positive. Then the proof of Theorem 3.1 is true. Suppose that $\alpha_1 > \alpha_2$ do not satisfy this assumption but the boundary condition $\beta_1 = \hat{y}(0)$, $\alpha_2 = \hat{y}(a)$ where $\alpha_1 > \beta_1 > \alpha_2$ satisfy it. Then by Theorem 3.1 the controlled trajectory $y(t, x)$ going out an arbitrary initial condition $y_0(x)$ can reach at a finite time moment T_1 the solution $\hat{y}_i(x)$ of (3.8) satisfying the boundary conditions: $\hat{y}_1(0) = \beta_1$, $\hat{y}_1(a) = \alpha_2$ if we would choose the appropriate boundary control. Thus, $y(T_1, x) = \hat{y}_1(x)$. Let $\hat{y}_2(x)$ be the solution of (3.8) with the boundary conditions $\hat{y}_2(0) = \beta_2 = \beta_1 + \varepsilon$, $y_2(a) = \alpha_2$ where $\varepsilon > 0$ is small enough. Applying Theorem 5.1 we can prolong the solution $y(t, x)$ behind $T_1 (t > T_1)$ such that at a time moment $T_2 > T_1$ the equality $y(T_2, x) = \hat{y}_2(x)$ holds. If $\hat{y}_2(0) = \beta_2 \geq \alpha_1$ then the proof is finished (we can take $\beta_2 = \alpha_2$). If $\hat{y}_2(0) = \beta_2 < \alpha_1$ we consider the solution $\hat{y}_3(x)$ of (3.8) with $\hat{y}_3(0) = \beta_2 + \varepsilon$ ($\varepsilon > 0$ is small), $\hat{y}_3(a) = \alpha_2$ and repeat the previous arguments. After several steps we will prove the Proposition. $\square$

REMARK 5.1.   *The methods of Sections 4,5 are general. Besides the Navier-Stokes system for which, surely, these methods can be generalized, there are a number of other systems in mathematical physics possessing nontrivial attractors on which a control by means of boundary values is interesting.*

In connection with control of motion on attractors, general problem on attractor's structure means very important and interesting.

REFERENCES

[1] A.V. FURSIKOV AND O.YU. IMANUVILOV, *On ε-controllability of the Stokes problem with distributed control concentrated in a subdomain*, Russian Math. Surveys, 47 N1 (1992), pp. 255–256.

[2] A.V. FURSIKOV AND O.YU. IMANUVILOV, *On approximate controllability of the Stokes system*, Annales de la Faculte des sciences de Toulous, 11,N2 (1993).

[3] A.V. FURSIKOV, *The Cauchy problem for a second-order elliptic equation in a conditionally well-posed formulation*, Trans. Moscow Math. Soc. (1990), pp. 139–175.

[4] A.V. FURSIKOV, *Lagrange principle for problems of optimal control of ill-posed or singular distributed systems*, J. Math. pure et appl., 71, N2 (1992), pp. 139–194.

[5] C. FABRE, J.-P. PUEL, AND E. ZUAZUA, *Contrôllabilité approchée de l'equation de la chaleur semi-linéaire*, C.R. Acad. Sci. Paris, 315 Serie 1 (1992), pp. 807–812.

[6] C. FABRE, J.-P. PUEL, AND E. ZUAZUA, *Approximate controllability for the semilinear heat equation*, (to appear).

[7] J. HENRY, *Etude de la contrôllabilité de certaines équations paraboliques*, Thèse d'Etat, Université Paris-VI, 1978.

[8] J.I. DIAZ, *Sur la contrôllabilité approche des inequations variationnelles et d'autres problèmes paraboliques non linéaires*, C.R. Acad. Sci. Paris, 312, serie I, 1991, pp. 519–522.

[9] J.L. LIONS, *Remarques sur la contrôllabilité approchée*, in Proceedings of "Jornalas Hispano-Francesas sobre Control de Sistemas Distribuidos", University of Malaga, Spain, October 1990.

[10] J.L. LIONS, *Are there connections between turbulence and controllability?*, In Analyse et Optimization des systems, Springer-Verlag Lecture Notes in Control and Informatic Series, 144 (1990).

[11] O. YU. IMANUVILOV, *Some problems of the optimization and exact controllability*, Thesis, Moscow 1991 (in Russian).

[12] O. YU. IMANUVILOV, *Exact controllability of the semilinear parabolic equation*, (To appear in Russian).

[13] O. YU. IMANUVILOV, *Exact boundary controllability of the parabolic equation*, Russian Math. Surveys 48 N3 (1993), pp. 211–212.

[14] J.L. LIONS, E. MAGENES, *Problemes aux limites non homogenes et applications*, Dunod. Paris, 1968.

[15] A.V. FURSIKOV, *Control problems and theorems concerning the unique solvability of a mixed boundary value problem for the three-dimensional Navier-Stokes and Euler equations*, Math. USSR Sbornik, 43(2) (1982), pp. 251–273.

[16] V.M. ALEKSEEV, V.M. TIKHOMIROV, S.V. FOMIN, *Optimal control*, Consultants Bureau, New-York, London, 1987.

[17] O.A. LADYZHENSKAYA, *The Mathematical Theory of viscous incompressible flow*, Gordon and Breach, New York, 1961, rev. 1969.

[18] R. TEMAM, *Navier-Stokes equations*, Theory and Numerical Analysis. A. North Holland, 1979.

[19] M.I. VISHIK AND A.V. FURSIKOV, *Mathematical Problems of Statistical hydromechanics*, Kluwer Ac. Publ. Dordrecht, Boston, London, 1988.

[20] O. YU. IMANUVILOV, *Optimal control of the inverse heat equation*, Siberian Math. Journal, N 1 (1993) pp. 204–211 (in Russian).

[21] L. HÖRMANDER, *Linear partial differential operators*, Springer-Verlag Berlin, Göttingen, Heidleberg, 1963.

[22] L. NIRENBERG, *Topics in nonlinear functional analysis*, New York, 1974.

[23] A.V. BABIN AND M.I. VISHIK, *Attractors of evolution equations*, North-Holland, 1992.

A PREHISTORY OF FLOW CONTROL AND OPTIMIZATION

MAX D. GUNZBURGER*

1. Introduction. Flow control and optimization is an ancient practice of man. For example any dam, sluice, canal, levee, irrigation ditch, valve, duct, pipe, pump, hose, vane, etc., is an exercise in flow control or optimization, i.e., and attempt to

> *control the mechanical state, e.g., the rate and direction of motion, and/or the thermodynamic state, e.g. the temperature, of a fluid in order to achieve a desired purpose.*

Even the animal kingdom has examples, e.g., beaver dams, of attempts at flow control.

However, until recently, flow control and optimization has been, for the most part, effected without the use of sophisticated fluid models and/or without the use of sophisticated optimization techniques. In spite of this, substantial successes have been achieved. On the other hand, sophisticated current and future uses of flow controls require a more systematic approach to these problems, and in particular, will require the use of sophisticated optimization techniques in conjunction with sophisticated flow models. Even the popular literature has recognized this need. For example, the January 1993 issue of *Popular Mechanincs* discusses the use injection of fluid near the nose of an aircraft in order to steer the aircraft in stall environments. Another example is the March 1, 1993 issue of *Aviation Week & Space Technology* in which the need for flow control theories involving thousands of degrees of freedom to replace current ones involving 10 degrees of freedom is discussed.

Here, our main goal is to briefly review some of the past successes in flow control and optimization. We also discuss why the time is now right for the incorporation of sophisticated fluid models and sophisticated optimization techniques into practical flow control and optimization methodology. Indeed, the purpose of this volume and of the meeting from which it emanates is to review some of the recent mathematical and engineering developments in this regard. We close with some remarks about the structure of flow control and optimization problems, and with some examples of interesting objective functionals and control mechanisms.

Lest one thinks that flow control and optimization is a recent quest among mathematicians, engineers, and scientists, consider the following drag minimization problem:

> *what is the shape that a surface of revolution moving at constant velocity in the direction of its axis must have if it*

---

* Department of Mathematics and Interdisciplinary Center for Applied Mathematics, Virginia Tech, Blacksburg, VA 24061-0531.

is to offer the least resistance to the motion?
The body is sketched in Figure 1.1.

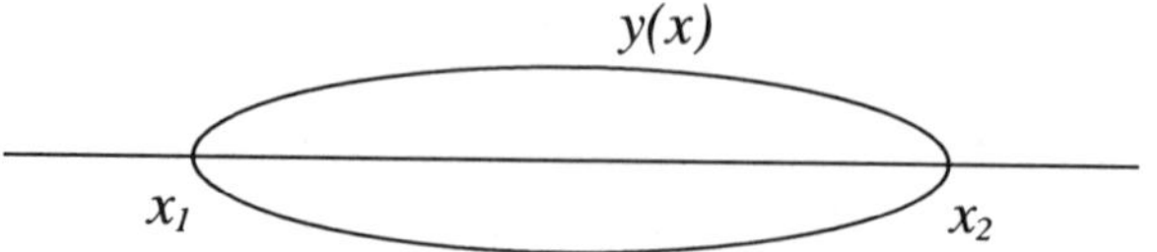

FIG. 1.1. *A body of revolution*

After making certain assumptions about the flow, one can show this problem is equivalent to finding a $y(x)$, $x_1 \le x \le x_2$, that minimizes

$$J(y) = \int_{x_1}^{x_2} \frac{y(x)[y'(x)]^3}{1 + [y'(x)]^2} \, dx \, .$$

This is the *first significant problem in the calulus of variations* and was posed (1687) and solved (1694) by *Newton*!

2. Flow control without fluids. By *flow control without fluids* we mean attempts to control a fluid flow and state without the utilization of sophisticated fluid models involving partial differential equations such as the Navier-Stokes equations, or the Euler equations, or the potential flow equations, etc., For the examples of dams, pumps, etc., mentioned above, flow control is effected without any attempt to solve such fluid equations.

An example of a very successful application of flow control without an accurate modeling of the fluid is the *design of the heating and cooling system in a building.* Here, one designs a system of ducts, fans, registers, vanes, sensors, actuators, heat pumps, furnaces, air conditioners, etc., so that the temperature in a building is close to a uniform, comfortable value and so that the heating/cooling bill is as low as possible. In the design process, the air flow is not computed using sophisticated models involving partial differential equations. Rather, one simply uses empirical rules for determining the flow rates necessary for carrying out the design. One also assumes that pumps, fans, furnaces, etc., move the air at constant flow rates through the ducts, registers, etc. Heat and temperature losses are also determined in an empirical manner.

Perhaps the most spectacular example of successful flow control without fluids is that of *aerodynamic controls.* Here, one determines a position of the rudder, wing flaps, elevators, airelons, throttle, etc., so that an aircraft executes a desired maneuver. To some extent, all modern aircraft employ automatic controls, i.e., controls that are not determined by the pilot, but perhaps by a computer. The extreme example in this regard is the Grumman X-29 airplane which uses such automatic control to keep the plane from going "unstable". Typically, aerodynamic controls are set by solving a small system of ordinary differential equations. The influence of the fluid flow on the controls appears as functions or constants in the differential equations. These functions and constants are determined a priori, very often using an empirical process. When the control settings are being determined, no attempt is made to solve partial differential equations for the fliud flow.

In these and numerous other examples, no attempts are made to employ sophisticated fluid models such as those involving partial differential equations. The flow of the fluid is modeled by a few constants, or at best functions of time, appearing in systems of ordinary differential equations that determine the optimal control settings, or by using a Bernoulli equation to relate mass flow and pressure, or, most often, by assuming constant mass flow rates. In this sense, one may view these efforts as constituting *flow control without fluids.*

3. Flow optimization without optimization. By *flow optimization without optimization* we mean attempts to control a fluid flow and state in order to meet a desired objective without the utilization of sophisticated optimization techniques such as Lagrange multiplier methods, quasi-Newton methods, etc. In many cases, including the ones described below, although sophisticated optimization algorithms are not involved, a detailed description of the fluid motion and state is employed. The latter are determined by experimental measurements, or analytical solutions, or computational simulations.

For the first example of flow optimization without optimization, we consider the large body of experimental work and somewhat smaller body of analytical work on *boundary layer control.* Here, the size, shape, formation, etc., of a boundary layer is to be affected, e.g., controlled, in order to meet a desired objective. Control mechanisms that have been considered are the movement of solid walls such as for a rotating cylinder, the injection or suction of fluid through orifices, shape variations such as camber, thickness, and flaps adjustments, etc. Objectives that have been considered are maximizing lift, minimizing drag, preventing separation, preventing or facilitating transition to turbulence, etc.

For example, consider the following question. Can the drag on a body be lowered by the suction of fluid through a narrow slit? Specifically, consider the sketch in Figure 3.1. Here, we have a cylinder in a uniform stream and we have fluid sucked through a slit on the back-side of the cylinder. This problem was the subject of Prandtl's first paper in 1904 [3]! What Prandtl found, through experimentation, is that indeed the drag on the cylinder could be reduced by sucking fluid out through the slit.

Another example is attempts towards the *cancellation of wave drag.* The Busemann biplane (1930) was an attempt to design a wing shape in order to reduce wave or shock drag; see Figure 3.2. The left-hand figure shows the shock waves under design conditions; the wedge angles are exactly those needed to cancel out the out-going waves. The right-hand figure

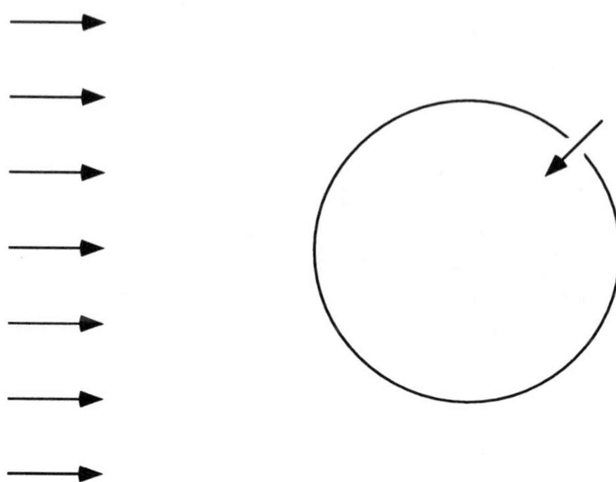

FIG. 3.1. *A cylinder in uniform flow with suction through a slit*

show off-design conditions for which the out-going waves are not completely cancelled.

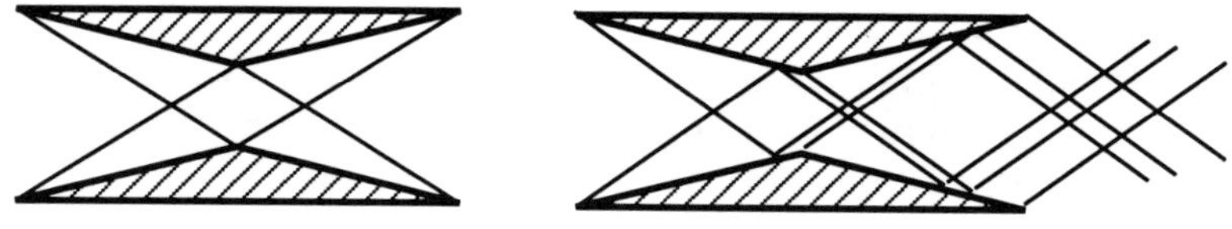

FIG. 3.2. *The Buseman biplane*

More recently, Garabedian and his co-workers [1], and others, have designed transonic airfoil shapes that generate shock-free flows. Again, under design conditions, there is no shock present at the back of the supersonic bubble on the upper side of the airfoil; at off-design conditions, a weak shock is present there.

In these examples, and many others as well, sophisticated flow models were used in experiments, analyses, or computations of optimal designs. However, no attempt was made to employ sophisticated optimization algorithms. Solutions were obtained by doing experiments or solving equations for a (small) set of configurations, and then comparing results. In essence, optimization, e.g., minimization, is effected by variants of the following algorithm (which for simplicity, we describe in the case of having only one design parmeter):

Given a functional $f(p)$ to be minimized with respect to the parameter p,

1.choose n distinct values $\{p_1, p_2, \ldots, p_n\}$ of the parameter;

2.evaluate $f(p_i)$ for $i = 1, \ldots, n$; and

3.examine the set $\{f(p_1), \ldots f(p_n)\}$ and choose a value p_j such that $f(p_j) \leq$ $f(p_i)$ for $i = 1, \ldots, n$.

For example, plot the values of $f(p_i)$, $i = 1, \ldots, n$, as in Figure 3.3, and then choose the parameter that yields the minimal value of f among the plotted values. One may view such efforts as *flow optimization without optimization*.

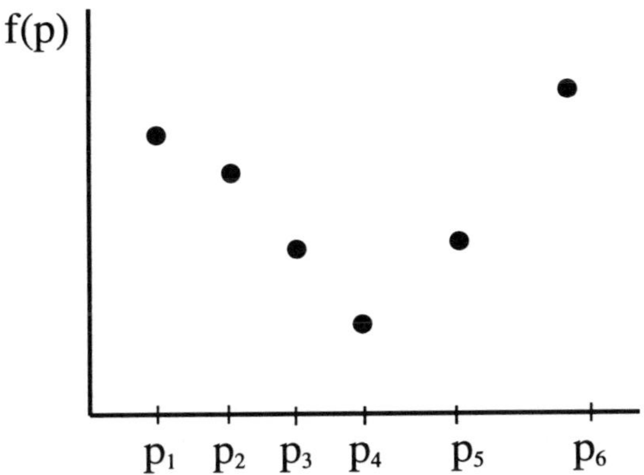

FIG. 3.3. *Graphical minimization of a functional*

4. Flow control without objectives. By *flow control without objectives* we mean attempts to use control and optimization ideas in a fluids setting, not to have the fluid flow meet some desired objective, but in order to meet some independent objective.

As an example, we consider the work reported in [2] and related papers on the use of optimization ideas to generate *incompressible computational fluid dynamics algorithms*. The connection between a CFD algorithm and flow optimization is made as follows. If $(\mathbf{u}, p)$ is a solution of the Navier-Stokes equations

$$-\nu \Delta \mathbf{u} + \mathbf{u} \cdot \nabla \mathbf{u} + \nabla p = \mathbf{f} \quad \text{in } \Omega,$$

$$\nabla \cdot \mathbf{u} = 0 \quad \text{in } \Omega,$$

and

$$\mathbf{u} = \mathbf{g} \quad \text{on } \partial\Omega,$$

in some region Ω, then $\mathbf{u}$ minimizes the functional

$$\mathcal{J}(\mathbf{v}) = \frac{\nu}{2} \int_\Omega |\nabla \mathbf{\Phi}(\mathbf{v})|^2 \, d\Omega$$

over a suitable function class, where, for given $\mathbf{v}$, $(\mathbf{\Phi}, \xi)$ is a solution of the Stokes problem

$$-\nu \Delta \mathbf{\Phi} + \nabla \xi = -\nu \Delta \mathbf{v} + \mathbf{v} \cdot \nabla \mathbf{v} - \mathbf{f} \quad \text{in } \Omega,$$

$$\nabla \cdot \mathbf{\Phi} = 0 \quad \text{in } \Omega,$$

and

$$\mathbf{\Phi} = \mathbf{0} \quad \text{on } \partial\Omega.$$

Moreover, if $\mathbf{u}$ is minimizer of $\mathcal{J}$, then

$$\mathbf{\Phi} = \mathbf{0}, \quad \xi = -p, \quad \text{and} \quad \mathcal{J}(\mathbf{u}) = \mathbf{0}.$$

The problem of minimizing $\mathcal{J}$ can be solved by a conjugate gradient algorithm having the property that at each iteration only a sequence of Stokes solves is required. Thus, and efficient CFD algorithm is generated. However, note that no intrinsic property of the flow is being optimized; hence, in this sense, we have *flow control without any objective.*

5. Timeliness of flow control problems. At this point it is natural to ask:

> can one put together *sophisticated flow models* and *sophisticated optimization techniques* in order to meet desired objectives?

An affirmative answer to this question depends on being able to obtain a like answer to the next question:

> has flow control and optimization become a subject ready for *rigourous mathematical treatment* and *systematic computational resolution* using *sophisticated fluid models* and *sophisticated optimization algorithms?*

An affirmative answer to the second question follows from the observations that there has recently been significant advances in the *theory of partial differential equations* for the equations of fluid mechanics, especially for the Navier-Stokes equations for incompressible flows, and there have also been significant advances in efficient and robust *algorithms for computational fluid dynamics* for all types of flow regimes that enable the analysis and approximation of flow control problems.

6. The structure of flow control problems. The structure of a flow control or optimization problem is similar to that of any such problem. First, one has an

objective, i.e., a reason why one wants to control the flow.

There are numerous objectives of interest in applications, e.g., flow matching, drag minimization, lift enhancement, preventing separation, preventing transition to turbulence, deterring temperature variations, enhancing mixing, deterring mixing, etc. Mathematically, such an objective is expressed as a *cost functional*.

Next, one has

constraints that must be imposed on candidate optimizers
that determine what type of flow one is interested in.

One must decide what type of fluid model is adequate for the flows one is interested in, i.e., is one satified with assuming the flow is a potential flow, an invsicid flow, a viscous flow, an incompressible flow, a compressible flow, a stationary flow, a time-dependent flow, etc. Mathematically, the type of flow is expressed in terms of a specific set of *partial differential equations.*

Finally, on has

controls or *design parmeters* at one's disposal in order to
meet the objective.

One can have *boundary value controls* such as injection or suction of fluid and heating or cooling or temperature controls, etc.; one could have *distributed controls* such as heat sources or magnetic fields, etc.; or, one could have *shape controls* such as leading or trailing edge flaps, movable walls, rudders, propeller pitch, surface roughness, or domain design, etc. Mathematically, controls are expressed in terms of *unknown data* in the problem specification.

Thus, the mathematical specification of a flow control or optimization problem involves:

state variables

$\phi = \mathbf{u}, \, p, \, T, \, e,$ etc. ,   the velocity, pressure, temperature, internal energy, etc.;

control variables or design parameters

g ,   e.g., the velocity on the boundary, the heat flux on the boundary, the shape of the boundary, etc.;

an objective or cost functional

$\mathcal{J}(\phi, g)$,   e.g., drag, temperature gradient, etc.

constraints

$F(\phi, g) = 0$,   i.e., flow equations.

The flow control problem is then simply stated as the following *minimization problem:*

find controls g and states ϕ such that $\mathcal{J}(\phi, g)$ is minimized,
subject to $F(\phi, g) = 0$.

The functionals to be minimized do not usually explicitly depend on the controls or design parameters; this may result in unbounded optimal controls. Thus, one must *limit the size of the control.* There are two ways to do this. One may place some a priori constraints on the size of admissible controls so that one looks for optimal controls within a bounded set, e.g., one could look for optimal controls g such that, for some suitable norm,

$$\|g\| \leq K \ .$$

A second method for limiting the size of the control is to penalize the objective functional with some norm of the control, i.e., instead of minimizing a functional $\mathcal{J}(\phi)$ one could minimize

$$\mathcal{J}(\phi) + \epsilon \|g\|^{\beta} \ .$$

By making judicious choices for the parameters ϵ and β and for the norm on g, one may at the same time effectively limit the size of the control and obtain states such that the value of $\mathcal{J}$ is small.

7. Sample objectives. We now give a short sample of the many possible objectives that arise in practical flow control and optimization problems. We emphasize that there are many other useful and interesting objective functionals that have or should be considered.

Flow tracking. Let $\mathbf{u}$ denote the velocity field and $\mathbf{U}_d$ denote a prescribed desired velocity field. We want to control the flow so that $\mathbf{u}$ is "close" to $\mathbf{U}_d$. It is natural to minimize some norm of the difference between $\mathbf{u}$ and $\mathbf{U}_d$. For example, one choice that has been considered is to minimize

$$\mathcal{J}(\mathbf{u}) = \frac{1}{4} \int_{\Omega} |\mathbf{u} - \mathbf{U}_d|^4 \, d\Omega \, ,$$

where Ω denotes the flow domain. (The particular choice of the L^4 norm is governed by technical considerations.) One can also try to match the flow on part of the flow domain, or even on some surface. For example, one may minimize

$$\mathcal{J}(\mathbf{u}) = \frac{1}{4} \int_{\Gamma_0} |\mathbf{u} - \mathbf{U}_d|^2 \, d\Gamma \, ,$$

where Γ_0 is some plane in the flow field.

Viscous drag minimization. An important objective in many applications is the minimization of drag. For some incompressible flows, the drag on a body can be computed from the integral of the dissipation function

$$\mathcal{J}(\mathbf{u}) = \frac{\mu}{2} \int_{\Omega} \left|(\nabla \mathbf{u}) + (\nabla \mathbf{u})^t\right|^2 \, d\Omega \, ,$$

where μ denotes the viscosity coefficient. Thus, if one wishes to minimize the drag on a body, one merely minimizes the above functional.

Avoiding hot spots. In many applications it is desirable to avoid "hot spots" along bounding surfaces, i.e., places where temperature peaks occur, since often such phenomena lead to meltdown or to flexural failures. Such difficulties may be avoided by minimizing the functional

$$\mathcal{J}(T) = \int_{\Gamma_T} |\nabla_s T|^2 \, d\Gamma \,,$$

where T denotes the temperature, ∇_s the surface gradient, and Γ_T the portion of the boundary along which one would like to avoid the above problems. Another candidate functional to be minimized is given by

$$\mathcal{J}(T) = \int_{\Gamma_T} |T - T_d|^2 \, d\Gamma \,,$$

where T_d denotes a desired temperature distribution.

Well-mixed flows. One common objective is to have two (or more) fluids become well-mixed at, for example, the outflow of some flow region. At the inflow, perhaps, the fluids are not well-mixed; we could have an air flow with fuel being injected through an orifice. By controlling the flow, we would like, by the time the fluids reach the outflow region, not to have high concentrations of either fluid present. One way to achieve this is to minimize

$$\mathcal{J}(c) = \int_{\Gamma_0} |\nabla_s c|^2 \, d\Gamma \,,$$

c denotes the mass fraction of fuel, ∇_s the surface gradient, and Γ_0 the outflow boundary. By minimizing the above functional we achieve a quasi-uniform concentration distribution at the outflow boundary.

Poorly mixed flows. In other applications one wants two or more fluids to mix as little as possible. For example, one would like one fluid to remain confined to a certain portion of the flow domain, and not penetrate into other portions of the flow domain. If one wants to exclude a particular species from the portion Ω_0 of the flow domain Ω we could, in this case, achieve our objective by minimizing

$$\mathcal{J}(c) = \|\chi c\|^\beta \,,$$

where χ denotes the characteristic function for Ω_0.

8. Sample control mecahnisms and design parameters. We now give a short sample of the many control mechanisms that arise in practical flow control and optimization problems. Again, we emphasize that there are many other useful and interesting control mecahnisms that have or should be considered.

Velocity along portions of the boundary. A very much used mechanism of control is to inject or suck fluid through orifices along bounding surfaces. Thus, if Γ_c denotes the portion of the boundary covered by the orifices, we would seek a control $\mathbf{g}$ such that one of the functionals is minimized, subject to the appropriate flow equations, and also

$$\mathbf{u} = \mathbf{g} \quad \text{on } \Gamma_c \,.$$

Temperature and heating controls. Another common control mechanism is to adjust the temperature, or even more often, the heat flux, along portions of the boundary of the flow domain in order to achieve one of the desired objectives. Within this class of controls we find "heating" and "cooling" controls. For example, one could seek a control q such that one of the functionals is minimized, subject to the appropriate flow equations, and also

$$\frac{\partial T}{\partial n} = q \quad \text{on } \Gamma_T \,,$$

where Γ_T denotes the portion of the boundary along which one allows the control to act and $\partial/\partial n$ denotes the normal derivative at the boundary.

Distributed controls. One could try to effect control through the body force in the Navier-Stokes equation. Thus, one would seek a control, defined on the flow domain Ω or on a portion of Ω, such that some functional is minimized and subject to the appropriate flow equations. Physically, one way to effect such control is by a magnetic field acting on an ionized fluid or an electrically conducting fluid. Another distributed control of interest is a heat source in the energy equation. Physically, one way to effect such a control is through radiation mechanisms, or through a targeted laser beam.

Shape controls. The control mechanisms discussed so far are collectively known as *value controls*; this refers to the fact that we try to effect control through the adjustment of the values of the data of the problem. Another class of controls are known collectively as *shape controls*; in this case control is effected by adjusting the shape of the flow domain. The shape of the flow domain may be changed in many ways. For example, one could use leading and/or trailing edge flaps, or movable walls, or rudders, or propeller pitch. A related problem is the *optimal design* problem. Here, we want to choose a flow domain, e.g., the exterior of an airfoil, so that some objective is achieved. Of course, the flow domain is determined by its boundary, e.g., the airfoil itself.

9. Acknowledgements. The author gratefully acknowledges the supprort of the Air Force Office of Scientific Research under grant number AFOSR-93-1-0061 and the Office of Naval Research under grant number N00014-91-J-1493.

REFERENCES

[1] F. Bauer, P. Garabedian, and D. Korn; *Supercritical Wing Sections*, Springer, Berlin, (1972).

[2] M. Bristeau, R. Glowinski, J. Perrier, O. Pironneau, and G. Poirer; Application of optimal control and finite element methods to the calculation of transonic flows and incompressible flows, *Numerical Methods in Applied Fluid Mechanics*, Academic, London, (1980), 203-312.

[3] H. Schlichting, *Boundary Layer Theory*, McGraw-Hill, New York, (1960).

MATHEMATICAL ISSUES IN OPTIMAL DESIGN OF A VAPOR TRANSPORT REACTOR

KAZUFUMI ITO*, HIEN T. TRAN* , AND JEFFERY S. SCROGGS*

Abstract. In this paper the optimal design of a vertical reactor for growing crystals and epitaxial layers by physical vapor transport technique is discussed. The transport phenomena involved in the deposition process is modeled by the gasdynamics equations and chemical kinematics. The problem is formulated as a shape optimization with respect to the geometry of the reactor and an optimal control problem by controlling the wall temperature. The material and shape derivatives of solutions to the so-called Boussinesq approximation are derived. Optimality condition and a numerical optimization method based on the augmented Lagrangian method are discussed for the boundary control of the Boussinesq flow. A numerical approximation based on the Jacobi polynomials for the axi-symmetric flow is developed along with a discussion of an iterative method based on GMRES for solving the resulting system of nonlinear equations.

1. Introduction. In this paper we discuss the mathematical issues involved in designing an optimal reactor for growing crystals and epitaxial layers by vapor transport techniques. The application of these materials in modern computers, communication systems, and other electronic and optical devices demand precisely controlled electrical and optical properties, and hence extremely high purity and uniformity. Our design effort is focused on the Scholz geometry depicted in Figure 1. The source material and the growing crystal are sealed in a fused silica ampoule that is heated by an isothermal furnace liner at its outer cylindrical surface. The substrate (the single crystal) is located on a fused silica window (W) which is cooled by a jet of helium gas from the outer surface. HPVT processes are based on physical vapor transport and can be described very roughly as proceeding via evaporation at the polycrystalline source and condensation at the surface of the cooler substrate.

Our effort on mathematical modeling of transport and growth process in the high pressure vapor transport (HPVT) arises from collaboration with Klaus Bachmann in a joint project between the Center for Research in Scientific Computing and the Material Research Laboratory, both at North Carolina State University. Preliminary studies in the laboratory have shown that crystals grown by HPVT of $ZnGeP_2$ exhibit superior properties than those grown by the existing techniques. We have begun to explore the conditions that favor these properties by modeling a vertical reactor along with a numerical simulation of 2-D axi-symmetric steady flow of a homogeneous P_2 gas at 1 and 10 atm pressure using the Boussinesq equation [TSB]. Numerical simulations were performed in [TBS] to study the flow dynamics and temperature distribution inside the reactor chamber and to illustrate the feasibility of an optimal reactor design study.

---

* Center for Research in Scientific Computation, North Carolina State University, Raleigh, North Carolina 27695-8205.

Our study is concerned with the transport mechanisms inside the reactor. We quantify the uniformity of the epitaxial layer and formulate the optimization problems in terms of the following performance indices:

(1) the variation of the temperature of the substrate
- less variation should increase uniformity and purity,

(2) variations in the relative fluxes of reactants
- this determines the stoichiometry which must be controlled to
- within 1 percent,

(3) the net flux of reactants onto the substrate
- this determines the growth rate,

(4) Absence of local recirculation flow.

The possible control variables consist of the shape of the reactor, aspect ratio, total pressure, orientation of the reactor with respect to the gravity vector, wall temperature distribution.

The mathematical model for the transport phenomena involved in the deposition process involves the gasdynamics equations (conservation of mass, momentum and energy) and the conservation of species equation for the reactants: i.e.,

$$\frac{\partial}{\partial t}\rho + \nabla \cdot (\rho\, u) = 0$$

$$\rho\left(\frac{\partial}{\partial t}u + u \cdot \nabla u\right) + \nabla p = \mu\left(\Delta u + \frac{1}{3}\nabla(\nabla \cdot u)\right) - \rho g\, e_3$$

(1.1)

$$\rho C_v\left(\frac{\partial}{\partial t}T + u \cdot \nabla T\right) + p(\nabla \cdot u) = \nabla \cdot (k\,\nabla T) + 2\mu\left(\epsilon_{ij} - \frac{1}{3}\epsilon_{kk}\right)^2,$$

$$\rho\left(\frac{\partial}{\partial t}c_i + u \cdot \nabla c_i\right) = \nabla \cdot (\rho D\,(\nabla c_i + \alpha\, c_i \nabla log T)) + r_i$$

where $\epsilon_{i,j} = \frac{1}{2}\left(\frac{\partial u_j}{\partial x_i} + \frac{\partial u_i}{\partial x_j}\right)$, the state function (ρ, u, T, c_i) includes the density $\rho(t, x)$, the mass-average velocity $u(t, x) \in R^d$, the temperature $T(t, x)$, and the mass fractions $c = (c_1, \cdots, c_m)$ of each specie i with $\sum_{i=1}^{m} c_i = 1$. In the equation r_i is the reaction rate of the i-th specie, α is the thermal diffusion factor (Soret coefficient), D is the solutal diffusivity, k is thermal diffusivity, and C_v is the specific heat. We assume the perfect gas law for all species; i.e.,

(1.2)
$$p_i = \frac{\rho c_i R_0 T}{m_i}$$

where R_0 is the universal gas law constant and m_i is the molecular weight of the i-th specie. We assume that the velocity field satisfies the non-slip boundary condition and the temperature distribution is given at the boundary Γ; i.e.,

$$(1.3) \qquad u = 0 \quad \text{and} \quad T = \theta \quad \text{on} \quad \Gamma$$

In the case of binary (carrier and reactant) gases we may consider the following boundary condition for the concentration, which models the surface reactions and deposition along the substrate [YHC]

$$(1.4a) \qquad n \cdot (\nabla c_i + \alpha\, c_i \nabla logT) - \epsilon c_i = 0$$

Hwhere n is the outward normal vector at the boundary and ϵ is the Dämkohler number. On the wall we assume

$$(1.4b) \qquad n \cdot (\nabla c_i + \alpha\, c_i \nabla logT) = 0$$

and at the source the concentration c_1 is assumed to be given.

The paper is organized as follows. In §2 existence of solutions to the steady problem is discussed. In §3 a numerical method based on Jacobi-polynomial based spectral (tau-) approximation is developed for the axisymmetric solution to the Boussinesq equation. An iterative method based on a preconditioned projection method and GMRES for solving the resulting system of nonlinear equations is developed. In §4 the shape optimization for the Boussinesq flow is formulated and the sensitivity equation based on the shape derivative is derived. In §5 the optimal control for the Boussinesq flow is discussed and the first order and second order optimality condition is established. A solution technique based on the augmented Lagrangian method with second order update is described.

2. Existence of solutions. In this section we discuss the existence of solutions to the steady problem. For simplicity of our discussions we consider the case when no reaction is taking place (i.e., involving only a single carrier gas). Then it is not difficult to show that if $T = \theta_0$ (a constant), then

$$(2.1) \qquad \bar{u} = 0, \;\; \bar{T} = \theta_0, \;\; \bar{\rho} = \rho_0 \exp\left(-\frac{mg}{R_0 T_0} z\right) \quad \text{and} \quad \bar{p} = \frac{R_0 T_0}{m}\bar{\rho}$$

is a solution to the steady equation of (1.1) where $p_0 = \frac{R_0 \rho_0}{m} T_0$. Then it is shown in [MN] that there exists a global unique (classical) solution to (1.1) provided that the initial condition is sufficiently close to $(\bar{\rho}, \bar{u}, \bar{T})$ in $H^3(\Omega)^5$.

The steady equation can be written as

$$\nabla \cdot u = -\frac{\nabla \rho}{\rho} \cdot u$$

$$(2.2) \qquad -\mu \left(\Delta u + \frac{1}{3}\nabla \nabla \cdot u\right) + \nabla p = -\rho g\, e_3 - \rho u \cdot \nabla u$$

$$-\nabla \cdot (k\nabla T) = \mu\, \Phi(Du) - p\, \nabla \cdot u - C_v \rho\, u \cdot \nabla T.$$

We will show the existence of solutions in a neighborhood of $(\bar{\rho}, \bar{u}, \bar{T})$ by applying the implicit function theory to (2.2). Consider the state space: $(\rho, u, T) \in W^{1,r}(\Omega) \times W^{1,r}(\Omega)^3 \times W^{2,r/2}(\Omega)$ for $r \geq 2$. Assume that $r \geq 4$. Then since $W^{1,r}(\Omega)$ is continuously embedded into $L^{\infty}(\Omega)$ the right hand side of (2.2) belongs to $L^r \times (L^r)^3 \times L^{r/2}$. Then it follows from [Gi] that the linear equation (2.2) has a unique solution $(u, p, T) = S(\rho, u, T)$ in $W^{1,r}(\Omega) \times L^r(\Omega) \times W^{2,r/2}(\Omega)$, given the right hand side in $L^r \times (L^r)^3 \times L^{r/2}$. However from the perfect gas law

$$p = \frac{R_0 \rho}{m} T$$

we must have $p \in W^{1,r}(\Omega)$. That is, we have a mismatch of the regularity for the pressure p. In order to overcome this difficulty we assume a bulk viscosity assumption [Sch]

$$(2.3) \qquad p_{th} = p - \frac{4}{3}\mu\, \nabla \cdot u \quad \text{and} \quad p_{th} = \frac{R_0 \rho}{m} T$$

where p_{th} is the thermodynamic pressure. Note that for the incompressible flow the thermodynamic pressure equals to the mechanical pressure and at the inviscid limit (assuming such a limit exists) (2.3) reduces to the perfect gas law.

From [FM] we have the vector field decomposition of $L^r(\Omega)^3$; that is,

$$L^r(\Omega)^3 = S_r(\Omega) \oplus G_r(\Omega)$$

where $S_r(\Omega)$ is the closure of C^{∞} solenoidal ($\nabla \cdot u = 0$) functions with compact support in Ω with respect to $L^r(\Omega)$ topology and $G_r(\Omega)$ is the gradient field $= \{\nabla \phi : \phi \in W^{1,r}(\Omega)\}$. Suppose $u = v + w$ with $v \in S_r(\Omega)$ and $w \in G_r(\Omega)$ then the left hand side of the second equation in (2.2) can be written as

$$-\mu\, \Delta v + \nabla \left(p - \frac{4}{3}\mu\, \nabla \cdot u\right)$$

The linearized equation of $(2.2) - (2.3)$ at $(\bar{\rho}, \bar{u}, \bar{T}, \bar{p})$ is given by

$$\nabla \cdot U - \frac{mg}{R_0 \theta_0} U_3 = F_1 \in L^r$$

$$-\mu \left(\Delta U + \frac{1}{3} \nabla \nabla \cdot U \right) + \nabla P + \Pi g\, e_3 = F_2 \in (L^r)^3$$

(2.4)

$$-\nabla \cdot (k \, \nabla \Theta) + \bar{p} \nabla \cdot U = F_3 \in L^{r/2} \quad \text{and} \quad \Theta|\Gamma = \theta$$

$$P - \frac{4}{3}\mu \nabla \cdot U - \frac{R_0}{m}(\bar{T}\Pi + \bar{\rho}\Theta) = F_4 \in W^{1,r}.$$

Hence by the implicit function theory if (2.4) has a unique solution $(\Pi, U, \Theta, P) \in X = W^{1,r} \times (W_0^{1,r})^3 \times W^{2,r/2} \times W^{1,r}$ that continuously depends on $F = (F_1, F_2, F_3, F_4) \in Y = L^r \times (L^r)^3 \times L^{r/2} \times W^{1,r}$ and $\theta \in \Sigma = $ the trace of $W^{2,r}$ on Γ, then there is a unique continuous solution mapping of equation $(2.2) - (2.3)$ defined in a neighborhood V of T_0 in Σ: $V \to (\rho, u, T, p) \in X$.

Assume the operator Q defined by $QU = \nabla \cdot U - \frac{mg}{R_0 \theta_0} e_3 \cdot U$ on $W^{1,r}$ is surjective. The first two equations of (2.4) has a unique solution $(U, P) \in (W^{1,r})^3 \times L^r$, where P is uniquely determined by the condition that the total thermal pressure, $= (p_{th}, 1)_\Omega$, is a constant. Moreover if $U = V + W$ with $V \in S_r(\Omega)$ and $W \in G_r(\Omega)$ then the second equation of (2.4) is written as

(2.5)
$$-\mu \, \Delta V + \nabla \left(P - \frac{4}{3}\mu \, \nabla \cdot U \right) = F_2 - \Pi g\, e_3$$

in the sense of distributions since $\mathrm{grad}\Delta = \Delta\mathrm{grad}$. Since $\mathrm{div}\Delta = \Delta\mathrm{div}$ in the sense of distributions it thus follows from the vector field decomposition of $(L^r)^3$ that $P - \frac{4}{3}\mu \nabla \cdot U \in L^r$ is continuously depend on Π and $F_2 \in (L^r)^3$. Then the third equation has a unique solution $\Theta \in W^{2,r/2}$ as a continuous function of $U \in (W^{1,r})^3$. Thus, the last equation of (2.4) can be equivalently written as

(2.6)
$$\frac{R_0 \theta_0}{m} \Pi + \Psi(\Pi) = F_4 \quad \text{in } W^{1,r}$$

where $\Psi \in \mathcal{L}(W^{1,r}, W^{1,r})$ is defined by the solution (U, Θ, P) to the first three equation of (2.4), described as above and is compact since $W^{1,r}$ is compactly embedded into L^r. It then follows from the Riesz-Shauder theory that if $-\frac{R_0 \theta_0}{m}$ is not an eigenvalue of the linear operator Ψ then (2.6) has a unique solution $\Pi \in W^{1,r}$ that in turn implies (2.4) has a unique solution. The range conditions on the operators Q, Ψ, which depend continuously on the total pressure, are generically satisfied.

3. Axi-symmetric flow and Jacobi polynomial based spectral method. In this section we consider a numerical approximation of the axi-symmetric flow of a homogeneous carrier gas. The so-called Boussinesq approximation of (1.1) assumes that the density ρ is constant. Thus, from the mass conservation we have $\nabla \cdot u = 0$. The buoyancy force in the presence of a gravitational force is modeled by

$$\rho_0 \left(1 - \frac{T - \theta_0}{\theta_0}\right)$$

which is obtained by the Taylor expansion of the perfect gas law (1.2); i.e.,

$$\rho \sim \rho_0 \left(1 + \frac{p - p_0}{p_0} - \frac{T - \theta_0}{\theta_0}\right)$$

where the pressure dependent term is neglected when no reaction is taking place. This results in

$$\rho_0 \left(\frac{\partial}{\partial t} u + u \cdot \nabla u\right) + \nabla p = \mu \, \Delta u + \frac{\rho_0}{\theta_0} (T - \theta_0) g e_3,$$

$$(3.1) \qquad \nabla \cdot u = 0, \quad u_\Gamma = 0,$$

$$\rho_0 C_v \left(\frac{\partial}{\partial t} T + u \cdot \nabla T\right) = \nabla \cdot (k \, \nabla T), \quad T|_\Gamma = \theta.$$

Furthermore, we consider the axi-symmetric flow; i.e., $\vec{u} = (u \cos\phi, u \sin\phi, w)$ where u, w, the radial and vertical component of the velocity field $\vec{u}$, satisfies

$$\rho_0 \left(\frac{\partial u}{\partial t} + u \frac{\partial u}{\partial r} + w \frac{\partial u}{\partial z}\right) + \frac{\partial p}{\partial r} = \mu \left(\Delta_r u - \frac{u}{r^2}\right)$$

$$\rho_0 \left(\frac{\partial w}{\partial t} + u \frac{\partial w}{\partial r} + w \frac{\partial w}{\partial z}\right) + \frac{\partial p}{\partial z} = \mu \, \Delta_r u + \frac{\rho_0}{\theta_0} (T - \theta_0) g$$

$$(3.2) \qquad \nabla_r \cdot (u, w) = \frac{1}{r} \frac{\partial}{\partial r} (ru) + \frac{\partial w}{\partial z} = 0$$

$$C_v \rho_0 \left(\frac{\partial T}{\partial t} + u \frac{\partial T}{\partial r} + w \frac{\partial T}{\partial z}\right) = k \, \Delta_r T$$

$$u, \ w = 0 \quad T = \theta \quad \text{on } \Gamma.$$

Here, $\Delta_r \phi = \frac{1}{r} (r \frac{\partial}{\partial r} \phi) + \frac{\partial^2}{\partial z^2} \phi$. The domain Ω_α can be parameterized by

$$(3.3) \qquad \Omega_\alpha = \{(r, z) : 0 < r < R \text{ and } 0 < z < \alpha(r)\}$$

where we assume that $\alpha \in C^2(0, R)$ is positive. Note the singularity appearing in the operator Δ_r is removable in the sense that

$$(3.4) \qquad -\int_\Omega \Delta_r \phi \psi \, r \, dr dz = \int_\Omega \left(\frac{\partial \phi}{\partial r} \frac{\partial \psi}{\partial r} + \frac{\partial \phi}{\partial z} \frac{\partial \psi}{\partial r}\right) r \, dr dz$$

for $\psi \in C^1(\Omega_\alpha)$ and $\psi_\Gamma = 0$. Similarly,

$$(3.5) \qquad -\int_\Omega \nabla_r \cdot (u, w) q\, r\, drdz = \int_\Omega (u\, \frac{\partial q}{\partial r} + w\, \frac{\partial q}{\partial z})\, r\, drdz$$

for $q \in C^1(\Omega_\alpha)$. Define the Hilbert spaces $H = L^2(\Omega, r\, drdz)$ and $V = \{\phi \in H : \nabla \cdot \phi \in H \text{ and } \phi_\Gamma = 0\}$. Then the right hand side of (3.4) defines a bounded, symmetric, coercive sesquiliner form a on $V \times V$. Thus Δ_r with $\operatorname{dom}(\Delta_r) = \{\phi \in V : \Delta_r \phi \in H\}$ is a self-adjoint operator on H (see, [Ta]). From (3.5) $-\nabla_r \cdot = \nabla^*$ with $\operatorname{dom}(\nabla_r \cdot) = V$. Hence the weak or variational formulation of (3.2) is given by

$$\rho_0 \left(\langle \frac{du(t)}{dt}, \phi \rangle + b_1((u, w), \phi)\right) + a_r(u(t), \phi) = 0$$

$$\rho_0 \left(\langle \frac{dw(t)}{dt}, \psi \rangle + b_2((u, w), \psi)\right) + a(u(t), \psi) = 0$$

$$(3.6)$$

$$C_v \rho_0 \left(\langle \frac{dT(t)}{dt}, \eta \rangle + b_3((u, w, T), \eta)\right) + k\, a(T(t), \eta) = 0$$

$$\nabla_r \cdot (u(t), w(t)) = 0, \quad (u(t), v(t), T(t)) \in V^3 + (0, 0, \theta),$$

for all $(\phi, \psi, \eta) \in V^3$ satisfying $\nabla_r \cdot (\phi, \psi) = 0$. The pressure dependent term is eliminated by the fact that $\langle \nabla p, (\phi, \psi) \rangle = 0$. Here the sesquilinear form a_r on $V \times V$ is defined by

$$(3.7) \qquad a_r(\phi_1, \phi_2) = a(\phi_1, \phi_2) + \int_\Omega \frac{1}{r^2} \phi_1 \phi_2\, drdz$$

and the tri-linear forms b_1, b_2 and b_3 are defined by

$$b_1((u, w), \phi) = \int_\Omega (\frac{\partial}{\partial r}(\frac{1}{2}(u^2 + w^2)) - w\, \operatorname{curl} \cdot (u, w))\phi\, rdrdz$$

$$(3.8) \qquad b_2((u, w), \psi) = \int_\Omega (\frac{\partial}{\partial z}(\frac{1}{2}(u^2 + w^2)) + u\, \operatorname{curl} \cdot (u, w))\psi\, rdrdz$$

$$b_3((u, w, T), \eta) = \frac{1}{2} \int_\Omega ((u, w) \cdot \nabla T + \nabla_r \cdot ((u, w)T))\eta\, r\, drdz$$

for ϕ, ψ, $\eta \in V$, where $\operatorname{curl} \cdot (u, w) = \frac{\partial w}{\partial r} - \frac{\partial u}{\partial z}$. Note that

$$b_1((u, w), u) + b_2((u, w), w)) = \frac{1}{2} \int_\Omega \nabla(u^2 + w^2) \cdot (u, w)\, r\, drdz = 0$$

and

$$(u, w) \cdot \nabla T = \nabla_r \cdot ((u, w)T)$$

for $(u, w) \in V^2$ satisfying $\nabla_r \cdot (u, w) = 0$.

Consider a boundary value problem

$$(3.9) \qquad -\frac{1}{r} \frac{d}{dr}\left(r \frac{d}{dr} u(r)\right) + \frac{m^2}{r^2} u(r) = f(r), \quad u(R) = 0.$$

in order to present the basic idea of the tau-method based on Jacobi-polynomials. A weak (or variational) form is given by

$$(3.10) \qquad \sigma(u, \psi) = \int_0^R \left(\frac{du}{dr} \frac{d\psi}{dr} + \frac{m^2}{r^2} u\psi\right) r \, dr = \int_0^R f\psi \, r \, dr.$$

for all $\psi \in C^1(0, R)$ with $\psi(0) = \psi(R) = 0$. Let W be the completion of $C^1(0, R)$ with $\psi(R) = 0$ with respect to the norm defined by $\sqrt{\sigma(\cdot, \cdot)}$; i.e., if $m > 0$ then

$$W = \{\phi \in AC_{\ell oc}(0, R) : \phi(R) = 0 \text{ and } r^{1/2} \frac{d}{dr}\phi, \ r^{-1/2} \phi \in L^2(0, R)\}.$$

Then σ defines a bounded coercive sesquilinear form on $W \times W$ and thus for $f \in W^*$ there exists a unique $u \in W$ that satisfies (3.10). Tau-approximation is based on representing an approximate solution u^n of u by

$$u^n = \sum_{k=0}^{n} u_k \, J_k((2r - R)/R)$$

where $J_k(\cdot)$ is the k-th Jacobi polynomial and satisfies the orthogonality [CHQZ]:

$$\int_{-1}^{1} J_k(x) J_l(x)(1 + x) \, dx = 0, \quad k \neq l.$$

Then $u^n \in Z^n \times W$ satisfies

$$(3.11) \qquad \int_0^R \left(\frac{d}{dr} u^n \frac{d}{dr}\psi + \frac{m^2}{r^2} u^n \psi\right) r \, dr = \int_0^R (P^{n-2}f)\psi \, r \, dr$$

for all $\psi \in Z^n \times W$, where Z^n is the space of polynomials of degree at most n on $(0, R)$ and P^{n-2} is the orthogonal projection of $L^2(0, R, r \, dr)$ onto Z^{n-2}. Note that $u^n \in Z^n \times W$ implies $u^n(0) = u^n(R) = 0$ and such conditions are forced on the approximate solution u^n (not on each element). The projection P^{n-2} reflects the fact that the dimension of the subspace $Z^n \times W$ is $n - 1$. Using the standard argument [CHQZ], one can show that $|u^n - u|_W \to 0$ as $n \to \infty$.

Similarly, the above outlined method can be applied to (3.6); i.e., an approximate solution (u^n, w^n, T^n) is represented as

$$u^n(t) = \sum_{k=0}^{n_1} \sum_{l=0}^{n_2} u_{k,l}(t) \, J_k((2r - R)/R) L_l((2z - \alpha(r))/\alpha(r))$$

and similarly for w^n and T^n where $L_l(\cdot)$ is the Legendre polynomial of degree l [CHQZ]. The divergence free condition $\nabla_r \cdot (u(t), w(t)) = 0$ is approximated by

$$(\nabla_r \cdot (u^n(t), w^n(t)), p)_r = ((u^n(t), w^n(t)), \nabla p)_r = 0 \quad \text{for all } p \in Z^{n_1-2, n_2-2}$$

where $(\cdot, \cdot)_r$ denotes the inner product of $H = L^2(\Omega, r\, dr dz)$ and

$$Z^{n_1, n_2} = \{\phi = \sum_{k=0}^{n_1} \sum_{l=0}^{n_2} \phi_{k,l} J_k((2r-R)/R) L_l((2z - \alpha(r))/\alpha(r))\}.$$

Let P^n be the orthogonal projection of H onto $Z^{n1-1, n2-2}$ and W^n be the divergence free subspace of $(Z^{n_1, n_2} \cap V)^2$, defined by

(3.12)
$$W^n = \{(u^n, w^n) \in (Z^{n_1, n_2} \times V)^2 : ((u^n(t), w^n(t)), \nabla p)_r = 0$$
$$\text{for all } p \in Z^{n1-2, n2-2}\}.$$

Then $(u^n(t), w^n(t)) \in W^n$ and $T^n(t) \in Z^{n_1, n_2} \cap V + \theta$ satisfies

$$\rho_0 \left((P^n \frac{du^n(t)}{dt}, \phi)_r + b_1^n((u^n, w^n), \phi)) + a_r^n(u^n(t), \phi) = 0 \right.$$

(3.13)
$$\rho_0 \left((P^n \frac{dw^n(t)}{dt}, \psi)_r + b_2^n((u^n, w^n), \psi)) + a^n(u^n(t), \psi) = 0 \right.$$

$$C_v \rho_0 \left((P^n \frac{dT^n(t)}{dt}, \eta)_r + b_3((u^n, w^n, T^n), \eta)) + k\, a(T(t), \eta) = 0 \right.$$

for all $(\phi, \psi) \in W^n$ and $\eta \in Z^{n_1, n_2} \times V$. Here, the approximate forms b_1^n, b_2^n and b_3^n are defined by
(3.14)
$$b_1^n((u, w), \phi) = (\frac{\partial}{\partial r}(\tfrac{1}{2} P^n(u^2 + w^2)) - P^n((P^n w) \operatorname{curl} \cdot (u, w)), \phi)_r$$

$$b_2^n((u, w), \psi) = (\frac{\partial}{\partial z}(\tfrac{1}{2} P^n(u^2 + w^2)) + P^n((P^n u) \operatorname{curl} \cdot (u, w)), \psi)_r$$

$$b_3^n((u, w, T), \eta) = \tfrac{1}{2}((u, w) \cdot \nabla T, \eta)_r - ((u, w) \cdot \nabla \eta, T)_r).$$

The sesquilinear forms a^n, a_r^n are given by

$$a^n(w, \psi) = (P^n \nabla w, \nabla \psi)_r$$

(3.15)
$$a_r^n(u, \psi) = a^n(u, \phi) + (P^n(u/r), \phi/r)_r.$$

Note that the above approximation is energy conservative in the sense that

$$b_1^n((u^n, w^n), u^n) + b_2^n((u^n, w^n), w^n)) = 0 \quad \text{and} \quad b_3^n((u^n, w^n, T^n), T^n) = 0$$

for $(u^n, w^n) \in W^n$ and $T^n \in Z^{n_1,n_2}$.

Next, we discuss an iterative method based on the Generalized Minimum Residual Method (GMRES) [SS] and the pre-conditioned projection method [Gi],[DI] for solving the steady problem of $(3.12) - (3.15)$. First note that $Z^{n_1,n_2} \cap V$ is isomorphic to Z^{n_1-2,n_2-2}. Consider the Stokes projection P_S onto the divergence free subspace W^n, defined by

$$(3.16) \qquad A^n x + B^n p = f \quad \text{and} \quad (B^n)^* x = 0$$

where A^n, B^n is the matrix representation of the tau-approximation of

$$\begin{pmatrix} -\Delta_r + \frac{1}{r^2} & 0 \\ 0 & -\Delta_r \end{pmatrix} \quad \text{and} \quad \begin{pmatrix} \frac{\partial}{\partial r} \\ \frac{\partial}{\partial z} \end{pmatrix},$$

respectively and

$$P_s f = \sum_{k=0}^{n_1-2} \sum_{l=0}^{n_2-2} x_{k,l} J_k((2r - R)/R) L_l((2z - \alpha(r))/\alpha(r)).$$

That is, in order to calculate the Stokes projection P_S onto W^n we require a solution to the (approximate) Stokes equation (3.16). An alternative and less expensive projection is the L_2-projection:

$$P_{L_2} = I - B^n((B^n)^* B^n)^{-1}(B^n)^*$$

which corresponds to (3.16) where A^n is replaced by I. Thus the preconditioned projection based on the L_2- projection P_{L_2} is defined by

$$P_2 = P_{L_2}(A^n)^{-1} P_{L_2}.$$

The preconditioning for the thermal equation may be given by the elliptic pre-conditioner $(-\Delta_r^n)^{-1}$. However, the elliptic preconditioner is less effective for the convective dominant flow (i.e., high density or high pressure flow). Hence the pre-conditioned problem of (3.13) is written as a constrained nonlinear equation:

$$(3.17) \qquad PF(y) = 0 \quad \text{and} \quad y \in \text{range}(P).$$

where y consists of the solution vectors for (u^n, w^n, T^n) and P represents the matrix representation of the pre-conditioning described above and symmetric positive definite. Since the nonlinearity in (3.13) is quadratic it is easy to calculate the Jacobian $J(y)$ of F. We extend the hybrid Krylov method for nonlinear equations in [BS] to (3.17). Set $J = J(y_c)$ at a current iterate y_c and $r = -PF(y_c)$. Let K_m be the Krylov subspace

$$K_m = \text{span}\{r, PJr, \cdots, (PJ)^{m-1}r\}.$$

We define an approximate solution $\delta^{(m)}$ to the Newton update $PJ\delta = r$ (i.e., the Newton iterate is given by $y_+ = y_c + \delta$) by the least square minimization:

$$(3.18) \qquad \text{minimize} \quad (J\delta + F(y_c))^* P(J\delta + F(y_c)) \quad \text{over } \delta \in K_m$$

The following algorithm is an extension of the nonlinear version of the GMRES algorithm developed in [BS] to equation of form (3.17), which involves the Gram-Schmitz orthogonalization of the Krylov subspace K_m.

Algorithm: Newton-GMRES

(1) Choose y_1 and m and set $k = 1$.

(2) Set $r = -PF$ where $F = F(y_k)$ and $J = J(y_k)$. Compute $\beta = -(r, F)$ and $v_1 = r/\beta$. For $j = 1, 2, \cdots, m$ do

$$h_{i,j} = (Jv_j, v_i), \ i = 1, 2, \cdots, j,$$

$$\hat{v}_{j+1} = PJv_j - \sum_{i=1}^{j} h_{i,j} v_i$$

$$h_{j+1,j} = (\hat{v}_{j+1}, Fv_j) - \sum_{i=1}^{j} |h_{i,j}|^2 \quad \text{and} \quad v_{j+1} = \hat{v}_{j+1}/h_{j+1,j}.$$

(3) Define H_m to be the $(m+1) \times m$ (Hessenberg) matrix whose nonzero entries are the coefficients $h_{i,j}$, $1 \le i \le j+1$, $1 \le j \le m$. Compute the least square solution

$$z = \beta((H_m)^* H_m)^{-1}(H_m)^* e_1 \quad \text{and set} \quad \delta^{(m)} = \sum_{i=1}^{m} z_i v_i.$$

(4) Set $y_{k+1} = y_k + \delta^{(m)}$. If convergence criterion is not satisfied then set $k = k + 1$ and go to (2).

Numerical implementation and convergence analysis of the proposed method will be reported in a forthcoming paper.

4. Shape optimization and shape derivative. In this section we discuss the shape derivative of solutions to the thermally coupled Navier-Stokes equations. For the sake of clarity of our presentation we consider the 2-D steady case (evolution, 3-D, axi-symmetric problems and a more general boundary condition can be treated as well); i.e., $(u, p, T) \in (H_0^1(\Omega(\alpha))^2 \times L^2(\Omega(\alpha)) \times (H_0^1(\Omega(\alpha)) + \theta)$ satisfies the Boussinesq equation

$$-\nu \Delta u + u \cdot \nabla u + \nabla p = gT e_2 + f, \quad \nabla \cdot u = 0$$
$$(4.1)$$
$$-k \Delta T + u \cdot \nabla T = 0$$

Throughout this section we assume that $\Omega(\alpha)$ is sufficiently smooth and θ is given as the trace of a function in $H^2(\Omega(\alpha))$. Here the solution (u, p, T)

depends on the shape of domain $\Omega(\alpha)$ which is parameterized by $\alpha \in Q_{ad}$. Consider the shape minimization problem [HN],[Pi]:

$$(4.2) \qquad \text{minimize} \quad J(u, T, \alpha) \quad \text{over } \alpha \in Q_{ad}$$

subject to (4.1). For example, the cost functional J is given as follows

$$J(u, \alpha) = \int_{\Omega(\alpha)} |u - u_d|^2 \, dx + \beta \, N(\alpha)$$

$$(4.3) \qquad J(u, \alpha) = \int_{\Omega(\alpha)} |\nabla u|^2 \, dx + \beta \, N(\alpha)$$

$$J(T, \alpha) = \int_{\Omega(\alpha)} |T - T_d|^2 \, dx + \beta \, N(\alpha)$$

where u_d, T_d is the target vector field and thermal distribution, respectively, $\beta \geq 0$ and $N(\alpha)$ denotes the regularization of the shape of domain $\Omega(\alpha)$. A successful numerical optimization method is commonly based on the gradient of the cost functional with respect to α. In order to calculate the gradient of J we will employ the so-called material derivative method. Material derivative concepts are well-known in continuum mechanics and have been applied to shape optimization problems in [Ce],[Zo],[HCK] and the references therein.

Let $\alpha \in Q_{ad}$ be fixed and for $|t|$ sufficiently small, let $\Omega_t(\alpha) = F_t(\Omega(\alpha))$ be the image of $\Omega(\alpha)$ obtained by the mapping $F_t : R^2 \to R^2$ defined as

$$F_t(x_1, x_2) = (x_1, x_2) + t \, h(x_1, x_2).$$

In the context of §3 we have

$$\Omega_t(\alpha) = \Omega(\alpha_t) = \Omega(\alpha + t \, v), \quad v \in C^2(0, R).$$

In what follows the dependency of $\Omega(\alpha)$ on α will be dropped. For $\varphi \in H^1(\Omega)$ and $\varphi_t \in H^1(\Omega_t)$ let us define

$$(4.4) \qquad \varphi^t = \varphi_t \circ F_t.$$

The material derivative of φ for field $h \in (H^1(\Omega))^2$ is given by

$$(4.5) \qquad \dot{\varphi}(x) = \lim_{t \to 0} \frac{\varphi_t(x + t \, h) - \varphi(x)}{t} \quad \text{for } x \in \Omega.$$

If φ_t has a regular extension to a neighborhood of $\bar{\Omega}_t$, then

$$(4.6) \qquad \varphi\prime(x) = \lim_{t \to 0} \frac{\varphi_t(x) - \varphi(x)}{t} = \dot{\varphi}(x) - h(x) \cdot \nabla \varphi(x), \ x \in \Omega$$

is called the shape derivative of φ. Define

$$I_t = \det(DF_t) \quad \text{and} \quad A_t = (DF_t^{-1})^*(DF_t^{-1})I_t.$$

where $*$ denotes the transpose of a matrix and DF_t is the Jacobian of F_t. It is then easy to verify that

$$\tfrac{d}{dt}F_t|_{t=0} = h, \quad \tfrac{d}{dt}(DF_t)|_{t=0} = Dh, \quad \tfrac{d}{dt}(DF_t^{-1})|_{t=0} = -(Dh)^*$$

(4.7)

$$\tfrac{d}{dt}I_t|_{t=0} = \operatorname{div} h \quad \text{and} \quad \tfrac{d}{dt}A_t|_{t=0} = \operatorname{div} h\, I - ((Dh)^* + Dh).$$

Moreover we have

Lemma 4.1 Let

$$E_t = \int_{\Omega_t} \varphi_t\, dx_t, \quad \varphi_t \in H^1(\Omega_t).$$

Then

$$(4.8) \qquad \dot{E} = \frac{d}{dt}E_t|_{t=0} = \int_{\Omega} \dot{\varphi} + \varphi \operatorname{div} h\, dx = \int_{\Omega} \varphi' + \operatorname{div}(h\varphi)\, dx.$$

Proof: Using Fubini's theorem we obtain

$$E_t = \int_{\Omega} \varphi^t I_t\, dx.$$

By differentiating E_t with respect to t we obtain

$$\frac{d}{dt}E_t = \int_{\Omega} \frac{d}{dt}I_t\, \varphi^t + I_t\, \frac{d}{dt}\varphi^t\, dx.$$

Since $I_0 = 1$ and $\frac{d}{dt}I_t|_{t=0} = \operatorname{div} h$ (4.8) follows by setting $t = 0$.

Note that $\nabla\varphi_t = (DF_t^{-1})\nabla\varphi^t$. Thus, $(4.1) - (4.3)$ is equivalently written as: (u^t, p^t, T^t) satisfies

(4.9)
$$(u^t \cdot (J_t \nabla u^t),\, \phi) + \nu\,(A_t \nabla u^t,\, \nabla\phi) + (J_t \nabla p^t,\, \phi) = (I_t\,(gT^t\, e_2 + f \circ F_t),\, \phi)$$

$$(u^t,\, J_t \nabla\psi) = 0$$

$$(u^t \cdot (J_t \nabla T^t),\, \eta) + k\,(A_t \nabla T^t,\, \eta) = 0$$

for $\phi \in (H_0^1(\Omega))^2$, $\psi \in H^1(\Omega)$ and $\eta \in H_0^1(\Omega)$, where $J_t = I_t\, DF_t^{-1}$ and $(\cdot,\cdot)$ denotes the $L_2(\Omega)$−inner product.

We will sketch a proof of the existence and regularity of solutions to (4.1). Let V_0 be the divergence free subspace of $(H_0^1(\Omega))^2$. Define the solution map S on $V_0 \times (H_0^1(\Omega) + \theta)$ by $S(u,T) = (\tilde{u}, \tilde{T})$ where $(\tilde{u}, \tilde{T})$ is a unique weak solution to

$$-\nu\,\Delta\tilde{u} + u \cdot \nabla\tilde{u} + \nabla p = g\tilde{T}\, e_2 + f, \quad \nabla \cdot \tilde{u} = 0$$

$$-k\,\Delta\tilde{T} + u \cdot \nabla\tilde{T} = 0$$

First we show that

$$\theta_1 = \min_{x \in \Gamma} \le \tilde{T} \le \max_{x \in \Gamma} = \theta_2 \quad \text{a.e. } x \in \Omega$$

Let $\psi(x) = \inf(\tilde{T}, \theta_1)$. Then $\psi \in H_0^1(\Omega)$ [Tr] and we have

$$k\,(\nabla \tilde{T}, \nabla \psi) + (u \cdot \nabla \tilde{T}, \psi) = 0.$$

Thus,

$$k\,(\nabla \psi, \nabla \psi) + \frac{1}{2}\,(u, \nabla |\psi|^2) = 0.$$

Since $\nabla \cdot u = 0$ we obtain $|\nabla \psi|^2 = 0$ which implies $\psi = 0$ an hence $\tilde{T} \ge \theta_1$. Similarly, one can prove that $\tilde{T} \le \theta_2$, choosing the test function $\psi = \sup(\tilde{T}, \theta_2)$. Next define a sesquiliner form σ on $V_0 \times V_0$ by

$$\sigma(w, v) = \nu\,(\nabla w, \nabla v) + (u \cdot \nabla w, v).$$

Then $\tilde{u}$ satisfies

$$(4.10) \qquad \sigma(\tilde{u}, \phi) = (g\tilde{T}\, e_2 + f,\, \phi) \quad \text{for all } \phi \in V_0.$$

Note that σ is bounded and coercive since $\sigma(v, v) = \nu\,|\nabla v|^2$ (e.g., see [Te]). Thus by Lax-Milgram theorem (4.10) possesses a unique solution $\tilde{u} \in V_0$ and we have

$$|\tilde{u}|_{V_0} \le \frac{M}{\nu}\,(|f|_{L_2} + g\,|T|_{L_2}) \quad \text{for some } M > 0.$$

Let C be a closed convex subspace of $V_0 \times (H_0^1(\Omega) + \theta)$ defined by

$$C = \{(u, T) : |u|_V \le M\,(|f|_{L_2} + \theta_{max}\,|\Omega|) \text{ and } \theta_1 \le T \le \theta_2,\ \text{a.e. } x \in \Omega\}.$$

where $\theta_{max} = \max(|\theta_1|, |\theta_2|)$. Then S maps from C into C. Note that

$$|(u \cdot \nabla w, v)| \le M_1\,|u|_{L_4}|w|_{H^1}|v|_{H^1} \quad \text{for } u, w, v \in V$$

for some $M_1 > 0$ and that $H^1(\Omega)$ is compactly embedded into $L_4(\Omega)$. Hence one can show (e.g., see [DI]) that the solution map S is compact. By Shauder fixed point theorem (e.g, see [Tr]) there exists at least one solution to (4.1). Define the Stokes operator Δ_S on $H = \{\phi \in (L_2(\Omega))^2 : \nabla \cdot \phi = 0$ and $n \cdot \phi = 0\}$ by

$$(-\Delta_S u, \phi) = (\nabla u, \nabla \phi) \quad \text{for } \phi \in V$$

with domain

$$\text{dom}(-\Delta_S) = \{u \in V : |(\nabla u, \nabla \phi)| \le c\,|\phi|_H \text{ for all } \phi \in V\}.$$

Then it is known [Ta] that $-\Delta_S$ is a positive self-adjoint operator on H, $\mathrm{dom}(-\Delta_S) \subset H^2(\Omega)$ and $V = \mathrm{dom}(-\Delta_S^{1/2}) = [H, \mathrm{dom}(-\Delta_S)]_{1/2}$. Moreover, we have $\phi \cdot \nabla\phi \in V_{-1/2} = \mathrm{dom}(-\Delta_S^{-1/4})$ for $\phi \in V$. Thus,

$$u = (-\Delta_S)^{-1}(gT\,e_2 + f - u\cdot\nabla u) \in \mathrm{dom}(-\Delta_S^{3/4}) = [V, \mathrm{dom}(-\Delta_S)]_{1/2} \subset H^{3/2}(\Omega).$$

Hence, $u \in L^\infty(\Omega)$ and $u\cdot\nabla u$, $u\cdot\nabla T \in L_2(\Omega)$. This implies that $(u, T) \in (H^2(\Omega))^3$.

Assume that $h \in C^{1,\infty}(R^2)$. Then there exists a solution $(u^t, p^t, T^t) \in (H_0^1(\Omega) \cap H^2(\Omega))^2 \times H^1(\Omega)/R \times H^2(\Omega)$ to (4.9) provided that $f^t \in L_2(\Omega)$. Assume that the linearized equation of (4.9) at (u, p, T) (i.e, $t = 0$)

$$-\nu\,\Delta\xi + u\cdot\nabla\xi + \xi\cdot\nabla u + \nabla q = g\theta\,e_2 + f_1, \quad \nabla\cdot\xi = f_2$$

(4.11)

$$u\cdot\nabla\theta + \xi\cdot\nabla T = k\,\Delta\theta + f_3$$

has a unique solution $(\xi, q, \theta) \in (H_0^1(\Omega) \cap H^2(\Omega))^2 \times H^1(\Omega)/R \times H_0^1(\Omega) \cap H^2(\Omega)$ which depends continuously on f_1, $f_3 \in L_2(\Omega)$ and $f_2 \in H^1(\Omega)$ with $(1, f_2) = 0$. Then, since F_t, A_t, I_t and J_t is continuously differentiable in t and Lipschitz in x it follows from the implicit function theory that for $|t|$ sufficiently small (4.9) has a (locally) unique solution (u^t, p^t, T^t). Moreover, one can argue that

$$\lim_{t\to 0} \frac{u^t - u}{t} = \dot{u} \quad \text{and} \quad \lim_{t\to 0} \frac{T^t - T}{t} = \dot{T}$$

exist in $H^2(\Omega)$ and $\dot{p}$ exists in $H^1(\Omega)$. Note that

$$\frac{d}{dt}(I_t\,f \circ F_t,\,\phi)|_{t=0} = (\mathrm{div}\,h\,f,\,\phi) + (h\cdot\nabla f) = (\mathrm{div}(f\,h),\,\phi) = -(f,\,h\cdot\nabla\phi).$$

Since $\frac{d}{d}I_t = \mathrm{div}\,h$, $\frac{d}{dt}A_t = A = \mathrm{div}\,h\,I - ((Dh)^* + Dh)$, and $\frac{d}{dt}J_t = \mathrm{div}\,h\,I - (Dh)^*$ it follows from (4.9) that $(\dot{u}, \dot{p}, \dot{T})$ satisfies
(4.12)

$$\nu\,(\nabla\dot{u},\,\nabla\phi) + \nu\,(A\,\nabla u,\,\nabla\phi) + (u\cdot\nabla\dot{u} + \dot{u}\cdot\nabla u + u\cdot(J\nabla u),\,\phi)$$

$$+(\nabla\dot{p},\,\phi) + (J\nabla p,\,\phi) = (g\dot{T}\,e_2,\,\phi) - (f,\,h\cdot\nabla\phi)$$

$$(\dot{u},\,\nabla\psi) + (u,\,J\nabla\psi) = 0$$

$$k\,(\nabla\dot{T},\,\nabla\eta) + k\,(A\,\nabla T,\,\nabla\eta) + (u\cdot\nabla\dot{T} + \dot{u}\cdot\nabla T + u\cdot(J\nabla T),\,\eta) = 0$$

for all $\phi \in H_0^1(\Omega)$, $\psi \in H^1(\Omega)$ and $\eta \in H_0^1(\Omega)$.

Next we derive an equation for the shape derivative (u', p', T'). Note

that
(4.13)
$$(\nabla(h \cdot \nabla u), \phi) + (A \nabla u, \nabla \phi) - (\Delta u, h \cdot \nabla \phi)$$

$$= \left\langle \left(\begin{array}{cc} (h_2)_{x_2} - (h_1)_{x_1} & -((h_1)_{x_2} + (h_2)_{x_1}) \\ -((h_1)_{x_2} + (h_2)_{x_1}) & (h_1)_{x_1} - (h_2)_{x_2} \end{array} \right) \left(\begin{array}{c} u_{x_1} \\ u_{x_2} \end{array} \right), \left(\begin{array}{c} \phi_{x_1} \\ \phi_{x_2} \end{array} \right) \right.$$

$$+ \left(\begin{array}{c} (h_1)_{x_1} u_{x_1} + h_1 u_{x_1 x_1} + (h_2)_{x_1} u_{x_2} + h_2 u_{x_1 x_2} - h_1(u_{x_1 x_1} + u_{x_2 x_2}) \\ (h_1)_{x_2} u_{x_1} + h_1 u_{x_1 x_2} + (h_2)_{x_2} u_{x_2} + h_2 u_{x_2 x_2} - h_2(u_{x_1 x_1} + u_{x_2 x_2}) \end{array} \right),$$

$$\left. \left(\begin{array}{c} \phi_{x_1} \\ \phi_{x_2} \end{array} \right) \right\rangle$$

$$= ((h_2 u_{x_1} - h_1 u_{x_2})_{x_2}, \phi_{x_1}) - ((h_2 u_{x_1} - h_1 u_{x_2})_{x_1}, \phi_{x_2})$$

$$= (\operatorname{curl}(h_2 u_{x_1} - h_1 u_{x_2}), \operatorname{grad}\phi) = 0,$$

where we assumed that $\phi \in H_0^1(\Omega) \cap H^2(\Omega)$ satisfies $n \cdot \nabla \phi = 0$. Similarly, we have
(4.14)
$$(\nabla(h \cdot \nabla p, \phi) + (J \nabla p, \phi) + (\nabla p, h \cdot \nabla \phi)$$

$$= \left\langle \left(\begin{array}{c} (h_2)_{x_2} p_{x_1} - (h_2)_{x_1} p_{x_2} \\ -(h_1)_{x_2} p_{x_1} + (h_1)_{x_1} p_{x_2} \end{array} \right), \left(\begin{array}{c} \phi_1 \\ \phi_2 \end{array} \right) + \left(\begin{array}{c} p_{x_1} \\ p_{x_2} \end{array} \right), \left(\begin{array}{c} h_1(\phi_1)_{x_1} + h_2(\phi_1)_{x_2} \\ h_1(\phi_2)_{x_1} + h_2(\phi_2)_{x_2} \end{array} \right) \right.$$

$$\left. - \left(\begin{array}{c} h_1 p_{x_1} + h_2 p_{x_2} \\ h_1 p_{x_1} + h_2 p_{x_2} \end{array} \right), \left(\begin{array}{c} (\phi_1)_{x_1} \\ (\phi_2)_{x_2} \end{array} \right) \right\rangle$$

$$= ((h_2 \phi_1)_{x_2}, p_{x_1}) - ((h_2 \phi_1)_{x_1}, p_{x_2}) - ((h_1 \phi_2)_{x_2}, p_{x_1}) + ((h_1 \phi_2)_{x_1}, p_{x_2})$$

$$= (\operatorname{curl}(h_1 \phi_2 - h_2 \phi_1), \operatorname{grad} p) = 0$$

For the convective term
(4.15)
$$((h \cdot \nabla u) \cdot \nabla u + u \cdot \nabla(h \cdot \nabla u), \phi) - (u \cdot (Dh^* \nabla u), \phi)$$

$$+ (\operatorname{div} h\,(u \cdot \nabla u, \phi) + (u \cdot \nabla u, h \cdot \nabla \phi)$$

$$= ((h \cdot \nabla u) \cdot \nabla u + u \cdot (h \cdot \nabla(\nabla u)), \phi) - (h \cdot \nabla(u \cdot \nabla u), \phi) = 0.$$

Moreover, we have

(4.16) $$(h \cdot \nabla T, \phi) + (\operatorname{div} h\, T, \phi) + (T, h \cdot \nabla \phi) = 0.$$

Since $\varphi' = \dot{\varphi} - h \cdot \nabla \varphi$ it follows from (4.1) and (4.12) $-$ (4.16) that

$$\nu(\nabla u', \nabla \phi) + (u \cdot \nabla u' + u' \cdot \nabla u, \phi) + (\nabla p', \phi) = (gT' e_2, \phi)$$

for all $\phi \in (H_0^1(\Omega) \cap H^2(\Omega))^2$ satisfying $n \cdot \nabla\phi = 0$. Using exactly the same arguments as above, we obtain

$$k\left(\nabla T', \nabla\eta\right) + \left(u \cdot \nabla T' + u' \cdot \nabla T, \eta\right) = 0$$

for all $\eta \in H_0^1(\Omega) \cap H^2(\Omega)$ satisfying $n \cdot \nabla\eta = 0$. Finally, for the divergence free equation we have

(4.17)
$$\left(h \cdot \nabla u, \nabla\psi\right) + \left(u, J\nabla\psi\right) + \left(\nabla \cdot u, h \cdot \nabla\psi\right)$$

$$= \left(h_1(u_1)_{x_1} + h_2(u_1)_{x_2}, \psi_{x_1}\right) + \left(h_1(u_2)_{x_1} + h_2(u_2)_{x_2}, \psi_{x_2}\right)$$

$$+\left(u_1, (h_2)_{x_2}\psi_{x_1} - (h_2)_{x_1}\psi_{x_2}\right) + \left(u_2, -(h_1)_{x_2}\psi_{x_1} + (h_1)_{x_1}\psi_{x_2}\right)$$

$$-\left((u_1)_{x_1} + (u_2)_{x_2}, h_1\psi_{x_1} + h_2\psi_{x_2}\right)$$

$$= \left(\mathrm{curl}(h_1 u_2 - h_2 u_1), \mathrm{grad}\,\psi\right) = 0$$

provided that $\psi \in H^2(\Omega)$. Thus, from (4.12),(4.17) we obtain

$$\left(u', \nabla\psi\right) = 0 \quad \text{for } \psi \in H^2(\Omega)$$

Hence one can conclude that if the assumption (4.11) holds, then for field $h \in C^{1,\infty}(R^2)$ the shape derivative $(u', p', T') \in (H^1(\Omega))^2 \times L_2(\Omega) \times H^1(\Omega)$ exists and satisfies

$$-\nu\,\Delta u' + u \cdot \nabla u' + u' \cdot \nabla u + \nabla p' = gT'\, e_2, \quad \nabla \cdot u' = 0$$

(4.18)

$$-k\Delta T' + u \cdot \nabla T' + u' \cdot \nabla T = 0$$

with boundary conditions

$$u' + h \cdot \nabla u = 0 \quad \text{and} \quad T' + h \cdot \nabla T = 0 \quad \text{on } \Gamma.$$

5. Augmented Lagrangian method with second-order update. In this section we discuss an application of the augmented Lagrangian method for constrained minimization problems that arise in flow control. Let H, U and Y be Hilbert spaces and set $X = H \times U$. Consider the constrained minimization problem:

(5.1)
$$\text{minimize} \quad f(u, \alpha) \quad \text{over } u \in H \text{ and } \alpha \in K$$

$$\text{subject to} \quad e(u, \alpha) = 0,$$

where K is a closed convex set in U. In practice, the control space U is of finite dimensional (i.e., which is parameterized or involves finite many inputs). For example, consider the following optimal control problem

(5.2)
$$\text{minimize} \quad J = \int_{\Omega} |T - T_d|^2\, dx \quad \text{over } \alpha \in K \subset R^m$$

subject to

$$\rho_0\, u \cdot \nabla u + \nabla p = \mu\, \Delta u - \frac{\rho_0}{\theta_0}\,(T - \theta_0)g\, e_3$$

(5.3)
$$\nabla \cdot u = 0 \quad u_\Gamma = 0,$$

$$\rho_0 C_v\, u \cdot \nabla T = \nabla \cdot (k\,\nabla T), \quad T|_\Gamma = \sum_{i=1}^{m} \alpha_i\, g_i,$$

where we assumed that g_i is the trace of Lipschitz continuous function θ_i on Γ and thus in $H^{1/2}(\Gamma)$. Using the same arguments as described in §4, given $\alpha \in K$ one can show that there exists at least one solution $(u, p, T) \in H^2(\Omega)^2 \times H^1(\Omega)/R \times H^1(\Omega) \cap L^\infty(\Omega)$ to (5.3). If we define a function $\hat{T}$ by $\hat{T} = T - \sum_{i=1}^{m} \alpha_i\, \theta_i$ then the third equation of (5.3) is equivalently written as

$$\rho_0 C_v\, u \cdot \nabla T = \nabla \cdot (k\,\nabla\hat{T}) + \nabla \cdot (k\,\nabla(\textstyle\sum_{i=1}^{m} \alpha_i\, \theta_i)), \quad \hat{T}|_\Gamma = 0.$$

Let V_0 be the divergence free subspace of $(H_0^1(\Omega))^3$, $H = V_0 \times H_0^1(\Omega)$ and $Y = V^* \times H^{-1}(\Omega)$. Then (5.3) can be written as $e((u, \hat{T}), \alpha) = 0$ where $e = (e_1(u, \hat{T}, \alpha), e_2(u, \hat{T}, \alpha))$ is defined by

(5.4)
$$\langle e_1(u, \hat{T}, \alpha), \phi \rangle = \mu\,(\nabla u,\, \nabla \phi) + \rho_0\,(u \cdot \nabla u,\, \phi) - \frac{\rho_0 g}{\theta_0}\,(T - \theta_0,\, \phi)$$

$$\langle e_2(u, \hat{T}, \alpha), \psi \rangle = (k\,\nabla\hat{T},\, \nabla\psi) + \rho_0 C_v\,(u \cdot \nabla T,\, \psi) + (k\,\nabla(\textstyle\sum_{i=0}^{m} \alpha_i\, \theta_i),\, \psi)$$

for $\phi \in V$ and $\psi \in H_0^1(\Omega)$, where $T = \hat{T} + \sum_{i=0}^{m} \alpha_i\, \theta_i$. The divergence free constraint is absorbed in the definition of V. Recall again that

$$|(u \cdot \nabla w,\, v)| \le M_1\, |u|_{L^4}|w|_{H^1}|v|_{H^1}$$

and that $H^1(\Omega)$ is compactly embedded into $L_4(\Omega)$. Thus, the cost functional J is sequentially weakly lower semi-continuous (e.g, see [DI]), and therefore (5.2)-(5.3) has at least one solution.

Assume the following hypotheses.

(H1)   there exists a solution $x^* = (u^*, \alpha^*)$ to (5.1).

(H2) f, e are twice continuously F−differentiable in a convex neighborhood of x^*.

(H3) x^* is a regular point in the sense [MZ] that

(5.5)
$$0 \in \text{int}\{e'(x^*)(v, h) : v \in H \text{ and } h \in K - \alpha^*\}.$$

Then it follows from [MZ] that there exists a Lagrange multiplier $\lambda^* \in Y^*$ such that

(5.6)
$$f'(x^*)(v, \alpha - \alpha^*) + \langle \lambda^*,\, e'(x^*)(v, \alpha - \alpha^*) \rangle \ge 0$$

for all $v \in X$ and $\alpha \in K$. The augmented Lagrangian method is based on an equivalent formulation of (5.1):

$$(5.7) \qquad \text{minimize} \quad f(u, \alpha) + \frac{c}{2} |e(u, \alpha)|_Y^2 \quad \text{over } u \in X \text{ and } \alpha \in K.$$

where $c \geq 0$. Then the augmented Lagrangian algorithm [Po],[He] is the multiplier method applied to (5.7); i.e., it involves a sequence of minimizations of the functional

$$L_{c_k}(u, \alpha, \lambda^k) = f(u, \alpha) + \langle \lambda^k, e(u, \alpha) \rangle + \frac{c}{2} |e(u, \alpha)|_Y^2$$

(5.8)

$$\text{subject to } \alpha \in K,$$

where the multiplier sequence $\{\lambda^k\}$ in Y^* is generated by the first order update

$$(5.9) \qquad \lambda^{k+1} = \lambda^k + (c_k - c_0) \, e(u_k, \alpha_k),$$

for $k \geq 1$. Here the pair (u_k, α_k) is a minimizer of $L_{c_k}(\cdot, \cdot, \lambda^k)$ and assume that $Y^* = Y$, otherwise each element in Y^* has its Riesz representation. To carry out this iterative a sequence of monotonically nondecreasing, positive real numbers $\{c_k\}$, $c_1 > c_0 \geq 0$ and a start up value λ^1 for the Lagrange multiplier for the equality constraint $e(u, \alpha) = 0$ need be chosen. The convergence results of the augmented Lagrangian method for the infinite dimensional optimization problem are established, for example, in [IK1],[PT]. The augmented Lagrangian method is a hybrid method of the penalty method (i.e., $\lambda^k = 0$) and the Lagrange multiplier method (i.e., $c_k = 0$) and combines good properties of the both methods. It overcomes the difficulty of the penalty method which requires to have a large value of c_k. The cost functional $L_{c_k}(x, \alpha, \lambda^k)$ is locally strictly convex provided that λ^k is sufficiently close to λ^* and the second order optimality condition

$$L_0''(u^*, \alpha^*, \lambda^*)((v, h), (v, h)) \geq \sigma \left(|v|_H^2 + |h|_U^2 \right)$$

(5.10)

$$\text{for all } (v, h) \in X \text{ satisfying } e'(u^*, \alpha^*)(v, h) = 0,$$

for some $\sigma > 0$, is satisfied. Here, $L_0''(u^*, \alpha^*, \lambda^*)$ denotes the bilinear form that characterizes the second derivative of $L_0(u, \alpha, \lambda) = f(u, \alpha) + \langle \lambda, e(u, \alpha) \rangle$ with respect to $x = (u, \alpha)$ at (x^*, λ^*). That is, the cost functional f is not necessary to be (locally) convex, which is required for convergence of the multiplier method. The algorithm (5.8) − (5.9) has been successfully applied to parameter estimation problems in elliptic PDEs [IK2],[IKK] and optimal control problems for 2-D incompressible Navier-Stokes [DI]. The first order update (5.9) provides q-linear convergence of the iterates (u_k, α_k) in X. In [IK3] we have investigated a second order update scheme for the augmented Lagrangian method. In what follows we assume that $\alpha^* \in \mathrm{int}\,(K)$. Thus, (H3) reduces to

$$(5.11) \qquad e'(x^*) \text{ is surjective.}$$

Hence the necessary condition (5.6) implies that

$$(5.12) \qquad L'_c(u^*, \alpha^*, \lambda^*) = 0 \quad \text{and} \quad e(u^*, \alpha^*) = 0,$$

for all $c \geq 0$. An algorithm proposed in [IK3] is to apply the Newton method to (5.12). Then the resulting algorithm is stated as: given a current iterate (x, λ) the next iterate (x_+, λ_+) satisfies

$$(5.13) \qquad \begin{pmatrix} L''_c(x, \lambda) & e'(x)^* \\ e'(x) & 0 \end{pmatrix} \begin{pmatrix} x_+ - x \\ \lambda_+ - \lambda \end{pmatrix} = - \begin{pmatrix} L'_c(x, \lambda) \\ e(x) \end{pmatrix}.$$

Note that

$$L'_c(x, \lambda) = L'_0(x, \lambda + c\,e(x))$$

and

$$(5.14) \qquad L''_c(x, \lambda) = L''_0(x, \lambda + c\,e(x)) + c\,\langle e'(x)(\cdot),\, e'(x)(\cdot)\rangle.$$

Consequently, suppose $|(x, \lambda) - (x^*, \lambda^*)|$ is sufficiently small then it follows from (5.10) [IK3] that $L''(x, \lambda)$ is coercive on $X \times Y$. Thus equation (5.13) can be regarded as a general Stokes equation. Following an argument due to Bertsekes one can avoid forming L''_c during the iteration. From the second equation of (5.13) we have $e'(x)(x_+ - x) = -e(x)$. Thus the first equation can be written as

$$L''_0(x, \lambda + c\,e(x))(x_+ - x) + e'(x)^*(\lambda_+ - (\lambda + c\,e(x))) = -L'_0(x, \lambda + c\,e(x))$$

and hence (5.13) is equivalent to

$$(5.15) \qquad \begin{pmatrix} L''_0(x, \hat{\lambda}) & e'(x)^* \\ e'(x) & 0 \end{pmatrix} \begin{pmatrix} x_+ - x \\ \lambda_+ - \hat{\lambda} \end{pmatrix} = - \begin{pmatrix} L'_0(x, \hat{\lambda}) \\ e(x) \end{pmatrix}$$

$$\text{where } \hat{\lambda} = \lambda + c\,e(x).$$

Note that $\hat{\lambda}$ is nothing but the first order update of the Lagrange multiplier if the current iterate x minimizes $L_c(x, \lambda)$. Equation (5.15) is more advantageous than (5.13) since the squaring term $c\,e'(x)^*e'(x)$ is absorbed and less calculation is involved. If we define a matrix operator S on $X \times Y$ by

$$S(x, \lambda) = \begin{pmatrix} L''_0(x, \lambda) & e'(x)^* \\ e'(x) & 0 \end{pmatrix},$$

then it follows from (5.10) that $S(x^*, \lambda^*)$ is boundedly invertible. Thus, suppose (x, λ) is sufficiently close to (x^*, λ^*) then equation (5.15) has a unique solution. We summarize our discussions as

Algorithm 5.1

(1) Choose $\lambda^1 \in Y$, $c > \bar{c} \geq 0$, and set $\hat{c} = c - \bar{c}$, $k = 1$.

(2) Determine $x = (u, \alpha) \in X \times K$ such that

$$L_c(u, \alpha, \lambda^k) \leq L_c(u^*, \alpha^*, \lambda) = f(x^*).$$

(3) Set $\hat{\lambda} = \lambda^k + \hat{c}\, e(x)$.

(4) Solve for $(x_+, \lambda_+) \in X \times Y$:

$$S(x, \hat{\lambda}) \begin{pmatrix} x_+ - x \\ \lambda_+ - \hat{\lambda} \end{pmatrix} = - \begin{pmatrix} L_0'(x, \hat{\lambda}) \\ e(x) \end{pmatrix}.$$

(5) Set $x_{k+1} = x_+$ and $\lambda^{k+1} = \lambda_+$. If the convergence criterion is satisfied then set $k = k + 1$ and go to (2).

Remark 5.2 A variant of Algorithm 5.1 is obtained by skipping step (2). Then it is reduced to the Newton method applied to equation (5.12). If $x = (u, \alpha)$ minimizes $L_c(\cdot, \lambda^k)$ over $H \times K$ then step (2) is completed. Step (2) implies a sufficient reduction of the merit functional (the augmented Lagrange functional). Let (H1),(H2) and (5,10),(5.11) hold. Then it is proved in [IK3] that if $|\lambda^1 - \lambda^*|_Y$ is sufficiently small then Algorithm 5.1 is well-posed and (x^k, λ^k) converges to (x^*, λ^*) q-quadratically.

REFERENCES

[BS] P.N.Brown and Y.Saad, Hybrid Krylov methods for nonlinear systems of equations, SIAM J. Sci. Statist. Comput.,11 (1990), 450-481.

[Ce] J.Céa, Conception optimale ou identification de formes, calcul rapide de la dérivée directionelle de la fonction coût, RAIRO Math.Mod.Num.Anal., 20 (1986), 371-402.

[CHQZ] C.Canuto, M. Y.Hussaini, A.Quarteroni and T.A.Zang, *Spectral Methods in Fluid Dynamics*, Springer-Verlag, Berlin, 1988.

[DI] M.C.Desai and K.Ito, Optimal Control of Navier-Stokes Equations, SIAM J. Control & Optim., to appear.

[FM] D.Fujiwara and H.Morimoto, An L_p-theory of the Helmholtz decomposition of vector field, J. Fac. Sci. Univ. Tokyo Sect. IA Math, 24 (1977), 685-700.

[Gi] Y.Giga, Domains of fractional powers of the Stokes operator in L_p spaces, Arch. Rational Mech. Anal., 89 (1985), 251-265.

[Gl] R.Glowinski, *Numerical Methods for Nonlinear Variational Problems*, Springer-Verlag, Berlin, 1984.

[GR] V.Girault and P.A. Raviart, *Finite Element Methods for Navier-Stokes Equations*, Springer-Verlag, Berlin, 1984.

[He] M.R.Hestenes, Multiplier and gradient methods, J. Opt. Theory Appl., 4 (1968), 303-320.

[HCK] E.J.Haug, K.K.Choi and V.Komkov, *Design Sensitivity Analysis of Structural Systems*, Academic Press, Orlando.

[HN] J.Haslinger and P.Neittaanmäki, *Finite Element Approximation for Optimal Shape Design: Theory and Applications*, John Wiley, New York, 1988.

[IK1] K.Ito and K.Kunisch, The augmented Lagrangian method for equality and inequality constraints in Hilbert spaces, Math. Programming.

[IK2] K.Ito and K.Kunisch, The augmented Lagrangian method for parameter estimation in elliptic systems, SIAM J. Control & Optm., 28 (1990), 113-136.

[IK3] K.Ito and K.Kunisch, The augmented Lagrangian method with second order update and its application to parameter estimation in elliptic systems, preprint.

[IKK] K.Ito, M.Kroller and K.Kunisch, A numerical study of an augmented Lagrangian method for the estimation of parameters in elliptic systems, SIAM J. Sci. Statist. Comput.,12 (1991), 884-910.

[MZ] H.Maurer and J.Zowe, First and second-order necessary and sufficient optimality conditions for infinite-dimensional programming problems, Math Programming, 16 (1979), 98-110.

[MN] A.Matsumura and T.Nishida, The initial value problem for the equations of motion of viscous and heat-conductive gases, J. Math. Kyoto Univ., 20 (1980), 67-104.

[Pi] O.Pironneau, *Optimal Shape Design for Elliptic Systems*, Springer-Verlag, New York 1984.

[Po] M.J.D.Powell, A method for nonlinear constraints in minimization problems, in "Optimization" (R.Fletcher, ed.), Academic Press, New York, (1969).

[PT] V.T.Polyak and Tret'yakov, The method of penalty estimates for conditional extremum problems, Z.Vychisl. Mat. i Mat. Fiz., 13 (1973), 34-46.

[Sch] H.Schlighting, *Boundary-Layer Theory*, 7th ed., McGraw-Hill, 1979.

[SS] Y.Saad and M.H.Schultz, GMRES: A generalized minimal residual algorithm for solving nonsymmetric linear systems, SIAM J. Sci. Statist. Comput.,7 (1986), 856-869.

[Ta] H.Tanabe, *Equations of Evolution*, Pitman, San Francisco, 1979.

[Te] R.Temam, *Navier-Stokes Equations: Theory and Numerical Analysis*, North Holland, Amsterdam, 1971.

[Tr] G.M. Troianiello, *Elliptic Differential Equations and Obstacle Problems*, Plenum Press, New York, 1987.

[TSB] H.Tran, J.S.Scroggs and K.J.Bachmann, Modeling of flow dynamics and its impact on the optimal reactor design problem, Proceeding of 1992 AMS-IMS-SIAM Summer Research Conference on Control and Identification of Partial Differential Equations.

[Yo] K.Yosida, *Functional Analysis*, Springer-Verlag, New York.

[YHC] G.W.Young, S.I.Hariharan and R.Carnahan, Flow effects in a vertical CVD reactor, SIAM J. Appl. Math., 52 (1992), 1509-1532.

[Zo] J.P.Zolesio, The material derivative (or spead) method for shape optimization, *Optimzation of Distributed Parameters, Part II*, (eds. E.J.Haung and J.Céa) NATO Advances Study Institute Series, Series E, Sijthoff & Noordhoff (1981),1089-1151.

MATHEMATICAL MODELING AND NUMERICAL SIMULATION IN EXTERNAL FLOW CONTROL

YUH-ROUNG OU*

Abstract. This paper presents an investigation of some active control problems for an external flow field. A series of numerical simulations are performed to investigate an unsteady viscous flow generated by a circular cylinder undergoing a combined rotary and rectilinear motion. By treating the rotation rate as a control variable, we present results of the time histories of forces acting on the cylinder surface and their time-averaged values under several types of rotations. The impact of changing rotation rate on the vortex formation, including the synchronization of cylinder and wake, is demonstrated. Based on the optimal control theory, an optimality system is formulated to determine the optimal rotation rates and the solution orbits. Though only the moving boundary mechanism is discussed, the results presented here add insight to the optimal design of control mechanism and may provide guidance to the formulation of other complex optimal flow control problems.

Key words. external flow, optimal control, rotating cylinder

AMS(MOS) subject classifications. 76D05, 49J20, 93C20

1. Introduction. Flow control has become a critical issue in aerodynamic improvement and design which may provide real-time effect for many important applications, such as highly instantaneous maneuvers for the super-maneuverable aircraft [15], and the optimum design of aerodynamic configurations [16]. It has been demonstrated in a number of experiments that the control mechanisms, such as moving surfaces, blowing, suction, injection of a different gas, etc, may provide useful tools in flow control. Considerable effort has been devoted to the improvement of control mechanisms. However, the principal progress to-date has been essentially accomplished by experimental investigations [11]. Most recently, the areas of both theoretical and computational approaches have received growing attention and become a subject of research focus [12,1,4,13,21,30,10,14,17,22,24,25,31,32,5].

This paper presents a systematic investigation on simulation and control of an external flow by using a moving surface mechanism. In order to keep the problem easy for analysis and simulation, we restrict our study to a simple geometry, i.e. a rotating cylinder. An unsteady flow generated by a circular cylinder undergoing a combined (steady or unsteady) rotary and rectilinear motion was studied. In this model, the rate of cylinder rotation

---

* Interdisciplinary Center for Applied Mathematics, and Aerospace and Ocean Engineering Department, Virginia Polytechnic Institute and State University, Blacksburg, VA 24061-0531. This work was supported by Air Force Office of Scientific Research under AFOSR Grant F-49620-92-J-0078. The author gratefully acknowledges Professors John Burns and S. S. Sritharan for many valuable discussions on various aspects of this project. Thanks are also due to Dr. M. Coutanceau for providing the experimental results.

is treated as a control parameter. Several specific flow control problems were formulated which depend on their corresponding objectives and constraints. The overall goal is to gain insight into the possible form of an optimal controller and demonstrate the feasibility of using time-dependent moving boundary mechanisms in external flow control.

Basically, this paper consists of two parts: numerical simulation and mathematical modeling. In §2, the problems of active control of flow around a circular cylinder are formulated. The governing equations and two types of flow control problems are described. In §3, a velocity/vorticity formulation of the governing equations and a computational algorithm used in this study are briefly described. All numerical results and discussion are presented in §4. The results demonstrate the feasibility of moving boundary mechanism in flow control. A mathematical theory in flow control associated with the problem of a rotating cylinder is formulated in §5. In §6, we outline the future directions in the area of external flow control. Although this investigation is mainly concentrated on the flow control problem of a rotating cylinder, we can extend the numerical algorithm and mathematical analysis into other types of flow geometry and control mechanism. For example, the utility of blowing/suction control mechanism in many investigations may only need little modification in both existing numerical algorithm and mathematical formulation [10,25,32].

2. Problems for a rotating cylinder. The most distinguishing feature of a rotating body traveling through a fluid is that the separation is eliminated on one side while the other side of the cylinder separation is continuously developed. In consequence, this asymmetry of flow development results in a transverse force acting on the cylinder surface in a direction perpendicular to that of flowing stream [33]. The research on the problem of a uniform stream past a cylindrical rotating body has been the subject of many experimental investigations and numerical simulations since the pioneered work of Prandtl [26,27]. See the papers by Taneda [35], Mo [19] and Tokumaru and Dimotakis [36] for a cylinder undergoing rotary oscillations, Taneda [34], Koromilas and Telionis [18], Coutanceau and Ménard [9], Badr and Dennis [3], Badr et al. [2], Chen, Ou and Pearlstein [8], Chang and Chern [6] and Ou and Burns [24] for a cylinder with a constant speed of rotation.

2.1. Governing equations. Let B denote a circular cylinder enclosed by an impermeable boundary Γ, while the two-dimensional exterior domain $D = \mathbf{R}^2 \backslash \{B \cup \Gamma\}$ is the region occupied by an incompressible viscous fluid. In this unbounded quiescent fluid, the circular cylinder is impulsively started with a translational velocity $U(\bar{t})\mathbf{e}_{\bar{x}}$ in the $\bar{x}$-direction normal to its generator and simultaneously a time-dependent angular velocity $\Omega(\bar{t})\mathbf{e}_{\bar{z}}$ about its axis. In an *inertial frame* fixed in space, the problem considered can be mathematically described by the Navier-Stokes equations:

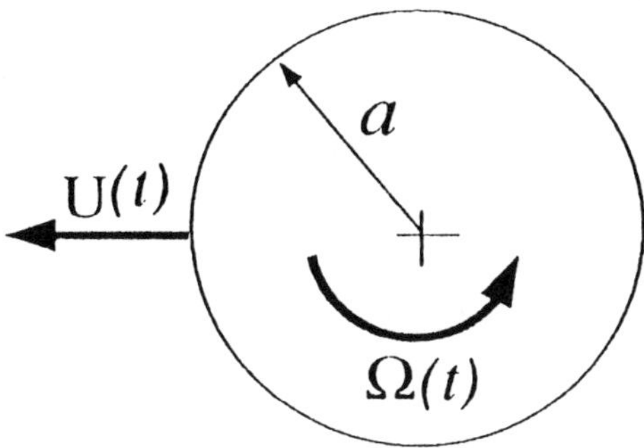

FIG. 2.1. *Schematic of a rotating cylinder in an inertial frame*

$$(2.1) \qquad \bar{\mathbf{u}}_t + (\bar{\mathbf{u}} \cdot \nabla)\bar{\mathbf{u}} = -\nabla p + \nu \nabla^2 \bar{\mathbf{u}} \quad \text{in} \quad D(\bar{t}),$$

$$(2.2) \qquad \nabla \cdot \bar{\mathbf{u}} = 0, \quad \text{in} \quad D(\bar{t}),$$

with the following boundary conditions and initial condition

$$(2.3) \qquad \bar{\mathbf{u}}(\bar{\mathbf{r}}, \bar{t})|_\Gamma = -U(\bar{t})\mathbf{e}_{\bar{x}} + \Omega(\bar{t})(-\bar{y}\mathbf{e}_{\bar{x}} + \bar{x}\mathbf{e}_{\bar{y}}),$$

$$(2.4) \qquad \bar{\mathbf{u}}(\bar{\mathbf{r}}, \bar{t}) = 0, \quad \text{as} \quad |\bar{\mathbf{r}}| \to \infty,$$

$$(2.5) \qquad \bar{\mathbf{u}}(\bar{\mathbf{r}}, 0) = 0, \quad \bar{\mathbf{r}} = (\bar{x}, \bar{y}) \in D(\bar{t}),$$

where $\bar{\mathbf{r}}$, $\bar{\mathbf{u}} = (\bar{u}, \bar{v})$, p and ν are, respectively, the position vector, the velocity field, the pressure field and the coefficient of kinematic viscosity. Also, $\mathbf{e}_{\bar{x}}$, $\mathbf{e}_{\bar{y}}$ and $\mathbf{e}_{\bar{z}}$ are denoted as the unit vector in the direction of $\bar{x}$-, $\bar{y}$- and $\bar{z}$-coordinate, respectively. Notice that in this coordinate frame, the exterior domain $D(\bar{t})$ is a time-varying region as shown in Figure 2.1.

In order that the region occupied by the fluid may be treated as a time-independent domain, it is necessary to recast these governing equations into a *non-inertial* reference frame attached to the body (i.e. the circular cylinder) without rotating of the reference frame. This can be done by introducing a new coordinate system (x, y) such that

$$\begin{cases} x = \bar{x} + \int_0^{\bar{t}} U(\tau)d\tau \\ y = \bar{y} \end{cases}.$$

Thus, the new velocity field $\mathbf{u} = (u, v)$ is given by

$$\begin{cases} u = \bar{u} + U(\bar{t}) \\ v = \bar{v} \end{cases}.$$

In this new non-rotating reference frame, the system of equations (2.1)-(2.5) can be rewritten as

$$(2.6) \qquad \mathbf{u}_t + (\mathbf{u} \cdot \nabla)\mathbf{u} = -\nabla p + \nu \nabla^2 \mathbf{u} + \frac{dU(t)}{dt}\mathbf{e}_x, \quad \text{in} \quad D \times [0, T]$$

$$(2.7) \qquad \nabla \cdot \mathbf{u} = 0, \quad \text{in} \quad D \times [0, T],$$

$$(2.8) \qquad \mathbf{u}(\mathbf{r}, t)|_\Gamma = \Omega(t)(-y\mathbf{e}_x + x\mathbf{e}_y),$$

$$(2.9) \qquad \mathbf{u}(\mathbf{r}, t) \to U(t)\mathbf{e}_x, \quad \text{as} \quad |\mathbf{r}| \to \infty,$$

$$(2.10) \qquad \mathbf{u}(\mathbf{r}, 0) = 0, \quad \mathbf{r} = (x, y) \in D.$$

The translational acceleration $dU(t)/dt$ of the body relative to the inertial frame appears as a fictitious body force in the equation of motion when written in the non-inertial frame. In this new reference frame, the domain occupied by the viscous fluid becomes *time-independent* $D(t) \equiv D$. Moreover, this formulation is equivalent to the problem of a uniform flow past a rotating cylinder. In all control problems considered in this study, the rotation rate $\Omega(t)$ will be varied while the rectilinear speed U is fixed to a constant value. In consequence, the fictitious body force is eliminated in the formulation.

2.2. Optimal control of flow field. From the standpoint of optimal control theory, various optimization problems may be formulated for a rotating cylinder that depend on the desired performances and control constraints. A simple example of optimal control problem is to drive the solution orbit $\mathbf{u}(t; \Omega)$ of system (2.6)-(2.10) to a desired flow field $\mathbf{z}_d$ by controlling the rotation rate $\Omega(t)$ with a minimum effort. Thus, one can define a cost functional as

$$(2.11) \qquad J(\Omega) = \int_0^T \int_D \|\mathbf{u}(t; \Omega) - \mathbf{z}_d\|^2 d\mathbf{r}dt.$$

For example, $\mathbf{z}_d$ is a desired equilibrium state in which no vortex shedding occurs. Then the problem is to find an optimal trajectory of $\Omega(t)$ such that it will drive the solution orbit $\mathbf{u}(t; \Omega)$ as closed as possible (in an appropriate working space) to the desired flow field $\mathbf{z}_d$ in a fixed time-interval with a minimum cost of (2.11).

In fact, the questions of possibility of suppressing vortex shedding by active control of rotation rate have been investigated by Taneda [35], Coutanceau and Ménard [9], Chen et al. [8] and Ou [23]. The studies of active control (or feedback control) of flow/structure interactions are of considerable practical important from the standpoint of wake modification and the reduction of flow-induced vibration [28]. In particular, the issues of suppressing the vortex shedding or tailoring the wake development have many potential applications in marine structure, civil engineering and advanced design of aero/hydro-maneuvering vehicles. However, the questions of whether cylinder rotation can destroy the Kármán vortex street and consequently suppress the vortex shedding have remained to be answered. Up to now, no previous attempt has been made in this area from the standpoint of optimization and control theory.

2.3. Optimal control of force coefficients. Although many experimental and numerical investigations have been conducted on the problems of rotating cylinder, most of previous works were primarily focused on the formation and development of vortices in cylinder wake. It appears that the effect of the rotation rate on the cylinder forces exerted by the fluid has received far less attention despite the fact that it has many important practical engineering applications. In this area, various problems of optimizing force performance can be formulated. For example, we can consider problems of finding an optimal control Ω^*, among a set of restricted control parameters, that will achieve the maximum value of the time-averaged lift functional

$$(2.12) \qquad J_1(\Omega) = \frac{1}{T_f} \int_0^{T_f} C_L(t, \Omega)dt,$$

or the minimum value of the time-averaged drag functional

$$(2.13) \qquad J_2(\Omega) = \frac{1}{T_f} \int_0^{T_f} C_D(t, \Omega)dt.$$

Here, T_f is the final time of motion after the cylinder impulsively started. Similarly, we can also formulate the optimization problems by seeking an optimal control that maximizes the following two important performance functionals

$$(2.14) \qquad J_3(\Omega) = \frac{1}{T_f} \int_0^{T_f} \left[\frac{C_L(t, \Omega)}{C_D(t, \Omega)} \right] dt,$$

and

$$(2.15) \qquad J_4(\Omega) = \frac{\int_0^{T_f} C_L(t, \Omega)dt}{\int_0^{T_f} C_D(t, \Omega)dt}.$$

All above performance functionals may provide us the valuable implication and insight to the optimal design of control mechanism.

In fact, the objective of optimal control of forces around the cylinder surface has a close relation to the objective mentioned in §2.2. From the fundamental theory of fluid mechanics, it is well known that there is no drag force on a circular cylinder which is immersed in a uniform potential flow. Thus, such control problem is to ask whether we can drive an arbitrary flow field to the potential flow (or at least as close as possible to the potential flow field) which no vortex shedding occurs.

3. Direct numerical simulation. In many practical numerical simulations for the laminar motion of a viscous incompressible fluid, both exterior as well as the interior flow domains, the formulation based on the velocity/vorticity variables would provide many advantages over the primitive-variable formulation of (2.6)-(2.10). The velocity/vorticity formulation is especially well suited to treating initial development of flow generated by an impulsively started body, in which the flow field is composed of a relatively small vortical viscous region embedded in a much large inviscid potential flow. In consequence, the computational domain may be restricted to a smaller region where all vorticity contributions are contained. In the numerical simulation part of this study, a velocity/vorticity formulation of governing equations was used in all computations.

3.1. Velocity/vorticity formulation. For a two-dimensional viscous flow, when the velocity field is rotational, the vorticity is defined by

$$(3.1) \qquad \omega \mathbf{e}_z = \nabla \times \mathbf{u}.$$

Here ω is the vorticity field. The vorticity transport equation is obtained by applying the curl operator to equation (2.6). The pressure term is thus eliminated when the continuity equation (2.7) and the definition of vorticity in (3.1) are used. The Cartesian coordinate form of the governing equation for the vorticity field can be expressed in the dimensionless form as

$$(3.2) \qquad \frac{\partial \omega}{\partial t} + \mathbf{u} \cdot \nabla \omega = \frac{2}{Re} \nabla^2 \omega.$$

In addition, the vector Poisson equation

$$(3.3) \qquad \nabla^2 \mathbf{u} = -\nabla \times (\omega \mathbf{e}_z)$$

again obtained from the continuity equation and the definition of vorticity, which can determine a velocity field from a known vorticity field. All the variables are made dimensionless by means of the characteristic quantities. The cylinder radius a is used as the length scale while a/U is used as the time scale. The Reynolds number $Re = 2Ua/\nu$ is based on the cylinder diameter $2a$ and the magnitude U of the rectilinear velocity.

In a non-rotating reference frame the dimensionless boundary conditions for a rotating cylinder can be written as

$$\mathbf{u} = -\alpha(t)y\mathbf{e}_x + \alpha(t)x\mathbf{e}_y, \quad \text{for} \quad (x,y) \in \Gamma$$

and

$$\mathbf{u} = \mathbf{e}_x, \quad \text{for} \quad \sqrt{x^2 + y^2} \to \infty.$$

Here, the ratio of speed of cylinder rotation to speed of translation is denoted as $\alpha(t) = \Omega(t)a/U$. This speed ratio is the primary control parameter throughout this work.

3.2. Computational procedure. The numerical approach is based on an explicit finite-difference/pseudo-spectral technique, and a new implementation of Biot-Savart law is used to produce accurate solutions to the governing equations (3.2)-(3.3) [7,8,23,5]. The vorticity transport equation (3.2) is first discretized by a second order central differences in the radial direction and a pseudospectral transform method in the circumferential direction for all spatial derivatives. This semi-discretization form of vorticity transport equation, consisting of a system of ordinary differential equations in time, can be written as

$$(3.4) \qquad \frac{d\hat{\omega}}{dt} = F(\hat{\omega}), \qquad \hat{\omega} = \left(\omega_{2,2}, \cdots, \omega_{M-1,N-1}\right)^T,$$

for all the interior grid points. Here M, N are denoted as the number of grid points in the circumferential and radial direction, respectively. The calculation procedure to advance the solution for any given time increment can be summarized as follows:

Step 1: Internal vorticity over the fluid region at each interior field point is calculated by solving the discretized vorticity transport equation. An explicit second-order rational Runge-Kutta marching scheme [37] is used to advance in time for (3.4):

$$\hat{\omega}^{n+1} = \hat{\omega}^n + \frac{2\hat{g_1}(\hat{g_1},\hat{g_3}) - \hat{g_3}(\hat{g_1},\hat{g_1})}{(\hat{g_3},\hat{g_3})},$$

and

$$\begin{cases} \hat{g_1} = F(\hat{\omega}^n)\Delta t \\ \hat{g_2} = F(\hat{\omega}^n + 0.5\hat{g_1})\Delta t \\ \hat{g_3} = 2\hat{g_1} - \hat{g_2} \end{cases}$$

where $(\hat{\cdot},\hat{\cdot})$ denotes the scalar product.

Step 2: Using known internal vorticity values at all the interior grid points from step 1, the generalized Biot-Savart law

$$(3.5) \qquad \begin{aligned} \mathbf{u}(\mathbf{r}_0,t) &= -\frac{1}{2\pi} \int\int_D \frac{\omega(\mathbf{r},t)\mathbf{e}_z \times (\mathbf{r}-\mathbf{r}_0)}{|\mathbf{r}-\mathbf{r}_0|^2} dA \\ &\quad -\frac{1}{2\pi} \int\int_B \frac{2\Omega(t)\mathbf{e}_z \times (\mathbf{r}-\mathbf{r}_0)}{|\mathbf{r}-\mathbf{r}_0|^2} dA + U\mathbf{e}_x \end{aligned}$$

is used to update the boundary vorticity values at all the surface nodes. Here $\mathbf{r}_0$ represents all grid points located on the solid boundary. This integral method proposed by Wu and Thompson [41] provides the basic link between velocity and vorticity fields throughout the numerical procedure.

Step 3: At this stage, all the vorticity values in the computational domain are known at the new time level. Then, the velocity at points on the outer perimeter of the computational domain is calculated by (3.5). In equation (3.5) now $\mathbf{r}_0$ denotes the points located on the outer perimeter of the computational domain.

Step 4: The new velocity field can be established by solving the Poisson equation (3.3) with prescribed solid boundary conditions and outer boundary conditions that have been calculated from step 3. The resulting discretized Poisson equation is then solved by a preconditioned biconjugate gradient routine. This step completes the computational loop for each time level.

One further important point to be noted in this integral approach is the determination of the initial flow field. In contrast to the special technique used by other methods, this integral approach enables the numerical code to generate the initial velocity field by executing one cycle of a solution procedure (from step 2 to step 4) rather than employing any additional treatments.

An important consequence of using the velocity/vorticity formulation is that the forces can be directly evaluated from the known vorticity values on the cylinder surface. In a viscous flow, it is well known that the total lift and drag forces are contributed by the pressure and skin friction due to the viscous effects. Hence, for known vorticity values on the cylinder surface, the lift and drag coefficients can be calculated in the r-θ coordinates by

$$(3.6) \qquad C_L(t) = C_{Lp}(t) + C_{Lf}(t) = -\frac{2}{Re} \int_0^{2\pi} \left(\frac{\partial \omega(t)}{\partial r} \right)_\Gamma \cos \theta \, d\theta \\ + \frac{2}{Re} \int_0^{2\pi} \omega(t)_\Gamma \cos \theta \, d\theta,$$

and

$$(3.7) \qquad C_D(t) = C_{Dp}(t) + C_{Df}(t) = \frac{2}{Re} \int_0^{2\pi} \left(\frac{\partial \omega(t)}{\partial r} \right)_\Gamma \sin \theta \, d\theta \\ - \frac{2}{Re} \int_0^{2\pi} \omega(t)_\Gamma \sin \theta \, d\theta,$$

where the subscript Γ denotes quantities evaluated on the cylinder surface. The subscripts p and f represent the contribution from pressure and skin friction, respectively. In particular, we denote the positive values of C_L in the negative y-direction (as noted in Figure 3.1).

4. Numerical results and discussion. In this section we present computational results for an unsteady flow around a rotating cylinder that undergoes a wide variety of steady and unsteady angular/rectilinear speed ratios at a Reynolds number of 200. In this model, the rectilinear speed is fixed to a constant value while the angular velocity is treated as a control variable. Although the choice of time-dependent rotation rates that may be used to control the rotating cylinder are unlimited, the computational results presented here are restricted to the following three types of rotation: 1) constant speed of rotation, $\alpha = $ constant; 2) time-harmonic rotary oscillation, $\alpha(t) = A \sin \pi F t$; 3) time-periodic rotation, $\alpha(t) = A|\sin \pi(F/2)t|$. All variables are normalized to the nondimensional forms in the formulation. For a type of time periodic rotation, $F = 2af/U$ is the reduced forcing

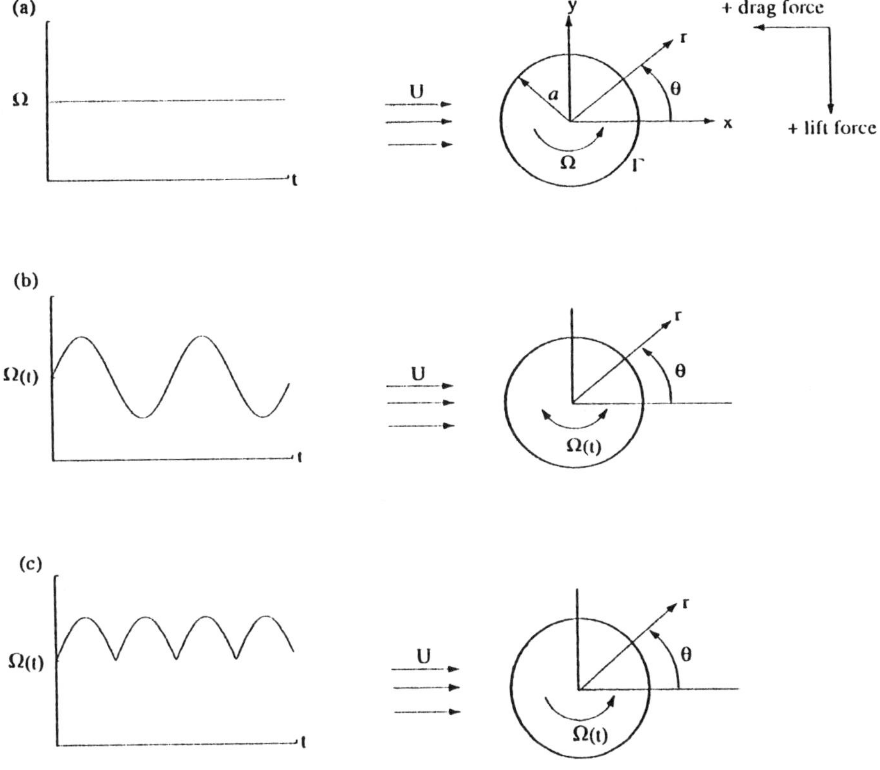

FIG. 3.1.   *Schematic of the rotating cylinder with three types of rotation: (a) $\alpha = \Omega a/U = constant$; (b) $\alpha(t) = A\sin \pi Ft$; (c) $\alpha(t) = A|\sin \pi(F/2)t|$.*

frequency and $A = \pi F\theta$ is the normalized maximum rotation rate of the forcing oscillations. Also, f, θ are denoted as the forcing frequency and the angular amplitude of rotation, respectively. In a non-rotating frame attached to the cylinder, the configurations for the different controls considered in the physical space are sketched in Figure 3.1, together with the corresponding time evolution of the angular velocity.

In the case of time-periodic rotation shown in Figure 3.1(c), the cylinder under control is rotated in the counterclockwise direction about its axis with a time-periodic angular speed. This particular type of rotation is expected to provide a substantial lift enhancement and drag reduction through a proper choice of both the angular amplitude (thus the normalized maximum rotation rate A) and forcing frequency (thus the reduced frequency F). This improvement can be demonstrated by comparing its respective force performances against the time-harmonic rotary oscillation. For a complete discussion of these performance improvements and comparisons, the reader is referred to Burns and Ou [5].

To assess the accuracy of the numerical algorithm, computations were first performed over a wide range of constant speed ratios up to 3.25 at a Reynolds number of 200. Several particular speed ratio parameters were chosen to allow for the comparison against the experimental work of Coutanceau and Ménard [9]. For a constant value of speed ratio $\alpha = 2.07$, Figure 4.1(a) shows an experimental flow visulization picture which is photographed by a camera representing an instantaneous streamline plot at time $t = 9.0$. The calculated result under the same conditions is shown in Figure 4.1(b). In the computation, the non-rotating reference frame is translating with the cylinder while the camera in the experiment is moving with the cylinder as well. Excellent agreement is obtained, despite the fact that a high velocity gradient is induced in the near wake due to the cylinder rotation. In Figures 4.2(a,b), a similar excellent agreement is also demonstrated at a greater speed ratio $\alpha = 3.25$. A detailed discussion of the accuracy of the numerical scheme can be found in [8].

4.1. Force performance: Constant speed of rotation. Figure 4.3 shows plots of the time histories of lift, drag and lift/drag coefficients at various values of speed ratios ($0 < \alpha \leq 3.25$) and for time in the interval $0 \leq t \leq 24$. As seen in Figure 4.3(a), when the speed ratio is increased to 2.07, the lift increases timewise proportionally. However, as the speed ratio further increases, lift appears to initially decrease then increases gradually at later times. Not surprisingly, the maximum value of C_L that can be achieved by rotation is also higher as the speed ratio grows. It is also observed that, at speed ratios lower than 2, the respective lift curves exhibit a well established periodic evolution. However, in the range of $\alpha > 2$, it is not known whether the nature of this periodicity will continue if the time of investigation is expanded. Apparently, as can be seen from Figure 4.3(a), the cylinder rotation (worked as a boundary moving control mechanism) does yield a substantial lift enhancement.

As illustrated by the drag curve in Figure 4.3(b), there is a substantial increase in drag when the speed ratio is increased. In all cases considered here, these drag curves seem to converge after a certain time and then oscillate under different amplitudes and frequencies thereafter. Detailed numerical results on the effect of the speed ratio to the resulting lift/drag curve are shown in Figure 4.3(c). In the range $0 < \alpha < 2.07$, the lift/drag performance appears to improve timewise (for $0 < t \leq 24$) with an increase of α. If a comparison is made between $\alpha = 2.07$ and $\alpha = 0.05$, a noticeable improvement of the lift/drag performance is observed. Although a higher lift/drag ratio is achieved by increasing the rotation rate in this range, the question arises whether any further increase of α will result in a continued improvement of the lift/drag ratio. Intuitively, it is natural to expect a monotonical increase in the lift/drag ratio as α increases to $\alpha = 3.25$. However, this is not the case as a comparison is made between

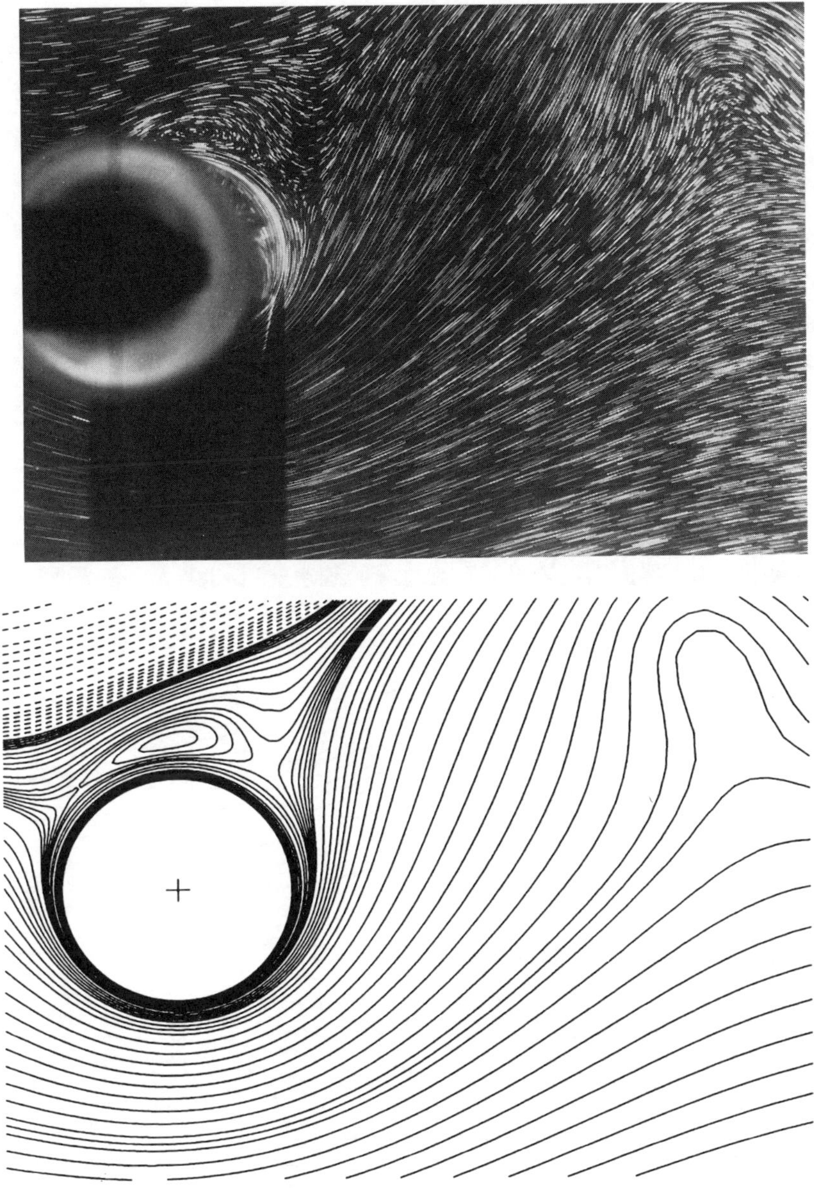

FIG. 4.1. *Instantaneous streamline plots for* $Re = 200$, $\alpha = 2.07$ *at* $t = 9.0$. *(a) flow-visulization picture, (b) computed result.*

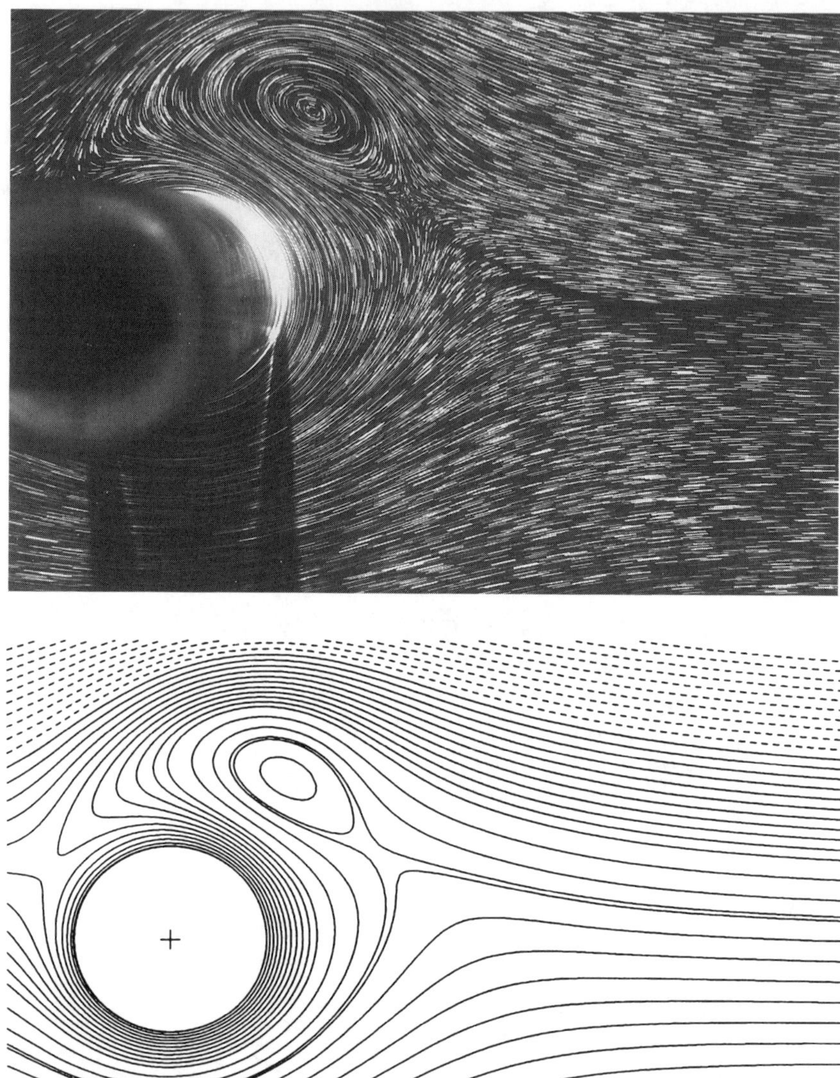

FIG. 4.2. *Instantaneous streamline plots for* $Re = 200$, $\alpha = 3.25$ *at* $t = 5.0$. (a) *flow-visulization picture,* (b) *computed result.*

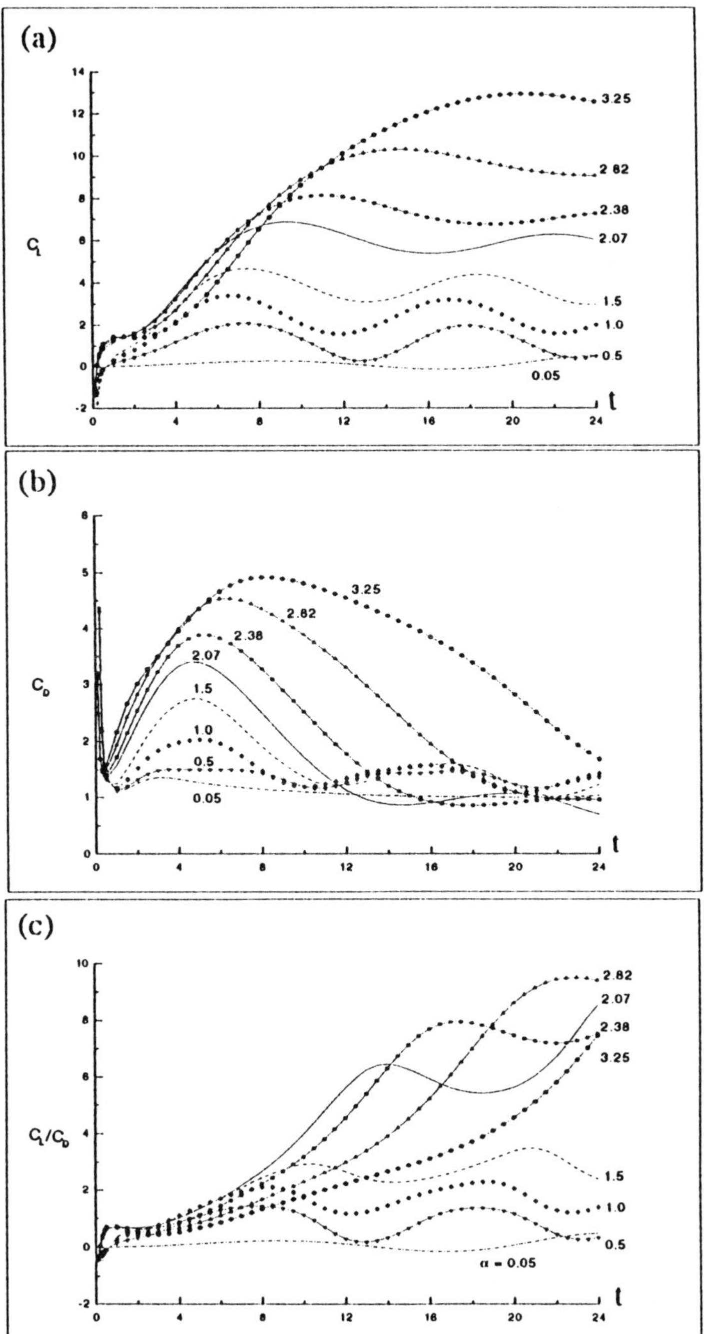

FIG. 4.3. *Temporal evolutions of the lift (a), drag (b) and lift/drag (c) coefficients at* $Re = 200$ *with various constant speed ratios* $(0.05 \leq \alpha \leq 3.25)$.

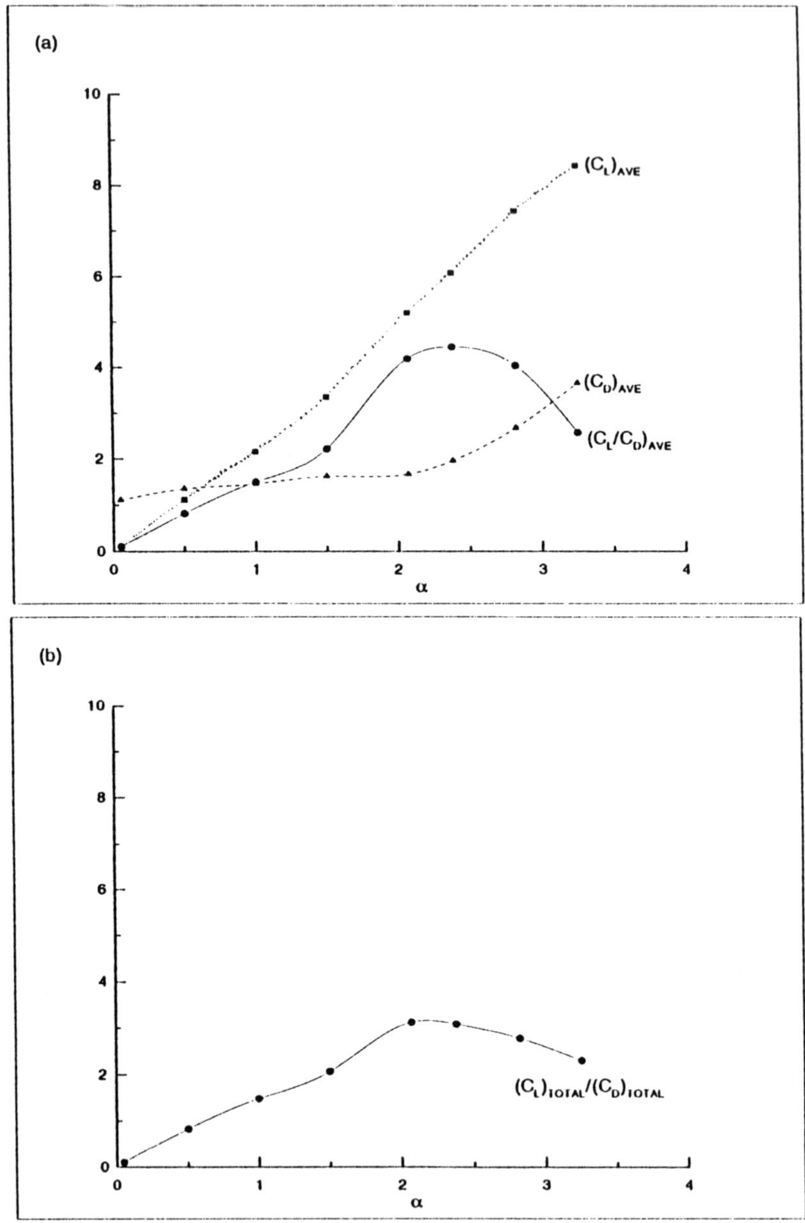

FIG. 4.4. *Effect of the speed ratio on time-averaged lift, drag and lift/drag coefficients (a) and on (total lift)/(total drag) force ratio (b) for $0 < \alpha \leq 3.25$.*

$\alpha = 3.25$ and $\alpha = 2.07$. In fact, the lift/drag curves illustrate a gradually decrease in performance over certain time interval when the speed ratio increases beyond 2. Moreover, this tendency toward lower lift/drag ratio becomes noticeable when α reaches the highest value ($\alpha = 3.25$) considered here. Nevertheless, for all α considered here, a significant increase in the maximum value of C_L/C_D can be obtained by increasing α. However, it is found that it will reach its maximum value at a much later time for higher values of α.

From the results of force improvement observed in Figure 4.3, it is interesting to examine these performance functionals described in (2.12)-(2.15) as the speed ratio is altered. Figure 4.4 illustrates the use of direct computation to calculate $J_i, i = 1, \cdots, 4$ under various constant speeds of rotation. These curves shown in Figure 4.4(a) represent the time-averaged lift, drag and lift/drag coefficients with respect to the speed ratio in the range $0 < \alpha \leq 3.25$ and for time in the interval $0 < t \leq 24$. It illustrates that the time-averaged lift J_1 is almost linearly proportional to the speed ratio, while the time-averaged drag J_2 remains as a constant value up to $\alpha = 2$, then monotonically increases with speed ratio thereafter. As shown in the figure, the optimal speed ratios corresponding to the maximum value of J_1 and the minimum value of J_2 are $\alpha_1^* = 3.25$ and $\alpha_2^* \approx 0$, respectively. Most importantly, the resulting time-averaged lift/drag is *not* linearly proportional to the speed ratio. As shown in the figure, the highest value of the speed ratio $\alpha = 3.25$ considered here is not the optimal constant rotation rate corresponding to the maximum value of J_3. The maximum value of J_3 occurs at a lower speed ratio, approximately $\alpha_3^* = 2.38$, and it represents a substantial increase of 440% over the lower speed ratio $\alpha = 0.5$.

In Figure 4.4(b), the variation of the (total lift)/(total drag) force ratio (i.e. J_4 in (2.15)) with respect to the speed ratio is shown for α in the range $0 < \alpha \leq 3.25$. Although the maximum value of J_4 is achieved at a value α_4^* between $\alpha = 2.0$ and $\alpha = 2.38$, it should be noted that this optimal speed ratio α_4^* is not necessarily the same optimal value α_3^* as shown in Figure 4.4(a). The results presented in Figures 4.4(a,b) demonstrate an effective way of improving performance by changing the rotation rate and illustrate the important of selecting a proper rotation rate in order to optimize the force performance.

4.2. Force performance: Time periodic inputs. The previous results only applied to constant rotation rates. In this section we consider time-varying rotations. Because the goal of this report is to demonstrate the feasibility of using a time-dependent moving surface mechanism for optimizing force performance, we shall restrict our simulations to two periodic inputs. That is, the time-harmonic rotary oscillation $\alpha(t) = \sin \pi F t$ and the time-periodic rotation $\alpha(t) = |\sin \pi (F/2)t|$.

It is well known that when a cylinder oscillates in a uniform flow, the associated forcing oscillating frequency and amplitude can influence the

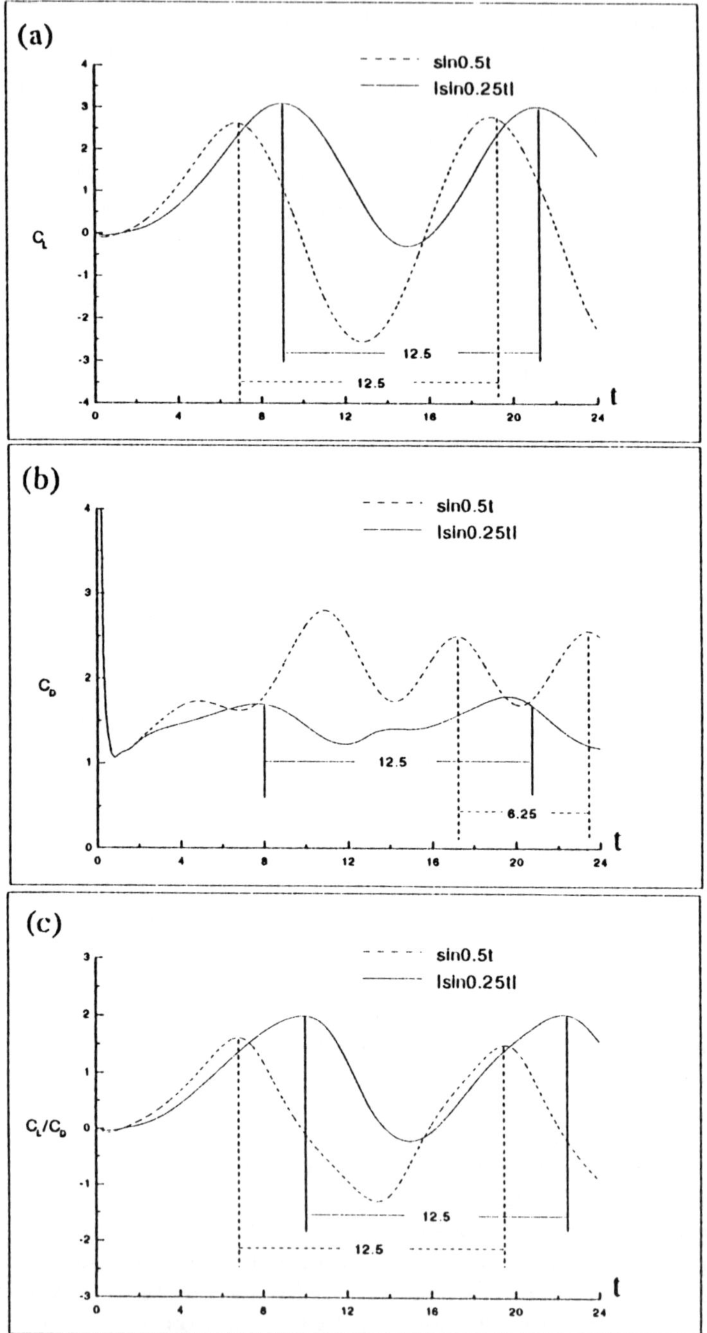

FIG. 4.5.　*Comparisons of temporal evolution of the lift (a), drag (b) and lift/drag (c) coefficients for a time-periodic rotation* $\alpha(t) = |\sin 0.25t|$ $(T = 12.5)$ *with a time-harmonic rotary oscillation* $\alpha(t) = \sin 0.5t$ $(T = 12.5)$ *at* $Re = 200$ *for* $0 \leq t \leq 24$.

vortex formulation and forces response substantially [38,40]. It has been experimentally shown that at $Re = 200$, the natural Strouhal frequency of a non-rotating circular cylinder ($\alpha = 0$) is approximately $F_n = 0.185$ [39]. It is of important to study the behavior of fluctuating forces at imposed forcing frequencies which lie in the neighborhood of the natural frequency. The temporal evolutions of lift, drag and lift/drag are shown separately in Figures 4.5(a,b,c) for a time-periodic rotation $\alpha(t) = |\sin 0.25t|$ and a time-harmonic rotary oscillation $\alpha(t) = \sin 0.5t$, respectively. Notice that these two types of rotation are employed by the same forcing frequency (i.e. $F = 0.16$) which lies in the neighborhood of the natural frequency. The numerical results clearly confirm the expected benefit of this time-periodic rotation for both lift and drag forces, as shown in Figure 4.5(a,b).

In comparing these two types of rotation, it should be noted that rotating in the same direction causes the lift curve to be shifted upwards due to the nature of rotation, while the drag curve is shifted downwards. In terms of performance, this corresponds to an increase of the time-averaged lift force in the time-span of the investigation, while in the same time interval, a substantial reduction of the time-averaged drag as well. The resulting improvement of the lift/drag ratio is shown in Figure 4.5(c). There is an interest in addressing the relationship between the force improvement and the vortex development around the cylinder surface. Although not shown here, one particular interesting feature is the phase difference between the maximum value of lift and the vortex sheet cutting time [23]. A thorough investigation regarding such issue may gain some insight into the possible form of an optimal controller.

To demonstrate the influence of time-varying rotation on the temporal development of these force coefficients, several additional values of forcing frequency were performed. Figures 4.6(a,b,c) show the comparisons of the time-averaged values of lift, drag and lift/drag coefficients (i.e. J_1, J_2 and J_3) between these two time periodic inputs with variation of the reduced forcing frequency in a range of $0.08 \leq F \leq 0.32$. These forces were averaged with respect to the time interval $0 \leq t \leq 24$.

In the case of time-harmonic rotary oscillation, the local maximum value of time-averaged lift, drag and lift/drag ratios correspond to the forcing frequency which lies in the neighborhood of the natural frequency, as shown in the Figure 4.6. This particular feature was also observed in the numerical results of Mo [19] where it was shown that the drag peak occurs at the forcing frequency equal to the natural frequency.

As for the cases of time-periodic rotation, it illustrates that a variation of forcing frequency in this range (i.e. $0.08 \leq F \leq 0.32$) has litter effect on the time-averaged forces. Although the differences in time-averaged forces are minor, the forcing frequency which lies in the neighborhood of the natural frequency ($F = 0.185$) corresponds to a larger time-averaged drag and a smaller time-averaged lift. In terms of performance, Figure 4.6 presents a clear improvement for the time-periodic rotation ($\alpha(t) = |\sin \pi(F/2)t|$)

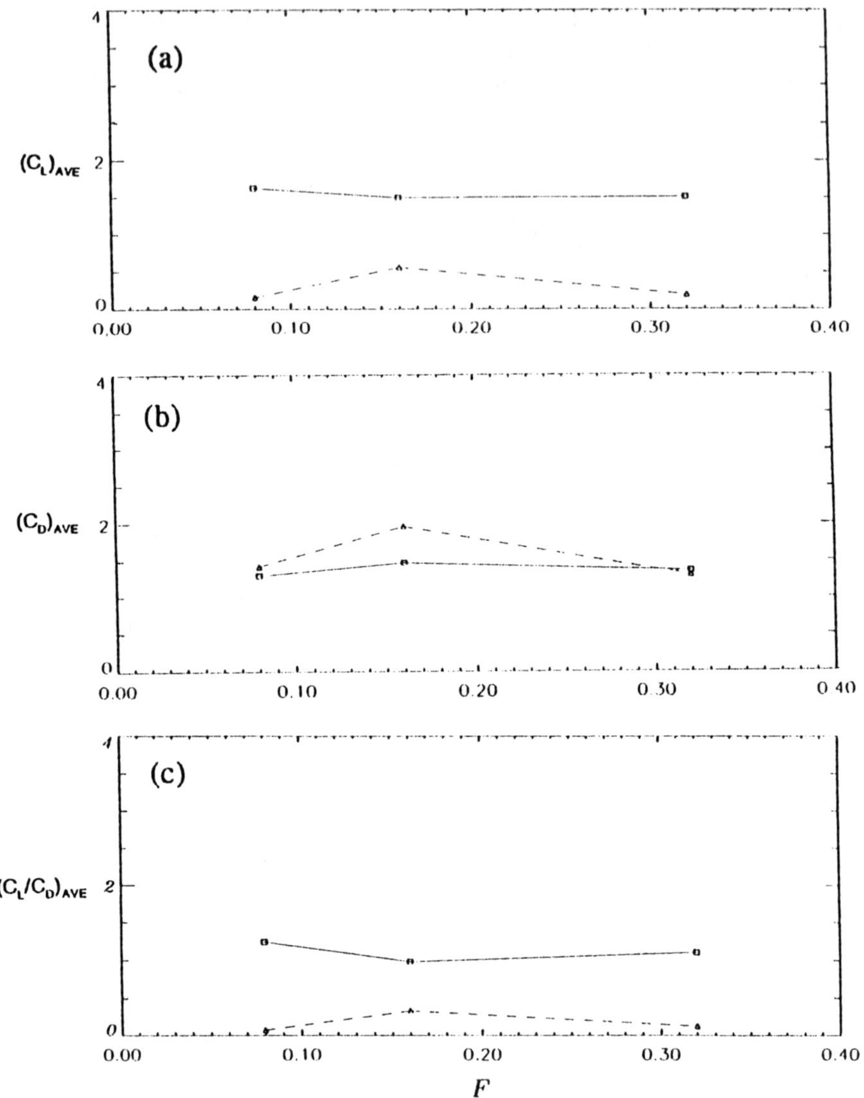

FIG. 4.6. *Time-averaged lift (a), drag (b) and lift/drag (c) coefficients for* $Re = 200$ *and* $0.08 \leq F \leq 0.32$. - - - - : $\alpha(t) = \sin \pi F t$; ———: $\alpha(t) = |\sin \pi (F/2)t|$.

when compared to the time-harmonic rotary oscillation ($\alpha(t) = \sin \pi F t$) at every tested forcing frequency. It appears that the forcing frequency which lies in the neighborhood of the natural frequency exhibits a smaller lift increase and larger drag reduction when compared to these frequencies lie outside the neighborhood of the natural frequency.

At this stage, we have demonstrated that the particular type of time-periodic rotation exhibits clear improvement of force performance. This motivated us to examine the effect of angular amplitude on the force development while the forcing frequency is fixed to a constant value. The parameter A now becomes the control variable in the optimal control calculations. Figure 4.7 shows that resulting forces on the cylinder can differ significantly at different angular amplitudes for $\alpha(t) = A|\sin 0.314t|$. This type of rotation corresponds to a forcing Strouhal number of 0.2 which is in the neighborhood of the natural Strouhal number of 0.185. The angular amplitudes considered here are $A = 1.0, 2.07$ and 3.25. Apparently, as can be seen from these figures, a larger angular amplitude definitely yields an incremental lift coefficient over the time-span of investigation ($0 \leq t \leq 36$). However, initially the drag increases with an increase of A, then after a certain time it oscillates with almost the same amplitude and frequency around an averaged value. Consequently, this leads to a substantial improvement in lift/drag with increasing A, as clearly shown in Figure 4.7(c).

The effect of angular amplitude on the time-averaged values of lift, drag and lift/drag coefficients is shown in Figure 4.8 for $\alpha(t) = A|\sin 0.314t|$ averaged over $0 \leq t \leq 36$. In a range of $1 \leq A \leq 3.25$, it illustrates that all the time-averaged values are almost linearly proportional to the angular amplitude. Significant increment in lift coefficients with increasing angular amplitude is particularly noticeable. This can be demonstrated by the comparison of $A = 3.25$ with $A = 1$. It represents a 240% increment of lift performance. However, a slight increment in drag coefficients with increasing angular amplitude is observed. A moderate improvement of time-averaged lift/drag ratio is also seen. Moreover, the effect of angular amplitude on these time-averaged forces is very noticeable when compared to the effect of the forcing frequency shown in Figure 4.6.

As noted in equations (3.6) and (3.7), the total lift and drag forces are contributed by the pressure and skin friction due to the viscous effect. In Figures 4.9(a,b), the pressure and skin friction contributions to the lift and drag are shown separated for $\alpha(t) = |\sin 0.283t|$ over the time-span of investigation ($0 \leq t \leq 36$). It appears that the pressure has larger contribution to the lift and drag at this particular Reynolds number (i.e. $Re = 200$). As a matter of fact, similar contributions are also observed for all frequencies and amplitudes considered in this study.

4.3. Synchronization of cylinder and wake. The synchronization of cylinder and wake has long been known to be an important component

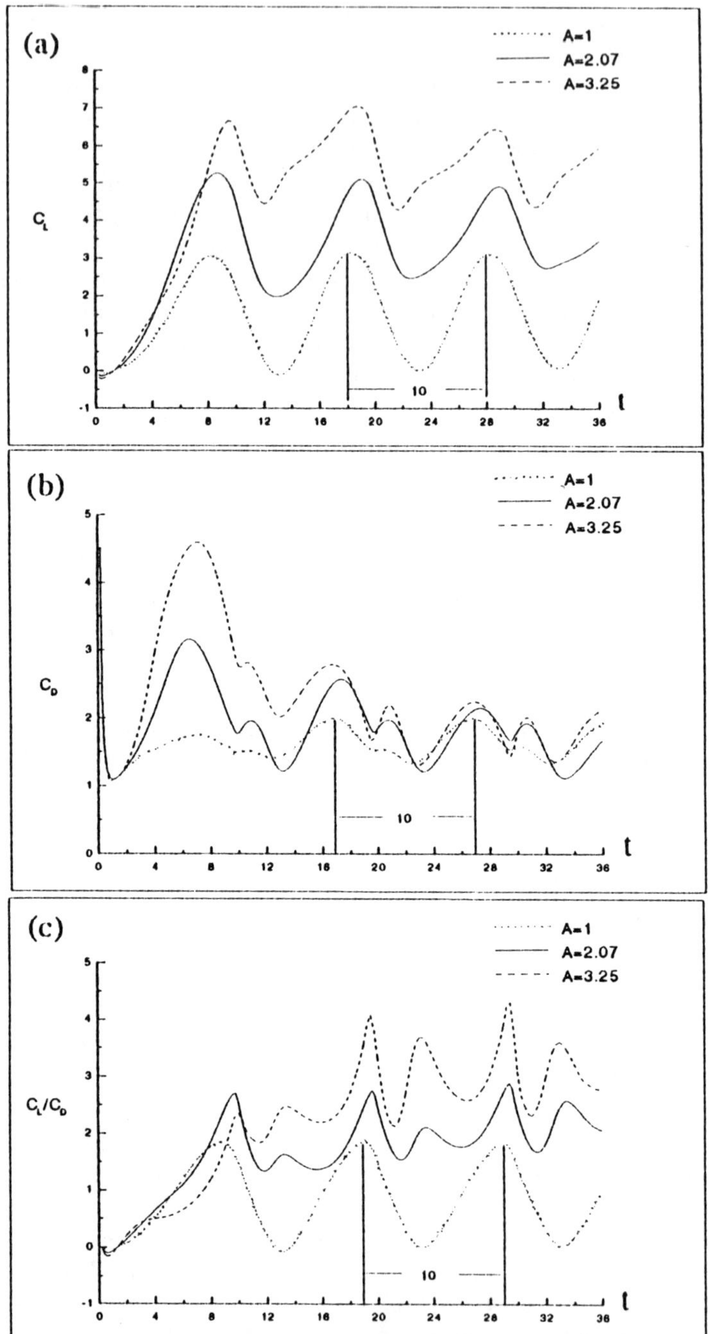

FIG. 4.7. *Temporal evolutions of the lift (a), drag (b) and lift/drag (c) coefficients for a time-periodic rotation* $\alpha(t) = A|\sin 0.314t|$ *at* $Re = 200$ *with amplitudes of* $A = 1.0, 2.07$ *and 3.25 for* $0 \leq t \leq 36$.

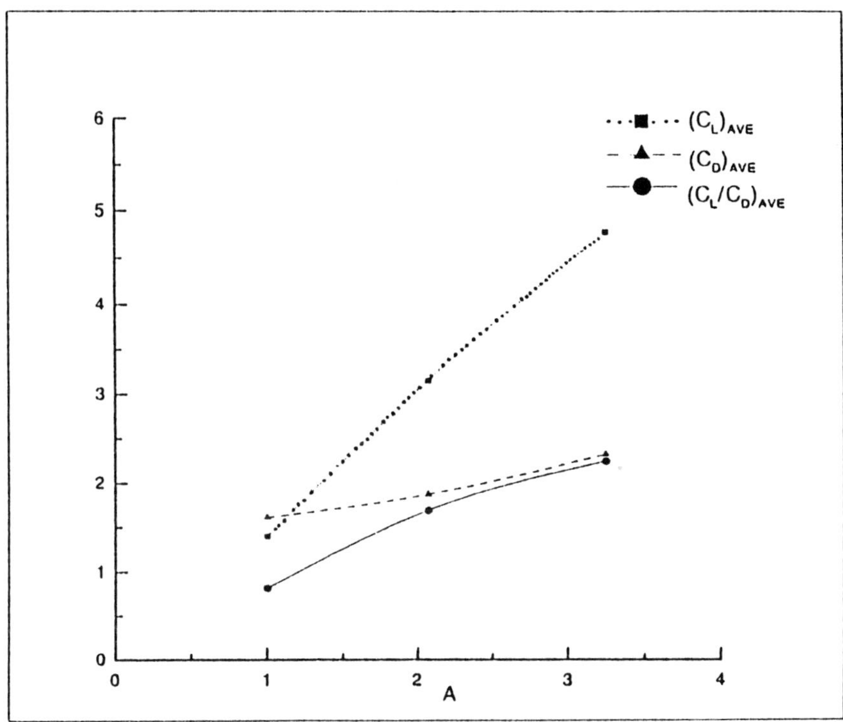

FIG. 4.8. *Variation of time-averaged force coefficients with respect to the angular amplitude for* $\alpha(t) = A|\sin 0.314t|$ *and* $1 \leq A \leq 3.25$.

of vortex-induced oscillations [28]. A detailed study of various types synchronization for a body oscillating transversely in a uniform stream can be found in Williamson and Roshko [40]. In the case of time-harmonic rotary oscillations, the effects of the forcing frequency and amplitude on a cylinder wake have been investigated experimentally by Tokumaru and Dimotakis [36]. Several vortex formations were observed in the wake. Their experiments dealt with a range of amplitudes and frequencies at a Reynolds number of $Re = 1.5 \times 10^4$. By fixing the reduced amplitude A in their experiments, four qualitatively different vortex shedding modes were identified when the forcing frequency was increased. For the case of time-periodic rotations considered here, it is natural to ask whether such synchronization can occur and how well the numerical results can predict the occurrence of this important phenomenon.

An examination of the responses in Figure 4.5, shows that the combined system of cylinder and wake will be "locked in" by an imposed forcing frequency. This synchronization of the cylinder and wake is due to the fact

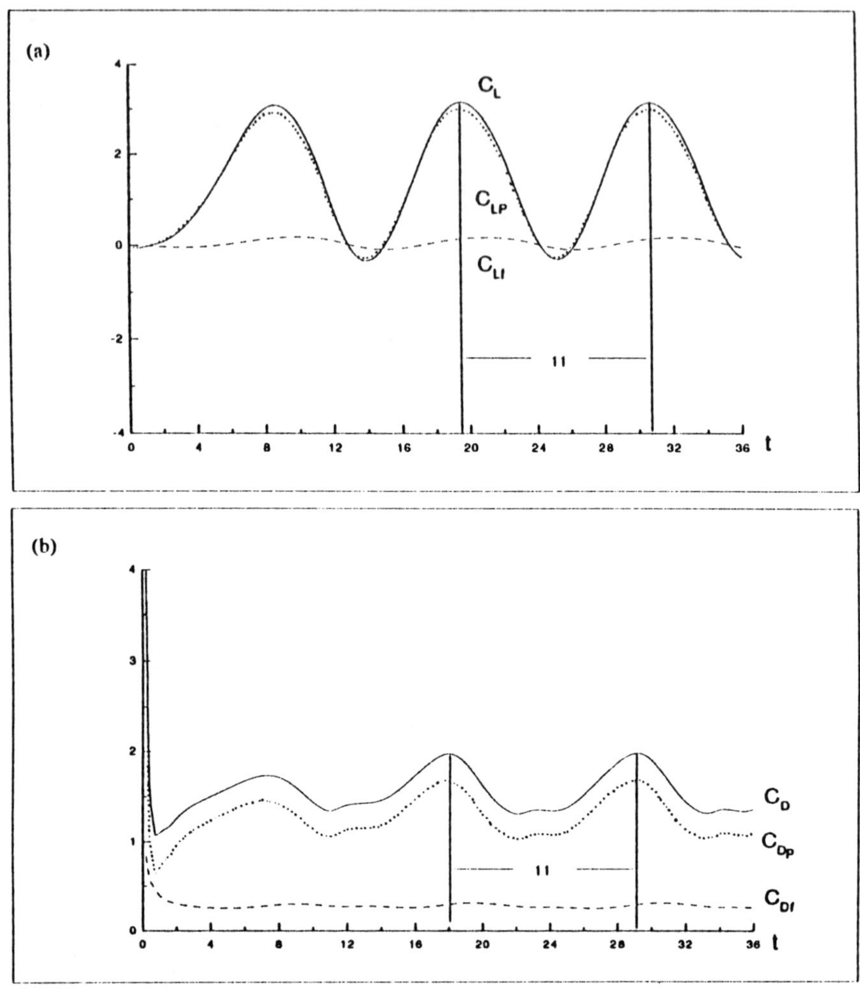

FIG. 4.9. *Contributions of pressure and skin friction to lift (a) and drag (b) for* $Re = 200$ *and* $\alpha(t) = |\sin 0.283t|$ *($F = 0.18$) for* $0 \leq t \leq 36$. - - - - : *skin friction;* $\cdots\cdots$: *pressure;* ———: *total.*

that the forcing frequency of rotation ($F = 0.16$) lies in the neighborhood of the natural frequency ($F_n = 0.185$). Notice that in the case of time-periodic rotation shown in Figure 4.5, both lift and drag curves oscillate with the forcing frequency (corresponding to a time period of $T = 12.5$), clearly exhibiting a periodic response. However, in the case of time-harmonic rotary oscillation, the lift curve oscillates with the same forcing frequency ($T = 12.5$) while drag curve oscillates with the period of $T/2$. Consequently, the lift/drag ratios oscillate at the same frequency ($T = 12.5$) for both types of rotation.

For the case of time-periodic rotation $\alpha(t) = A|\sin 0.314t|$ presented in Figure 4.7, we extended our observation to a relatively longer time. For $0 < t \leq 36$, an examination of these force curves for $A = 1.0$ exhibits a periodic response with a frequency ($F = 0.2$) precisely equal to the input forcing frequency (i.e. $T = 10$). Although this periodic behavior is not established for $A = 2.07$ and 3.25, the corresponding curves are almost periodic in time. In order to confirm this periodicity, a sequence of instantaneous streamlines plots are shown in Figure 4.10. In Figure 4.10, each plot is separated with an interval of one time period. These streamlines are plotted in a frame fixed with the undisturbed fluid. The periodicity of the flow is clearly noticeable. Two opposite-sign vortices are shed alternately on opposite sides of the cylinder at each cycle of rotation. The vortex formation in the wake is similar to the case of a non-rotating cylinder ($\alpha = 0$). However, the midline of the vortex street has been displaced slightly upwards due to the nature of rotation (in the counterclockwise direction). These results indicate that rotation may provide an effective control of the cylinder wake.

4.4. Controlling of vortex shedding. In the case of constant rotation, the most complete investigation by experiment regarding the issue of vortex shedding was accomplished by Coutanceau and Ménard [9]. In their experiments, it was concluded that a Kármám vortex street disappears as speed ratio increases to a limiting value $\alpha_L \approx 2$. Their experiments indicate that there is no formation of any eddy after the first eddy created. One particular interesting feature is the difference between the experimental work and our calculated observation regarding the conclusion of suppressing of vortex shedding at high speed ratios.

Figures 4.11(a,b,c,d) show the calculated equi-vorticity contours for various constant speed ratios at $t = 24$. These calculated plots indicate that vortex shedding continues to occur even at high rotation rates ($\alpha \geq 2.07$). However, at these high α, the observed formation of the vortex street behind a rotating cylinder seems to contradict the experimental conclusion described in [9]. This difference is due to the fact that the experimental apparatus was such that only 10 dimensionless time units of data could be collected and in part by the flow visualization techniques used in their experiments. For a detailed discussion of this contradictory result, the reader is referred to [8].

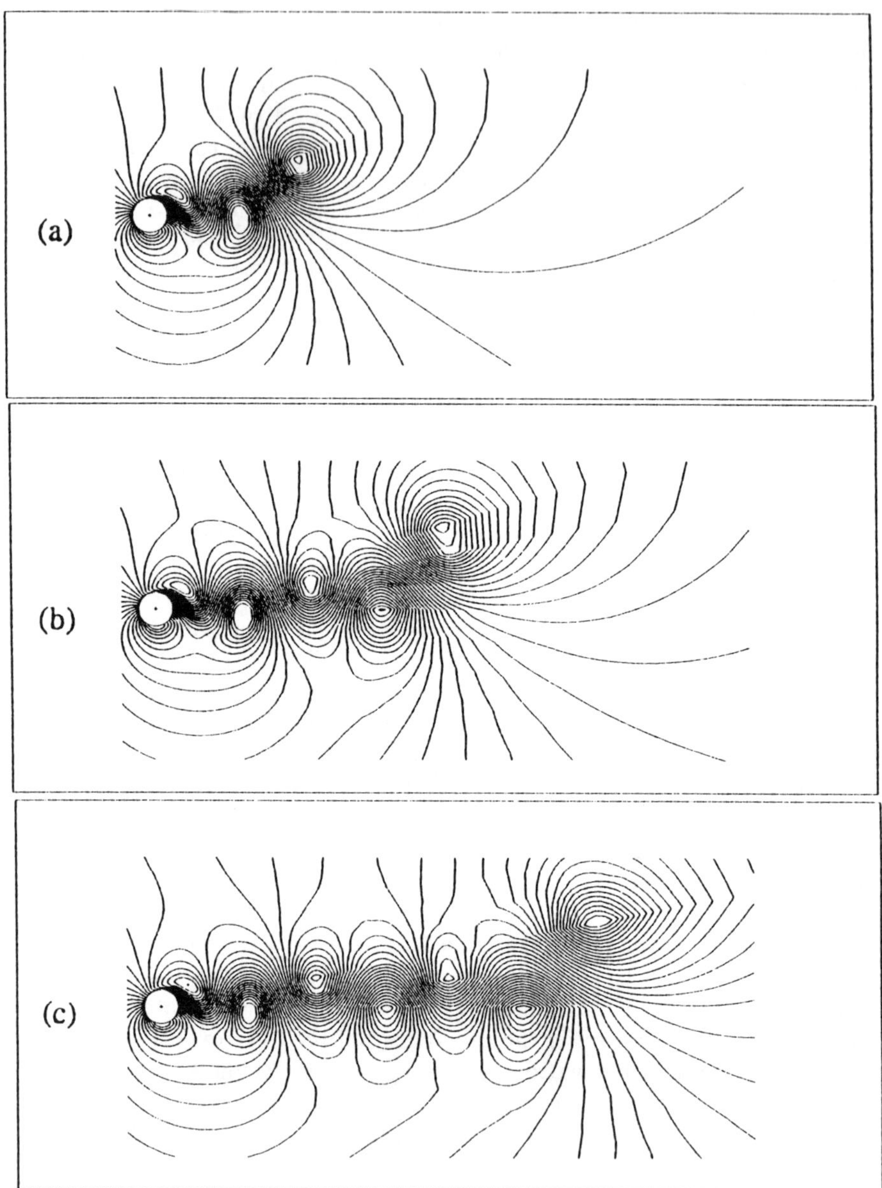

FIG. 4.10. *Instantaneous streamlines plots for $Re = 200$, $\alpha(t) = |\sin 0.314t|$ $(F = 0.2)$, viewed from a frame fixed with the undisturbed fluid.* (a) $t = 16$, (b) $t = 26$, (c) $t = 36$.

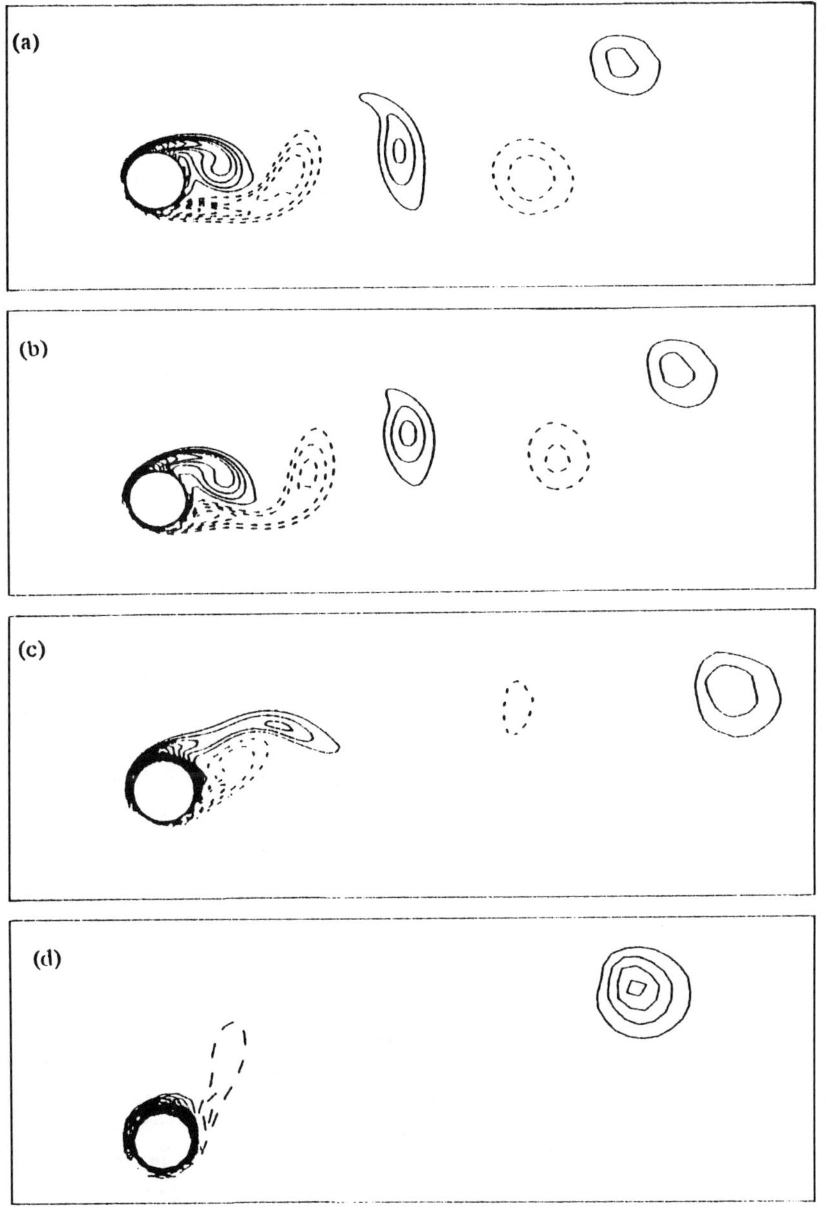

FIG. 4.11. *Vortex shedding patterns for various constant speed ratios at* $Re = 200$ *and* $t = 24.0$. *(a)* $\alpha = 0.5$, *(b)* $\alpha = 1.0$, *(c)* $\alpha = 2.07$, *(d)* $\alpha = 3.25$.

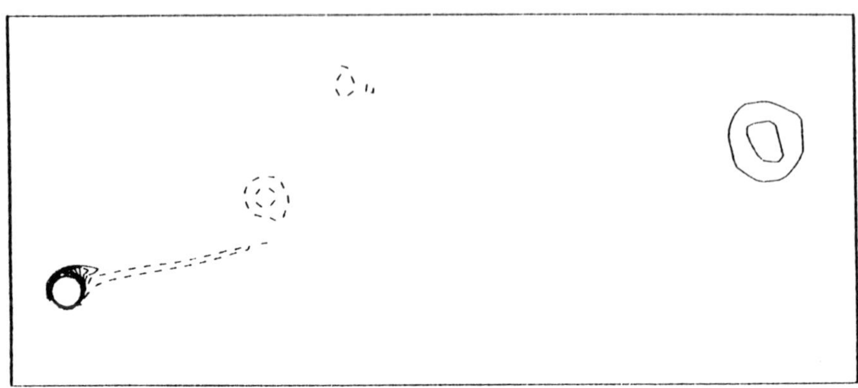

FIG. 4.12. *Equi-vorticity contours for Re* = 200 *and* α = 3.25 *at* t = 54.0.

To better elucidate the continuation of vortex formation and consequently its evolution in the wake, we extend the computation to a relative large time-span for a high speed ratio $\alpha = 3.25$. Figure 4.12 shows the computed equi-vorticity contour at $t = 54$. It illustrates that votices are continuously created and shed to the downstream. However, the vortex shedding process and flow pattern are qualitatively different from that of lower speed ratios. In order to confirm the continuous existence of vortex shedding even at higher speed ratios, Figure 4.13 shows the core trajectory of the first vortex up to $t = 24$ for various α. It appears that the vortex core moves further away the centerline ($y = 0$) of cylinder motion when the speed ratio is increased.

It is important to note a recent investigation by Badr et al. [2] regarding the issues of vortex formation and shedding. Their observations were performed both experimentally and numerically at Reynolds numbers of $Re = 10^3$ and $Re = 10^4$. For a rotation rate at $\alpha = 3$ and $Re = 10^3$, they show that no other eddy is created after the shedding of two vortices. In addition, the temporal evolutions of the lift and drag coefficients imply that a steady state is indeed approached. However, for a fixed Reynolds number the issue of whether there exists a limiting value of speed ratio α_L beyond which vortex shedding completely disappears in the wake remains to be determined. If such a critical value does indeed exist, then it is of interest to know its dependence on the Reynolds number.

Although the suppression of vortex shedding may be achieved at certain Reynolds number under a constant high speed ratio, this does not

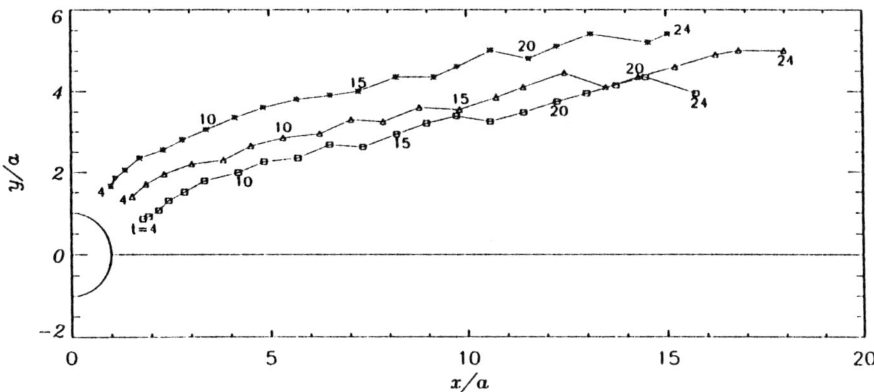

FIG. 4.13. *Trajectory of the core of the first vortex up to time $t = 24.0$. $\square$: $\alpha = 1.0$; $\triangle$: $\alpha = 2.07$; $\star$: $\alpha = 3.25$.*

immediately imply that the constant rotation is the most effective way to suppress the vortex shedding among all possible forms of rotation. In fact, the effect of a time-harmonic rotary oscillation on the vortex shedding process had been studied experimentally by Taneda [35]. At several values of Reynolds number, his experiments demonstrates that vortex shedding can be eliminated under a sufficiently large value of forcing amplitude and frequency. As motivated by his experimental observations, we have tested a similar case of high amplitude and frequency by using our computational algorithm. As shown in Figure 4.14, the time development of equi-vorticity contours indicates that there is no vortex shedding in the wake at least in the time-span of investigation. There are only two attached eddies created on both side of the cylinder surface. Moreover, these eddies grow and elongate toward a tongue shape around the cylinder as time evolves. Nevertheless, the disappearance of these vortices at large time has not yet been determined due to the computational time limitation.

Figures 4.15(a,b,c) show the calculated equi-vorticity contours for time-harmonic rotary oscillations under three values of forcing frequency at time $t = 24$. For the forcing frequency $F = 0.16$ (i.e. $\alpha(t) = \sin 0.5t$) which lies in the neighborhood of the Natural frequency as shown in Figure 4.15(b), the process of vortex formation and shedding is qualitatively similar to the non-rotating case (i.e. $\alpha = 0$). However, when a forcing frequency moves away the natural frequency, the vortex shedding patterns are changed significantly as illustrated in Figure 4.15(a) and (c). The calculated equi-vorticity contours for the time periodic rotation $\alpha(t) = |\sin \pi(F/2)t|$ are

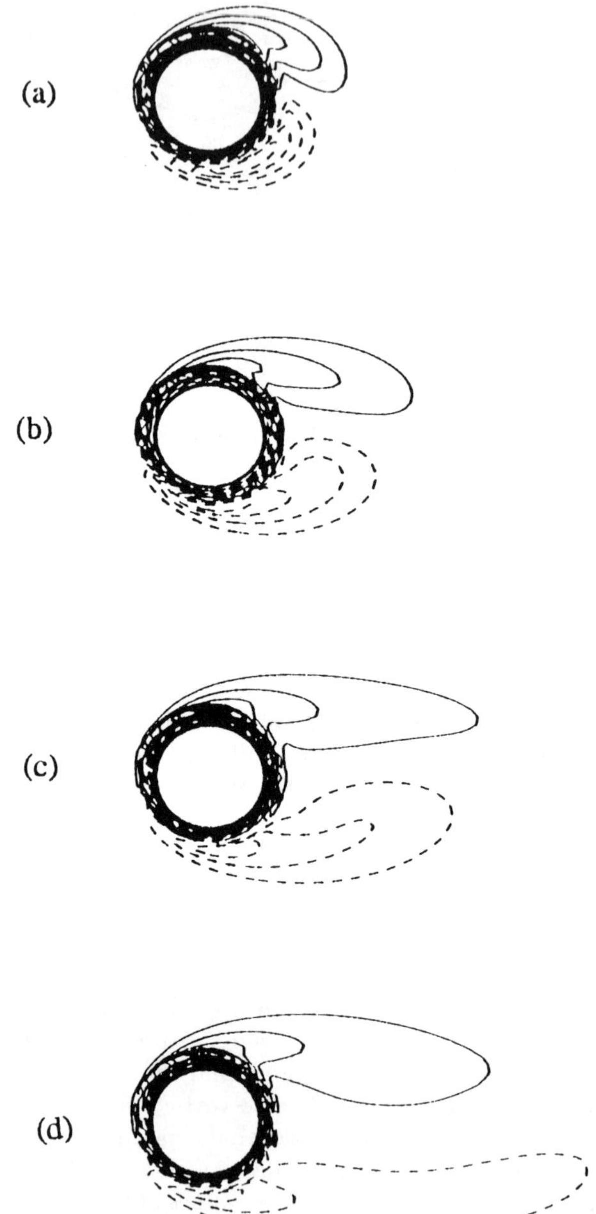

FIG. 4.14. *Equi-vorticity contours for a time-harmonic rotary oscillation* $\alpha(t) = A\sin\pi Ft$ *with* $A = 15.0$, $F = 6.0$ *and* $Re = 35$. *(a)* $t = 3.0$, *(b)* $t = 6.6$, *(c)* $t = 12.0$, *(d)* $t = 23.0$.

shown in Figure 4.16(a,b,c). Although the vortex shedding process is similar to the case of $\alpha = 0$, the midline of vortex street has been displaced upward away the centerline ($y = 0$). This demonstrates that the type of rotation can influence the formation of vortex street.

Now, the next question is to ask whether a time periodic input or any constant rotations will produce the most effective way to suppress the vortex shedding. If we treat the rotation rate as a control variable, it is of interest to find an optimal control such that vortex shedding will be suppressed with a minimum effort. This leads us to consider the following more challenging and practical control problem. That is, to find the optimal trajectory of the rotation rate that will drive the solution to a desired flow field over a fixed time interval mentioned in §2.2. Notice that there are many control mechanisms other than moving surface can be applied for controlling flow field around a circular cylinder. For example, the blowing/suction on the cylinder surface may produce a similar result [25,32].

5. Mathematical theory . A precise understanding of time-varying moving surfaces in boundary layer control may provide an effective way for lift enhancement and drag reduction. By treating the rotation rate as a control variable, we will eventually be interested in finding the optimal control (i.e. a time history of the rotation rate) based on optimal control theory. Although here the optimal control problem associated with the constant rotation rate was solved by direct computations, it is still important to explore the possible implementation of a computational algorithm to calculate the optimal solution for the more general problems. In order to construct a systematic computational algorithm for practical designs, a mathematical approach is proposed which is based on the mathematical works described in [31,32]. The detail of mathematical analysis of generalized solutions for the Navier-Stokes equations associated with external flows can be found in [29].

The following discussion is specifically formulated for control and optimization problems of a rotating cylinder. However, for other types flow control problem encountered in incompressible viscous flows, such adjoint method may be also used.

5.1. Existence theorem of optimal controls. In this section, we will establish an existence theorem of optimal controls. Firstly, the system of equations (2.6)-(2.10) is recast into an evolutionary equation with *homogeneous* boundary conditions. Namely, one need to construct two solenoidal vector fields $\boldsymbol{\Psi}(\mathbf{r})$ and $\boldsymbol{\Phi}(\mathbf{r})$ such that

$$\boldsymbol{\Psi}(\mathbf{r}) = \nabla \times \left[-\frac{\Theta_\epsilon(\mathbf{r})}{2}(x^2 + y^2)\mathbf{e}_z \right],$$

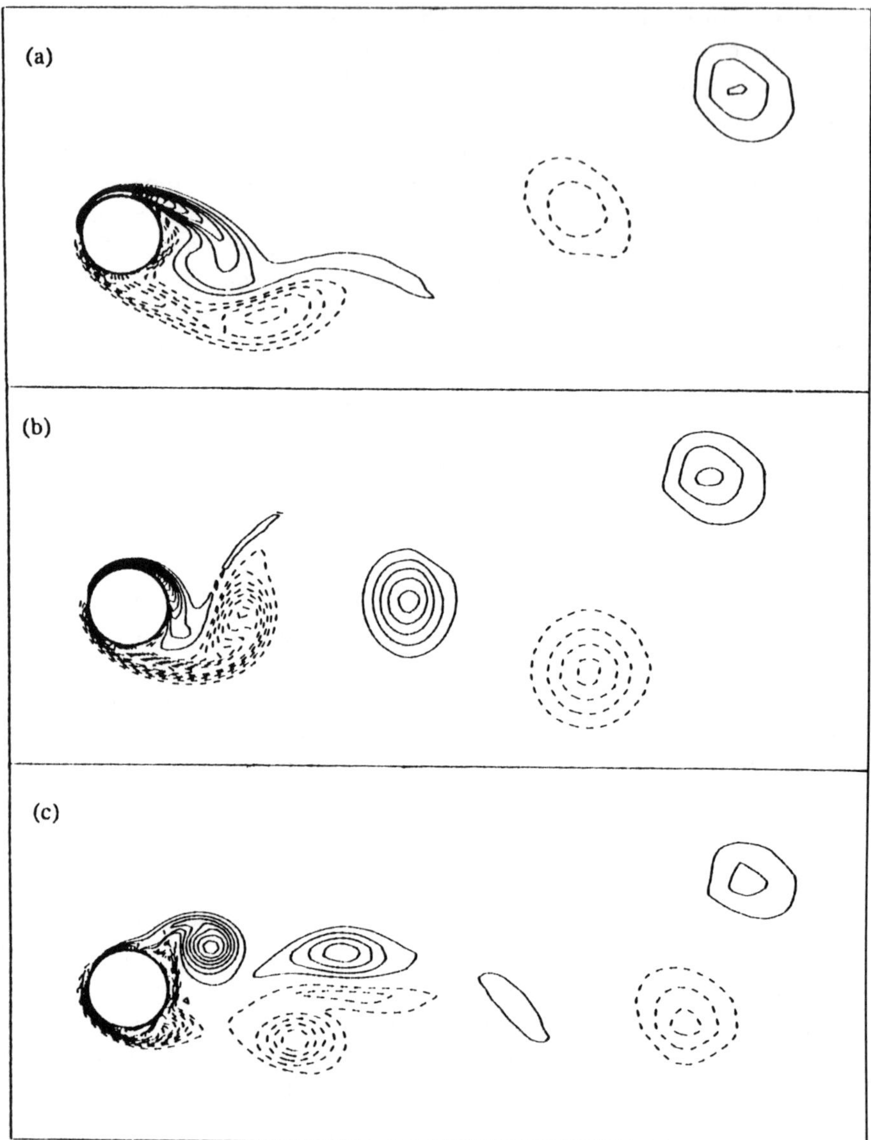

FIG. 4.15. *Vortex shedding patterns of the time-periodic rotations* $\alpha(t) = \sin \pi F t$ *with various forcing frequencies for* $Re = 200$, $t = 24.0$. *(a)* $F = 0.08$, *(b)* $F = 0.16$, *(c)* $F = 0.32$.

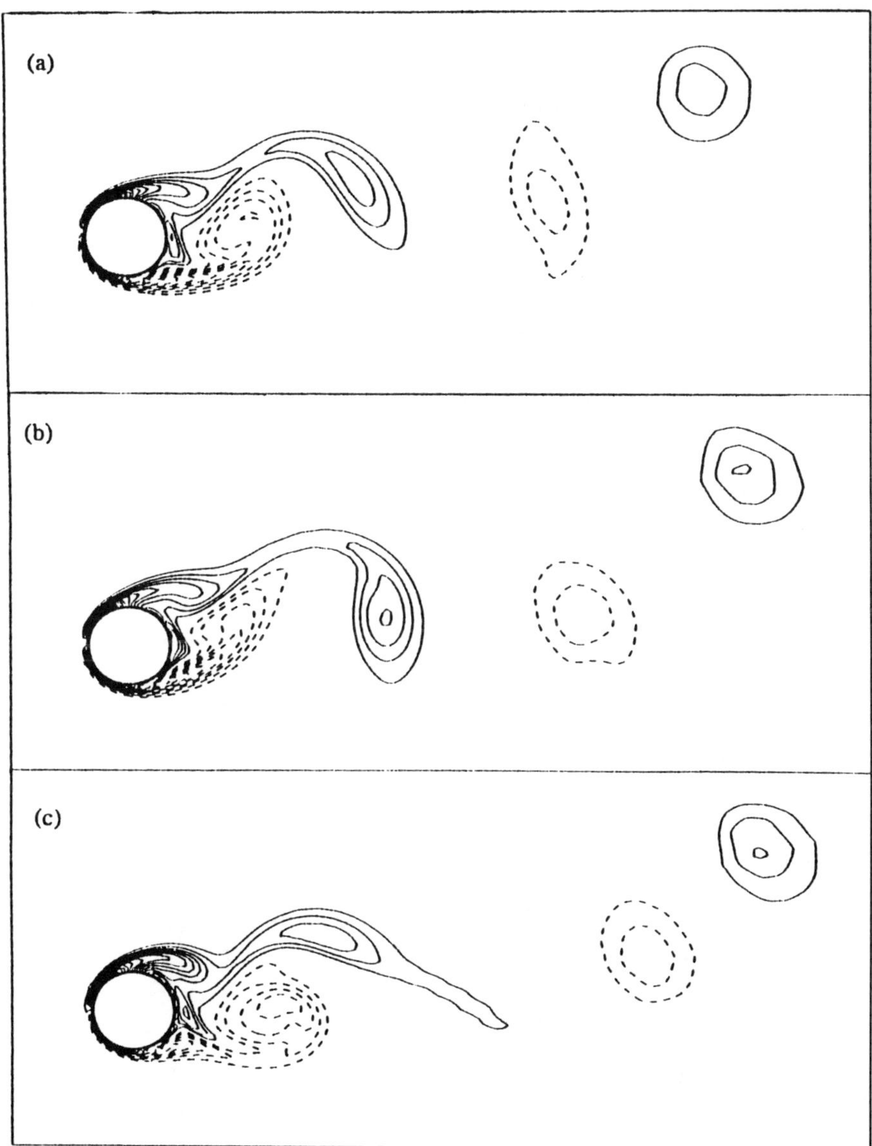

FIG. 4.16. *Vortex shedding patterns of the time-periodic rotary oscillations* $\alpha(t) = |\sin \pi(F/2)t|$ *with various forcing frequencies for* $Re = 200$, $t = 24.0$. *(a)* $F = 0.08$, *(b)* $F = 0.16$, *(c)* $F = 0.32$.

$$\begin{cases} \nabla \cdot \boldsymbol{\Psi} = 0 \\ \boldsymbol{\Psi}|_{\Gamma} = -y\mathbf{e}_x + x\mathbf{e}_y \\ \boldsymbol{\Psi} = 0, \ \text{for} \ \rho(\mathbf{r}) \geq 2e^{-1/\epsilon} \end{cases}$$

and

$$\boldsymbol{\Phi}(\mathbf{r}) = \nabla \times \left[y \left(1 - \Theta_\epsilon(\mathbf{r}) \right) \mathbf{e}_z \right],$$

$$\begin{cases} \nabla \cdot \boldsymbol{\Phi} = 0 \\ \boldsymbol{\Phi}|_{\Gamma} = 0 \\ \boldsymbol{\Phi} = \mathbf{e}_x, \ \text{for} \ \rho(\mathbf{r}) \geq 2e^{-1/\epsilon} \end{cases}$$

These two vector fields $\boldsymbol{\Psi}(\mathbf{r})$ and $\boldsymbol{\Phi}(\mathbf{r})$ would carry the non-homogeneous boundary values at the solid surface and far field, respectively. Here $\Theta_\epsilon(\mathbf{r})$ is a positive scalar cut-off smooth function such that for $\epsilon > 0$,

$$\begin{cases} \Theta_\epsilon(\mathbf{r}) = 1, \quad \mathbf{r} \in N(\Gamma, \epsilon), \quad \text{neighborhood of } \Gamma \\ \Theta_\epsilon(\mathbf{r}) = 0, \quad \rho(\mathbf{r}) \geq 2e^{-1/\epsilon} \\ \left| \frac{\partial \Theta_\epsilon(\mathbf{r})}{\partial \mathbf{r}^k} \right| \leq \frac{\epsilon}{\rho(\mathbf{r})}, \quad \rho(\mathbf{r}) \leq 2e^{-1/\epsilon}, \quad k = 1, 2 \end{cases}$$

where $\rho(\mathbf{r}) = \text{dist}(\Gamma, \mathbf{r}), \mathbf{r} \in D$.

Let us now introduce a change of variable such that,

$$\mathbf{u}(\mathbf{r}, t) = \mathbf{v}(\mathbf{r}, t) + U\boldsymbol{\Phi}(\mathbf{r}) + \Omega(t)\boldsymbol{\Psi}(\mathbf{r}).$$

A system of equations with homogeneous boundary values is obtained:

$$\mathbf{v}_t + (\mathbf{v} \cdot \nabla)\mathbf{v} + U(\mathbf{v} \cdot \nabla\boldsymbol{\Phi}) + \Omega(t)(\mathbf{v} \cdot \nabla\boldsymbol{\Psi}) + U(\boldsymbol{\Phi} \cdot \nabla\mathbf{v}) + \Omega(t)(\boldsymbol{\Psi} \cdot \nabla\mathbf{v})$$
$$= -\nabla p + \nu\nabla^2\mathbf{v} + \mathbf{f}_{\boldsymbol{\Phi\Psi}} \quad \text{in } D \times [0, T]$$

$\nabla \cdot \mathbf{v} = 0, \ \text{in } D \times [0, T],$

$\mathbf{v}|_{\Gamma} = 0,$

$\mathbf{v} \to 0, \ \text{as } |\mathbf{r}| \to \infty,$

$\mathbf{u}(\mathbf{r}, 0) = 0, \quad \mathbf{r} = (x, y) \in D,$

where $\mathbf{f}_{\boldsymbol{\Phi\Psi}} = f(U, \Omega, \Omega_t, \boldsymbol{\Psi}, \boldsymbol{\Phi})$ and $supp\{\mathbf{f}_{\boldsymbol{\Phi\Psi}}\} \subset\subset D$.

In consequence, this system of equations is then projected to the solenoidal subspace H by the orthogonal projector $P_H : L^2(D) \to H(D)$, we get

$$(5.1) \quad \begin{cases} \partial_t \mathbf{v}(t; \Omega) + \nu A\mathbf{v}(t; \Omega) + N(\boldsymbol{\Phi}, \boldsymbol{\Psi}, \mathbf{v}(t; \Omega)) = F(\boldsymbol{\Phi}, \boldsymbol{\Psi}, \Omega) \\ \mathbf{v}(0) = 0, \end{cases}$$

where $H = \{\mathbf{v} : D \to R^2; \mathbf{v} \in L^2(D), \nabla \cdot \mathbf{v} = 0, \text{ and } \mathbf{v} \cdot \mathbf{n}|_{\Gamma} = 0\}$. In (5.1), A is the Stokes operator and F is all known quantities, while N

includes the inertial term of original equations. Also, $\Omega(t)$ is the angular speed and is treated as a control variable in the formulation. The proof of existence theorem for the system (5.1) is analogous to the procedure outlined in [29,31], we will omit the proof here.

A simple example of optimal control problems as mentioned in §2.2 is to drive a solution orbit $\mathbf{u}(t;\Omega)$ to a desired flow field $\mathbf{z}_d$ by using the rotation rate $\Omega(t)$ as a control parameter. Hence the *optimal control* problem is to find an *admissible pair* $(\mathbf{v},\Omega)$ such that minimizes the cost functional

$$(5.2) \quad J(\mathbf{v},\Omega) = \int_0^T \|\mathbf{v}(t;\Omega) + U\boldsymbol{\Phi} + \Omega(t)\boldsymbol{\Psi} - \mathbf{z}_d\|^2_{L^2(D)} dt + \lambda \int_0^T |\Omega_t|^2 dt,$$

over an *admissible set* $\mathcal{U}_{ad}$. Here $\mathcal{U}_{ad}$ is the set of all admissible pair $(\mathbf{v},\Omega)$ that satisfy equation (5.1) and

> (i) $(\mathbf{v},\Omega) \in L^2(0,T;V) \times H^1(0,T)$;
>
> (ii) $J(\mathbf{v},\Omega) < \infty$.

Notice that the cost functional in (5.2) is penalized by the control, which is necessary in the proof of existence of an optimal control. Also, V is a subspace of H. The existence theorem can be stated as follows:

Existence Theorem. *There exists an optimal solution* $(\mathbf{v}^*,\Omega^*) \in \mathcal{U}_{ad}$ *such that the corresponding value of the cost functional achieves the absolute minimum. i.e.*

$$J(\mathbf{v}^*,\Omega^*) = \inf_{(\mathbf{v},\Omega)\in\mathcal{U}_{ad}} J(\mathbf{v},\Omega).$$

5.2. An optimality system. The next question is to ask how to determine the optimal controls. This can be accomplished by introducing an adjoint state which corresponding to the adjoint of a linearized version of state equation system (5.1). The *optimal control* Ω^* is thus determined by the solution of the optimality system:

$$\begin{cases} \partial_t \mathbf{v}(\Omega^*) + \nu A\mathbf{v}(\Omega^*) + N(\mathbf{v}(\Omega^*),\Omega^*) = F(\Omega^*) \\ -\partial_t \mathbf{p}(\Omega^*) + \nu A\mathbf{v}(\Omega^*) + N'(\mathbf{v}^*)\mathbf{p}(\Omega^*) = \mathbf{v}(\Omega^*) + U\boldsymbol{\Phi} + \Omega^*\boldsymbol{\Psi} - \mathbf{z}_d, \\ \mathbf{v}(0) = 0, \mathbf{p}(T) = 0, \end{cases}$$

(5.3)

and

$$(5.4) \quad \int_0^T \left[(\mathbf{p}(t,\Omega^*), F'(t,\Omega^*))_{L^2(D)} + \lambda\Omega_t^*(\Omega_t - \Omega_t^*) \right] dt \geq 0,$$

where $\mathbf{p}$ is the adjoint state and $N'(\mathbf{v}^*)$ is the Fréchet derivative of $N(\cdot)$ at $\mathbf{v}^*$. This optimality system consists of the evolutionary Navier-Stokes equations, the adjoint equation and an optimality condition (5.4) that relates the optimal control Ω^* with the optimal state $\mathbf{v}^*$. However, the resulting optimality system is complex and formidable. Therefore, the next step is to implement an efficient numerical algorithm to solve the equations (5.3)-(5.4) computationally.

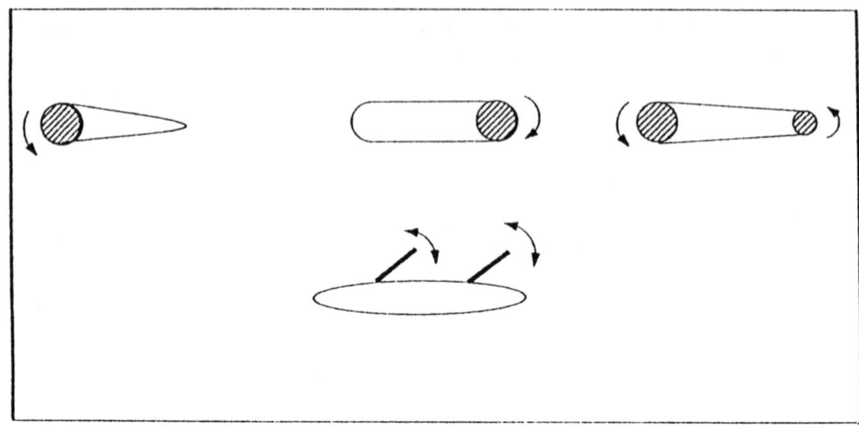

FIG. 6.1. *Applications of boundary moving surfaces mechanism in external flow control.*

6. Conclusion. The objectives to demonstrate the feasibility of a time-dependent moving surface as a control mechanism in enhancing force performance and changing the flow field were achieved. Although all optimal control problems for a rotating cylinder in this study were directly computed by trail and error variation of controls, the numerical results are significant because they show a proper choice of the rotation rate can lead to improved flow fields. For the case of a constant speed of rotation, several optimal control problems were considered and solved computationally. Using time-periodic rotations leads to a considerable improvement in the force coefficients and was shown to be very effective, especially compared to time-harmonic rotary oscillations. Very precise periodicity of the force for certain cases was established, and this periodic behavior has considerable impact on controlling the vortex formation in the cylinder wake. The possible form of controller to suppress vortex shedding was discussed. Based on the theoretical approach described in previous section, a computational algorithm may be implemented to seek an optimal control such that it will suppress the vortex shedding with a minimum cost.

The rotating cylinder mechanism (as a moving surface control) developed here can also be used to investigate fundamental question regarding unsteady separation control. For example, the moving surfaces mechanism has been successfully applied to boundary-layer control in a number of experiments by Modi et al. [20]. In their experiments, the boundary-layer flow is controlled by an application of two rotating cylinders located at the leading and trailing edges of an airfoil. It has been shown that this mechanism can prevent flow separation by retarding the initial growth of the boundary layer, with the important consequences of lift enhancement and stall delay. In spite of the fact that considerable aerodynamic benefits were gained by changing the cylinder speed ratio, in their experiments the speed of rotation was performed merely with constant values. However, it should be noted that if the rotating cylinder mechanism is applied to a region of flow domain in which time-dependent separations occurred, a constant rotation rate may not correspond to the optimal performance when an airfoil is undergoing a rapid maneuver. Such observation provide the motivation for us to consider problems of unsteady flow control by means of a time-dependent moving surface mechanism. Figure 6.1 shows some possible flow geometries for future investigations. Using such mechanisms as a controller allows us to formulate a wide variety of practical control problems in real engineering applications. Modifications of existing numerical algorithms needed for these types of control problems depend on performance and design constraints. It is our hope that this investigation will represent a step toward control of external flow, and serve as a guide on the formulation of many practical optimal flow control problems.

REFERENCES

[1] F. Abergel and R. Temam, *On some control problems in fluid mechanics*, Theoret. Comput. Fluid Dynamics, 1 (1990), pp. 303-325.

[2] H. M. Badr, M. Coutanceau, S. C. R. Dennis and C. Ménard, *Unsteady flow past a rotating circular cylinder at Reynolds numbers 10^3 and 10^4*, J. Fluid Mech., 220 (1990), pp. 459-484.

[3] H. M. Badr and S. C. R. Dennis, *Time-dependent viscous flow past an impulsively started rotating and translating circular cylinder*, J. Fluid Mech., 158 (1985), pp. 447-488.

[4] J. A. Burns and S. Kang, *A control problem for Burgers' equation with bounded input/output*, Nonlinear Dynamics, 2 (1991), pp. 235-262.

[5] J. A. Burns and Y.-R. Ou, *Effect of rotation rate on the forces of a rotating cylinder: simulation and control*, submitted to Phys. fluids A, (1993).

[6] C.-C. Chang and R.-L. Chern, *Vortex shedding from an impulsively started rotating and translating cylinder*, J. Fluid Mech., 235 (1992), pp. 265-298.

[7] Y.-M. Chen, *Numerical Simulation of the Unsteady Two-dimensional Flow in a Time-dependent Doubly-connected Domain*, PhD thesis, University of Arizona, 1989.

[8] Y.-M. Chen and Y.-R. Ou and A. J. Pealstein, *Development of the wake behind a circular cylinder impulsively started into rotatory and rectilinear motion: intermediate rotation rates*. ICASE Report 91-10 (1991), J.

Fluid Mech., 253 (1993), pp. 449-484.

[9] M. Coutanceau and C. Ménard, *Influence of rotation on the near-wake development behind an impulsively started circular cylinder*, J. Fluid Mech., 158 (1985), pp. 399-446.

[10] H. Fattorini and S. S. Sritharan, *Existence of optimal controls for viscous flow problems*, Proc. R. Soc. Lond. A, 439 (1992), pp. 81-102.

[11] M. Gad-el-Hak, *Flow control*, Appl. Mech. Rev., 42 (1989), pp. 261-293.

[12] M. D. Gunzburger, L. S. Hou and T. P. Svobodny, *Numerical approximation of an optimal control problem associated with the Navier-Stokes equations*, Appl. Math. Lett., 2 (1989), pp. 29-31.

[13] ————, *Analysis and finite element approximation of optimal control problems for stationary Navier-Stokes equations with distributed and Neuman controls*, Mathematics of Computations, 57 (1991), pp. 123-151.

[14] ————, *Boundary velocity control of incompressible flow with an application to viscous drag reduction*, SIAM J. Control and Optim., 30 (1992), pp. 167-181.

[15] W. B. Herbst, *Supermaneuverability*, Workshop on Unsteady Separated Flow, Sponsored by AFOSR, FJSRL, U. of Colorado (1983).

[16] A. Jameson, *Automatic design of transonic airfoil to reduce the shock induced pressure drag*, the 31st Israel Annual Conference on Aviation and Aeronautics, (1990).

[17] S. Kang And K. Ito, *A control problem for fluid flow*, Proc. 31st IEEE Conference on Decision and Control, Tucson, AZ (1992), pp. 3393-3398.

[18] C. A. Koromilas and D. P. Telionis, *Unsteady laminar separation: an experimental study*, J. Fluid Mech., 97 (1980), pp 347-384.

[19] J. Mo, *An Investigation on the Wake of a Cylinder with Rotational Oscillations*, PhD thesis, U. of Tennessee Space Institute 1989.

[20] V. J. Modi, F. Mokhtarian and T. Yokomizo, *Effect of moving surfaces on the airfoil boundary-layer control*, J. Aircraft., 27 (1990), pp 42-50.

[21] Y.-R. Ou, *Control of oscillatory forces on a circular cylinder by rotation*, Proc. 4th International Symposium CFD, U. of California. Davis, CA (1991), pp. 897-902.

[22] ————, *Active flow control relative to a rotating cylinder*, Proc. 31st IEEE Conference on Decision and Control, Tucson, AZ (1992), pp. 3399-3404.

[23] ————, *Active control of exterior hydrodynamics - computational results*, to appear in *Optimal Control of Viscous Flow*, SIAM Frontiers in Applied Mathematics, ed. S.S.Sritharan, Society for Industrial and Applied Mathematics, Philadelphia, PA, 1993.

[24] Y.-R. Ou and J. A. Burns, *Optimal Control of lift/drag ratios on a rotating cylinder*, Appl. Math. Lett., 5 (1992), pp. 57-62.

[25] D. S. Park, D. M. Ladd, E. Hendricks, *Feedback control of Kármán vortex shedding*, Symposium on Active Control of Noise and Vibration, ASME Winter Annual Meering, Anaheim, CA (1992).

[26] L. Prandtl, *Über flüssigkeitsbewegung bei sehr kleiner reiburg*, Proc. 3rd Int. Math. Congr, Heidelberg, Germany (1904), pp. 484-491.

[27] ————, *The Magnus effect and windpowered ships*, Naturwissenschaften, 13 (1925), pp. 93-108.

[28] T. Sarpkaya, *Vortex-induced oscillations*, J. Appl. Mech., 46 (1979), pp.

241-258.

[29] S. S. Sritharan, *Invariant Manifold Theory for Hydrodynamic Transition*, Pitman Research Notes in Mathematics Series 241, (1990).

[30] ————, *Dynamic programming of the Navier-Stokes equations*, Systems and Control Letters, 16 (1991), pp. 299-307.

[31] ————, *An optimal control problem in exterior hydrodynamics*, Proc. of the Royal Society of Edinburgh, 121A (1992), pp. 5-32.

[32] S. S. Sritharan, Y.-R. Ou, J. A. Burns, D. Park, D. Ladd, E. Hendrick and N. Nossier, *Optimal control of viscous flow past a cylinder: mathematical theory, computation and experiment,,* to be published (1992).

[33] W. M. Swanson, *The Magnus effect: a summary of investigations to date*, ASME J. Basic Engrg., 83 (1961), pp. 461-470.

[34] S. Taneda, *Visual study of unsteady separated flows around bodies*, Prog. Aero. Sci., 17 (1977) pp. 287-348.

[35] ————, *Visual observations of the flow past a circular cylinder performing a rotatory oscillation*, J. Phys. Soc. Japan, 45 (1978), pp. 1038-1043.

[36] P. T. Tokumaru and P. E. Dimotakis, *Rotary oscillation control of a cylinder wake*, J. Fluid Mech., 224 (1991), pp. 77-90.

[37] A. Wambecq, *Rational Runge-Kutta methods for solving systems of ordinary differential equations*, Computing, 20 (1978), pp. 333-342.

[38] C. H. K. Williamson, *Sinusoidal flow relative to circular cylinders*, J. Fluid Mech., 155 (1985) pp. 141-174.

[39] ————, *Defining a universal and continuous Strouhal-Reynolds number relationship for the laminar vortex shedding of a circular cylinder*, Phys. Fluids, 31 (1988), pp. 2742-2744.

[40] C. H. K. Williamson and A. Roshko, *Vortex formation in the wake of an oscillating cylinder*, J. Fluids and Structures, 2 (1988), pp. 355-381.

[41] J. C. Wu And J. F. Thompson, *Numerical solutions of time-dependent incompressible Navier-Stokes equations using an integral-differential formulation*, Computers & Fluids, 1 (1973), pp. 197-215.

OPTIMAL FEEDBACK CONTROL OF HYDRODYNAMICS: A PROGRESS REPORT

S.S. SRITHARAN*

Abstract. In this article we review some of the recent results in the mathematical theory of optimal feedback control of viscous flow. Main results are existence of ordinary and chattering controls, Pontryagin maximum principle and feedback synthesis using infinite dimensional Hamilton-Jacobi equation of dynamic programming. Some preliminary results on stochastic control also presented.

AMS(MOS) subject classifications. 49,93,76,60,49,35,46

1. Introduction. Optimal feedback control of viscous flow has many applications in engineering sciences. In this context, both deterministic as well as stochastic control of Navier-Stokes equations are of interest. During the past few years several fundamental advances were made in the deterministic case. Main questions addressed were existence theorem for ordinary optimal controls [21,32,28,19], existence of chattering controls [17,18], necessary conditions for free terminal state problem [21,30,1,23] as well as the full Pontryagin maximum principle for problems with terminal constraint [16] and feedback synthesis using Hamilton-Jacobi -Bellman equation [30,16,31]. Finite element methods for the maximum principle with free end state are analyzed in [22]. See also the forthcoming book [33] for reports of progress by various authors of this field. The concepts used in these works have their origins in the classical works of Euler, Lagrange, Hamilton, Jacobi and Weierstrass and also in the modern works of Caratheodori, Tonelli, Young, Pontryagin and Bellman. In this article we will review some of these developments. Some initial thoughts on stochastic optimal control theory will also be presented.

As shown in the above papers of Sritharan and of Fattorini and Sritharan, time dependent flow problems with *boundary control* can be reduced to infinite dimensional semilinear evolution equations of the following type in a separable Hilbert space $\boldsymbol{X}$:

$$(1.1) \qquad v_t + \nu \mathcal{A} v + \mathcal{B}(v) = \mathcal{N}(v, U)$$

$$(1.2) \qquad v(\tau) = \zeta \in \boldsymbol{X}.$$

Here $\mathcal{A}$ and $\mathcal{B}$ are respectively the well known Stokes operator and the inertia term. $\mathcal{N}$ is the control operator whose form is dictated by the type

---

* Code 574, Naval Command, Control and Ocean Surveillance Center, San Diego, CA 92152-5000. Supported by the Mathematical Sciences and Mechanics Divisions of ONR.

of forcing (boundary forcing, distributed forcing etc). The special cases of the control operator

$$(1.3) \qquad\qquad \mathcal{N}(\boldsymbol{v}, U) = L_N U$$

and

$$(1.4) \qquad\qquad \mathcal{N}(\boldsymbol{v}, U) = U + \boldsymbol{f}$$

where L_N is a bounded linear operator and $\boldsymbol{f}$ is a given element of $\boldsymbol{X}$ are also of interest. As shown in [19], for a large class of flow control problems including exterior hydrodynamics and flow through water tunnels the control operator appears as a linear term similar to (1.3). When the boundary control is distributed [17] we obtain a nonlinear control operator similar to (1.1). Similar models have also been proposed by experimentalists [25,26]. The simple control operator of the type (1.4) was proposed by Fursikov [21].

When we do not have adequate convexity, the controls will be taken as probability measures (chattering controls) $\boldsymbol{\mu}$ defined on the control set $\boldsymbol{U}$ and the control operator $\mathcal{N}(\cdot, \cdot)$ will be formally replaced by

$$\boldsymbol{N}(\boldsymbol{v})\boldsymbol{\mu} = \int_{\boldsymbol{U}} \mathcal{N}(\boldsymbol{v}, U)\mu(dU).$$

Then $\boldsymbol{N}(\boldsymbol{v})\boldsymbol{\mu} \in \overline{\text{conv}}\mathcal{N}(\boldsymbol{v}, \boldsymbol{U})$ with closure in the weak topology of $\boldsymbol{X}$. The corresponding trajectories will be called relaxed trajectories. In such situations, as discussed below, a similar relaxation should also be introduced in the cost functionals.

In some of our problems the state will have a terminal constraint of the type

$$\boldsymbol{v}(T) \in \boldsymbol{Y} \subseteq \boldsymbol{X}$$

where $\boldsymbol{Y}$ is a closed subset.

We will consider two classes of cost functionals:

(1) *Finite Horizon*:

$$(1.5) \qquad\qquad \phi(\boldsymbol{v}(T))) + \int_0^T \mathcal{L}(t, \boldsymbol{v}(t), U(t))dt \to \inf.$$

(2) *Infinite Horizon*:

$$(1.6) \qquad\qquad \int_0^\infty e^{-\lambda t}\mathcal{L}(t, \boldsymbol{v}(t), U(t))dt \to \inf.$$

where $\lambda > 0$ is some discount factor and $\mathcal{L}(\cdot, \cdot)$ is the Lagrangian (for specific forms of the Lagrangian see the papers quoted above). The corresponding relaxed functionals will be,

$$\phi(\boldsymbol{v}(T))) + \int_0^T \int_{\boldsymbol{U}} \mathcal{L}(t, \boldsymbol{v}(t), U)\mu(t, dU)dt \to \inf$$

and

$$\int_0^\infty e^{-\lambda t} \int_{\boldsymbol{U}} \mathcal{L}(t, \boldsymbol{v}(t), U)\mu(t, dU)dt \to \inf$$

respectively.

Let $(\Upsilon, \Sigma_\Upsilon, m)$ be a complete probability space, where Υ be a set of elementary events, Σ_Υ is a sigma algebra of subsets of Υ and $m(\cdot) : \Sigma_\Upsilon \to [0, 1]$ is a complete probability measure. For stochastic control we will consider the random evolution system on $(\Upsilon, \Sigma_\Upsilon, m)$ with a white noise forcing,

$$(1.7) \qquad d\boldsymbol{v} = \mathcal{F}(\boldsymbol{v}, U)dt + d\boldsymbol{W}$$

where the "drift" term is given by

$$(1.8) \qquad \mathcal{F}(\boldsymbol{v}, U) = -\nu \mathcal{A}\boldsymbol{v} - \mathcal{B}(\boldsymbol{v}) + \mathcal{N}(\boldsymbol{v}, U),$$

and $\boldsymbol{W}$ is the $\boldsymbol{X}$-valued Wiener process with covariance $Q \in \mathcal{L}(\boldsymbol{X}; \boldsymbol{X})$ being a symmetric, nonnegative, nuclear operator (ie. of finite trace $\mathrm{Tr}\, Q < +\infty$).

We will consider three classes of cost functionals:

(1) *Finite Horizon*:

$$(1.9) \qquad \mathcal{E}\left[\phi(\boldsymbol{v}(T))) + \int_0^T \mathcal{L}(t, \boldsymbol{v}(t), U(t))dt\right] \to \inf.$$

(2) *Infinite Horizon*:

$$(1.10) \qquad \mathcal{E}\left[\int_0^\infty e^{-\lambda t}\mathcal{L}(t, \boldsymbol{v}(t), U(t))dt\right] \to \inf.$$

(3) *Ergodic Control*

$$(1.11) \qquad \liminf_{T \to \infty} \frac{1}{T} \int_0^T \mathcal{L}(t, \boldsymbol{v}(t), U(t))dt \to \inf \quad \text{almost surely}.$$

In the above, $\mathcal{E}[\cdot]$ represents the expectation,

$$(1.12) \qquad \mathcal{E}\left[\psi(\boldsymbol{v})\right] := \int_\Upsilon \psi(\boldsymbol{v}(\omega))m(d\omega).$$

2. Definition of trajectories. Let us first recall some of the properties of the operators $\mathcal{A}$, $\mathcal{B}$ and $\mathcal{N}$ from earlier papers. In this paper we will restrict ourselves to fluid flow in bounded domains and refer the readers to the literature quoted for analysis in unbounded domains. The Stokes operator $\mathcal{A}$ satisfies the following well known properties [7].

PROPOSITION 2.1. *$\mathcal{A}$ is self-adjoint and positive definite.*

These results have the following consequences. $-\mathcal{A}$ generates a *holomorphic* semigroup $S(t) = \exp(-t\mathcal{A})$. The fractional powers $\mathcal{A}^\alpha, \alpha \in \mathbf{R}$ are well defined and $\mathcal{A}^\alpha$ for $\alpha \leq 0$ are bounded. For $\alpha \geq 0$ we write $\mathbf{X}_\alpha = D(\mathcal{A}^\alpha)$ and equip this space with the natural inner product $(\boldsymbol{v}, \boldsymbol{u})_\alpha = (\mathcal{A}^\alpha \boldsymbol{v}, \mathcal{A}^\alpha \boldsymbol{u})$, corresponding to the norm $\|\boldsymbol{v}\|_\alpha = \|\mathcal{A}^\alpha \boldsymbol{v}\|$. For $\alpha < 0$, $\mathbf{X}_\alpha$ is the completion of $\mathbf{X}$ under $\|\cdot\|_\alpha$.

The inertia term $\mathcal{B}(\cdot)$ satisfies the following

PROPOSITION 2.2. *There exists β, $0 \leq \beta < 1/2$ such that $\mathcal{B}(\cdot)$ maps $\mathbf{X}_{1/2}$ into $\mathbf{X}_{-\beta}$. Moreover, $\mathcal{B}(\cdot) : \mathbf{X}_{1/2} \to \mathbf{X}_{-\beta}$ is continuous, locally bounded, and locally Lipschitz continuous, i.e. for every $C > 0$ there exist constants $K_B = K_B(C)$, $L_B = L_B(C)$ such that*

$$(2.1) \qquad \|\mathcal{B}(\boldsymbol{v})\|_{-\beta} \leq K_B, \ for \ \boldsymbol{v} \in \mathbf{X}_{1/2} \ and \ \|\boldsymbol{v}\|_{1/2} \leq C,$$

$$\|\mathcal{B}(\boldsymbol{v}) - \mathcal{B}(\boldsymbol{u})\|_{-\beta} \leq L_B \|\boldsymbol{v} - \boldsymbol{u}\|_{1/2},$$

$$(2.2) \qquad for \ \boldsymbol{v}, \boldsymbol{u} \in \mathbf{X}_{1/2} \ and \ \|\boldsymbol{v}\|_{1/2}, \|\boldsymbol{u}\|_{1/2} \leq C.$$

The control $U(\cdot)$ takes its values in the control set $\mathbf{U}$ which is a normal topological space. The control operator $\mathcal{N}(\boldsymbol{v}, U)$ is defined in $\mathbf{X}_{1/2} \times \mathbf{U}$.

PROPOSITION 2.3. *$\mathcal{N}(\cdot, \cdot)$ continuously maps $\mathbf{X}_{1/2} \times \mathbf{U}$ into $\mathbf{X}$. There exists a continuous function $\kappa(\cdot) : \mathbf{U} \to \mathbf{R}$, $\kappa(U) \geq 1$ such that, for every $C > 0$ there exist $K_N = K_N(C)$ and $L_N = L_N(C)$ such that*

$$(2.3) \quad \|\mathcal{N}(\boldsymbol{v}, U)\| \leq K_N \kappa(U), \ for \ \boldsymbol{v} \in \mathbf{X}_{1/2}, \|\boldsymbol{v}\|_{1/2} \leq C, U \in \mathbf{U},$$

$$\|\mathcal{N}(\boldsymbol{v}, U) - \mathcal{N}(\boldsymbol{u}, U)\| \leq L_N \|\boldsymbol{v} - \boldsymbol{u}\|_{1/2} \kappa(U),$$

$$(2.4) \qquad for \ \boldsymbol{v} \in \mathbf{X}_{1/2}, \boldsymbol{u} \in \mathbf{X}_{1/2}, U \in \mathbf{U} \ and \ \|\boldsymbol{v}\|_{1/2}, \|\boldsymbol{u}\|_{1/2} \leq C.$$

The space $\mathcal{U}_{\mathrm{ad}}(0, T; \mathbf{U}, \kappa)$ of admissible controls consists of all $\mathbf{U}$-valued functions defined almost everywhere and satisfying,

$$(2.5) \qquad \kappa(U(\cdot)) \in L^2(0, T).$$

This implies that [17], $\forall \boldsymbol{v}(\cdot) \in C([0, T]; \mathbf{X}_{1/2})$, *the control operator $\mathcal{N}(\boldsymbol{v}(\cdot), U(\cdot))$ is strongly measurable.* In fact, we have $\mathcal{N}(\boldsymbol{v}(\cdot), U(\cdot)) \in L^2(0, T; \mathbf{X})$.

2.1. Ordinary trajectories. By definition, *solutions or trajectories* of the initial value problem (1.1)-(1.2) in an interval $0 \leq t \leq T'$ are $\boldsymbol{X}_{1/2}$-valued functions $\boldsymbol{v}(\cdot)$ continuous in the norm of $\boldsymbol{X}_{1/2}$ and satisfying

$$\boldsymbol{v}(t) = S(t)\zeta - \int_0^t \mathcal{A}^\beta S(t-r)\mathcal{A}^{-\beta}\mathcal{B}(\boldsymbol{v}(r))dr$$

$$(2.6) \qquad + \int_0^t S(t-r)\mathcal{N}(\boldsymbol{v}(r), U(r))dr, \qquad 0 \leq t \leq T'.$$

The following results defines the trajectory for our control system.

THEOREM 2.1. *Let $\zeta \in \boldsymbol{X}_{1/2}$ and $U(\cdot) \in \mathcal{U}_{ad}(0,T;\boldsymbol{U},\kappa)$. Then (2.6) possesses a unique solution $\boldsymbol{v}(\cdot) \in C([0,T'];\boldsymbol{X}_{1/2})$ for some $T' \leq T$. Moreover, suppose that*

$$(2.7) \qquad \|\boldsymbol{v}(t)\|_{1/2} \leq C, \quad 0 \leq t < T'.$$

Then, if $[0,T_m)$ is the maximal interval of existence of $\boldsymbol{v}(\cdot)$, we have $T_m > T'$.

This implies that, if $\boldsymbol{v}(t)$ is a solution of (2.6) in a closed interval $0 \leq t \leq T'$, then $\boldsymbol{v}(t)$ can always be extended to a larger interval $0 \leq t \leq T''$, $T'' > T'$ solving the equation in $t \geq T'$ with $\boldsymbol{v}(T')$ as initial condition. This implies that each solution $\boldsymbol{v}(\cdot)$ of (2.6) either exists in $0 \leq t \leq T$ or possesses a maximal interval of existence $[0,T_m)$, $T_m < T$ with $\limsup_{t \to T_m} \|\boldsymbol{v}(t)\|_{1/2} = +\infty$.

2.2. Chattering controls and relaxed trajectories. We will begin with the the class $\mathcal{V}_{bc}(0,T;\boldsymbol{U},\kappa)$ of chattering controls [17]. Here the control set $\boldsymbol{U}$ is required to be a normal topological space and the instantaneous values of the chattering controls are regular finitely additive probability measures $\Sigma_{rba}(\boldsymbol{U},\Phi_c)$ defind on an algebra Φ_c of subsets of $\boldsymbol{U}$. This Banach space of measures coincides with the strong dual of the Banach space $C_b(\boldsymbol{U})$ of bounded continuous functions.

DEFINITION 2.1. **Chattering Controls**: *The space $\mathcal{V}_{bc}(0,T;\boldsymbol{U},\kappa)$ of chattering controls consists of all*

$$\boldsymbol{\mu}(\cdot) \in (L^1(0,T;C_b(\boldsymbol{U})))^* = L_w^\infty(0,T;\Sigma_{rba}(\boldsymbol{U},\Phi_c))$$

that satisfy

$$(2.8) \qquad (i) \qquad \|\boldsymbol{\mu}(\cdot)\|_{L_w^\infty(0,T;\Sigma_{rba}(\boldsymbol{U},\Phi_c))} \leq 1.$$

$$(ii) \qquad \boldsymbol{f}(\cdot) \in L^1(0,T;C_b(\boldsymbol{U}))$$

is such that for $U \in \boldsymbol{U}$, $f(t,U) \geq 0$, a.e. in $0 \leq t \leq T$ then

$$(2.9) \qquad \int_0^T \int_{\boldsymbol{U}} f(t,U)\mu(t,dU)dt \geq 0.$$

(iii)   if $\mathcal{X}_e(\cdot)$ is the characteristic function of a measurable set $e \subseteq [0, T]$ and $\mathcal{X}_U(\cdot)$ is the function identically 1 in U then

$$(2.10) \qquad \int_0^T \int_U (\mathcal{X}_e(t) \otimes \mathcal{X}_U(U)) \mu(t, dU) dt = meas\ (e).$$

$$(2.11) \qquad (iv) \qquad \int_U \kappa(U)^2 \mu(\cdot, dU) \in L^1(0, T).$$

Note that ordinary controls in the space $\mathcal{U}_{\mathrm{ad}}(0, T; U, \kappa)$ can be duplicated by chattering controls. If $V(\cdot) \in \mathcal{U}_{\mathrm{ad}}(0, T; U, \kappa)$, we define a chattering control by the Dirac measure concentrated on $V(t)$: $\boldsymbol{\mu}(t) = \boldsymbol{\delta}_{V(t)}$. It is obvious that $\boldsymbol{\mu}(t)$ satisfies conditions (i),(ii),(iii); as to (iv),

$$\int_U \kappa(U)^2 \mu(\cdot, dU) = \kappa(V(\cdot))^2$$

which belongs to $L^1(0, T)$ by the condition (2.5).

We proceed to the definition of the relaxed system. It will be of the form (1.1)-(1.2), but with different control set and control operator. The relaxed counterparts (of the control set U and control operator $\mathcal{N}$) will be denoted by $\Re U$ and N respectively.

Let $\boldsymbol{\Sigma}_{\mathrm{rba}}(U, \boldsymbol{\Phi}_{\mathrm{c}}, \kappa)$ be the subspace of $\boldsymbol{\Sigma}_{\mathrm{rba}}(U, \boldsymbol{\Phi}_{\mathrm{c}})$ whose elements satisfy

$$\kappa(\boldsymbol{\mu})^2 := \int_U \kappa(U)^2 |\mu|(dU) < \infty.$$

We will also denote $\kappa(\boldsymbol{\mu}) = \|\boldsymbol{\mu}\|_\kappa$.

The **chattering (or relaxed) control set** $\Re U$ corresponds to all $\mu \in \boldsymbol{\Sigma}_{\mathrm{rba}}(U, \boldsymbol{\Phi}_{\mathrm{c}}, \kappa)$ that satisfy

$$\mu(A) \geq 0, \ \forall A \in \boldsymbol{\Phi}_{\mathrm{c}} \text{ and } \mu(U) = 1.$$

Chattering controls take values in $\Re U$ and satisfy the control space hypothesis (2.5) with $\kappa(\boldsymbol{\mu})$ playing the part of $\kappa(U)$. Concerning the structure of chattering controls we have

PROPOSITION 2.4. *Let D be the set of all* Dirac measures *defined as:*

$$D = \{\mu \in \Re U; \mu = \delta_U, U \in U\}$$

and let $\overline{conv}(D)$ be its closed convex hull with closure taken in the $C_b(U)$-weak topology of $\Sigma_{rba}(U; \boldsymbol{\Phi}_c)$. Then

$$(i) \qquad \Re U = \overline{conv}(D).$$

(ii) Elements of D are exactly the **extremal points** *of $\Re U$.*

The relaxed control operator $N : X_{1/2} \times \Re U \to X$ will be denoted $N(v)\mu$ to emphasize the linearity in μ and is defined in the following way: $N(v)\mu$ is the unique element of X satisfying

$$(z, N(v)\mu) = \int_U (z, \mathcal{N}(v, U))\mu(dU), \quad \forall z \in X.$$

In fact we can show that,

PROPOSITION 2.5. $N(v)\mu$ *is continuous in* $X_{1/2} \times \Re U$ *and locally Lipschitz continuous: if* $K_N = K_N(C)$ *and* $L_N = L_N(C)$ *are the constants in (2.3-2.4) then*

$$(2.12) \qquad \|N(v)\mu\| \le K_N \kappa(\mu), \quad for\ v \in X_{1/2}, \|v\|_{1/2} \le C.$$

$$(2.13) \qquad \|N(v)\mu - N(z)\mu\| \le L_N \|v - z\|_{1/2}\kappa(\mu),$$

$$for\ v \in X_{1/2}, z \in X_{1/2}, \mu \in \Re U\ and\ \|v\|_{1/2}, \|z\|_{1/2} \le C.$$

Moreover, $\forall v(\cdot) \in C(0, T; X_{1/2})$ *and* $\forall \mu(\cdot) \in \mathcal{V}_{bc}(0, T; U, \kappa)$, $N(v(\cdot))\mu(\cdot)$ *is strongly measurable.*

The **relaxed system** corresponding to $\mathcal{V}_{bc}(0, T; U, \kappa)$ is

$$(2.14) \qquad v_t(t) + \mathcal{A}v(t) + \mathcal{B}(v(t)) = N(v(t))\mu(t)$$

and the unique solvability can be deduced from Theorem 1.

THEOREM 2.2. *Let* $\zeta \in X_{1/2}$ *and* $\mu(\cdot) \in \mathcal{V}_{bc}(0, T; U, \kappa)$. *Then (2.14) possesses a unique solution* $v(\cdot) \in C([0, T']; X_{1/2})$ *for some* $T' \le T$. *Moreover, suppose that*

$$(2.15) \qquad \|v(t)\|_{1/2} \le C, \quad 0 \le t < T'.$$

Then, if $[0, T_m)$ *is the maximal interval of existence of* $v(\cdot)$, *we have* $T_m > T'$.

2.3. Relaxation theorem. Let us now describe certain interesting approximation results for relaxed trajectories [18]. First result is a continuous dependence theorem for the relaxed problem (2.14),(1.2).

PROPOSITION 2.6. *Let* $\mu(\cdot) \in \mathcal{V}_{bc}(0, T; U, \kappa)$ *be such that the trajectory* $v(t, \mu)$ *for (2.14) exists in* $0 \le t \le T$. *Let* $\{\mu_\alpha(\cdot)\} \subseteq \mathcal{V}_{bc}(0, T; U, \kappa)$ *be a generalized sequence with*

$$\int_0^T \|\mu_\alpha(t) - \mu(t)\|_\kappa^2 dt \to 0.$$

Then there exists α_0 *and a constant* C *such that, if* $\alpha \ge \alpha_0$ *then* $v(t, \mu_\alpha)$ *exists in* $0 \le t \le T$ *and*

$$\|v(t, \mu_\alpha) - v(t, \mu)\|_{1/2} \le C$$

$$(2.16)$$

$$\left[\int_0^T \|\mu_\alpha(r) - \mu(r)\|_\kappa^2 dr \right]^{1/2}, \quad \forall t \in [0, T].$$

The main result below assures that the chattering control space is not too large and each relaxed trajectory (of (2.14), (1.2)) can be uniformly approximated by an ordinary trajectory (of (1.1), (1.2)).

THEOREM 2.3. *Let* $\mu(\cdot) \in \mathcal{V}_{bc}(0, T; \boldsymbol{U}, \kappa)$ *be such that the trajectory* $v(t, \mu)$ *for (2.14) exists in* $0 \leq t \leq T$ *and let* $\epsilon > 0$. *Then there exists a countably valued* **ordinary control** $U(\cdot) \in \mathcal{U}_{bc}(0, T; \boldsymbol{U}, \kappa)$ *satisfying*

$$ess.\ sup_{t \in [0,T]} \kappa(U(t)) < \infty$$

such that the trajectory $v(t, U)$ *exists in* $0 \leq t \leq T$ *and*

$$(2.17) \qquad \|v(t, \mu) - v(t, U)\|_{1/2} \leq \epsilon, \quad \forall t \in [0, T].$$

2.4. Trajectories for stochastic Navier-Stokes equation. Solvability theorem for *uncontrolled* stochastic Navier-Stokes equation (ie (1.2), (1.4 and (1.7)) with $U = 0$) with additive and multiplicative noise has been established by many authors[4,34,3,6]. Here we will present a solvability theorem for the *controlled* system (1.2)-(1.4) and (1.7) for the case of *two dimensions*. Proof of this result is only slightly different and will be given in [33].

Let us denote by Σ_t^W the σ-algebra generated by $\{\boldsymbol{W}(s), s \leq t\}$. We will define the class of *admissible controls* $\mathcal{U}_{ad}^W(\boldsymbol{R}^+; \boldsymbol{X})$ as $\boldsymbol{X}$-valued stochastic processes $U(t, \omega)$ which satisfy the following two conditions.

(1) $U(t, \omega)$ is Brownian adapted. That is, for each $t \geq 0, \omega \to U(t, \omega)$ is measurable from $(\Omega, \Sigma_t^W) \to (\boldsymbol{X}, \boldsymbol{B}(\boldsymbol{X}))$ where $\boldsymbol{B}(\boldsymbol{X})$ is the Borel algebra generated by the closed sets of $\boldsymbol{X}$.

(2) $\mathcal{E}\left[\int_0^T \|U(t)\|^2 dt\right] < \infty, \forall T > 0$.

THEOREM 2.4. *Let* $U \in \mathcal{U}_{ad}^W(\boldsymbol{R}^+; \boldsymbol{X})$, $\boldsymbol{f} \in L_{loc}^2(\boldsymbol{R}^+; \boldsymbol{X})$, $\zeta \in \boldsymbol{X}$ *and let* $\boldsymbol{W}$ *be an* $\boldsymbol{X}$-*valued Wiener process with trace class covariance* $\mathcal{Q}$. *Then there exists a unique solution* v *to (1.2)-(1.4) and (1.7) such that*

$$(2.18) \quad v(\cdot, \omega) \in L_{loc}^2(\boldsymbol{R}^+; \boldsymbol{X}_{1/2}) \cap L_{loc}^\infty(\boldsymbol{R}^+; \boldsymbol{X}) \cap C^{-s, \kappa}, \quad \omega \ a.s$$

and

$$(1)\ \mathcal{E}\left[\|v(t)\|^2 + \nu\|v\|_{L^2(0,t; X_{1/2})}^2\right] \leq \|\zeta\|^2 + \frac{1}{\nu}\|\boldsymbol{f}\|_{L^2(0,t; X)}^2$$

$$(2.19) \qquad\qquad + \frac{1}{\nu}\mathcal{E}\left[\|U\|_{L^2(0,t; X)}^2\right] + t \cdot Tr\mathcal{Q}, \quad \forall t \geq 0,$$

$$(2.20) \qquad\qquad (2)\ \mathcal{E}\left[\|v\|_{L^\infty(0,t; X)}^2\right] \leq C(T) < \infty, \quad \forall T \geq 0,$$

and $\forall s > 1, 0 < \kappa < 1/2$,

$$(2.21) \qquad\qquad (3)\ \mathcal{E}\left[\|v\|_{C_T^{-s, \kappa}}\right] \leq C(T) < \infty, \quad \forall T \geq 0.$$

Here $\mathcal{C}_T^{-s,\kappa}$ denote the class of continuous functions from $[0,T] \to \boldsymbol{X}_{-s}$ such that

$$\|v\|_{\mathcal{C}_T^{-s,\kappa}} = \sup_{0 \leq t_1 < t_2 \leq T, |t_1 - t_2| \leq 1} \frac{\|v(t_2) - v(t_1)\|_{-s}}{|t_2 - t_1|^\kappa} < \infty,$$

and $\mathcal{C}^{-s,\kappa} = \bigcap_{T < \infty} \mathcal{C}_T^{-s,\kappa}$.

3. Existence of optimal controls. In this section we will give a flavor of the type of existence theorems established in our papers. More general theorems can be found in those papers. Results of this section rely on global unique solvability of (1.1)-(1.2).

3.1. Tonelli Type: Ordinary Controls. When adequate form of convexity is available, optimal controls can be sought within the class of ordinary controls. Such existence theorems were established in [21,32,19]. Simplest of this type of theorems can be formulated as follows. Let us consider the case of free terminal condition (i.e $\boldsymbol{Y} = \boldsymbol{X}$) and linear control operator $\mathcal{N}(\boldsymbol{v}, U) = L_N U$. Control set $\boldsymbol{U} \subseteq \boldsymbol{F}$ is a closed convex subset of a Hilbert space $\boldsymbol{F}$. We will consider a special cost functional of the form,

$$(3.1) \qquad \int_0^T \mathcal{L}(r, \boldsymbol{v}(r), U(r)) dr \to \inf$$

where $\mathcal{L}(t, \boldsymbol{v}, \cdot) : \boldsymbol{U} \to \boldsymbol{R}^+$ is a convex function.

THEOREM 3.1. *Let the Lagrangian satisfies the growth condition,*

$$(3.2) \quad \mathcal{L}(t, \boldsymbol{v}, U) \geq \alpha \|\boldsymbol{v}\|_{1/2}^2 + \beta \|U\|_F^2 - \gamma, \quad \forall \boldsymbol{v} \in \boldsymbol{X}_{1/2}, \forall U \in \boldsymbol{F},$$

with $\beta > 0$ and $\alpha, \gamma \geq 0$. Then there exists an optimal control $U \in L^2(0, T; \boldsymbol{F})$.

3.2. Young type: Chattering controls. For general flow control problems existence theorems using Young measures can be established without requiring the convexity of $\mathcal{L}(\boldsymbol{v}, \cdot)$ and of $\boldsymbol{U}$. Moreover, it is possible to handle the nonlinear form of the control operator $\mathcal{N}(\boldsymbol{v}, U)$. We will work with the relaxed evolution system (2.14) and the relaxed cost functional,

$$(3.3) \qquad \int_0^T \int_{\boldsymbol{U}} \mathcal{L}(r, \boldsymbol{v}(r), U) \mu(r, dU) dr \to \inf.$$

The following theorem is a special case of the results in [17].

THEOREM 3.2. *Let the Lagrangian satisfies the growth condition,*

$$(3.4) \quad \mathcal{L}(t, \boldsymbol{v}, U) \geq \alpha \|\boldsymbol{v}\|_{1/2}^2 + \beta \kappa(U)^2 - \gamma, \quad \forall \boldsymbol{v} \in \boldsymbol{X}_{1/2}, \forall U \in \boldsymbol{U},$$

with $\beta > 0$ and $\alpha, \gamma \geq 0$. Then there exists an optimal chattering control $\mu(\cdot) \in \mathcal{V}_{bc}(0, T; \boldsymbol{U}, \kappa)$.

4. Pontryagin maximum principle. In this section, we will present the Pontryagin maximum principle which provides the necessary conditions for optimal controls. In its complete form, this theorem is proven in [16], using the Ekeland's variational principle. A special case of this theorem was proven in [30] using viscosity solution technique for the Hamilton-Jacobi equation.

4.1. Pontryagin maximum principle for ordinary controls. Let us consider the control system (1.1)-(1.2) with control operator (1.3), cost functional (1.5) and target condition $v(T) \in Y \subseteq X$. Let us define the *Pseudo-Hamiltonian,*

$$(4.1) \qquad \tilde{\mathcal{H}}(t, v, p, U) = <p, \mathcal{F}(v, U)> - \mathcal{L}(t, v, U),$$

where $\mathcal{F}(v, U)$ is given by (1.8) and (1.3). Note that we can now write (1.1) as

$$(4.2) \qquad v_t = \nabla_p \tilde{\mathcal{H}}(t, v, p, U).$$

THEOREM 4.1. *Let $\hat{U} \in L^2(\tau, T; U)$ be an optimal control and $\hat{v}(t, \tau, \zeta, \hat{U})$ be the corresponding trajectory with initial data $\zeta \in X_1$ at $t = \tau$. Then there exists an adjoint state $p \in C([\tau, T]; X) \cap L^2(\tau, T; X_{1/2})$ such that*

$$(4.3) \qquad p_t = -\nabla_v \tilde{\mathcal{H}}(t, \hat{v}, \hat{U}, p),$$

with final data,

$$(4.4) \qquad -p(T) - \nabla\phi(\hat{v}(T)) \in N_Y(\hat{v}(T)),$$

where $N_Y(\hat{v}(T))$ is the Clarke normal cone to Y at $\hat{v}(T)$. Moreover,

$$(4.5) \qquad \tilde{\mathcal{H}}(t, \hat{v}, p, \hat{U}) = \max_{U \in U} \tilde{\mathcal{H}}(t, \hat{v}, p, U).$$

4.2. Pontryagin maximum principle for chattering controls. Let us consider the relaxed control system (2.14) with general control operator, initial data (1.2), cost functional (1.5) and target condition $v(T) \in Y \subseteq X$. Let us define the *Relaxed Pseudo-Hamiltonian,*

$$(4.6) \qquad \tilde{\mathcal{H}}_R(t, v, p, \mu) = <p, F(v, \mu)> - L(t, v)\mu,$$

where

$$(4.7) \qquad F(v, \mu) = -\nu A v - B(v) + \int_U N(v, U)\mu(t, dU)$$

and

$$(4.8) \qquad L(t, v)\mu = \int_U \mathcal{L}(t, v, U)\mu(t, dU).$$

Hence also,

$$(4.9) \qquad \tilde{\mathcal{H}}_R(t, \boldsymbol{v}, \boldsymbol{p}, \boldsymbol{\mu}) = \int_{\boldsymbol{U}} \tilde{\mathcal{H}}(t, \boldsymbol{v}, \boldsymbol{p}, U) \mu(t, dU).$$

Note that we can now write (2.14) as

$$(4.10) \qquad \boldsymbol{v}_t = \nabla_p \tilde{\mathcal{H}}_R(t, \boldsymbol{v}, \boldsymbol{p}, \boldsymbol{\mu}).$$

The following generalization of the Pontryagin maximum principle can be proven by methods analogous to those used in the previous theorem. Details of the proof is given in [33].

THEOREM 4.2. *Let* $\hat{\boldsymbol{\mu}} \in \mathcal{V}_{bc}(0, T; \boldsymbol{U}, \kappa)$ *be an optimal control and* $\hat{\boldsymbol{v}}(t, \tau, \boldsymbol{\zeta}, \hat{\boldsymbol{\mu}})$ *be the corresponding trajectory with initial data* $\boldsymbol{\zeta} \in \boldsymbol{X}_1$ *at* $t = \tau$. *Then there exists an adjoint state* $\boldsymbol{p} \in C([\tau, T]; \boldsymbol{X}) \cap L^2(\tau, T; \boldsymbol{X}_{1/2})$ *such that*

$$(4.11) \qquad \boldsymbol{p}_t = -\nabla_v \tilde{\mathcal{H}}_R(t, \hat{\boldsymbol{v}}, \hat{\boldsymbol{\mu}}, \boldsymbol{p}),$$

with final data,

$$(4.12) \qquad -\boldsymbol{p}(T) - \nabla \phi(\hat{\boldsymbol{v}}(T)) \in \boldsymbol{N}_Y(\hat{\boldsymbol{v}}(T)),$$

where $\boldsymbol{N}_Y(\hat{\boldsymbol{v}}(T))$ *is the Clarke normal cone to* $\boldsymbol{Y}$ *at* $\hat{\boldsymbol{v}}(T)$. *Moreover,*

$$(4.13) \qquad \tilde{\mathcal{H}}_R(t, \hat{\boldsymbol{v}}, \boldsymbol{p}, \hat{\boldsymbol{\mu}}) = \max_{\boldsymbol{\mu} \in \Re \boldsymbol{U}} \tilde{\mathcal{H}}_R(t, \hat{\boldsymbol{v}}, \boldsymbol{p}, \boldsymbol{\mu}).$$

5. Dynamic programming and feedback analysis: Deterministic case. In this section we will describe the results on feedback control theory originated in [30] and elaborated in [16,31]. Let us define the *value function* for the control problem (1.1),(1.2) and the relaxed cost corresponding to (1.5) as

$$\mathcal{V}(\tau, \boldsymbol{\zeta}) = \min \left\{ \phi(\boldsymbol{v}(T))) + \int_\tau^T \boldsymbol{L}(t, \boldsymbol{v}(t)) \boldsymbol{\mu}(t) dt; \quad \boldsymbol{\mu}(\cdot) \in \mathcal{V}_{bc}(\tau, T; \boldsymbol{U}, \kappa) \right\},$$

with a similar definition for the infinite horizon case. From this definition it is possible to establish the *Bellman principle of optimality*:

$$\mathcal{V}(\tau, \boldsymbol{\zeta}) = \inf \left\{ \int_\tau^t \boldsymbol{L}(r, \boldsymbol{v}(r)) \boldsymbol{\mu}(r) dr + \mathcal{V}(t, \boldsymbol{v}(t)); \quad \boldsymbol{\mu}(\cdot) \in \mathcal{V}_{bc}(\tau, t; \boldsymbol{U}, \kappa) \right\}.$$

For the case of infinite horizon this principle takes the form,

$$\mathcal{V}(\boldsymbol{\zeta}) = \inf \left\{ \int_0^t e^{-\lambda t} \boldsymbol{L}(r, \boldsymbol{v}(r)) \boldsymbol{\mu}(r) dr + \mathcal{V}(\boldsymbol{v}(t)); \quad \boldsymbol{\mu}(\cdot) \in \mathcal{V}_{bc}(0, t; \boldsymbol{U}, \kappa) \right\}.$$

From this, if the value function is differentiable we obtain the infinite dimensional, first order, **Hamilton-Jacobi-Bellman** equation,

$$(5.1) \qquad \partial_\tau \mathcal{V} - \mathcal{H}_R(\tau, \zeta, -\partial_\zeta \mathcal{V}) = 0,$$

with

$$\mathcal{V}(T, \zeta) = \phi(\zeta), \quad \forall \zeta \in \boldsymbol{X}.$$

Here the **true Hamiltonian** $\mathcal{H}_R(\cdot, \cdot, \cdot)$ is defined as

$$(5.2) \qquad \mathcal{H}_R(t, \boldsymbol{v}, \boldsymbol{p}) = \max_{\mu \in \Re \boldsymbol{U}} \tilde{\mathcal{H}}_R(t, \boldsymbol{v}, \boldsymbol{p}, \boldsymbol{\mu}).$$

For the infinite Horizon case we get

$$(5.3) \qquad \lambda \mathcal{V} + \mathcal{H}_R(\zeta, -\partial_\zeta \mathcal{V}) = 0.$$

We will note the following important result. If we define a true Hamiltonian $\mathcal{H}(\cdot, \cdot, \cdot)$ using ordinary controls,

$$(5.4) \qquad \mathcal{H}(t, \boldsymbol{v}, \boldsymbol{p}) = \sup_{U \in \boldsymbol{U}} \tilde{\mathcal{H}}(t, \boldsymbol{v}, \boldsymbol{p}, U),$$

then, we have

PROPOSITION 5.1.

$$(5.5) \qquad \mathcal{H}_R(t, \boldsymbol{v}, \boldsymbol{p}) = \mathcal{H}(t, \boldsymbol{v}, \boldsymbol{p}).$$

Proof: Note that in (4.9), taking a Dirac measure $\boldsymbol{\mu} = \delta_U$ we obtain,

$$(5.6) \qquad \mathcal{H}(t, \boldsymbol{v}, \boldsymbol{p}) \leq \mathcal{H}_R(t, \boldsymbol{v}, \boldsymbol{p}).$$

Now,

$$\tilde{\mathcal{H}}_R(t, \boldsymbol{v}, \boldsymbol{p}, \boldsymbol{\mu}) = \int_{\boldsymbol{U}} \tilde{\mathcal{H}}(t, \boldsymbol{v}, \boldsymbol{p}, U)\mu(t, dU)$$

$$\leq \int_{\boldsymbol{U}} \sup_{U \in \boldsymbol{U}} \tilde{\mathcal{H}}(t, \boldsymbol{v}, \boldsymbol{p}, U)\mu(t, dU)$$

$$(5.7)$$

$$= \mathcal{H}(t, \boldsymbol{v}, \boldsymbol{p}) \int_{\boldsymbol{U}} \mu(t, dU) = \mathcal{H}(t, \boldsymbol{v}, \boldsymbol{p}).$$

(5.2), (5.7) and (5.6) give us (5.5).

In general the value function does not have sufficient differentiability. In fact as presented in Theorem 9 below, we only know that it is locally

Lipschitz continuous. Therefore, equation (5.1) needs to be interpreted either in the sense of nonsmooth analysis [8,9] or in the sense of viscosity solutions [11,10,12,13,14]. We will begin by recalling certain usful notions of derivatives.

DEFINITION 5.1. *Let* $f : \boldsymbol{X} \to \boldsymbol{R}$ *be locally Lipschitz function.*
(i) The superdifferential of f at a point $\boldsymbol{x} \in \boldsymbol{X}$ is the subset of $\boldsymbol{X}$ defined as

$$(5.8) \quad \partial^{+} f(\boldsymbol{x}) = \left\{ \zeta \in \boldsymbol{X}; \limsup_{\boldsymbol{y} \to \boldsymbol{x}} \left[\frac{f(\boldsymbol{y}) - f(\boldsymbol{x}) - (\zeta, \boldsymbol{y} - \boldsymbol{x})}{\|\boldsymbol{y} - \boldsymbol{x}\|} \right] \leq 0 \right\}.$$

(ii) The subdifferential of f at a point $\boldsymbol{x} \in \boldsymbol{X}$ is the subset of $\boldsymbol{X}$ defined as

$$(5.9) \quad \partial^{-} f(\boldsymbol{x}) = \left\{ \zeta \in \boldsymbol{X}; \liminf_{\boldsymbol{y} \to \boldsymbol{x}} \left[\frac{f(\boldsymbol{y}) - f(\boldsymbol{x}) - (\zeta, \boldsymbol{y} - \boldsymbol{x})}{\|\boldsymbol{y} - \boldsymbol{x}\|} \right] \geq 0 \right\}.$$

(iii) The Clarke generalized gradient of f at a point $\boldsymbol{x} \in \boldsymbol{X}$ is the subset of $\boldsymbol{X}$ defined as

$$(5.10) \qquad \partial f(\boldsymbol{x}) = \left\{ \zeta \in \boldsymbol{X}; f^{0}(\boldsymbol{x}; \boldsymbol{v}) \geq (\zeta, \boldsymbol{v}), \quad \forall \boldsymbol{v} \in \boldsymbol{X} \right\},$$

where $f^{0}(\boldsymbol{x}; \boldsymbol{v})$ denotes the directional derivative,

$$(5.11) \qquad f^{0}(\boldsymbol{x}; \boldsymbol{v}) = \limsup_{t \downarrow 0, \boldsymbol{y} \to \boldsymbol{x}} \frac{f(\boldsymbol{y} + t\boldsymbol{v}) - f(\boldsymbol{y})}{t}.$$

PROPOSITION 5.2. *: Let $f : \boldsymbol{X} \to \boldsymbol{R}$ be a locally Lipschitz function. Then for any point $\boldsymbol{x} \in \boldsymbol{X}$,*
(i) the sets $\partial^{+} f(\boldsymbol{x})$, $\partial^{-} f(\boldsymbol{x})$ and $\partial f(\boldsymbol{x})$ are closed convex,

$$(5.12) \qquad (ii) \quad -\partial^{-}(-f)(\boldsymbol{x}) = \partial^{+} f(\boldsymbol{x}),$$

$$(5.13) \qquad (iii) \quad -\partial(-f)(\boldsymbol{x}) = \partial f(\boldsymbol{x}),$$

$$(5.14) \qquad and \quad (iv) \quad \partial^{-} f(\boldsymbol{x}) \bigcup \partial^{+} f(\boldsymbol{x}) \subseteq \partial f(\boldsymbol{x}) \subseteq \boldsymbol{X}.$$

We now return to the Hamilton-Jacobi-Bellman equation (5.1) and define generalized solutions.

DEFINITION 5.2. **(Viscosity solutions)** *Let $\mathcal{V} : [0, T] \times \boldsymbol{X} \to \boldsymbol{R}$ be a locally Lipschitz function. Then $\mathcal{V}$ is called a viscosity subsolution to the equation (1.1) if for each $(t, \boldsymbol{y}) \in (0, T) \times \boldsymbol{X}_{1}$,*

$$(5.15) \qquad -\zeta + \mathcal{H}(t, \boldsymbol{y}, -\boldsymbol{\xi}) \leq 0, \quad \forall(\zeta, \boldsymbol{\xi}) \in \partial^{+} \mathcal{V}(t, \boldsymbol{y})$$

and viscosity supersolution if for each $(t, \boldsymbol{y}) \in (0, T) \times \boldsymbol{X}_{1}$,

$$(5.16) \qquad -\zeta + \mathcal{H}(t, \boldsymbol{y}, -\boldsymbol{\xi}) \geq 0, \quad \forall(\zeta, \boldsymbol{\xi}) \in \partial^{-} \mathcal{V}(t, \boldsymbol{y}).$$

If $\mathcal{V}$ satisfies (5.15) and (5.16) then it is called a viscosity solution.

DEFINITION 5.3. (**Clarke generalized solutions**) *Let $\mathcal{V} : [0,T] \times X \to R$ be a locally Lipschitz function. Then $\mathcal{V}$ is called a Clarke generalized solution to the equation (5.1) if for each $(t, y) \in (0, T) \times X_1$,*

$$(5.17) \qquad \max\{-\zeta + \mathcal{H}(t, y, -\xi); \quad (\zeta, \xi) \in \partial \mathcal{V}(t, y)\} = 0.$$

Using the continuity properties of the cost functional and continuous dependence theorems for the state with respect to the data τ, ζ and control U, the following theorem is proved in [30,29,16] for the control problem (1.1), (1.2), (1.3) and (1.5) with quadratic $\phi(\cdot) : X \to R$ and quadratic Lagrangian $\mathcal{L}(\cdot, \cdot) : X_{1/2} \times F \to R$.

THEOREM 5.1. **Verification Theorem** *The value function $\mathcal{V}(\cdot, \cdot) \in C([0,T] \times X)$. For each $t \in [\tau, T]$, $\mathcal{V}(t, \cdot)$ is locally Lipchitz in X and for each $\zeta \in X_1$, $\mathcal{V}(\cdot, \zeta)$ is absolutely continuous in $t \in (\tau, T)$. Moreover, $\mathcal{V}(\cdot, \cdot) : [\tau, T] \times X \to R$ is a viscosity solution of the Hamilton-Jacobi-Bellman equation and $\forall y \in X_1$,*

$$-\zeta + \mathcal{H}(t, y, -\xi) = 0, \; for \; \mathrm{some} \; (\zeta, \xi) \in \partial^+ \mathcal{V}(t, y),$$

for t a.e in $[0, T]$.

A major open problem in feedback control theory of Navier-Stokes equations is the uniqueness of viscosity solution.

For the above class of flow control problems, we have the following connection between the two types of generalized solutions. A similar result for the finite dimensional case was proven by Frankowska [20]. Our case, however, is considerably more involved because of the infinite dimentionality and the *unboundedness* of the Hamiltonian. The following hypothesis is motivated by a weaker result of Preiss [27].

DIFFERENTIABILITY HYPOTHESIS 5.2. *Let $\partial \mathcal{V}(t, y)$ be the Clarke derivative of $\mathcal{V}$ at some arbitrary point $(t, y) \in [0, T] \times X_1$. Then*

$$\partial \mathcal{V}(t, y) = \bigcap_{\epsilon > 0} \overline{co} \{\nabla \mathcal{V}(r, x) \in R \times X; \quad (r, x) \in B_{R \times X_1}((t, y); \epsilon)\}$$

where the closure here is in the weak topology.

THEOREM 5.3. *Let $\mathcal{V} : [0, T] \times X \to R$ be locally Lipschitz.*
(I) Suppose that $\mathcal{V}$ is a viscosity solution, satisfies $\forall (t, y) \in (0, T) \times X_1$,

$$(5.18) \qquad -\zeta + \mathcal{H}(t, y, -\xi) = 0, \qquad \mathrm{for \; some} \; (\zeta, \xi) \in \partial \mathcal{V}(t, y)$$

and satisfies the differentiability hypothesis I. Then $\mathcal{V}$ is also a Clarke generalized solution and $\forall (t, y) \in (0, T) \times X_1$,

$$(5.19) \qquad -\zeta + \mathcal{H}(t, y, -\xi) = 0, \; \forall (\zeta, \xi) \in \partial^- \mathcal{V}(t, y).$$

On the other hand,
(II) if $\mathcal{V}$ is a Clarke generalized solution and satisfies (5.19) then $\mathcal{V}$ is also a viscosity solution.

NONSMOOTH ANALYSIS AND FREE BOUNDARY PROBLEMS FOR POTENTIAL FLOW

SRDJAN STOJANOVIC*

Abstract. New approach to some Free boundary problems, is introduced. Those problems are studied first by Alt and Caffarelli [2] in the case of a potential flow. Their approach seem not to be possible to extend to the case of a Stokes flow. In this paper, the variable domain problem is relaxed so that it becomes a nonsmooth optimization problem on the fixed domain for the somewhat singular state equation. State equation is considered, and the multivalued generalized gradient of the variational functional is studied. Here, we considered Potential flow.

1. Statement of the problem. In this paper we introduce a new approach to some Free boundary problems in fluid mechanics. Here, we shall consider only the case of a Potential flow, but the same method can be adopted for the Stokes flow, which will be the subject of the fortcoming paper.

Consider the set of admissible shapes, i.e., the control set

$$(1.1) \qquad U = \left\{ u \in H_0^3(-1, 1); 0 \le u(x) \le 1, -1 < x < 1 \right\}.$$

Denote

$$(1.2) \qquad \Gamma_u = \{(x, u(x)); -1 < x < 1\},$$

and extend $u \in U$ as zero outside of $(-1, 1)$. Define the domain

$$(1.3) \qquad \Omega_u = \{(x, y); |x| < a, u(x) < y < 2\}.$$

Also, let $\Omega \supset \Omega_u$ be a domain defined by

$$(1.4) \qquad \Omega = \{(x, y); |x| < a, 0 < y < 2\}.$$

Now, consider inviscid, incompressible, irrotational flow in a finite channel Ω with an immersed obstacle Γ_u with shape $u \in U$. So, the flow actually takes place in Ω_u.

It is well known that such a flow can be described by the stream function $w = w^u$, which is a solution of (to fix ideas, we take flux to be equal to one)

$$\Delta w = 0 \text{ in } \Omega_u$$
$$w = 0 \text{ in } \{(x, 0); -a < x < -1 \text{ or } 1 < x < a\}$$
$$w = 0 \text{ in } \Gamma_u$$
$$w = 2 \text{ in } \{(x, 2); -a < x < a\}$$
$$(1.5) \qquad w_x = 0 \text{ in } \{(\pm a, y); 0 < y < 1\}.$$

---

* Supported in part by the NSF Grant DMS-91-11794.

Department of Mathematical Sciences, University of Cincinnati, Cincinnati, OH 45221-0025.

We could also take $a = \infty$ in (1.5), i.e., consider a flow in an infinite channel. Then the last condition in (1.5) is substituted by the requirement that w is bounded.

If the stream function w is known then, of course, the velocity vector field $\mathbf{v}$ can be computed easily as $\mathbf{v} = < w_y, -w_x >$.

The problem we propose is the following:

For given $g = g(x, y) \in W^{2,q}(\Omega)$, $q > 2$, (occasionally we will not have to assume that much) and such that

$$(1.6) \qquad\qquad\qquad\qquad\qquad\qquad g \geq 0,$$

$$(1.7) \qquad\qquad\qquad\qquad g = 0 \text{ in } \Omega \cap \{|x| > 1\},$$

find (if possible) $u \in U$ (the shape of the immersed obstacle) such that, if w^u is the corresponding solution of (1.5), then also

$$(1.8) \qquad\qquad\qquad\qquad |\mathbf{v}| = |\nabla w^u| = g \text{ in } \Gamma_u.$$

By Bernoulli's law

$$(1.9) \qquad\qquad\qquad\qquad P + \frac{1}{2}|\nabla w^u|^2 = const.$$

throughout the fluid (here P denotes the pressure). Hence, we see that the requesting specific velocity profile on the immersed obstacle is equivalent to requesting the specific pressure profile. Obviously, this is a problem with wide possibilities for applications.

We note that this problem is closely related to the following, by now, well known variational problem (see [2,3,6]; see also [11] for numerical considerations):

Find $w \in H^1(\Omega)$ satisfying the boundary conditions

$$(1.10) \qquad \begin{aligned} w &= 0 \text{ in } \{(x,0); -a < x < a\} \\ w &= 2 \text{ in } \{(x,2); -a < x < a\} \end{aligned}$$

and such that the variational functional

$$(1.11) \qquad\qquad \mathbf{J}(w) = \int_\Omega \left[|\nabla w|^2 + g^2 \mathbf{I}_{\{w > 0\}} \right]$$

is minimized. Here $\mathbf{I}_D$ is a characteristic function of the set D, i.e.,

$$(1.12) \qquad\qquad \mathbf{I}_D(x) = \begin{cases} 1 & \text{if } x \in D \\ 0 & \text{if } x \notin D \end{cases}.$$

The reason to develop the method introduced here is that it extends to other equations for which there is no known analog of $\mathbf{J}$ (for example, Stokes problem, to be discussed in Part 2).

2. A relaxation of the problem. In this section we assume that

$$\text{(2.1)} \qquad\qquad g \in H^2(\Omega).$$

Suppose that there exists an $u \in U$ such that corresponding w^u solves (1.5,1.8). We shall say then that u is an *exact control*. Now, extend w^u from Ω_u to Ω as z^u:

$$\text{(2.2)} \qquad\qquad z^u = \begin{cases} w^u & \text{on } \Omega_u \\ 0 & \text{on } \Omega \backslash \Omega_u \end{cases}.$$

It follows

LEMMA 2.1. *If $u \in U$ is an exact control, then $z^u \in H^1(\Omega)$, and it is a solution of the following elliptic boundary value problem (with singular right hand side)*

$$\Delta z^u = \xi_u \ in \ \Omega$$
$$z^u = 0 \ in \ \{(x,0); -a < x < a\}$$
$$z^u = 2 \ in \ \{(x,2); -a < x < a\}$$
$$\text{(2.3)} \qquad (z^u)_x = 0 \ in \ \{(\pm a, y); 0 < y < 1\}$$

where $\xi_u \in H^{-1}(\Omega)$ is a measure given by

$$\text{(2.4)} \qquad\qquad \xi_u(\varphi) = \int_{\Gamma_u} g\varphi d\sigma.$$

Proof: Obviously, since by elliptic estimates w^u is regular in Ω_u, $z^u \in C^{0,1}(\bar{\Omega})$ (regarding regularity near corners see the begining of the proof of the Theorem 3.1), and in particular $z^u \in H^1(\Omega)$.

By the Trace Theorem,

$$|\xi_u(\varphi)| = \left| \int_{\Gamma_u} g\varphi d\sigma \right| \leq \|g\|_{L^2(\Gamma_u)} \|\varphi\|_{L^2(\Gamma_u)}$$
$$\text{(2.5)} \qquad\qquad\qquad \leq c_u \|g\|_{H^1(\Omega)} \|\varphi\|_{H^1(\Omega)}.$$

So, in particular, $\xi_u \in H^{-1}(\Omega)$. Also, since $g \geq 0$, ξ_u is a measure.

Now, more explicitly, (2.3) can be written as: Find $z \in H^1(\Omega)$ such that

$$z^u = 0 \ \text{in} \ \{(x,0); -a < x < a\}$$
$$\text{(2.6)} \qquad z^u = 2 \ \text{in} \ \{(x,2); -a < x < a\}$$

and

$$\text{(2.7)} \qquad\qquad -\int_\Omega \nabla z^u \cdot \nabla \varphi = \int_{\Gamma_u} g\varphi d\sigma$$

for all $\varphi \in H^1(\Omega)$ such that

$$(2.8) \qquad \varphi = 0 \text{ in } \{(x,0); -a < x < a\} \cup \{(x,2); -a < x < a\}.$$

To check (2.7), we note that by the maximum principle, a solution of (1.5) is positive. Hence (1.8) and the boundary condition in (1.5) imply that

$$(2.9) \qquad \frac{\partial w^u}{\partial \nu_u} = -g \text{ in } \Gamma_u,$$

where ν_u is the exterior unit normal to $\partial \Omega_u$. Hence,

$$-\int_\Omega \nabla z^u \cdot \nabla \varphi = -\int_{\Omega_u} \nabla w^u \cdot \nabla \varphi =$$

$$(2.10) \qquad = \int_{\Omega_u} (\Delta w^u)\varphi - \int_{\partial \Omega_u} \frac{\partial w^u}{\partial \nu_u} \varphi d\sigma = \int_{\Gamma_u} g\varphi d\sigma,$$

which completes the proof of the Lemma. $\qquad \square$

LEMMA 2.2. *Let z^u be a solution of (2.6-2.8). If it happens that $z^u|_{\Gamma_u} = 0$, then $z^u|_{\Omega_u}$ is a solution of (1.5,1.8), i.e., u is an exact control.*

Proof: In the next section we shall prove that z^u is regular enough so that calculations performed here are legitimate. More precisely, by (3.6) below, it suffices to assume that $\varphi \in C_0^1(\Omega)$. We have

$$\int_{\Gamma_u} g\varphi d\sigma = -\int_{\Omega_u} \nabla z^u \cdot \nabla \varphi - \int_{\Omega \setminus \Omega_u} \nabla z^u \cdot \nabla \varphi$$

$$= \int_{\Omega_u} (\Delta z^u)\varphi + \int_{\Omega \setminus \Omega_u} (\Delta z^u)\varphi$$

$$(2.11) \qquad - \int_{\partial \Omega_u} \frac{\partial z^u}{\partial \nu} \varphi d\sigma - \int_{\partial(\Omega \setminus \Omega_u)} \frac{\partial z^u}{\partial \nu} \varphi d\sigma.$$

Let ν be exterior to Ω_u, and let

$$z^{u,int} \overset{def}{=} z^u|_{\Omega \setminus \Omega_u}$$

$$(2.12) \qquad z^{u,ext} \overset{def}{=} z^u|_{\Omega_u}.$$

Then (2.11) implies that

$$(2.13) \qquad \int_{\Gamma_u} g\varphi d\sigma = \int_{\Gamma_u} \left(\frac{\partial z^{u,int}}{\partial \nu} - \frac{\partial z^{u,ext}}{\partial \nu} \right) \varphi d\sigma, \ \forall \varphi \in C_0^1(\Omega).$$

So,

$$(2.14) \qquad g = \frac{\partial z^{u,int}}{\partial \nu} - \frac{\partial z^{u,ext}}{\partial \nu} \text{ on } \Gamma_u.$$

We observe that (2.14) always holds for the solution of (2.3).

Now, if $z^u|_{\Gamma_u} = 0$, then $z^u|_{\Omega \setminus \Omega_u} = 0$, so that $\frac{\partial z^{u,int}}{\partial \nu} = 0$, and then

$$(2.15) \qquad g = -\frac{\partial z^{u,ext}}{\partial \nu} \text{ on } \Gamma_u,$$

i.e.,

$$(2.16) \qquad g = |\nabla (z^u|_{\Omega_u})| \text{ on } \Gamma_u,$$

i.e., (1.8) holds. $\qquad\qquad\qquad\qquad\qquad\qquad\qquad\qquad\qquad\qquad\qquad\square$

Lemma 2.2 motivates the following

DEFINITION 2.1. *$u^* \in U$ is said to solve the* relaxed shape optimization problem *if the corresponding z^u defined by (2.3), is such that*

$$(2.17) \qquad \Phi(u) = \frac{1}{2} \int_{\Gamma_u} (z^u)^2 d\sigma$$

is minimized, i.e., that there exists an $u^* \in U$ such that

$$(2.18) \qquad \Phi(u^*) = \min_{u \in U} \Phi(u).$$

Of course, an exact control is a minimizer, i.e., a solution of (2.18). On the other hand, a solution of (2.18) is an exact control provided an exact control exists.

We do not consider the exact controllability. Rather, we shall study the relaxed problem introduced in Definition 2.1.

3. The state equation. It will be convenient to state the regularity theorem for the general boundary value. So let ψ be a given function on Ω such that $z^u = \psi$ on $\partial_0 \Omega \subset \partial \Omega$. We assume that the boundary and ψ are sufficiently regular (see [12] for details; also we shall give some details in the case of the boundary and boundary values in our case). For any $z \in H^1(\Omega)$, we define $\|z\|_{L^\infty(\partial\Omega)}$ as

$$(3.1) \qquad \|z\|_{L^\infty(\partial\Omega)} \stackrel{\text{def}}{=} \inf \left\{ m \geq 0; -m \leq z \leq m \text{ on } \partial\Omega \text{ in } H^1(\Omega) \right\},$$

where inequalities in $H^1(\Omega)$ are defined in e.g. [9]. Also, we define

$$(3.2) \qquad W^{2,q}_{\{y>0\}-loc}(\Omega) \stackrel{\text{def}}{=} \cap_{\epsilon>0} W^{2,q}(\Omega \cap \{y > \epsilon\}).$$

We have

THEOREM 3.1. *For any $u \in U$ the state equation (2.3) has a unique weak solution. Let q be such that $2 \leq q < \infty$. If $g \in W^{1,q}(\Omega)$ then*

$$(3.3) \qquad z^u \in W^{1,q}(\Omega) \cap C^\infty(\bar{\Omega} \setminus \Gamma_u),$$

and the apriori estimate

$$(3.4) \qquad \|z^u\|_{W^{1,q}(\Omega)} \leq c \left(1 + \|u\|_{C^{0,1}(-1,1)}\right) \left(\|g\|_{W^{1,q}(\Omega)} + \|\psi\|_{W^{1,q}(\Omega)}\right).$$

holds. If in addition $q > 2$ then also

$$(3.5) \qquad \|z^u\|_{L^\infty(\Omega)} \leq c \left(1 + \|u\|_{C^{0,1}(-1,1)}\right) \left(\|g\|_{W^{1,q}(\Omega)} + \|\psi\|_{L^\infty(\partial\Omega)}\right).$$

Moreover, if $g \in W^{2,q}(\Omega)$, and (1.7) holds, then (see (2.12))

$$(3.6) \qquad z^{u,ext} \in W^{2,q}(\Omega_u), \quad z^{u,int} \in W^{2,q}_{\{y>0\}-loc}(\Omega \setminus \bar\Omega_u).$$

and the apriori estimates

$$(3.7) \qquad \|z^{u,ext}\|_{W^{2,q}(\Omega_u)} \leq c \left(\|u\|_{H^3(-1,1)}, \|g\|_{W^{2,q}(\Omega)}, \|\psi\|_{W^{2,q}(\Omega)}\right),$$

and

$$\|z^{u,int}\|_{W^{2,q}((\Omega\setminus\Omega_u)\cap\{y>\epsilon\})} \leq$$
$$(3.8) \qquad \leq c \left(\epsilon, \|u\|_{H^3(-1,1)}, \|g\|_{W^{2,q}(\Omega)}, \|\psi\|_{W^{2,q}(\Omega)}\right).$$

hold.

Proof: Since $\xi_u \in H^{-1}(\Omega)$ existence and uniqueness of a weak solution z^u of (2.3) is trivial. Also, since z^u is harmonic in $\Omega \setminus \Gamma_u$, it follows that $z^u \in C^\infty(\bar\Omega \setminus \Gamma_u)$. Few words are needed here due to the presence of corners in Ω. To prove regularity of z^u in the neighborhood of corners, say in the neighborhood of $(-a, 0)$, one can extend z^u in $\{x < -a, 0 < y < 2\}$ as $\widetilde{z^u}$ by the formula

$$(3.9) \qquad \widetilde{z^u}(x, y) \overset{\text{def}}{=} \begin{cases} z^u(-2a - x, y) & \text{if } x < -a \\ z^u(x, y) & \text{if } x \geq -a. \end{cases}$$

Then since $\widetilde{z^u}$ is continuous on $\{x = -a\}$ and $\widetilde{z^u}_x = 0$ on $\{x = -a\}$, it is elementary to show that $\widetilde{z^u}$ is harmonic across $\{x = -a\}$. Indeed, let $B_\rho(A) = B_1 \cup (B_\rho(A) \cap \{x = -a\}) \cup B_2 \subset \{0 < y < 2\}$ be a ball centered at $A \in \{x = -a\}$ with radius ρ. Here, $B_1 = B_\rho(A) \cap \{x > -a\}$ and $B_2 = B_\rho(A) \cap \{x < -a\}$. Then,

$$\int_{B_\rho(A)} \widetilde{z^u} \Delta\varphi = \int_{B_1} \widetilde{z^u} \Delta\varphi + \int_{B_2} \widetilde{z^u} \Delta\varphi =$$
$$(3.10) \qquad \int_{\{x=-a\}\cap B_\rho(A)} \left[-\varphi_x \widetilde{z^u} + \varphi \widetilde{z^u}_x + \varphi_x \widetilde{z^u} - \varphi \widetilde{z^u}_x\right] dy = 0,$$

for all $\varphi \in C_0^\infty(B_\rho(A))$, so that $\widetilde{z^u}$ is harmonic across $\{x = -a\}$ as clamed. Henceforth $\widetilde{z^u}$ is as regular in the neighborhood of $(-a, 0)$ as the (extended) boundary data is. In particular, in our case $\psi = 0$ there, so that (3.3) follows.

Set

$$(3.11) \qquad \varphi = \psi - z^u$$

in (2.7). It easily follows

$$\int_\Omega |\nabla z^u|^2 = \int_{\Gamma_u} g(\psi - z^u)d\sigma - \int_\Omega \nabla z^u \cdot \nabla \psi$$

$$\leq \left(\int_{\Gamma_u} g^2 d\sigma\right)^{\frac{1}{2}} \left[\left(\int_{\Gamma_u} \psi^2 d\sigma\right)^{\frac{1}{2}} + \left(\int_{\Gamma_u} (z^u)^2 d\sigma\right)^{\frac{1}{2}}\right]$$

$$(3.12) \qquad\qquad\qquad\qquad\qquad + \left|\int_\Omega \nabla z^u \cdot \nabla \psi\right|.$$

Now since $z^u = (z^u - \psi) + \psi$, using Poincaré inequality, we have

$$\|z^u\|_{H^1(\Omega)} \leq c \left(\|\nabla(z^u - \psi)\|_{L^2(\Omega)} + \|\psi\|_{H^1(\Omega)}\right)$$

$$(3.13) \qquad\qquad\qquad \leq c \left(\|\nabla z^u\|_{L^2(\Omega)} + \|\psi\|_{H^1(\Omega)}\right)$$

Combining (3.12,3.13) we get

$$\|z^u\|_{H^1(\Omega)}^2 \leq c \left(1 + \|u\|_{C^{0,1}(-1,1)}\right) \left[\|g\|_{H^1(\Omega)} \left(\|\psi\|_{H^1(\Omega)} + \|z^u\|_{H^1(\Omega)}\right)\right]$$

$$(3.14) \qquad\qquad\qquad + c\|z^u\|_{H^1(\Omega)}\|\psi\|_{H^1(\Omega)} + c\|\psi\|_{H^1(\Omega)}^2.$$

In (3.14) the inequality follows from the proof of the Trace Theorem (see e.g. [10], or [5]). Indeed, one can see ([5], p. 132) that for $1 \leq q < \infty$ one has

$$(3.15) \qquad \|z^u\|_{L^q(\Gamma_u)}^q \leq c \left(1 + \|u\|_{C^{0,1}(-1,1)}^2\right)^{\frac{1}{2}} \|z^u\|_{W^{1,q}(\Omega)}^q,$$

which implies

$$(3.16) \qquad \|z^u\|_{L^q(\Gamma_u)} \leq c \left(1 + \|u\|_{C^{0,1}(-1,1)}\right)^{\frac{1}{q}} \|z^u\|_{W^{1,q}(\Omega)},$$

¿From (3.14) we easily conclude that (3.4) holds for $q = 2$.

Proceeding, (we assume $\frac{1}{q} + \frac{1}{q'} = 1$)

$$|\xi_u(\varphi)| \leq \|g\|_{L^q(\Gamma_u)}\|\varphi\|_{L^{q'}(\Gamma_u)}$$

$$\leq c \left(1 + \|u\|_{C^{0,1}(-1,1)}\right)^{\frac{1}{q}} \|g\|_{W^{1,q}(\Omega)} \left(1 + \|u\|_{C^{0,1}(-1,1)}\right)^{\frac{1}{q'}} \|\varphi\|_{W^{1,q'}(\Omega)}$$

$$(3.17) \qquad\qquad = c \left(1 + \|u\|_{C^{0,1}(-1,1)}\right) \|g\|_{W^{1,q}(\Omega)}\|\varphi\|_{W^{1,q'}(\Omega)}.$$

So, $\xi_u \in \left(W^{1,q'}(\Omega)\right)^*$ (here X^* represents the dual space of the space X) and

$$(3.18) \qquad \|\xi_u\|_{(W^{1,q'}(\Omega))^*} \leq c \left(1 + \|u\|_{C^{0,1}(-1,1)}\right) \|g\|_{W^{1,q}(\Omega)}.$$

We know (see e.g. [1]) that ξ has a representation

$$(3.19) \qquad \xi(\varphi) = \int_\Omega [f_0\varphi + f_1\varphi_x + f_2\varphi_y],$$

for some $f_i \in L^q(\Omega), i = 0, 1, 2$, and

$$(3.20) \qquad \|\xi_u\|_{(W^{1,q'}(\Omega))^*} = \sum_{i=0}^{2} \|f_i\|_{L^q(\Omega)}.$$

Now from elliptic regularity (see [12], p. 179) we have

$$(3.21) \qquad \|z^u\|_{W^{1,q}(\Omega)} \le c \left(\sum_{i=0}^{2} \|f_i\|_{L^q(\Omega)} + \|\psi\|_{W^{1,q}(\Omega)} + \|z^u\|_{H^1(\Omega)} \right).$$

¿From (3.18,3.20,3.21) and since we already proved (3.4) in the case $q = 2$, we conclude that (3.4) holds.

To prove (3.5), we recall (see e.g. [12] p. 103) that if $q > 2$ and if $z^u \le 0$ on $\partial_0\Omega$ in the sense of $H^1(\Omega)$, then

$$(3.22) \qquad \operatorname*{ess\,sup}_\Omega z^u \le c \left(\sum_{i=0}^{2} \|f_i\|_{L^q(\Omega)} + \|z^u\|_{L^2(\Omega)} \right).$$

Hence,

$$\operatorname*{ess\,sup}_\Omega \left(z^u - \|z^u\|_{L^\infty(\partial\Omega)} \right) \le$$

$$(3.23) \qquad \le c \left(\sum_{i=0}^{2} \|f_i\|_{L^q(\Omega)} + \|z^u\|_{L^2(\Omega)} + \|z^u\|_{L^\infty(\partial\Omega)} \right),$$

and similarly for $-z^u + \|z^u\|_{L^\infty(\partial\Omega)}$. This easily implies (3.5).

Now, we shall consider further regularity of $z^u|_{\Omega_u}$ and $z^u|_{\Omega\backslash\Omega_u}$. Since the singular set is on Γ_u, we expect higher regularity in the tangential direction. To prove that this is the case we flatten the Γ_u first, since then it is easier to differentiate.

Define v, $\tilde{g}$ and $\tilde{\varphi}$ by

$$(3.24) \qquad v(x,y) = z^u(x, y + u(x)),$$

$$(3.25) \qquad \tilde{g}(x,y) = g(x, y + u(x))\sqrt{1 + u'^2(x)},$$

$$(3.26) \qquad \tilde{\varphi}(x,y) = \varphi(x, y + u(x))$$

and operator L by

$$(3.27) \qquad Lv = \Delta v + v_{yy}(u_x)^2 - 2v_{xy}u_x - v_y u_{xx}.$$

Of course, L is uniformly elliptic, since the matrix

$$(3.28) \qquad [l_{ij}] = \begin{bmatrix} 1 & -u_x \\ -u_x & 1 + u_x^2 \end{bmatrix}$$

is positive definite. Indeed,

$$(3.29) \qquad l_{ij}\xi_i\xi_j = (\xi_1 - u_x\xi_2)^2 + \xi_2^2.$$

So, if c is such that $|u_x| \le c$, then if $|\xi_2| < \frac{1}{2c}|\xi_1|$ then

$$(3.30) \qquad (\xi_1 - u_x\xi_2)^2 > \frac{1}{4}\xi_1^2.$$

On the other hand if $|\xi_2| \ge \frac{1}{2c}|\xi_1|$ then

$$(3.31) \qquad \xi_2^2 \ge \frac{1}{4c^2}\xi_1^2.$$

So, it is easy to see that if we take $\alpha = \min\left(\frac{1}{4}, \frac{1}{8c^2}\right)$ then

$$(3.32) \qquad l_{ij}\xi_i\xi_j \ge \alpha|\xi|^2.$$

Let, also, Ξ_u be the map with the image Ω given by the formula

$$(3.33) \qquad \Xi_u(x,y) = (x, y + u(x)).$$

Then,

$$(3.34) \qquad \Delta z^u \circ \Xi_u = Lv,$$

and since $|det\, D\Xi_u| = 1$ (here $D\Xi_u$ is the gradient matrix of the map Ξ_u so that $|det\, D\Xi_u|$ is the Jacobian)

$$(3.35) \qquad (Lv)(\tilde\varphi) = (\Delta z^u)(\varphi).$$

Hence

$$(Lv)(\tilde\varphi) = \int_{\Gamma_u} g\varphi d\sigma$$

$$= \int g(x, u(x))\varphi(x, u(x))\sqrt{1 + u'^2(x)}\,dx$$

$$(3.36) \qquad = \int_{\{y=0\}} \tilde g\tilde\varphi dx \stackrel{\text{def}}{=} \tilde\xi(\tilde\varphi).$$

So,

$$(3.37) \qquad Lv = \tilde\xi$$

in the sense of distributions. Since the singular set is now on $\{y = 0\}$, we expect higher regularity in x-direction. To prove that, we want to differentiate (or more precisely, difference) equation (3.37) with respect to x. Somewhat more precisely, define the standard difference operator (in the x-direction) δ_h^1 as

$$(3.38) \qquad \left(\delta_h^1 u\right)(x) = \frac{1}{h}\left(u(x+h,y) - u(x,y)\right), \ h \neq 0.$$

Then from (3.37) we get

$$(3.39) \qquad (Lv)\left(\delta_{-h}^1 \tilde{\varphi}\right) = \tilde{\xi}\left(\delta_{-h}^1 \tilde{\varphi}\right).$$

We shall discuss in some details only the right-hand side. We have

$$\tilde{\xi}\left(\delta_{-h}^1 \tilde{\varphi}\right) = \int_{\{y=0\}} \tilde{g}\delta_{-h}^1 \tilde{\varphi}\,dx$$

$$(3.40) \qquad = -\int_{\{y=0\}} \left(\delta_h^1 \tilde{g}\right)\tilde{\varphi}\,dx \longrightarrow -\int_{\{y=0\}} \tilde{g}_x\tilde{\varphi}\,dx,$$

as $h \to 0$. We conclude that

$$(3.41) \qquad (Lv)_x(\tilde{\varphi}) = \tilde{\xi}_x(\tilde{\varphi}),$$

and hence

$$(3.42) \qquad L_1 v_x = \tilde{\xi}_x - v_{yy}2u_x u_{xx} + v_y u_{xxx},$$

where

$$(3.43) \qquad L_1 w = \Delta w + (u_x)^2 w_{yy} - 2u_x w_{xy} - 3u_{xx}w_y,$$

and where

$$(3.44) \qquad \tilde{\xi}_x(\tilde{\varphi}) \stackrel{\text{def}}{=} \int_{\{y=0\}} \tilde{g}_x\tilde{\varphi}\,dx.$$

We observe that the differencing performed above is legitimate, since

$$(3.45) \qquad \tilde{\xi}_x - v_{yy}2u_x u_{xx} + v_y u_{xxx} \in \left(W^{1,q'}\right)^*.$$

Indeed, $g \in W^{2,q}(\Omega)$, and also observe that $u_{xxx} \in L^2$, and that $L^2 \hookrightarrow \left(W^{1,q'}\right)^*$. Also, since L_1 has the same principal part as L, L_1 is uniformly elliptic, as well.

Now we can conclude from (3.42,3.45) that

$$(3.46) \qquad v_x \in W^{1,q}.$$

This implies, by the Trace Theorem, that $v_x|_{\{y=0\}} \in W^{1-\frac{1}{q},q}$, so that

$$(3.47) \qquad v|_{\{y=0\}} \in W^{2-\frac{1}{q},q}.$$

We observe that because of (1.1,1.7), the preceding analysis is true also in the $\{y > 0\}$-neighborhood of (the pre-image of) $(\pm 1, 0)$, so that (3.47) holds up to the initial and terminal points of (the pre-image of) Γ_u. Elliptic regularity then yields

$$(3.48) \qquad v|_{\{y \geq 0\}} \in W^{2,q}.$$

Unfortunately, we can not claim the same global result for $v|_{\{y \leq 0\}}$ because of the nonsmoothness of $\partial(\Omega \setminus \Omega_u)$, i.e., we have to localize in $\{y < 0\}$. This concludes the proof of (3.6). Now, regarding estimates (3.7,3.8) we have

$$\|z^{u,ext}\|_{W^{2,q}(\Omega_u)} \leq c \left(\|u\|_{H^3(-1,1)}\right) \|v|_{\{y \geq 0\}}\|_{W^{2,q}(\Xi_u^{-1}(\Omega_u))}$$

$$(3.49) \qquad \leq c \left(\|u\|_{H^3(-1,1)}, \|g\|_{W^{2,q}(\Omega)}, \|\psi\|_{W^{2,q}(\Omega)}\right),$$

and similarly (after localization in $\{y < 0\}$) for $z^{u,int}$, which completes the proof of the Theorem. $\qquad\qquad\square$

COROLLARY 3.1. *If $g \in W^{2,q}(\Omega)$ for some $q > 2$, and if (1.7) holds, then $z^u \in C^{0,1}_{\{y>0\}-loc}(\bar{\Omega})$ and the following apriori estimate holds*

$$(3.50) \qquad \|z^u\|_{C^{0,1}(\bar{\Omega} \cap \{y \geq \epsilon\})} \leq c \left(\epsilon, \|u\|_{H^3(-1,1)}, \|g\|_{W^{2,q}(\Omega)}, \|\psi\|_{W^{2,q}(\Omega)}\right),$$

for any $\epsilon > 0$.

Proof: ¿From (3.7,3.8) and by the Imbedding theorem (see e.g. [7]), we have

$$\|z^{u,ext}\|_{C^1(\bar{\Omega}_u)} + \|z^{u,int}\|_{C^1(\overline{\Omega \setminus \Omega_u} \cap \{y \geq 0\})} \leq$$

$$(3.51) \qquad \leq c \left(\epsilon, \|u\|_{H^3(-1,1)}, \|g\|_{W^{2,q}(\Omega)}, \|\psi\|_{W^{2,q}(\Omega)}\right).$$

This implies (3.50). $\qquad\qquad\square$

COROLLARY 3.2. *If $g \in W^{2,q}(\Omega)$ for some $q > 2$, and if (1.7) holds, then*

$$(3.52) \qquad z_y^{u,int} \in C^{0,1-\frac{2}{q}}\left(\Gamma_u\right),$$

and the following apriori estimate

$$(3.53) \qquad \|z_y^{u,int}\|_{C^{0,1-\frac{2}{q}}(\overline{\Gamma_u})} \leq c \left(\|u\|_{H^3(-1,1)}, \|g\|_{W^{2,q}(\Omega)}, \|\psi\|_{W^{2,q}(\Omega)}\right)$$

holds.

The interest in this Corollary is due to the lack of $(\Omega \setminus \Omega_u)$-global regularity of z^u.

Proof: Let τ and ν be unit tangent and unit normal to Γ_u. More precisely, set

$$(3.54) \qquad \tau = \frac{1}{\sqrt{1 + u'^2}} \langle 1, u' \rangle,$$

and

$$(3.55) \qquad \nu = \frac{1}{\sqrt{1 + u'^2}} \langle u', -1 \rangle.$$

It is elementary to compute that then

$$(3.56) \qquad z_y^{u,int} = \frac{u'}{\sqrt{1 + u'^2}} z_\tau^{u,int} - \frac{1}{\sqrt{1 + u'^2}} z_\nu^{u,int}.$$

But since, by Theorem 3.1 $z_\tau^{u,int} = z_\tau^{u,ext}$ and (also, by Lemma 2.2) $z_\nu^{u,int} = g + z_\nu^{u,ext}$ on Γ_u, we have

$$(3.57) \qquad z_y^{u,int}\Big|_{\Gamma_u} = \left(\frac{u'}{\sqrt{1 + u'^2}} z_\tau^{u,ext} - \frac{1}{\sqrt{1 + u'^2}} [g + z_\nu^{u,ext}] \right)\Big|_{\Gamma_u}.$$

The Corollary follows due to the Ω_u-*global* regularity of $z^{u,ext}$, and by the Imbedding theorem. Indeed,

$$\left\| z_y^{u,int} \right\|_{C^{0,1-\frac{2}{q}}(\overline{\Gamma_u})} = \left\| \frac{u'}{\sqrt{1 + u'^2}} z_\tau^{u,ext} - \frac{1}{\sqrt{1 + u'^2}} [g + z_\nu^{u,ext}] \right\|_{C^{0,1-\frac{2}{q}}(\overline{\Gamma_u})}$$

$$\leq c \|u\|_{H^3(-1,1)} \left[\|z^{u,ext}\|_{C^{1,1-\frac{2}{q}}(\overline{\Gamma_u})} + \|g\|_{C^{0,1-\frac{2}{q}}(\overline{\Gamma_u})} \right]$$

$$\leq c \|u\|_{H^3(-1,1)} \left[\|z^{u,ext}\|_{W^{2,q}(\Omega_u)} + \|g\|_{W^{1,q}(\Omega_u)} \right]$$

$$\leq c \left(\|u\|_{H^3(-1,1)}, \|g\|_{W^{2,q}(\Omega)}, \|\psi\|_{W^{2,q}(\Omega)} \right).$$

$$(3.58)$$

$\square$

4. Existence of a minimizer. In order to claim existence of a minimizer, i.e., existence of a solution of the relaxed problem, one needs compactness. One way of introducing compactness would be to bound the set of admissible controls to

$$(4.1) \qquad U_b = \{ u \in U; \|u\|_{H^3(-1,1)} \leq b \}$$

where b is some prescribed (large) positive constant.

PROPOSITION 4.1. *Let* $g \in W^{1,q}(\Omega)$, *for some* $q > 2$. *Then, there exists an* $u^* \in U_b$ *such that*

$$(4.2) \qquad \Phi(u^*) = \min_{u \in U_b} \Phi(u).$$

Proof: Let $(u_n)_{n=1,2,\dots} \subset U_b$ be a minimizing sequence. By Theorem 1. we know that

$$\tag{4.3} \|z^{u_n}\|_{H^1(\Omega)} + \|z^{u_n}\|_{C^0(\bar{\Omega})} \le c.$$

By taking subsequences, if necessary, we can assume without loss of generality that there exist $u^* \in U_b$ and $z^* \in H^1(\Omega)$ such that

$$\tag{4.4} u^n \to u^* \text{ in } H^2(-1,1)$$

$$\tag{4.5} z^{u_n} \rightharpoonup z^* \text{ weakly in } H^1(\Omega)$$

$$\tag{4.6} z^{u_n} \to z^* \text{ in } C^0(\bar{\Omega}).$$

Recall that

$$\tag{4.7} -\int_\Omega \nabla z^{u_n} \cdot \nabla \varphi = \int_{\Gamma_{u_n}} g\varphi d\sigma$$

for all $\varphi \in H^1(\Omega)$ such that $\varphi|_{\{y=0\}} = \varphi|_{\{y=2\}} = 0$. If, in addition, $\varphi \in C^1(\bar{\Omega})$ then it is easy to see that

$$\tag{4.8} \lim_{n\to\infty} \int_{\Gamma_{u_n}} g\varphi d\sigma = \int_{\Gamma_{u^*}} g\varphi d\sigma.$$

Hence, for such φ we can pass $n \to \infty$ in (4.7) to conclude

$$\tag{4.9} -\int_\Omega \nabla z^* \cdot \nabla \varphi = \int_{\Gamma_{u^*}} g\varphi d\sigma$$

for any $\varphi \in C^1(\bar{\Omega})$ such that $\varphi|_{\{y=0\}} = \varphi|_{\{y=2\}} = 0$. But then, by the density, (4.9) holds for all $\varphi \in H^1(\Omega)$ such that $\varphi|_{\{y=0\}} = \varphi|_{\{y=2\}} = 0$. We conclude, by uniqueness, that

$$\tag{4.10} z^* = z^{u^*}.$$

Now since

$$\tag{4.11} \Phi(u_n) = \frac{1}{2} \int_{\Gamma_{u_n}} (z^{u_n})^2 d\sigma$$

(4.4,4.6) imply that

$$\tag{4.12} \lim_{n\to\infty} \Phi(u_n) = \Phi(u^*).$$

This completes the proof of the Proposition. $\square$

5. Differentiability properties of the variational functional Φ.
Our goal is to derive information about the *multivalued* generalized gradient of Φ. To make our results more precise we shall introduce several definitions.

Let Φ be a real-valued function on the subset U of the Banach space X.

DEFINITION 5.1. *Φ is said to be* directionally differentiable *at $u \in U$ if the limit*

$$(5.1) \qquad \lim_{\lambda \downarrow 0} \frac{\Phi(u + \lambda v) - \Phi(u)}{\lambda}$$

exists for any $v \in X$ such that $u + \lambda v \in U$, for small enough $\lambda > 0$. If that is the case, then the limit in (5.1) is called directional derivative *and it is denoted by $\Phi'(u; v)$.*

DEFINITION 5.2. *Φ is said to be* subdifferentiable *at u, if there exists an $f \in X^*$ such that*

$$(5.2) \qquad \Phi'(u; v) \geq f(v)$$

for every $v \in X$ such that $u + \lambda v \in U$, for small enough $\lambda > 0$. Set of all such f's is called subdifferential, *and it is denoted by $\partial_* \Phi(u)$.*

DEFINITION 5.3. *Φ is said to be* superdifferentiable *at u, if there exists an $f \in X^*$ such that*

$$(5.3) \qquad \Phi'(u; v) \leq f(v)$$

for every $v \in X$ such that $u + \lambda v \in U$, for small enough $\lambda > 0$. Set of all such f's is called superdifferential, *and it is denoted by $\partial^* \Phi(u)$.*

If Φ is both sub- and superdifferentiable at $u \in int(U)$, and moreover $\partial_* \Phi(u) \cap \partial^* \Phi(u) \neq \emptyset$, then $\partial_* \Phi(u) \cap \partial^* \Phi(u)$ is a singleton and Φ is Gâteaux differentiable.

We go back now to our problem. Of course, $X = H_0^3(-1, 1)$, U is defined in (1.1).

Proceeding, define the adjoint variable p^u, as a solution of the (adjoint) equation

$$\Delta p^u = \eta_u \text{ in } \Omega$$
$$p^u = 0 \text{ in } \{(x, 0); -a < x < a\} \cup \{(x, 2); -a < x < a\}$$
$$(5.4) \qquad p_x^u = 0 \text{ in } \{(\pm a, y); 0 < y < 2\}$$

where $\eta_u \in H^{-1}(\Omega)$ is a (signed) measure given by

$$(5.5) \qquad \eta_u(\varphi) = \int_{\Gamma_u} z^u \varphi d\sigma.$$

Obviously, (5.4) is the same type of equation as (2.3).

In this section, as before, $z^{u,ext} = z^u|_{\Omega_u}$ and $z^{u,int} = z^u|_{\Omega\backslash\Omega_u}$; also, below we shall use the notation $p^{u,ext} = p^u|_{\Omega_u}$ and $p^{u,int} = p^u|_{\Omega\backslash\Omega_u}$. That is essential in this calculation, since z^u and p^u are *not* differentiable *across* the Γ_u.

LEMMA 5.1. *Let* $g \in W^{2,q}(\Omega)$, *for some* $q \geq 2$. *Then*

$$(5.6) \qquad p^{u,ext} \in W^{2,q}(\Omega_u), \quad p^{u,int} \in W^{2,q}_{\{y>0\}-loc}(\Omega \backslash \bar{\Omega}_u).$$

and the apriori estimates

$$(5.7) \qquad \|p^{u,ext}\|_{W^{2,q}(\Omega_u)} \leq c\left(\|u\|_{H^3(-1,1)}, \|g\|_{W^{2,q}(\Omega)}, \|\psi\|_{W^{2,q}(\Omega)}\right),$$

and

$$\|p^{u,int}\|_{W^{2,q}((\Omega\backslash\Omega_u)\cap\{y\geq\epsilon\})} \leq$$
$$(5.8) \qquad \leq c\left(\epsilon, \|u\|_{H^3(-1,1)}, \|g\|_{W^{2,q}(\Omega)}, \|\psi\|_{W^{2,q}(\Omega)}\right).$$

hold.

Proof: Comparing (2.3) and (5.4) we see that the only difference is in right-hand sides. Namely, in (5.5), $z^u \notin W^{2,q}(\Omega)$. Nevertheless, for example, $z^{u,ext} \in W^{2,q}(\Omega_u)$, and since η_u depends on z^u only through the trace on Γ_u, and since z^u and $z^{u,ext}$ have same traces on Γ_u we easily conclude the proof of the Lemma. $\qquad\square$

We shall use the usual notation: $v^+ \overset{\text{def}}{=} v\mathbf{I}_{\{v>0\}}$, and $v^- \overset{\text{def}}{=} -v\mathbf{I}_{\{v<0\}}$. So, $v = v^+ - v^-$.

Now we are ready to state the following

THEOREM 5.1. *Let* $g \in W^{2,q}(\Omega)$, *for some* $q > 2$. *Then* Φ *is directionally differentiable at any* $u \in U$ *such that* $u(x) > 0$ *for* $-1 < x < 1$, *and*

$$\Phi'(u;v) =$$
$$= \int_{-1}^{1}\left(z^u(z_y^{u,ext}v^+ - z_y^{u,int}v^-)\sqrt{1+u'^2} + (z^u)^2\frac{u'v'}{\sqrt{1+u'^2}}\right)dx$$
$$(5.9) \qquad + \int_{\Gamma_u}\left((gp^{u,ext})_y v^+ - (gp^{u,int})_y v^-\right)d\sigma + \int_{-1}^{1}gp^u\frac{u'v'}{\sqrt{1+u'^2}}dx.$$

Moreover, if

$$(5.10) \quad z^u z_y^{u,int} + (gp^{u,int})_y \leq z^u z_y^{u,ext} + (gp^{u,ext})_y \quad a.e. \ in \ (-1,1),$$

then Φ *is subdifferentiable at* u *and*

$$\partial_*\Phi(u) =$$
$$= \left[\left(z^u z_y^{u,int} + (gp^{u,int})_y\right)\sqrt{1+u'^2}, \left(z^u z_y^{u,ext} + (gp^{u,ext})_y\right)\sqrt{1+u'^2}\right]$$
$$- \left(\frac{u'}{\sqrt{1+u'^2}}\left((z^u)^2 + gp^u\right)\right)'$$
$$(5.11) \qquad \overset{\text{def}}{=} [l\partial_*\Phi(u), r\partial_*\Phi(u)] \subset L^\infty(-1,1).$$

On the other hand, if

$$(5.12) \quad z^u z_y^{u,int} + (gp^{u,int})_y \geq z^u z_y^{u,ext} + (gp^{u,ext})_y \quad a.e. \ in \ (-1,1),$$

then Φ is superdifferentiable at u and

$$\partial^*\Phi(u) =$$

$$= \left[\left(z^u(z_y^{u,ext} + (gp^{u,ext})_y) \right) \sqrt{1+u'^2}, \left(z^u(z_y^{u,int} + (gp^{u,int})_y) \right) \sqrt{1+u'^2} \right]$$

$$\left(\frac{u'}{\sqrt{1+u'^2}} \left((z^u)^2 + gp^u \right) \right)'$$

$$(5.13) \qquad\qquad\qquad \overset{\text{def}}{=} [l\partial^*\Phi(u), r\partial^*\Phi(u)] \subset L^\infty(-1,1)$$

Proof: We attempt to differentiate Φ. To this end, for given $u \in U$ and a suitable direction $v \in H_0^3(-1,1)$ (suitable in a sense that $u + \lambda v \in U$ for small enough $\lambda > 0$) we try to compute the (one sided) directional derivative $\Phi'(u;v)$. Using the regularity result (Theorem 3.1, and Corollary 3.2), we compute

$$\Phi'(u;v) = \lim_{\lambda \downarrow 0} \frac{\Phi(u+\lambda v) - \Phi(u)}{\lambda}$$

$$= \lim_{\lambda \downarrow 0} \frac{1}{2\lambda} \left(\int_{\Gamma_{u+\lambda v}} (z^{u+\lambda v})^2 d\sigma - \int_{\Gamma_u} (z^u)^2 d\sigma \right)$$

$$= \int_{-1}^1 \left(z^u(z_y^{u,ext} v^+ - z_y^{u,int} v^-)\sqrt{1+u'^2} + (z^u)^2 \frac{u'v'}{\sqrt{1+u'^2}} \right) dx +$$

$$(5.14) \qquad\qquad\qquad + \lim_{\lambda \downarrow 0} \frac{1}{2\lambda} \int_{\Gamma_u} \left((z^{u+\lambda v})^2 - (z^u)^2 \right) d\sigma,$$

Before proceeding with the proof, we shall need the following Lemma (more precisely, its Corollary).

LEMMA 5.2. *Under previous assumptions on u, and v, and for any $\alpha < 1$ the following estimate holds*

$$(5.15) \qquad\qquad\qquad \|z^{u+\lambda v} - z^u\|_{C^0(\bar{\Omega})} \leq c\lambda^\alpha.$$

Proof: We need to compare $z^{u+\lambda v}$ and z^u. This is difficult to do in the original domain Ω since (singular) right hand sides of the equations that they satisfy act on disjoint sets, so that there is no obvious cancelation. So, the idea of the proof is to map the original domain into different domains in such a way that the cancelation does take place.

Let, as before, Ξ_u be the map with the image Ω given by the formula

$$(5.16) \qquad\qquad\qquad \Xi_u(x,y) = (x, y + u(x)).$$

Then

$$(5.17) \qquad\qquad\qquad \Xi_u^{-1}(x,y) = (x, y - u(x)),$$

and (set $A = (x, y)$)

$$(5.18) \qquad dist\left(\Xi_{u+\lambda v}^{-1}(A) - \Xi_u^{-1}(A)\right) \leq c\lambda.$$

Now consider $\tilde{z}^{u+\lambda v}$ and $\tilde{z}^u$ defined as

$$
\begin{aligned}
\tilde{z}^{u+\lambda v} &= z^{u+\lambda v} \circ \Xi_{u+\lambda v}, \\
\tilde{z}^u &= z^u \circ \Xi_u,
\end{aligned}
$$
(5.19)

and operators L_u and $L_{u+\lambda v}$ defined by

$$(5.20) \qquad L_u w = \Delta w + w_{yy}(u_x)^2 - 2w_{xy}u_x - w_y u_{xx},$$

$$
\begin{aligned}
L_{u+\lambda v} w &= \Delta w + w_{yy}(u_x + \lambda v_x)^2 - 2w_{xy}(u_x + \lambda v_x) - w_y(u_{xx} + \lambda v_x) = \\
(5.21) \qquad &= L_u w + \lambda\left[w_{yy}(2u_x v_x + \lambda v_x^2) - 2w_{xy}v_x - w_y v_{xx}\right].
\end{aligned}
$$

Then $\tilde{z}^{u+\lambda v} - \tilde{z}^u$ satisfies the equation

$$
L_u\left(\tilde{z}^{u+\lambda v} - \tilde{z}^u\right) = \gamma -
$$
$$(5.22) \qquad \lambda\left[\tilde{z}_{yy}^{u+\lambda v}(2u_x v_x + \lambda v_x^2) - 2\tilde{z}_{xy}^{u+\lambda v}v_x - \tilde{z}_y^{u+\lambda v}v_{xx}\right]
$$

in $\Xi_{u+\lambda v}^{-1}(\Omega) \cap \Xi_u^{-1}(\Omega)$, where

$$(5.23) \qquad \gamma(\varphi) \overset{\text{def}}{=} \int_{\{y=0\}} (G_1 - G_2)\,\varphi dx$$

and where

$$
\begin{aligned}
G_1(x, y) &\overset{\text{def}}{=} g(x, y + u(x) + \lambda v(x))\sqrt{1 + (u'(x) + \lambda v'(x))^2}, \\
(5.24) \qquad G_2(x, y) &\overset{\text{def}}{=} g(x, y + u(x))\sqrt{1 + (u'(x))^2}.
\end{aligned}
$$

Observe that

$$(5.25) \qquad \|G_1 - G_2\|_{W^{1,q}\left(\Xi_{u+\lambda v}^{-1}(\Omega)\cap\Xi_u^{-1}(\Omega)\right)} \leq c\lambda.$$

Now since

$$(5.26) \qquad dist\left(\partial\left(\Xi_{u+\lambda v}^{-1}(\Omega)\right), \partial\left(\Xi_u^{-1}(\Omega)\right)\right) \leq c\lambda$$

and because of the Hölder continuity of $z^{u+\lambda v}$ and z^u, we conclude that

$$(5.27) \qquad \|\tilde{z}^{u+\lambda v} - \tilde{z}^u\|_{C^0\left(\partial\left(\Xi_{u+\lambda v}^{-1}(\Omega)\cap\Xi_u^{-1}(\Omega)\right)\right)} \leq c\lambda^\alpha.$$

Then (5.22,5.25,5.27) imply that

$$(5.28) \qquad \|\tilde{z}^{u+\lambda v} - \tilde{z}^u\|_{C^0\left(\overline{\Xi_{u+\lambda v}^{-1}(\Omega)\cap\Xi_u^{-1}(\Omega)}\right)} \leq c\lambda^\alpha.$$

Then we have (set $A = (x, y)$)

$$\left| z^{u+\lambda v}(A) - z^u(A) \right| =$$
$$= \left| \tilde{z}^{u+\lambda v}\left(\Xi_{u+\lambda v}^{-1}(A)\right) - \tilde{z}^u\left(\Xi_u^{-1}(A)\right) \right| \leq$$
$$\leq \left| \tilde{z}^{u+\lambda v}\left(\Xi_{u+\lambda v}^{-1}(A)\right) - \tilde{z}^u\left(\Xi_{u+\lambda v}^{-1}(A)\right) \right| +$$
$$+ \left| \tilde{z}^u\left(\Xi_{u+\lambda v}^{-1}(A)\right) - \tilde{z}^u\left(\Xi_u^{-1}(A)\right) \right| \leq$$
$$(5.29) \qquad\qquad c\lambda^\alpha + c\lambda^\alpha = c\lambda^\alpha.$$

In (5.29) we also used Hölder continuity of $\tilde{z}^u$. This completes the proof of the Lemma. $\qquad\square$

COROLLARY 5.1.

$$(5.30) \qquad \lim_{\lambda \downarrow 0} \frac{1}{\lambda} \int_{\Gamma_u} \left(z^{u+\lambda v} - z^u \right)^2 d\sigma = 0.$$

Proof: Take $\alpha > \frac{1}{2}$ in the Lemma. Then

$$(5.31) \qquad \frac{\left\| z^{u+\lambda v} - z^u \right\|_{C^0(\bar{\Omega})}^2}{\lambda} \leq c\lambda^\beta, \ \beta = 2\alpha - 1 > 0.$$

$$\square$$

Now, we can proceed with the proof of the Theorem. We compute the last term in (5.14).

$$\lim_{\lambda \downarrow 0} \frac{1}{2\lambda} \int_{\Gamma_u} \left((z^{u+\lambda v})^2 - (z^u)^2 \right) d\sigma$$
$$= \lim_{\lambda \downarrow 0} \frac{1}{\lambda} \int_{\Gamma_u} (z^{u+\lambda v} - z^u) z^u d\sigma + \lim_{\lambda \downarrow 0} \frac{1}{2\lambda} \int_{\Gamma_u} \left(z^{u+\lambda v} - z^u \right)^2 d\sigma$$
$$= \lim_{\lambda \downarrow 0} \frac{1}{\lambda} \int_{\Gamma_u} (z^{u+\lambda v} - z^u) z^u d\sigma$$
$$= -\lim_{\lambda \downarrow 0} \frac{1}{\lambda} \int_{\Omega} \nabla p^u \cdot \nabla(z^{u+\lambda v} - z^u)$$
$$= \lim_{\lambda \downarrow 0} \frac{1}{\lambda} \left(\int_{\Gamma_{u+\lambda v}} g p^u d\sigma - \int_{\Gamma_u} g p^u d\sigma \right)$$
$$= \int_{\Gamma_u} \left((g p^{u,ext})_y v^+ - (g p^{u,int})_y v^- \right) d\sigma + \int_{-1}^1 g p^u \frac{u' v'}{\sqrt{1 + u'^2}} dx.$$
$$(5.32)$$

Now from (5.14, 5.32) we conclude that Φ is directionally differentiable, and that (5.9) holds. Furthermore, if (5.10) holds, then

$$\Phi'(u; v) =$$
$$= \int_{-1}^1 \left(z^u(z_y^{u,ext} v^+ - z_y^{u,int} v^-)\sqrt{1 + u'^2} + (z^u)^2 \frac{u' v'}{\sqrt{1 + u'^2}} \right) dx$$

$$+ \int_{\Gamma_u} \left((gp^{u,ext})_y \, v^+ - (gp^{u,int})_y \, v^- \right) d\sigma + \int_{-1}^{1} gp^u \, \frac{u'v'}{\sqrt{1+u'^2}} dx$$

$$(5.33) \qquad \geq \int_{-1}^{1} \left(\tau - \left(\frac{u'}{\sqrt{1+u'^2}} \left((z^u)^2 + gp^u \right) \right)' \right) v \, dx$$

for all

$$\tau \in \left[\left(z^u z_y^{u,int} + (gp^{u,int})_y \right) \sqrt{1+u'^2}, \right.$$

$$(5.34) \qquad \left. \left(z^u z_y^{u,ext} + (gp^{u,ext})_y \right) \sqrt{1+u'^2} \right].$$

This proves that Φ is subdifferentiable at u and that (5.11) holds. Similarly, one can consider superdifferentiability of Φ. So, the Theorem follows. $\square$

6. Remarks. The above suggests the numerical algorithm (*the steepest descent method*) for minimization of Φ, i.e., for the numerical solution of the relaxed shape optimization problem:

Choose $u_0 \in U$. If $u_n \in U$ is already known, then u_{n+1} is determined by:

- compute z^{u_n} as a solution of (2.3);
- compute p^{u_n} as a solution of (5.4);
- if (5.10) holds, compute an u_{n+1} such that

$$(6.1) \qquad u_{n+1} \in \left(u_n - \rho_n A^{-1} \left(\partial_* \Phi(u_n) \right) \right) \cap U, \ \rho_n > 0,$$

and if (5.12) holds, compute an u_{n+1} such that

$$(6.2) \qquad u_{n+1} \in \left(u_n - \rho_n A^{-1} \left(\partial^* \Phi(u_n) \right) \right) \cap U, \ \rho_n > 0.$$

Here, A is the isomorphism between $H_0^3(-1,1)$ and its dual. So we see that it would be much better to work on $H_0^1(-1,1)$ instead, since then A would be a second order operator $-\frac{d^2}{dx^2}$ instead of the sixth order operator.

If neither (5.10) nor (5.12) holds, i.e., if Φ is neither convex nor concave at the point u_n, then it is more delicate to determine the steep(est) descent direction.

The actual *choice* of u_n in (6.1) or (6.2) is an interesting question. Somewhat formal considerations suggest that the following rules should be adopted:

- if (5.10) holds and $\partial_* \Phi(u_n) \geq 0$ a.e. in $(-1,1)$, then

$$(6.3) \qquad u_{n+1} = u_n - \rho_n A^{-1} \left(l \partial_* \Phi(u_n) \right), \ \rho_n > 0;$$

- if (5.10) holds and $\partial_* \Phi(u_n) \leq 0$ a.e. in $(-1,1)$, then

$$(6.4) \qquad u_{n+1} = u_n - \rho_n A^{-1} \left(r \partial_* \Phi(u_n) \right), \ \rho_n > 0;$$

- if (5.12) holds and $\partial^*\Phi(u_n) \geq 0$ a.e. in $(-1, 1)$, then

$$(6.5) \qquad u_{n+1} = u_n - \rho_n A^{-1}\left(r\partial^*\Phi(u_n)\right), \quad \rho_n > 0;$$

- if (5.12) holds and $\partial^*\Phi(u_n) \leq 0$ a.e. in $(-1, 1)$, then

$$(6.6) \qquad u_{n+1} = u_n - \rho_n A^{-1}\left(l\partial^*\Phi(u_n)\right), \quad \rho_n > 0.$$

One can show that if u is a local minimizer for Φ then (5.10) does hold. Also, we observe that in terms of Clarke's nonsmooth analysis (5.34) implies that (if (5.10) holds)

$$(6.7) \qquad \partial\Phi(u) \supset \partial_*\Phi(u)$$

where $\partial\Phi$ is the generalized gradient of Φ (observe that Φ is nonsmooth, i.e., $\partial\Phi$ is *multivalued*).

Finally, we note that the method introduced here is an unexpected follow-up of the research in the completely different context (electrophotography, see [4]; see also [8]). The difference is that in [4], instead of (2.3), the state equation is (up to nonessential details)

$$\Delta z^u = \mathbf{I}_{D_u} \text{ in } \Omega$$
$$z^u = 0 \text{ in } \{(x, 0); -a < x < a\}$$
$$z^u = 1 \text{ in } \{(x, 1); -a < x < a\}$$
$$(6.8) \qquad z^u_x = 0 \text{ in } \{(\pm a, y); 0 < y < 1\}$$

where D_u is the set enclosed by Γ_u, and the functional to minimize is, instead of (2.17),

$$(6.9) \qquad \Psi(u) = \frac{1}{2}\int_{\Gamma_u}\left(\frac{\partial z^u}{\partial \nu}\right)^2 d\sigma.$$

Observe that in (2.3) the right hand side, i.e., the measure ξ_u is, *essentially*, "derivative" of $\mathbf{I}_{D_u}$, the right hand side in (6.8). On the other hand in (6.9), z^u is under derivative. So, in the final balance those two problems have the same level of smoothness (which happens to be a kind of Lipschitz continuity), and hence, the analogous general ideas apply.

REFERENCES

[1] R. A. Adams, *Sobolev Spaces*, Academic Press, New York, 1975.

[2] H. W. Alt and L. A. Caffarelli, Existence and regularity for a minimum problem with free boundary, *J. Reine Angew. Math.* **105**, 105-144 (1981).

[3] H. W. Alt, L. A. Caffarelli, and A. Friedman, A free boundary problem for quasilinear elliptic equations, *Ann. Scuola Norm. Sup. Pisa*, (4) **11**, 1-44 (1984).

[4] V. Barbu and S. Stojanovic, A variational approach to a free boundary problem arising in electrophotography, IMA Preprint Series #815, *Numer. Funct. Analysis & Optimiz.*, to appear.

[5] L. C. Evans and R. F. Gariepy, *Measure Theory and Fine Properties of Functions*, Studies in Advanced Mathematics, CRC Press, Boca Raton, 1992.

[6] A. Friedman, *Variational Principles and Free-Boundary Problems*, Wiley-Interscience, New York, 1982

[7] D. Gilbarg and N. S. Trudinger, *Elliptic Partial Differential Equations of Second Order*, Springer-Verlag, Berlin, 1983.

[8] F. He and S. Stojanovic, A nonsmooth PDE optimization problem, submitted.

[9] D. Kinderlehrer and G. Stampacchia, *An Introduction to Variational Inequalities and their Applications*, Academic Press, New York, 1980.

[10] J. Nečas, *Les Methodes Directes en Theorie des Equations Elliptiques*, Masson, Paris, 1967.

[11] S. Stojanovic, Parallel computations for variational free boundary problem modeling injection of fluid from a slot into a stream, *Theoretical Aspects of Industrial Design*, D.A. Field and V. Komkov, Editors, SIAM, 1992.

[12] G. M. Troianiello, *Elliptic Differential Equations and Obstacle Problems*, Plenum Press, New York, 1987.

COMPUTATIONAL FLUID DYNAMICS ANALYSIS OF THE FLOW IN AN APCVD APPLICATOR SYSTEM

GARY S. STRUMOLO*

Abstract. Application of Atmospheric Pressure Chemical Vapor Deposition (APCVD) to the production of coated glass is addressed in this study. Several layers of thin films are deposited on the surface of the glass as it moves underneath the APCVD applicator system at high temperature. A memory effect in the form of film thickness streaks, corresponding to the location of the inlet holes located upstream in the upper manifold feed channel, is evident on the glass. This nonuniform film across the glass causes a color variation of the coating. Effective mixing of the gas streams is required to treat the hole memory problem. However, a premature reaction is to be avoided. Optimum design parameters to correct this problem include the geometry of the applicator and the sensitivity of the flow field to boundary conditions is of major interest. The Computational Fluid Dynamics (CFD) simulation and analysis package FIRE is used to predict the flow. The flow of gases involved is treated as that of a steady, viscous, incompressible fluid. Results for both two- and three-dimensional cases demonstrate that the deposition process can be improved by injecting the flow at an angle counter to the direction of glass motion, and that CFD techniques can be successfully used to predict the flow behavior of an APCVD applicator system and help optimize its design.

TABLE OF CONTENTS

---

* Ford Motor Company, Dearborn, MI 48124

1. Introduction. Atmospheric Pressure Chemical Vapor Deposition (APCVD)applicators are used in the production of thin-film, coated glass products like architectural glass where, for example, low emissivity thin coatings such as tin oxide are applied to the interior surface for the purpose of reducing the heat loss from buildings. They can also play a significant role in the development of automotive parts such as car windshields (thin films sandwiched in laminated glass to seve as transparent heaters) and sidelights (privacy glass with solar load reduction [20-30% transmission]). The design of applicators to deposit these films is crucial to the quality of the end product. In addition, APCVD applicators may be employed in tandem to lay down a sequence of coatings. An effective APCVD applicator system must keep the operation of adjacent applicators independent of each other.

Presently, many on-line glass coatings are performed using powder spray applicators. This process is open to the atmosphere and, therefore, susceptible to air currents leading to imperfections in the final product. One of the problems associated with this process is the mottle/haze created on the tinted glass product. On the other hand, APCVD applicators are placed inside the tin bath where gas currents are minimized. In present APCVD applicators design, the deposition gases are fed through a narrow channel at its lower exit; this is shown schematically in Figure 1.1. The glass ribbon underneath is moving at a speed of 400 in/min (0.169 m/s) at a temperature of about $605^\circ C (1120^\circ F)$. The gases are then extracted through two exhaust manifolds positioned at opposite sides of the applicator. The exhaust design must remove reaction by-products without inhibiting the reaction or interfering with the reducing tin bath atmosphere.

In the architectural glass example, a tin oxide/silicon dioxide four-layer stack could be used as an interference filter to reduce the color from a thick tin oxide film coated on top of the stack. The tin oxide reflects heat, while the four-layer stack underneath suppresses the unwanted color of the tin and acts as a passive diffusion barrier to insulate and protect it from the soda lime glass. Any non-uniformity in the gas flow across the glass ribbon would lead to film thickness non-uniformities that would become evident through a dramatic discoloration on the glass surface. With the APCVD applicators being considered the velocity field retains a "memory" of the holes corresponding their locations upstream in the feed manifold. The effect of the applicator feed holes on the applied coatings is evident by concentration "peaks", as depicted in Figure 1.2. It is difficult to erase this hole memory effect if the mixing process of the gas streams is inefficient. However, maintaining a simple and yet robust applicator system is essential to the manufacturing process.

Computer modelling and simulation of APCVD applicators is attractive, since it is cost effective, versatile and flexible. The result is an enhanced ability to visualize the flow and monitor gas mixing within the applicator environment, as functions of the geometry and boundary condi-

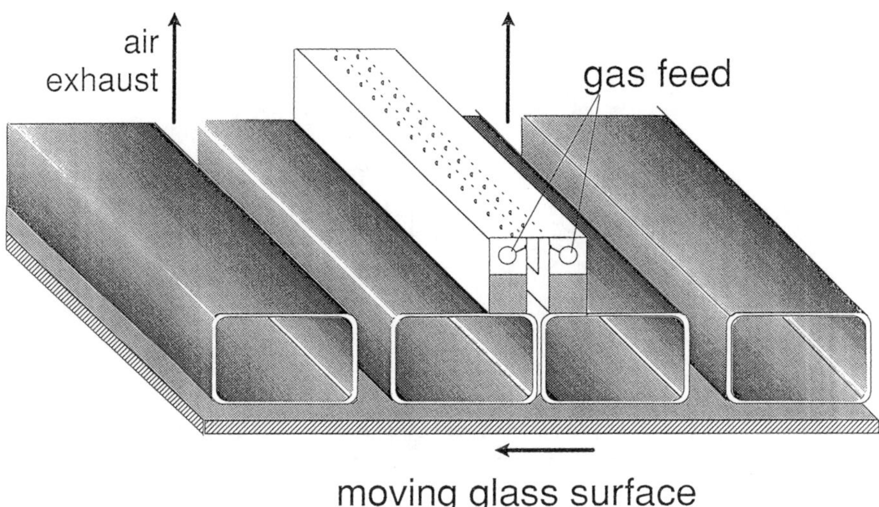

FIG. 1.1. *Schematic representation of a typical APCVD applicator system*

tions, through a variety of flow parameters such as velocities, temperatures, etc. The primary objective of the present study is to model and simulate the steady, viscous, incompressible gas flow in a APCVD applicator system. The Computational Fluid Dynamics (CFD) package FIRE is used for this purpose. Two- and three-dimensional models are investigated.

Knowledge of the flow field within the applicator is necessary to alleviate the problems outlined earlier and suggest possible design modification; mainly to eliminate any deposition hole memory effect, improve the film thickness uniformity across the ribbon width, reduce haze due to gas phase nucleating particulates, and generally improve film deposition efficiency.

2. Process and apparatus description. APCVD is a process that combines Chemical Vapor Deposition (CVD) with a conveyor operated furnace at atmospheric pressure. It originated in the microelectronics industry as a way to manufacture printed circuit boards. Today, it is principally used to produce thin films[1] for different coating processes without the use of a vacuum. It is considered to be a production-oriented, cost-effective means for providing high quality coatings (Gralenski, 1984). CVD usually involves the delivery of more than one gaseous chemical to a heated surface where a reaction occurs. The reaction can also happen before the chemicals reach the surface, although this is not often desirable. Reaction by-products are vented out through exhausts chimneys. Multiple coatings are also possible

---

[1] The thickness of these films is on the order few hundred Angstroms (1 Angstrom $= 10^{-10}$ m).

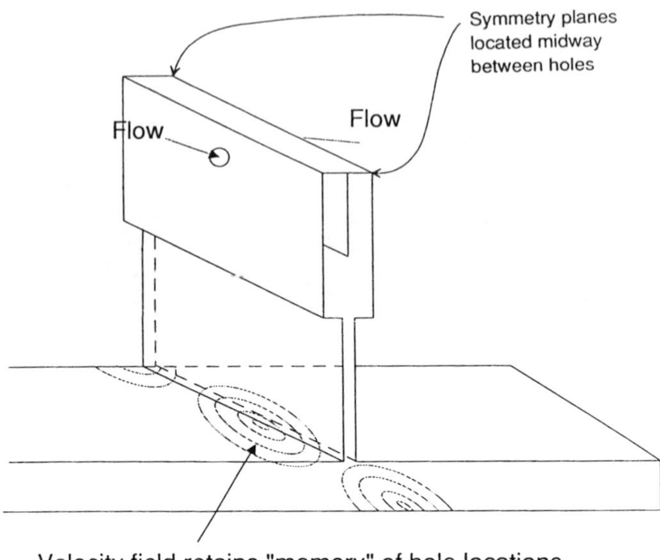

FIG. 1.2. *Schematic representation of surface streaks corresponding to hole memory*

through furnaces supplied with multistage APCVD systems.

Each APCVD applicator system consists of a gas feed and two exhaust chimneys; a two-dimensional representation of one being studied is given in Figure 2.1. It is designed to distribute the mixed gases across a 5-ft wide glass ribbon passing under the coating applicator system. The gas feed side consists of an upper manifold in the form of a 0.5 in.-wide and 2.5 in.-long channel that contains the gas inlets, and a narrower lower channel 0.125 in. in width and 3.0 in. in length that operates as an injector with an exit in the deposition area facing the glass top surface. The function of the injector is to effectively deliver the gaseous chemicals to the heated glass.

There are two separate streams of gases which are introduced upstream from opposite sides of the upper channel through two arrays of 0.067-in.-diameter distribution holes (see Figure 1.1). The holes are distributed 0.5 in. apart and positioned 0.25 in. below the top of the channel in the 28-in. middle segment of the 60-in. applicator span. The two fluids flow through the side holes in parallel streams at 30° angle with the normal to each surface, creating two streams counter to one another. The gases are supplied to the holes from one end of the 0.5-in.-diameter horizontal tube of the feed chamber. The total flow rate is 200 liter/min over a 2 ft width. There is a splitter plate separating the two gas streams as they enter the upper manifold channel; this plate extends along half the length of the channel. The function of the splitter plate is to delay the mixing process in order to prevent any possible precipitation resulting from premature reaction between

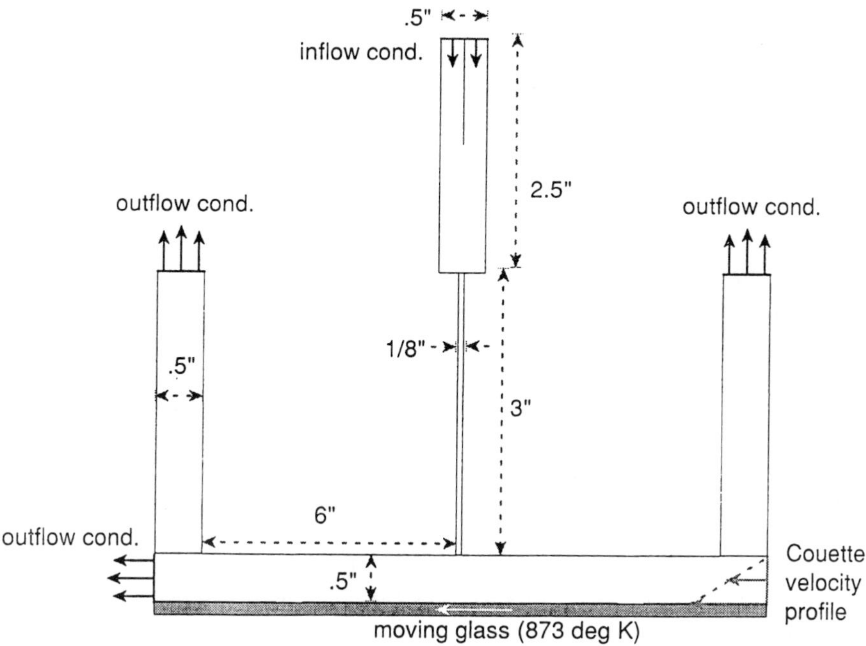

FIG. 2.1. *Two-dimensional representation of an APCVD applicator system*

the two gas streams; spontaneous low-temperature reaction may occur in some cases.

The two gases are then passed through to the second narrower channel to eventually impact the surface of the moving glass ribbon. This injector channel exit is 0.5 in. above the glass top surface. The lower channel of the applicator is internally cooled with water. However, the temperature of the flowing gas is maintained around or over 150°C in the case of tin oxide reactor. Most of the mixing between the two streams occur prior to the entry to the narrow channel. As stated earlier, the glass substrate is moving underneath the applicators at an approximate speed of 400 in/min (0.169 m/s). The coating gas is applied to the glass while it is moving inside the molten tin bath. At this point the glass surface is at an approximate temperature of 605°C (1120°F). The upper surface of the applicator system facing the glass is maintained at a controlled temperature.

In the silica reactors, the silicon dioxide (SiO_2) layers are formed from the reaction of silane (SiH_4)(0.5-1 %) with oxygen (> 50 %). The reactants are pre-mixed (1-2% silane) in the applicator manifold and maintained at room temperature (20°C). Premature reaction may occur if the reactant temperature and flow are not kept under control. In the tin oxide reactor the tin oxide (SnO_2) layers are formed from the reaction of tin tetrachloride ($SnCl_4$)(2-3 %) and water vapor (2% H_2O) in nitrogen at about 110°C. These reactants can be pre-mixed before entry to the feed manifold if desired.

The two exhaust chimneys are in the form of two vertical rectangular channels 0.5 in. in width and located at a distance of 6.0 in. from either side of the injector channel. Their task is to remove the by-products of the reaction without inhibiting it. The exit ports of the exhaust manifolds are maintained at a controlled temperature and pressure near vacuum. This ensures that the flow from two adjacent applicator systems stays separated. Makeup air, if needed for the chemical reaction and/or to prevent the reaction deposits from fouling the exhaust manifolds, is supplied from ports located between the adjacent applicators. The ambient air is nitrogen with 3-5 % hydrogen. Because of the motion of the glass, there is a strong flow of gas along the ribbon towards the sting-out (at the end of the tin bath). The excess gas flow rate due to entrainment is ignored since it does not substantially contribute the main flow.

There are several problems associated with this process. Foremost is the development of non-uniformity on the glass surface in the form of streak lines, due to the hole memory effect that is created as the flow propagates downstream from the holes. Any non-uniformity or streaking due to the gas flow shows up as a discoloration of the coating on the glass surface. it is important to note that in thin film technology, thickness variations often produce appreciable variations in physical, chemical, electrical, or optical properties (Gralenski, 1984).

3. Flow characteristics. It is helpful to have an idea of the basic features of the flow so that we can evaluate our numerical predictions for reasonableness. The velocity distribution of the flow at the inlet to the feed manifold is considered uniform. Examining the Reynolds number of the flow based on average velocity and channel width, we obtain

$$R_e \equiv \frac{UH}{V} \approx 400,$$

in the upper channel; room conditions are assumed for the fluid, i.e., room temperature of 20°C and atmospheric pressure. In the above definition, U is the average velocity, H is the channel width, and v is the kinematic viscosity.

The low Reynolds number suggests that the flow is well into the laminar range. For steady, two-dimensional, incompressible, isothermal flow of a Newtonian, isotropic, homogeneous, viscous fluid between two fixed parallel flat plates the critical Reynolds number at which transition from laminar to turbulent flow occurs is approximately 1500 (Potter and Foss, 1982). At the end of the narrow injector channel, the flow exceeds its laminar entry length L_e given by the relationship (Schlichting, 1979)

$$\frac{L_e}{H} = 0.04 R_e$$

and is fully-developed with Poiseuille's parabolic velocity profile; $L_e = 1.84$ in. whereas the length of the narrow injector channel is 3 in.

In the absence of turbulence, there is no effective mechanism for mixing between the two gas streams. It is clear that the laminar mixing (primarily due to molecular diffusion) between the two streams occur in the upper manifold channel, after the splitter plate prohibits any mixing that may otherwise take place. Further downstream, we would expect two small separated regions to exist on both sides of the upper channel in the region of contact with the lower narrow channel, where the area is reduced significantly. Massive separation is not expected in this region due to the low Reynolds number. However, we expect to see a more pronounced separation occurs in the vicinity of the exit of the injector channel where the flow meets the moving glass. There should also be recirculating regions on either side of the exiting jet, with different pattern of recirculating fluid due to the motion of the glass ribbon underneath.

A vortex structure can be identified near where the deposition occurs. As the exiting jet approaches the moving glass, a vortex loop forms from the action of the jet velocity profile. This loop moves toward the stagnation streamline, and reorients its path to diffuse into the boundary-layer fluid. Inside the boundary layer the loop is stretched and its vorticity is increased as the flow spreads along the glass.

Away from the separated flow zone, the velocity profile above the moving glass surface would be that of general Couette flow between two parallel flat walls (Schlichting, 1979), with decreased pressure in the direction of wall motion (i.e., negative pressure gradient, $dp/dx < 0$). From the no-slip boundary condition, the velocity on the lower wall is identical to that of the moving glass and becomes zero at the upper fixed wall; a simple Couette flow with linear distribution will result in the case of zero pressure gradient. Actually, the velocity distribution should be a superposition of the simple Couette flow and the parabolic profile of a steady parallel flow in a straight channel with two parallel fixed walls. The flow should not be evenly split between the two sides of the chamber due to the motion of the glass. This would result in a lower velocity in the region where the flow is moving in the direction opposite to the motion of the glass. Moreover, the buoyancy effect due to the temperature differential between the lower and upper walls should play a role in the dynamics of the flow here.

At the far ends of the moving glass, both upstream and downstream, the velocity profile would be the same as that described above. However, an assumption of simple Couette flow is used in the computer model due to the negligible effect of these profiles on the computation. A repeated boundary condition, which is presently not an available feature in FIRE, would have been more appropriate. The outflow conditions at the exit plane of both exhaust manifolds are assumed to resemble fully-developed channel flow, with a near parabolic velocity distribution.

It is certain that the velocity field within the applicator system described above will be influenced by boundary conditions. These include the velocity profiles at the inlets and outlets, speed of the moving glass, pressure considerations at the exhaust ports, as well as the temperature distributions within the flow and among the boundaries. The velocity field does not change significantly as the fluid enters the narrow injector channel, and retains memory of the holes corresponding the their upstream locations in the feed manifold. Additionally, it is interesting to note that we observe hole memory effects even when the glass is moving slowly or at a standstill. These streaks are more pronounced in the case of tin oxide rather than the silica.

Regarding the fluid properties, the gases involved are essentially pressurized nitrogen ($>$ 95%) which is passed through liquid chemicals to create the desired gaseous solution. Practically, the fluid flow is considered as that of an incompressible air, and the fluid density and viscosity are the same as that of air at atmospheric pressure.

4. Project goal. Our goal is the design of an APCVD applicator that exhibits optimum coating performance. This implies creating a film thickness across the entire ribbon width that is uniform and devoid of any deposition hole memory. Additionally, adjacent applicator systems must operate independently. The geometry and dimensions of the applicator are of particular interest. These include lengths and widths of the upper manifold channel and lower injector channel of the feed system, height of injector channel above the glass surface, position of both upstream and downstream exhaust manifolds, location of makeup air inlets, as well as the separating distance between adjacent applicators. Also, the influence of various hole shapes, sizes, spacing, distribution patterns, and the angle of the flow through the holes into the feed manifold. While these are important parameters, they will not be the subject of analysis in this paper.

In deference then to the above geometrical parameters, we are concerned with investigating the effect of the following on the gas flow pattern and mixing efficiency:

- Gas flow rates and velocities, including the inflow and the outflow ports (which are affected by the speed of the glass ribbon creating an unbalanced exhaust flow).
- Entrainment air flow above the glass ribbon (which is drafted from the surroundings at the edge of the applicator).
- Temperature gradients and buoyancy effects.
- Boundary-layer flow, separation, and stagnation region formation in the near vicinity of the exit of the narrow injector channel (as the flow hits the moving glass); these phenomena may enhance the formation of undesirable particulates.
- Manner of gas introduction (e.g., through holes, slits) and the angle of the injector channel.

Another motivation for our effort is the development of a clear understanding of the kinetics of the chemical reaction, primarily to identify the reaction time and reaction zone length. This is also related to the available flow rates and exhaust manifolds design, which must allow the removal of reaction by-products without inhibiting film deposition. The chemical reactions involved in the APCVD process take place on an atomic scale are by no means trivial to assess. We will not address these issues, however, in this study.

5. Simulation software. Solutions to extremely large and/or complex flow problems are increasingly more feasible due to continuing advancements in computing power. The CFD code FIRE, developed by AVL, Austria, is used to solve this flow problem. FIRE (Bachler et al., 1992) is a general purpose finite-volume based computational fluid dynamics analysis package, used to solve incompressible and compressible, viscous fluid flow problems. It is a menu-driven, fully interactive (with built-in graphics capabilities), multidimensional software that can simulate steady and unsteady flows that contain fixed or moving boundaries. It can handle both laminar or turbulent flows, Newtonian or non-Newtonian fluids, and non-isothermal flows as well.

We performed calculations on both the Apollo DN10000 and HP730 workstations. Run times varied according to the number of volume grids, time step size, and convergence criterion. As one might expect, the choice of a suitable time step was critical to the convergence characteristics and and validity of the end results.

6. Results and discussion. We calculated velocity components, pressures, and temperatures and present plots of these variables along with contours of a quantity called "passive scalar." The passive scalar represent a trace of fluid particles as the calculation advances in time. Think of it as injecting colored dye into the flow. Expressed as a number between 0 and 1, it represents the fraction of new fluid present in a computational cell. The results can be divided as two-dimensional or three-dimensional. The latter is critical to understanding the hole memory effect while the former becomes relevant once this effect is minimized.

6.1. Two-dimensional flow test cases. The starting point for our analysis is the consideration of the two-dimensional flow problem. It is important to thoroughly analyze this case because once the hole memory effect is eliminated the flow will indeed become two-dimensional.

We assume the flow to enter the top of the upper manifold channel at a uniform velocity of 0.431 m/s per unit depth. The two gases in the upper channel start to form a parabolic velocity profile after they pass the splitter plate. Two small recirculation regions form in the bottom corners prior to entering the narrow injector channel, as expected. Two distinct separated regions with recirculating flow are present below the exit of the injector. The size of these separated regions depends on the velocity, inclination, and height above the glass surface of the jet issuing from the exit of the injector, as well as the speed of the glass surface and any thermal gradients present. In general, the separated region upstream is smaller than that aft of the jet. The different cases used to investigate the influence these factors on the flow field are presented below.

6.1.1. Injector flow normal to the glass motion. This section considers the geometry where the injector channel meets the deposition chamber at a right angle, as shown in Figure 2.1. The velocity profiles, passive scalar contours, and temperature distribution are given, respectively, in Figures 6.1, 6.2 and 6.3 at time t = 0.61 sec. There is a large separated region just downstream of the jet near the upper surface of the deposition chamber, as well as a smaller, but still significant, recirculation region just upstream. A large downstream separated region aids in increasing the gas velocity near the glass surface by effectively acting as a barrier around which the gas jet must go. However, the upstream separation counteracts this effect somewhat since it is located near the glass surface and causes the fluid to lift up. The passive scalar indicates that, as one might expect, the fluid has a strong tendency to move in the direction of glass motion. Although the glass surface is heated, we initially maintained both the injected gas and the remaining applicator walls at room temperature. From the temperature distribution in Figure 6.3, it is evident that the jet, due to its high velocity, causes a local cooling in the deposition zone, and that the temperature gradient in the upstream segment of the deposition chamber is almost uniform. The same flow pattern described above is also demonstrated at $t = 3.0$ sec in the plots of the velocity profiles (Figure 6.4) and passive scalar (Figure 6.5). The latter shows a near total flush of the old fluid inside the applicator system by this time. The total pressure distribution, exhibited in Figure 6.6 at $t = 2.98$ sec, indicates a pressure loss as the flow moves down the injector channel, as well as a relatively high pressure in the deposition zone located next to the low pressure separated flow.

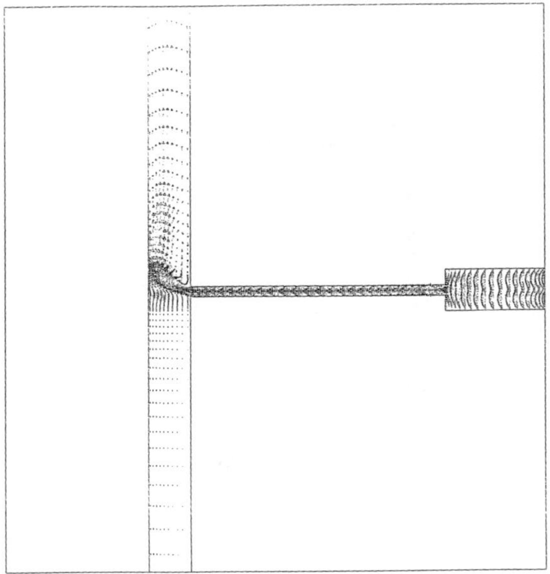

FIG. 6.1. *Velocity profiles for the two-dimensional flow case with normal injector; t =* 0.61 *sec.*

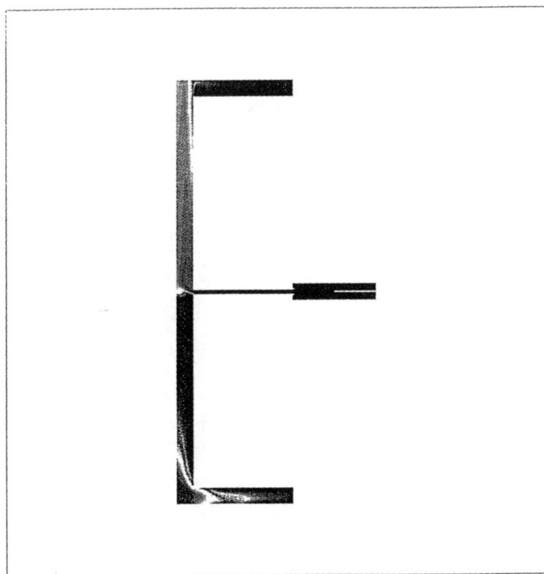

FIG. 6.2. *Passive scalar contours for the two-dimensional flow case with normal injector; t =* 0.61 *sec.*

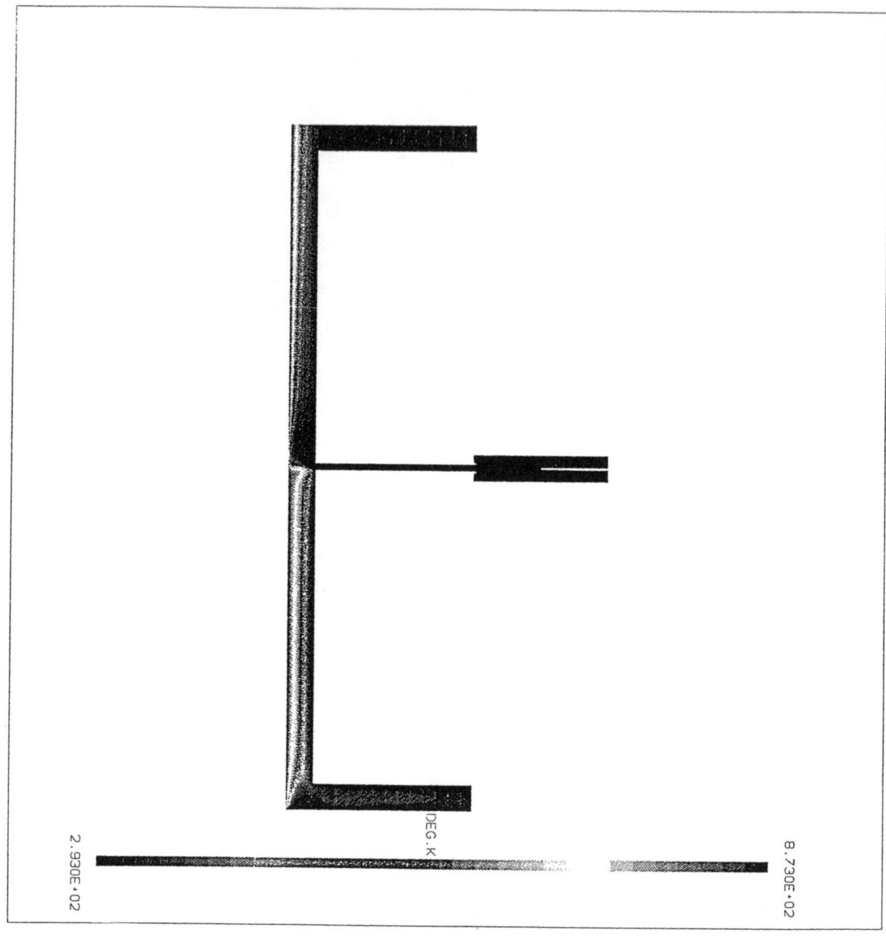

FIG. 6.3. *Temperature distribution for the two-dimensional flow case with normal injector; $t = 0.61$ sec.*

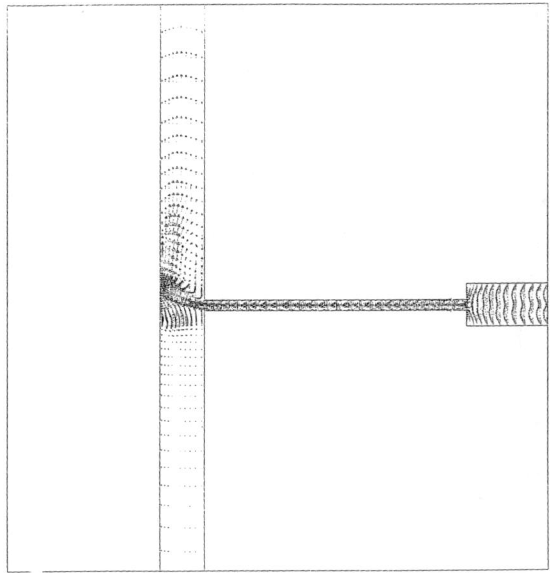

FIG. 6.4. *Velocity profiles for the two-dimensional flow case with normal injector; t = 3.0 sec.*

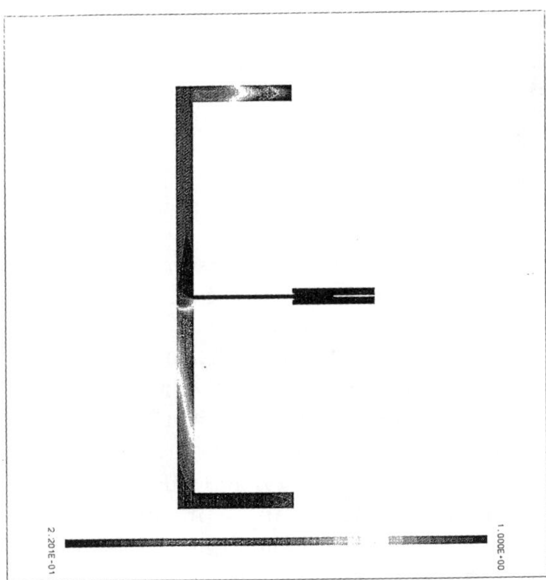

FIG. 6.5. *Passive scalar contours for the two-dimensional flow case with normal injector; t = 3.0 sec.*

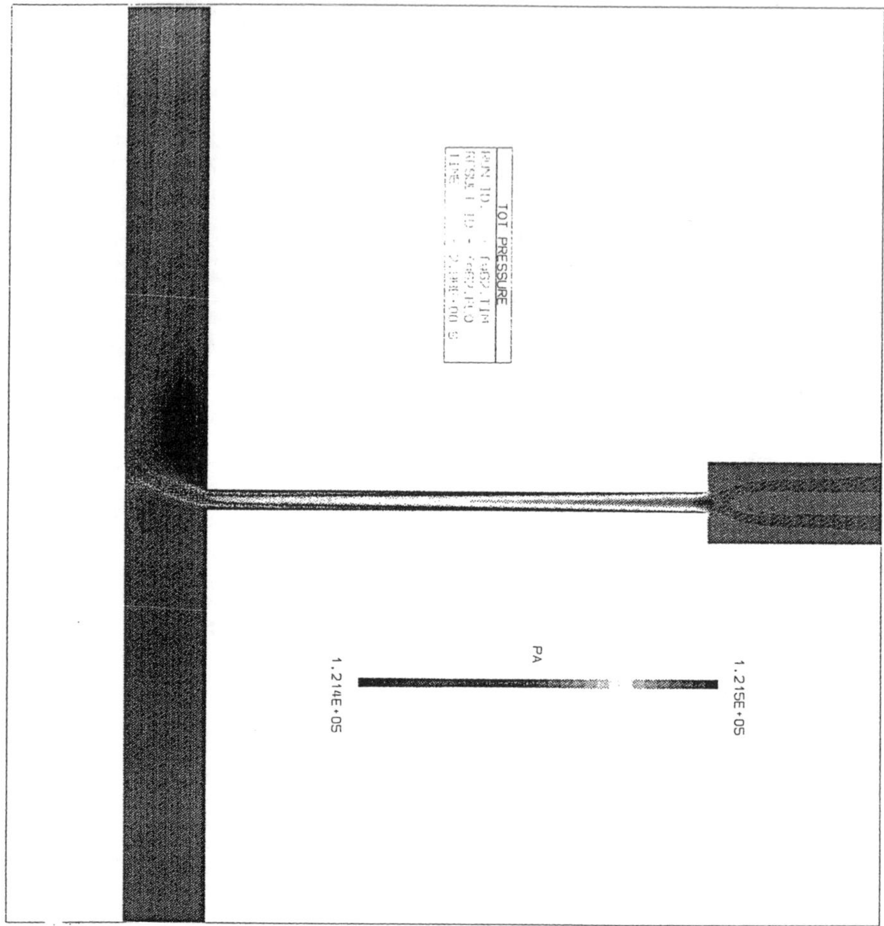

FIG. 6.6. *Total pressure distribution for the two-dimensional flow case with normal injector; $t = 2.98$ sec.*

To examine the influence that the efficiency of the exhaust manifold of an upstream applicator has on the flow field of the next applicator, we imposed the outlet velocity on the left as a boundary condition on the upstream (right) end of the deposition chamber. This would model the case where the previous applicator was allowed to flow freely into the next one. The velocity profiles and passive scalar are presented in Figures 6.7 and 6.8 respectively, at $t = 0.37$ sec. It is clear that the jet flow is dominated by the high velocity upstream incoming flow, producing low velocities and a lift up of the jet from the glass surface in the deposition zone. This results in low applicator efficiency, and implies that we have to insulate adjacent applicators from each other.

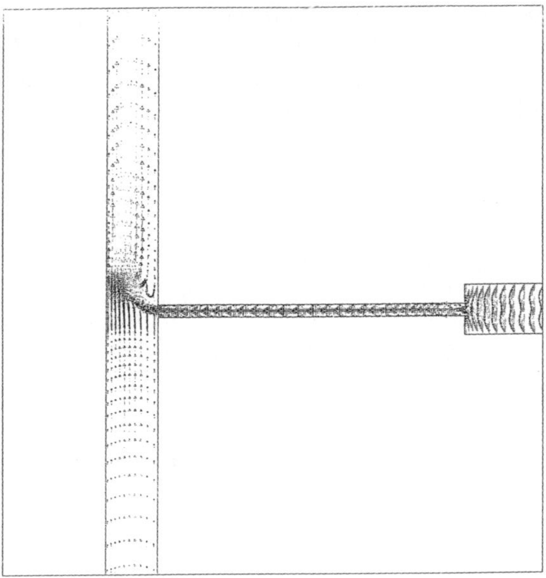

FIG. 6.7. *Velocity profiles for the two-dimensional flow case with normal injector and forced upstream flow; t = 0.37 sec.*

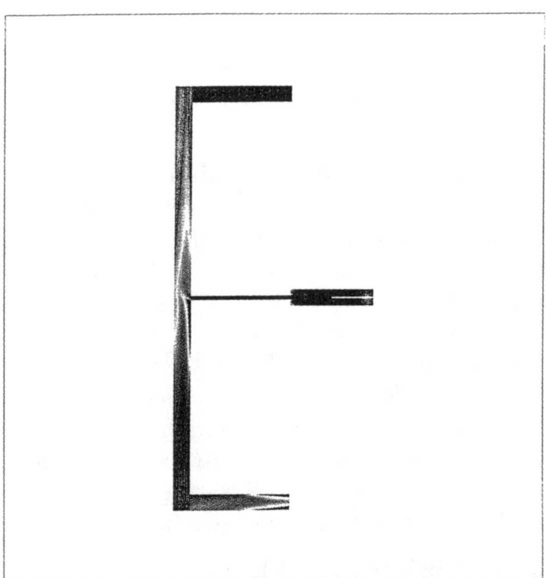

FIG. 6.8. *Passive scalar contours for the 2D flow case with normal injector and forced upstream flow; t = 0.37 sec.*

Figure 6.9 depicts the velocity profiles of the flow at $t = 0.127$ sec when the glass ribbon is slowly moving at 12 in/min $(0.0051m/s)^2$. It shows a considerably larger upstream separated region, more flow moving upstream, and lower velocity by the glass surface.

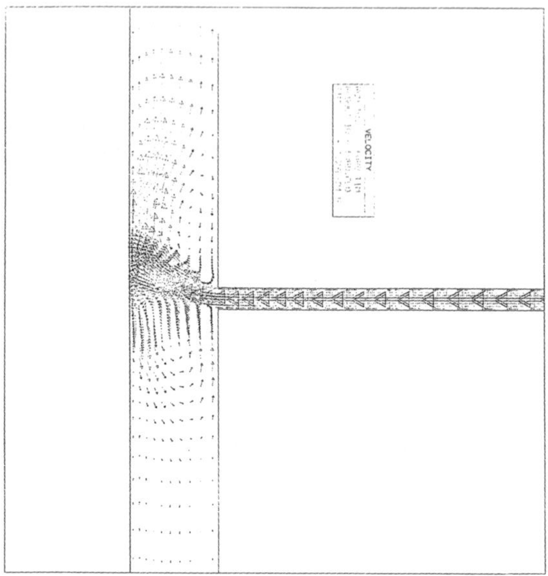

FIG. 6.9. *Velocity profiles for the 2D flow case with normal injector and glass ribbon moving at 12 in/min; $t = 0.127$ sec.*

6.1.2. Effect of temperature gradient. We next reduced the temperature difference across the height of the deposition chamber by increasing the upper surface from room temperature to $350°$ C. The velocity profiles are presented in Figure 6.10 at $t = 8.5$ sec. It shows a rather diminished upstream recirculation as the flow becomes less buoyant. The velocity near the glass in the deposition zone is decreased with decreased temperature differential; its maximum value is 1.21 m/s compared to 1.47 m/s for the case of high thermal gradient. This poses a delicate problem. On the one hand, too much recirculation created by a high thermal gradient could prohibit the chemical reaction and/or cause unwanted particulates to deposit on the glass surface. On the other, elevating the temperature along the upper walls could promote a premature reaction.

6.1.3. Effect of injection angle: flow at $30°$, $-15°$ and $-45°$. To study the effect of injection angle on the jet impact on the glass and the size of the separated regions, we modeled the jet with different injection angles.

---

[2] We allowed the glass ribbon to move at this speed to model commercial APCVD devices

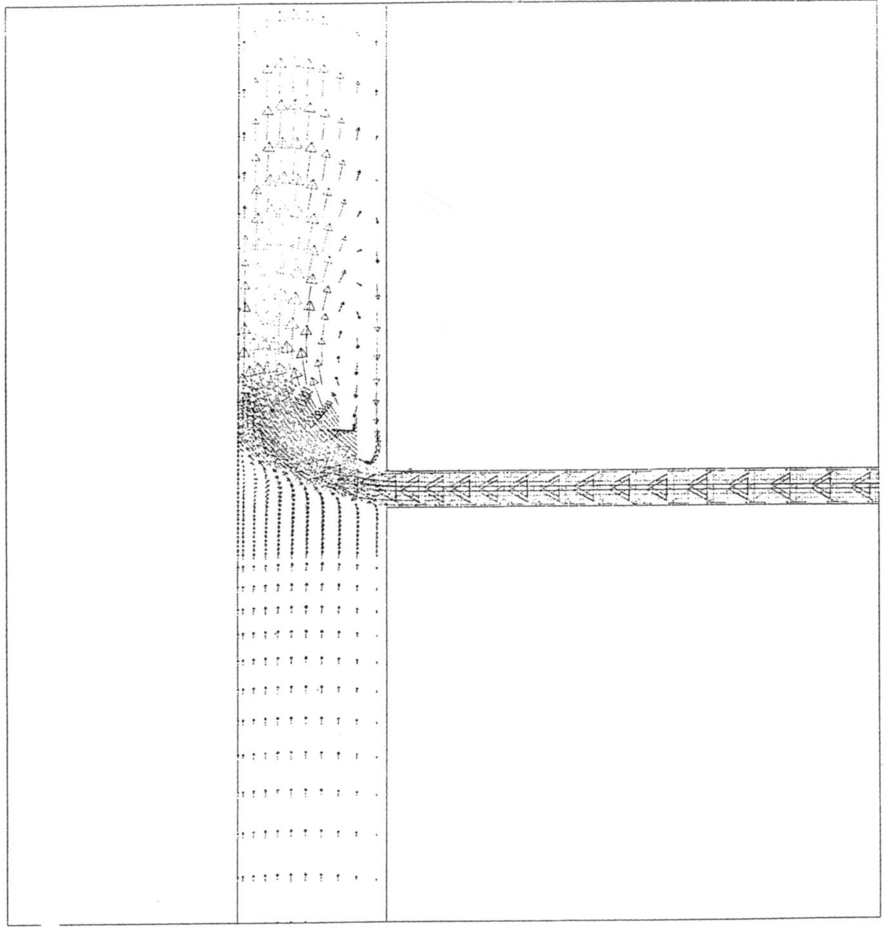

FIG. 6.10. *Velocity profiles for the 2D flow case with normal injector and upper deposition chamber surface maintained at 350°C; t = 8.491 sec.*

The injection angle θ is measured from the vertical axis perpendicular to the glass surface, and defined to be positive in the clockwise direction. Thus positive angles have the injection channel pointing in the direction of glass motion, while negative angles have the channel pointing counter to glass motion. The velocity profile for these flow are displayed in Figure 6.11 for $\theta = 30°$ and $t = 4.0$ sec, Figure 6.12 for $\theta = -15°$ and $t = 8.47$ sec, and Figure 6.13 for $\theta = -45°$ and $t = 0.127$ sec. Also the total pressure distribution for the case with $\theta = -15°$ is shown in Figure 6.14 at $t = 4.1$ sec. For all of these cases, the temperature of the upper surface of the deposition chamber is maintained at 350°C.

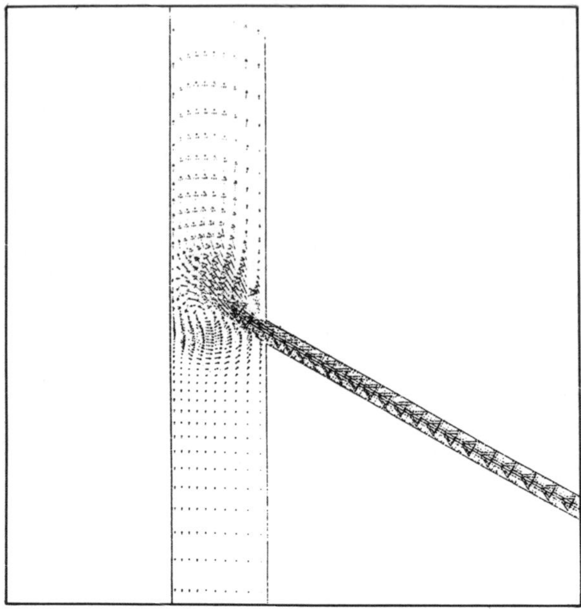

FIG. 6.11. *Velocity profiles for the two-dimensional flow case with injection angle* $\theta =$ 30°; $t = 4.0$ *sec.*

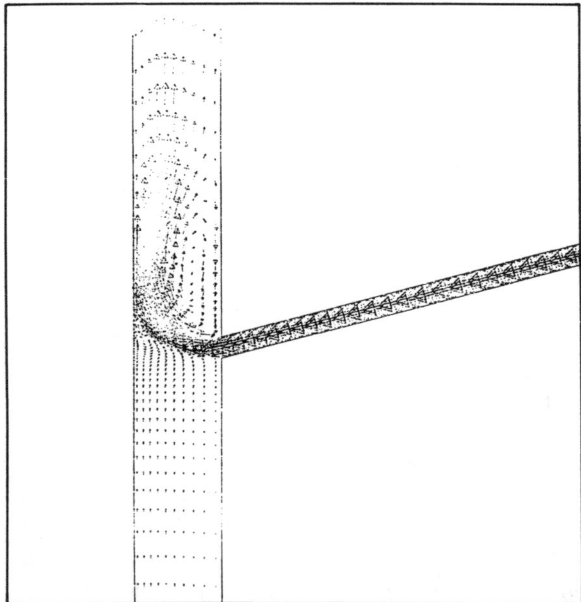

FIG. 6.12. *Velocity profiles for the two-dimensional flow case with injection angle* $\theta =$ −15°; $t = 8.47$ *sec.*

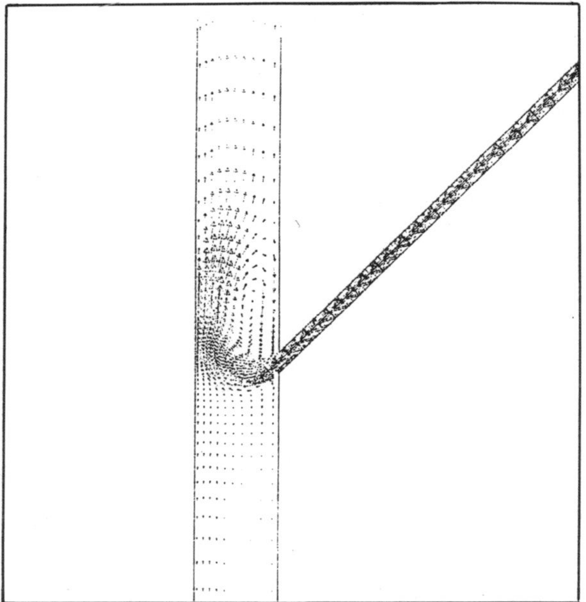

FIG. 6.13. *Velocity profiles for the two-dimensional flow case with injection angle* $\theta = -45°$; $t = 0.127$ *sec.*

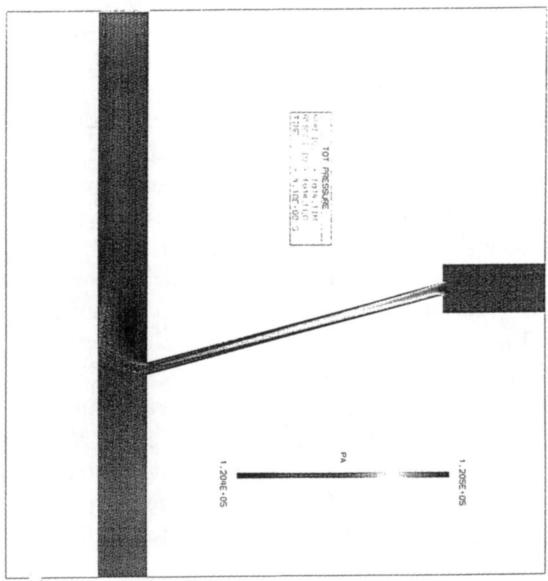

FIG. 6.14. *Total pressure distribution for the two-dimensional flow case with injection angle* $\theta = -15°$; $t = 8.47$ *sec.*

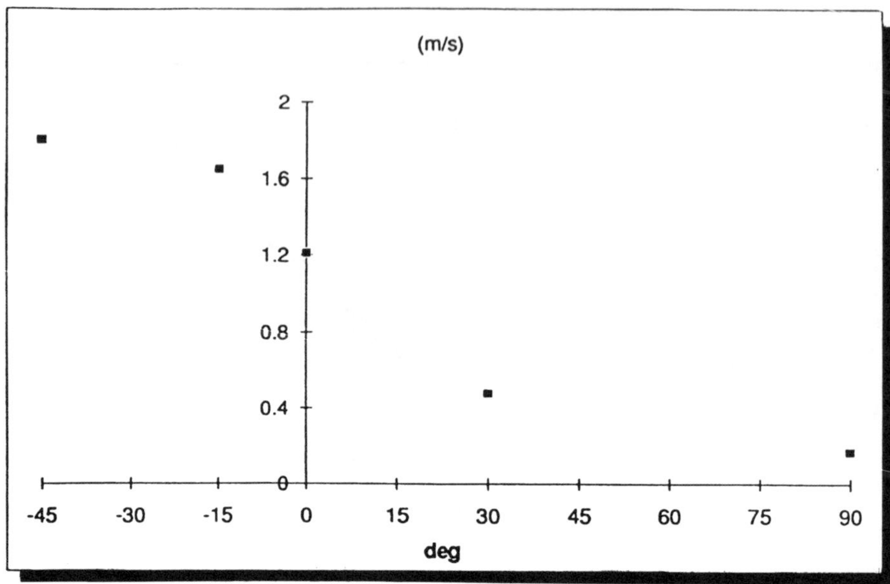

FIG. 6.15. *Maximum deposition velocity as a function of jet angle.*

In the case of $\theta = 30^o$, we clearly observe a reduced downstream separated region and an increased upstream recirculation. This upstream zone "lifts" the jet upward and away from the glass surface, causing the velocities by the glass surface to be smaller in magnitude. For $\theta = -15^o$, with the jet direction counter to the glass motion, there is a larger downstream separation compared to the standard case of injection at a right angle ($\theta = 0^o$, see Figure 6.10). The reduction in flow area due the presence of the larger downstream separated region pushes the jet further down toward the glass surface, and also accelerates the flow in the proximity of the glass surface. This effect is accentuated in the case of $\theta = -45^o$ in that it forces the jet even further down. The recirculation upstream almost disappears and the incoming flow near the upper surface is slowed down; but there is a bigger separated region downstream. From the preceding observations, it is evident that the deposition velocity[3] decreases as a function of the jet angle θ, as shown in Figure 6.15. The asymptotic value depicts the limiting case of a jet moving parallel to the glass ($\theta = 90^o$).

It thus appears that by angling the injection channel in a direction opposite to the glass motion we can suppress the upstream separated region and move it further upstream. This causes the maximum velocity and the total pressure near the glass surface in the immediate vicinity of the deposition zone to increase, which is desirable for higher quality deposition,

---

[3] Defined as the maximum velocity among the row of grid cell just above the glass surface

and keeps the jet on or near the glass surface for a longer distance. This is an interesting upshot and rather counterintuitive, since present APCVD systems have injector devices jetting the fluid either normally or at an angle in the direction of glass motion. We are currently developing an invention disclosure on this new approach.

6.2. Three-Dimensional flow test case.

6.2.1. Injection via side holes.
We need three-dimensional modelling to both detect and correct for the effect of hole memory on the flow field. This test case corresponds to the original design of the actual experimental model. Since the number of volume grids required is large resulting in extensive computation, only a section of the applicator is selected. This section contains two holes, one on each side, feeding the gases at 30° angles normal to the channel sides (see diagram below). The area of each hole is 0.002841 in^2 (dictated by the computational grid) and the magnitude of the gas velocity through the holes is 15.95 m/s (resulting in an x-component velocity of 13.813 m/s and a y-component velocity of 7.975 m/s). Although the two flows are in opposite directions, symmetry planes were assumed to exist midway between adjacent holes. In the absence of a repeated boundary condition feature in FIRE, this choice saves considerable computation time since the next option is to consider a model with few rows of holes (possible three), which can make the number of volume grids prohibitive for practical computation on a workstation.

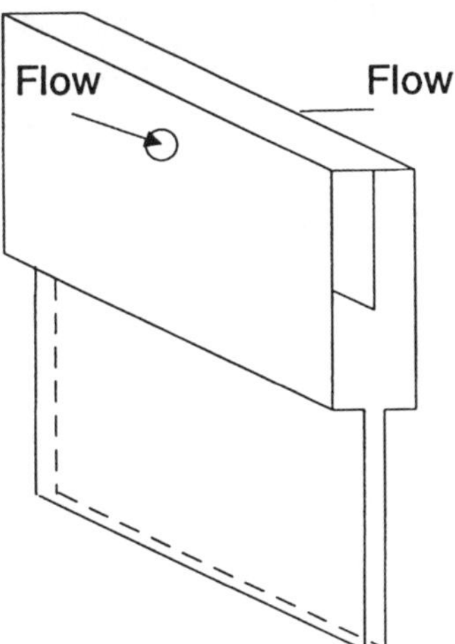

Plots of the passive scalar are shown in Figures 6.16 and 6.17 at different cross sections are in the flow, for t=0.11 sec and t=0.162 sec, respectively. The velocity field retains memory of the holes corresponding to their locations upstream in the feed manifold, which persists as the flow propagates downstream. This is evident by the clustering at the center to form an elliptical pattern. Now that we have verified the hole memory effect computationally, let's examine design alternatives aimed at alleviating it.

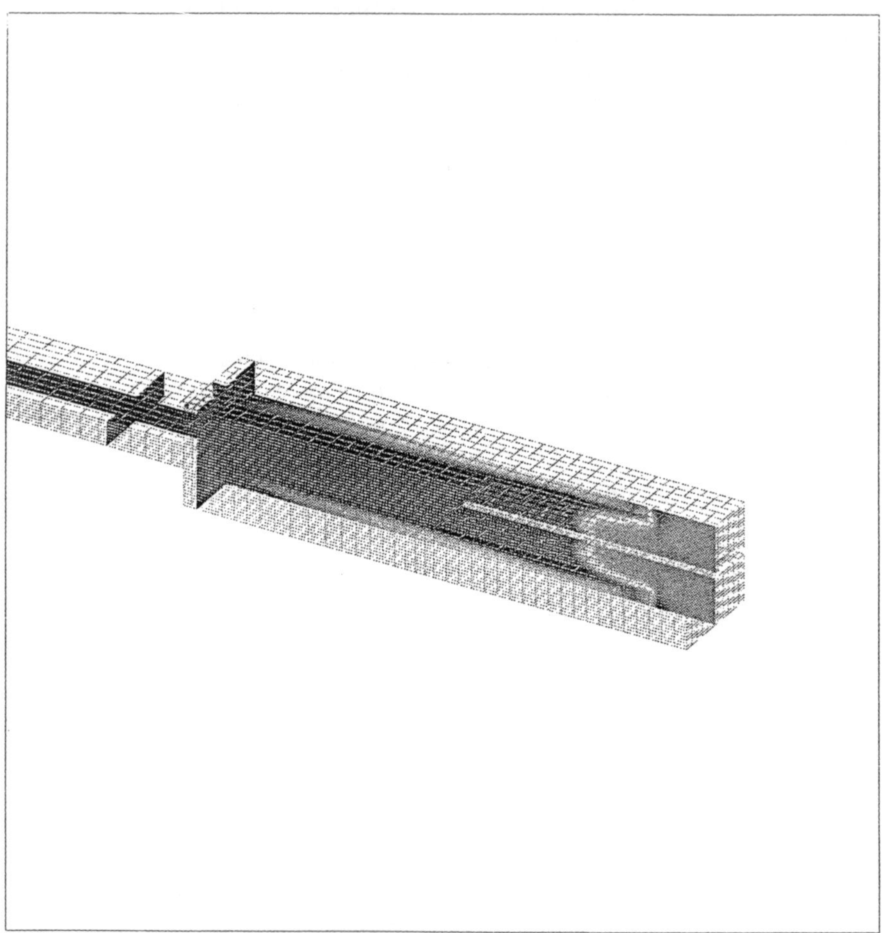

FIG. 6.16. *Vertical and horizontal passive scalar contours for the 3D flow case with gas inlet through side holes; $t = 0.11$ sec.*

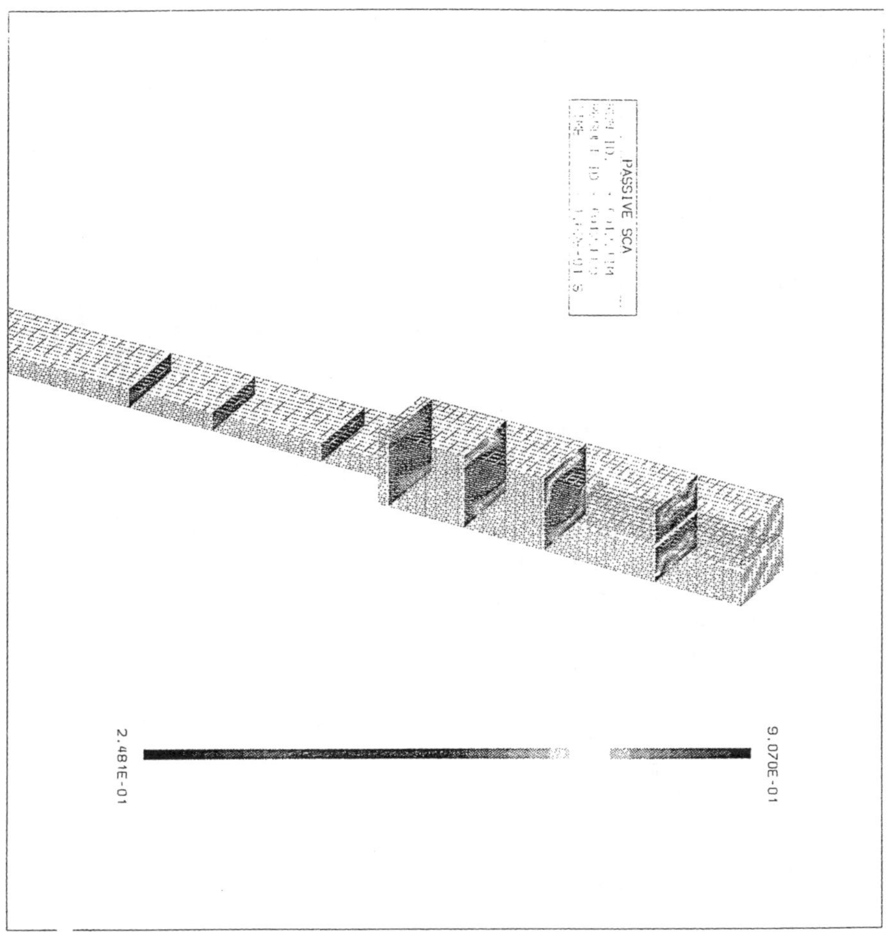

FIG. 6.17. *Horizontal passive scalar contours for the 3D flow case with gas inlet through side holes;* $t = 0.162$ *sec.*

6.2.2. Injection via the top: holes & slits. Our first approach was to modify the channel by replacing the hole gas inlets with slits. This configuration resembles the laboratory model without the splitter plate and the side holes replaced by slits on top of the feed manifold (see diagram on next page). As before, we simplified this model to that of three-dimensional channel flow with 0.5 in. × 0.5 in. square across section and symmetry planes. First, the flow from two slits with no overlap is simulated. The slit dimensions are 0.03125×0.5 in. with a flow velocity of 5.904 m/s.

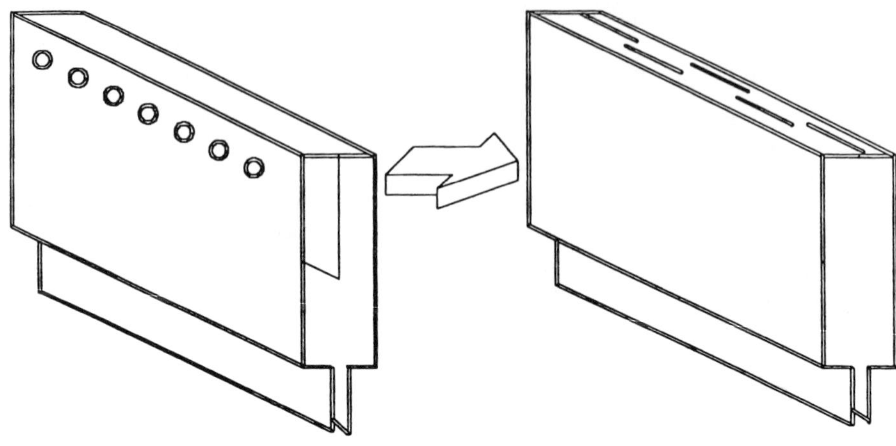

In Figure 6.18 the passive scalar at various cross sections for the flow at $t = 0.25$ sec shows that the flow from the two slits, separated at a distance of 0.0625 in., interacts and twists with a high concentration region at the center, and lower concentration on the sides of the channel. As the separating distance between the slits is increased to 0.1875 in., a more uniform distribution of the flow across the cross-sectional area of the channel is observed, as demonstrated in Figure 6.19 for the passive scalar at $t = 0.244$ sec.

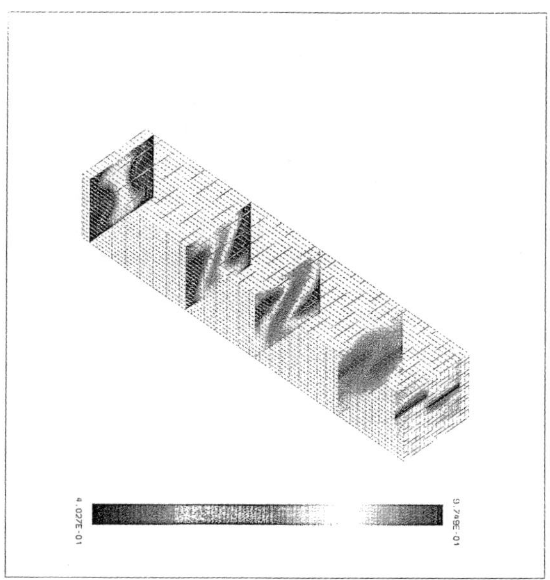

FIG. 6.18. *Passive scalar contours for the 3D square channel flow case with gas inlet through 0.03125 × 0.5 in. slits separated by 0.0625 in.; t = 0.25 sec.*

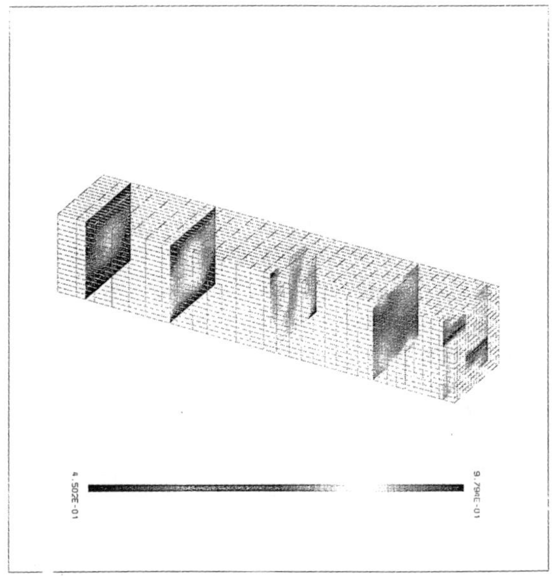

FIG. 6.19. *Passive scalar contours for the 3D square channel flow case with gas inlet through 0.03125 × 0.5 in. slits separated by 0.1875 in.; t = 0.244 sec.*

Additional calculations are planned using different arrangements of slits and holes to obtain a more homogeneous distribution of gases.

7. Concluding remarks. We have observed a number of problems in applying APCVD technology successfully to produce high quality coated glass. Some of the these relate to the memory of the gas feeds, the shape of the deposition jet, separation and recirculation zones, stagnation region where particulates can form, and exhaust efficiency. To predict the flow and design improved APCVD applicators to help alleviate the aforementioned problems, the gas feed system and injection angle must be modified based on the observation of their effect on the flow in the deposition zone near the glass surface. Parameters affecting the deposition process include: injection velocity and angle, height of injection jet exit above the glass surface, and speed of the moving glass. The preceding variables, except for the glass speed, can be individually altered for each applicator in order to achieve the desired performance.

The results of this study show that the velocity field does not change significantly as the laminar flow moves into the narrow injector channel, and retains memory of the hole locations upstream in the feed manifold. The memory problem is a result of the manner of injection and is created due to insufficient mixing in the upper manifold channel. Therefore, a mechanism is necessary to force the gases to turn and mix after flowing through the holes into the feed manifold, and before entering the injector. Turbulence can serve as a vehicle for that purpose. However, perturbations created to trigger turbulence will be dampened due to the low Reynolds number of the flow.

From experimental observations high velocity in the vicinity of the deposition zone is required for better coating. This translates into a higher total pressure on the surface of the glass. In the case of low velocities, the deposition film is vulnerable to outside disturbances. The effect of the injector channel angle is to accelerate the gas in the direction of the drawn glass. This results in a stronger impact for the gases with the glass in the proximity of the deposition zone. It is interesting to observe that due to the temperature differential between the lower and upper surfaces of the deposition chamber, the buoyancy effect is responsible for reducing the size of the upstream separated region. The role of temperature in enhancing any buoyancy effect will become insignificant if the upstream flow rate is increased, as it would then be dominated by the inertia of the flow.

REFERENCES

[1] Bachler, G. et al., *FIRE Instruction Manual*, Version 3.3, AVL LIST Gmbh, Graz, Austria, 1992.

[2] Gralenski, N.M., *Thin Films from a Thick Film Firing Furnace*, Hybrid Circuit Technology, May 1984.

[3] Potter, M.C., and Foss, J.F., *Fluid Mechanics*, Great Lakes Press, Okemos, MI, 1982.

[4] Schlichting, H., *Boundary-Layer Theory*, 7th Edn., McGraw-Hill, New York, 1979.

SHAPE OPTIMIZATION AND CONTROL OF SEPARATING FLOW IN HYDRODYNAMICS

THOMAS SVOBODNY*

Abstract. A model for computing flows with specified separation characteristics is presented. This is based on a shape optimization method for constructing a surface with a given tangential vorticity field.

1. Introduction. The Dirichlet problem for the Stokes operator is well-posed. That is, if we specify precisely the velocity on the boundary, then there exists a unique solution to the boundary value problem. Since the Stokes operator is the principal operator for the equations of viscous flow, the same considerations apply vis-a-vis the boundary conditions. The specification of velocity on the boundary is the relation that expresses the phenomenon of the fluid's adherence to a solid surface due to intermolecular forces. In some situations, one would perhaps want to model a boundary interaction by giving some other quantity on the surface, such as surface stress, or pressure, or vorticity ([1],[6]). In the present article we present a situation where one would like to specify both a surface vorticity while adhering to the requirement that the velocity be zero on the boundary. Clearly, something must give; what gives is the boundary: we specify vorticity and then the boundary velocity is a *cost* that we wish to drive to zero by finding the right surface. Even in the case where zero is unattainable, we can interpret the solution to this minimization problem in a physical way. The method described in this article can be used not only to construct surfaces with prescribed flow properties but also to compute flows with free surfaces.

In the next section the model of flow separation that motivates the use of the vorticity boundary condition is explained. In §3 we put everything in the context of shape optimization and compute the gradient of the relevant functional. In §4 we show that the optimization problem has a solution and how to define the gradient in a variational manner. In the concluding section, we discuss briefly the numerical computation.

2. Flow separation. The stall of an airplane wing is a familiar phenomenon: as the angle of attack is slowly increased, the form of the wing relative to the mean stream is no longer such that a streamwise pressure gradient on the lee-side of the wing invokes a favorable circulation over the wing to ensure the required lift. At a sufficiently high angle of attack this pressure is so reduced that there is a region on the lee-side where the flow is reversed near the surface; the streamwise velocity turns away from the surface and circumnavigates a "bubble" of the reversed flow or joins in a wake

---

* Supported in part by the Office of Naval Research Grant N00014-91-1494.
Department of Mathematics and Statistics, Wright-State University, Dayton, OH 45435.

behind the wing. In either case we say that the flow has separated from the surface. Thus, separation is typically defined ([3],[2]) as the departure of a streamline from the surface or as the occurence of a singularity so as to render invalid the boundary layer approximation. We make no direct use of this definition in the present work, nor do we consider the boundary layer approximation except to make some intuitive remarks to motivate interest in the role of the surface vorticity field in the separation phenomenon.

The separation at the surface plays a major part in the development of the global flow picture. Particularly, when separation occurs, there is usually formed a vortex-like structure or structures. In typical examples, we have the large vortical rolls that develop at the ends of wings of large transport planes: here there is no separation until close to the trailing edge of the wing; on the other hand, in swept-wing fighters separation occurs on the forebody and vortices form which can effect the flow over the aft portion of the wing as well as serve as dynamical drivers for structures such as vertical stabilizers. Worthy of mention in this context is the concept of vortex-lift, whereby the vortical structure over a delta wing induces favorable-to-lift pressure field. Actually, the story of lift for separated flow is not in good theoretical voice and is presently being told mainly through experimental and observational studies.

We should also make reference to the significance of separation to other engineering problems such as drag, pressure recovery, and noise generation. When one considers the wide range of flows that can occur in nature and indeed of which man could make use, one sees that attached flows with stable boundary layers form a very restricted class (this is analogous to the situation in systems theory vis a vis linear systems); yet, these flows are the only ones understood.

To observe separated flows and the attendant surface action, experimentalists can coat the surface of a wing or hydrodynamic surface with a viscous material such as paint, dye, or oil [23]. In the observed flow, the coating forms streaks along the surface; these follow the field line of the tangential surface shear. This vector field is known as the skin-friction, and it is observed that separation is characterized by the appearance of critical conditions in this vector field. (Hirsute individuals can do this very cheaply in the bath tub.) Mathematically, it is more convenient to work with the tangential vorticity, to which the skin-friction is closely related. Let us consider coordinates (ξ_1, ξ_2, η) in a region of the flow domain near a portion of the (smooth) boundary. The coordinates (ξ_1, ξ_2) refer to the bounding surface and η to the normal (into the flow domain), which latter we take to be euclidean distance from the boundary, so that every point near the boundary has the representation $\mathbf{R} = \mathbf{r}(\xi_1, \xi_2) + \eta \mathbf{n}$, where $\mathbf{r}$ is the surface parametrization and $\mathbf{n}$ is the unit normal. We use an orthonormal frame in a neighborhood of the boundary: $\mathbf{t}_1 = \frac{\partial \mathbf{R}}{\partial s_1}, \mathbf{t}_2 = \frac{\partial \mathbf{R}}{\partial s_2}, \mathbf{n} = \mathbf{t}_1 \times \mathbf{t}_2$, where s_k is arclength in the direction of increasing ξ_k: $ds^k = h_k d\xi^k$. Assuming the linear constitutive law of the Navier-Stokes equations, the surface stress is $\mathbf{T} = \mu\gamma^{31}\mathbf{t}_1 + \mu\gamma^{32}\mathbf{t}_2 + (\mu\gamma^{33} - p)\mathbf{n}$, where μ is viscosity and γ^{ij} are shear strain rate components. The skin-friction is the tangential component, $\boldsymbol{\tau}_s$,

$$
\begin{aligned}
\boldsymbol{\tau}_s &= \mu(\frac{\partial w}{\partial s_1} + \frac{\partial u}{\partial n})\mathbf{t}_1 + \mu(\frac{\partial w}{\partial s_2} + \frac{\partial v}{\partial n})\mathbf{t}_2 \\
&= \mu\frac{\partial u}{\partial n}\mathbf{t}_1 + \mu\frac{\partial v}{\partial n}\mathbf{t}_2
\end{aligned}
$$

(2.1)

Here, the flow velocity is $\mathbf{u} = u\mathbf{t}_1 + v\mathbf{t}_2 + w\mathbf{n}$ and we have used the condition $\mathbf{u} = 0$ for adherence to the boundary. The expression for the vorticity, $\boldsymbol{\omega} = \operatorname{curl}\mathbf{u}$, is, in these coordinates,

$$
\boldsymbol{\omega} = (\frac{\partial w}{\partial s_2} - \frac{1}{h_2}\frac{\partial(h_2 v)}{\partial n})\mathbf{t}_1 + (\frac{1}{h_1}\frac{\partial(h_1 u)}{\partial n} - \frac{\partial w}{\partial s_1})\mathbf{t}_2 + (\frac{1}{h_2}\frac{\partial(h_2 v)}{\partial s_1} - \frac{1}{h_1}\frac{\partial(h_1 u)}{\partial s_2})\mathbf{n},
$$

and so

$$
\mathbf{n} \times \boldsymbol{\omega} = (\frac{\partial w}{\partial s_2} - \frac{1}{h_2}\frac{\partial h_2 v}{\partial n})\mathbf{t}_2 + (\frac{\partial w}{\partial s_1} - \frac{1}{h_1}\frac{\partial h_1 u}{\partial n})\mathbf{t}_1,
$$

which on the surface reduces to

(2.2)
$$
\boldsymbol{\omega} \times \mathbf{n}|_{\eta=0} = \frac{\partial u}{\partial n}\mathbf{t}_1 + \frac{\partial v}{\partial n}\mathbf{t}_2,
$$

which is just $\mu^{-1}\boldsymbol{\tau}_s$ (2.1). (Notice that this is still the correct expression even if $\mathbf{u}|_\Gamma \neq 0$. The tangential vorticity field, $\boldsymbol{\omega} \times \mathbf{n}$, (actually this is vorticity rotated a right-angle about $\mathbf{n}$), will be the surface vector field of interest to us throughtout the work. From 2.2 , we see that $\mathbf{u} = (\boldsymbol{\omega} \times \mathbf{n})\eta + o(\eta)$, and so, for small η, the velocity field is tangent to the surface, except at critical points of $\boldsymbol{\omega} \times \mathbf{n}$. To see what the normal component is we can integrate the incompressibility condition, $\operatorname{div}\mathbf{u} = 0$; again, assuming that η is small, we have

$$
\begin{aligned}
\frac{\partial w}{\partial n} &= -\operatorname{div}_{tan}(\mathbf{u}_{tan}) \\
&= -\operatorname{div}_{tan}(\boldsymbol{\omega} \times \mathbf{n})\eta,
\end{aligned}
$$

or, upon integration,

$$
w = -\operatorname{div}_{tan}(\boldsymbol{\omega} \times \mathbf{n})\frac{\eta^2}{2} = -(\operatorname{curl}\boldsymbol{\omega} \cdot \mathbf{n})\frac{\eta^2}{2}.
$$

The tangential divergence, div_{tan}, can be defined without reference to coordinates in the following way. Let A denote a small surface patch centered at s_0 with area $|A|$, then

$$
\operatorname{div}_{tan}\mathbf{u}(s_0) = \lim_{|A| \to 0} |A|^{-1} \oint (\mathbf{n} \times \mathbf{u}) \cdot d\mathbf{l}.
$$

Thus,

$$\operatorname{div}_{tan}(\boldsymbol{\omega} \times \mathbf{n}) = \lim_{|A| \to 0} |A|^{-1} \oint \boldsymbol{\omega} \cdot d\mathbf{l} = (\operatorname{curl} \boldsymbol{\omega} \cdot \mathbf{n}).$$

The vector field, $\operatorname{curl} \boldsymbol{\omega} = \operatorname{curl} \operatorname{curl} \mathbf{u}$, which can be seen to be of major importance near critical points of $\boldsymbol{\omega} \times \mathbf{n}$, and which appears as a term in the Navier-Stokes equations and is seen to measure the rotation and stretching of vortex lines, is known as the flexion-field [22].

In summary, then, streamlines will be expected to be parallel to the surface for small η, as long as $\boldsymbol{\omega} \times \mathbf{n} \neq 0$. When $\boldsymbol{\omega} \times \mathbf{n} = 0$ which generally happens at isolated points (one needs special symmetry for $\boldsymbol{\omega} \times \mathbf{n} = 0$ to hold on a curve), we have [11] (i) a point of separation if $\operatorname{curl} \boldsymbol{\omega} \cdot \mathbf{n} < 0$, or (ii) a point of attachment if $\operatorname{curl} \boldsymbol{\omega} \cdot \mathbf{n} > 0$. If separation happens then the streamlines will tend away from the surface; following Lighthill [12], we can see that this is characterized by the convergence of near-surface streamlines toward a separating surface determined by a $\boldsymbol{\omega} \times \mathbf{n}$ field line. Look at the volume flow through a streamtube Σ, whose base is on the surface between two skin friction lines and the height of the tube is η:

$$\text{volume flow} = \iint_{\Sigma} \mathbf{v} \cdot d\mathbf{S} = \int_0^h \int_0^\eta |\boldsymbol{\omega} \times \mathbf{n}| z \, dz \, ds = \frac{1}{2} |\boldsymbol{\omega} \times \mathbf{n}| \eta^2 h$$

If $h \to 0$, then η^2 gets big, i.e., streamlines diverge from the surface. Thus, a necessary condition for separation is that skin-friction lines converge on a limiting line. What role, then, do the surace vorticity and flexion fields fulfill in forming the character of the mean outer flow? In particular, how are " vortices" generated at the surface, and how are their characteristics determined by what happens at the surface? Engineers are particularly interested in how to control forebody vortices on swept-wing planes ([18]).These questions will be dealt with in a future work; what is clear, from the above analysis is that the surface vorticity field plays an important part in any flow field and particularly in those flows which are said to be separating. Thus an important first step in the control theory of vortical and/or separating flows is to be able to have some control over this surface vorticity. In this work we are interested in the problem of how to achieve a prescribed tangential vorticity field by use of either a geometric control (shape of bounding surface) or boundary control (tangential blowing). In the next section, we discuss this in the context of shape optimization.

3. Shape optimization with a prescribed surface vorticity. We consider a body $\mathcal{B}$ in a viscous incompressible fluid moving with respect to the far-fluid at a uniform velocity $\mathbf{h}$. The flow is considered in a bounded region Λ containing $\mathcal{B}$. The boundary of the region Λ will be denoted as $\partial\Lambda$. The boundary of the body $\partial\mathcal{B}$ includes a connected component Γ, that we will consider to be *variable* or subject to design. Since the body can be parametrized by the variable part of the boundary Γ, we shall write also $\mathcal{B} = \mathcal{B}_\Gamma$, and we denote by Ω the actual flow region

$$(3.1) \qquad \Omega \overset{\text{def}}{=} \Lambda \setminus \bar{\mathcal{B}}_\Gamma,$$

so that we suppress the use of the parameter Γ for Ω.

The flow velocity $\mathbf{u}$ and the pressure field p, are assumed to satisfy the Navier-Stokes equation in the flow domain:

$$(3.2) \qquad \rho(\mathbf{u}_t + \mathbf{u} \cdot \nabla \mathbf{u}) = -\text{grad}\,(p + \psi) - \text{curl}\,(\mu\text{curl}\,\mathbf{u}),$$

where the mass density, ρ, and the viscosity, μ, are both constant parameters. The mathematical problem that we wish to consider is : Given a smooth vector field $\mathbf{f}$ (given parametrically on the unknown surface), find the surface Γ that minimizes

$$(3.3) \qquad J(\Gamma) = \frac{1}{2} \int_\Gamma |\mathbf{u} \times \mathbf{n}|^2 \, d\sigma,$$

where $\mathbf{u}$ satisfies the Navier-Stokes equations 3.2, with the boundary conditions

$$(3.4) \qquad \mathbf{u}|_{\partial\Omega\setminus\Gamma} = \mathbf{h},$$
$$(3.5) \qquad \mathbf{u} \cdot \mathbf{n}|_\Gamma = 0,$$
$$(3.6) \qquad (\nabla \times \mathbf{u}) \times \mathbf{n}|_\Gamma = \mathbf{f} \times \mathbf{n}.$$

The first condition gives the motion with respect to the far-fluid; the second condition implies that the surface is not to be penetrated by the fluid; the third condition fixes the tangential vorticity. In this work, we will only consider the case that none of the boundary surfaces are deformed in time and that the vector fields $\mathbf{h}$ and $\mathbf{f}$ are not time-varying. The outer flow of a separated flow is typically non-stationary and we will eventually consider this possibility; moreover, we will have to allow for this possibility that $\mathbf{u} \times \mathbf{n}$ may not be constant on the boundary. For the time-being however, our main goal is to study the shape derivative of this optimization problem and it will be convenient to first couch this study in the context of a steady flow. Notice that the condition (3.5) implies that the functional

$$\bar{J}(\Gamma) = \frac{1}{2} \int_\Gamma |\mathbf{u} \times \mathbf{n}|^2 \, d\sigma$$

could be used. The optimality systems for these two functionals are the same. An equivalent weak formulation of the governing equations can be found by partial integration. Defining

$$X = \{\mathbf{u} \in \mathbf{H}^1(\Omega) : \nabla \cdot \mathbf{u} = 0, \quad \mathbf{u}|_{\partial\Omega\setminus\Gamma} = \mathbf{0}, \quad \mathbf{u} \cdot \mathbf{n}|_\Gamma = 0\},$$

the equivalent problem is to find $\mathbf{u}$ so that $\mathbf{u} - \mathbf{h} \in X$ and satisfies

$$(3.7) \quad \int_\Omega \nu(\nabla \times \mathbf{u}) \cdot (\nabla \times \boldsymbol{\tau}) \quad + \quad (\nabla \times \mathbf{u}) \times \mathbf{u} \cdot \boldsymbol{\tau} \, dx$$

$$= \int_\Gamma \nu(\mathbf{f} \times \mathbf{n}) \cdot \boldsymbol{\tau} \, d\Gamma, \quad \forall \boldsymbol{\tau} \in X.$$

Here, $\nu = \mu \rho^{-1}$, is the kinematic viscosity or momentum diffusion coefficient. Now we have a framework for our problem. We want to minimize the functional $\mathcal{J}$ where the $\mathbf{u}$ in the integrand is constrained by (3.7). These state equations are well-posed; we will delineate suitable hypotheses under which there exists a minimum to the functional. For mininimization in this context one naturally considers an invesigation of the gradient. We will show that such an object exists. Moreover, it is straightforward to see that the solution to the optimization problem gives us some kind of a solution to our original problem: we are given a surface Γ^* upon which the adherence condition does not necessarily hold; thus $\mathbf{u}^* \times \mathbf{n}|_\Gamma = \mathbf{g} \times \mathbf{n}$ where one now considers $\mathbf{g}$ to be the (Dirichlet) control (tangential blowing or suction) [7]. This is one method of hybrid-control.

We want to compute an expression for the shape derivative of the functional J with respect to variations of the surface Γ. For this purpose let us gather together a few facts from the theory of shape optimization ([17],[19],[8],[15]). Let $\mathbf{V}$ denote a vector field (in R^3) defined in a normal-neighborhood of Γ, and vanishing on $\partial\Omega \setminus \Gamma$. For example, $\mathbf{V}$ can be given on Γ as $\mathbf{V} = V^1\mathbf{t}_1 + V^2\mathbf{t}_2 + V^n\mathbf{n}$ and then extended into Ω in some way; i.e., if $\mathbf{x} = \mathbf{r} + \eta\mathbf{n}$, then $\mathbf{V}$ can be extended in a constant way, $\mathbf{V}(\mathbf{x}) = \mathbf{V}(\mathbf{r})$, or perhaps as $\mathbf{V}(\mathbf{x}) = \mathbf{V}(\mathbf{r})h(\eta)$, where h is a cutoff function, etc., . A deformation of the boundary and thus the domain will be given by

$$F^\lambda(\mathbf{x}) = \mathbf{x} + \lambda\mathbf{V}(\mathbf{x})$$

The deformed control surface is $\Gamma^\lambda = F^\lambda(\Gamma)$. We will now define the material derivative of a functional defined on the domain or boundary. Let ϕ_λ(or ζ_λ) be a function defined on $F^\lambda(\Omega)$(or on $F^\lambda(\Gamma)$). Standard notation in shape optimization ([8],[19]) for the pullback to a function defined on Ω(or Γ) is

$$\phi^\lambda = (F^\lambda)^*\phi_\lambda = \phi_\lambda \circ F^\lambda$$

$$(\text{and, of course, } \zeta^\lambda = (F^\lambda)^*\zeta_\lambda = \zeta_\lambda \circ F^\lambda).$$

The *material derivative* is

$$\dot{\phi} = \frac{d}{d\lambda}\phi^\lambda|_{\lambda=0} = \lim_{\lambda \to 0} \lambda^{-1}((F^\lambda)^*\phi_\lambda - \phi);$$

we pull back to form the difference quotient in the fixed domain. The material derivative of a function on Γ is defined in an analogous way. Now suppose that $\mathbf{x} \in \Omega \cap \Omega^\lambda$, for some λ, then

$$
\begin{aligned}
\dot{\phi}(\mathbf{x}) = \frac{d}{d\lambda}\phi^\lambda(\mathbf{x})|_{\lambda=0} &= \frac{d}{d\lambda}\phi_\lambda(F^\lambda(\mathbf{x})) \\
&= \lim_{\lambda \to 0} \lambda^{-1}((\phi_\lambda)(\mathbf{x}) - \phi(\mathbf{x})) + \mathbf{V}(\mathbf{x}) \cdot \nabla\phi.
\end{aligned}
$$
(3.8)

The first term on the right side of the last equation is known as the *shape derivative*; it is denoted by ϕ' and can be shown to depend only on $\mathbf{V}|_{\eta=0}$, and thus we can analogously define the shape derivative of a function defined on Γ:

$$
\zeta'(\mathbf{r}) = \dot{\zeta}(\mathbf{r}) - \nabla_{tan}\zeta(\mathbf{r}) \cdot \mathbf{V},
$$

where $\nabla_{tan} = \nabla - \nabla_n = \nabla - \mathbf{n}\frac{\partial}{\partial\eta}$. The shape derivative measures the change in a function on a domain due to changes in the domain. For example, the function could be the solution to a differential equation to be solved on a domain whose shape may be subject to change. It is clear that the shape derivative is the object that appears to first order in an expansion of the solution in powers of λ. Notice that if f is defined everywhere on R^3, independently of Γ or Ω, then its shape derivative, $f' = 0$, since, in that case, for λ small enough, $f_\lambda(x) = f(x)$. The shape derivative can be shown to depend only on the component of $\mathbf{V}$ normal to the boundary.

To calculate the derivative of J, we will change variables to write $J(\Gamma^\lambda)$ as an integral over the fixed Γ; for this we will need to calculate the Jacobian of the resulting change of variables, at least to first order in λ. This Jacobian is defined as

$$
Jac(\lambda)d\Gamma = (F^\lambda)^* d\Gamma^\lambda.
$$
(3.9)

Using the notation of the previous section, we let subscripts denote derivatives of the vectors $\mathbf{r}$ and $\mathbf{v}$, i.e.,

$$
\mathbf{r}_k = \frac{\partial\mathbf{r}}{\partial\xi_k} = h_k\mathbf{t}_k.
$$

Then

$$
\begin{aligned}
(F^\lambda)^* d\Gamma^\lambda(\mathbf{r}_1, \mathbf{r}_2) &= d\Gamma^\lambda(F_*^\lambda\mathbf{r}_1, F_*^\lambda\mathbf{r}_2) \\
&= d\Gamma^\lambda(\mathbf{r}_1 + \lambda\mathbf{V}_1, \mathbf{r}_2 + \lambda\mathbf{V}_2) \\
&= |(\mathbf{r}_1 + \lambda\mathbf{V}_1) \times (\mathbf{r}_2 + \lambda\mathbf{V}_2)|d\xi_1 d\xi_2 \\
&= \left\{ 1 + \lambda\left(\frac{(\mathbf{r}_1 \times \mathbf{r}_2) \cdot (\mathbf{V}_1 \times \mathbf{r}_2 + \mathbf{r}_1 \times \mathbf{V}_2)}{|\mathbf{r}_1 \times \mathbf{r}_2|^2}\right) + \circ(\lambda) \right\} \\
&\qquad\qquad |\mathbf{r}_1 \times \mathbf{r}_2|^2 d\xi_1 d\xi_2 \\
&= Jac(\lambda)|\mathbf{r}_1 \times \mathbf{r}_2|^2 d\xi_1 d\xi_2;
\end{aligned}
$$

and we have the change of variables formula:

$$\int_{\Gamma^\lambda} f d\Gamma^\lambda = \int_\Gamma (F^\lambda)^* (f d\Gamma^\lambda) = \int_\Gamma (f \circ F^\lambda) Jac(\lambda) d\Gamma.$$

An expression for the derivative of the Jacobian can be found as follows: [1]

$$
\begin{aligned}
\frac{d}{d\lambda} Jac(\lambda)|_{\lambda=0} &= \frac{(\mathbf{r}_1 \times \mathbf{r}_2) \cdot (\mathbf{V}_1 \times \mathbf{r}_2 + \mathbf{r}_1 \times \mathbf{V}_2)}{|\mathbf{r}_1 \times \mathbf{r}_2|^2} \\
&= (h_1 h_2)^{-1} [\mathbf{V}_1 \cdot (\mathbf{r}_2 \times \mathbf{n}) + (\mathbf{V}_2 \cdot (\mathbf{n} \times \mathbf{r}_1))] \\
&= h_1^{-1} (\mathbf{V}_1 \cdot \mathbf{t}_1) + h_2^{-1} (\mathbf{V}_2 \cdot \mathbf{t}_2) \\
&= h_1^{-1} V_1^1 + h_2^{-1} V_2^2 + (h_1 h_2)^{-1} (V^2 \frac{\partial h_1}{\partial \xi_2} + V^1 \frac{\partial h_2}{\partial \xi_1}) \\
&\qquad - V^n (\mathbf{n} \cdot (\mathbf{r}_{11} + \mathbf{r}_{22})) \\
&= h_2^{-1} \frac{\partial}{\partial s_1} (h_2 V^1) + h_1^{-1} \frac{\partial}{\partial s_2} (h_1 V^2) + 2H V^n \\
&= \mathrm{div}_{tan} \mathbf{V}^{tan} + 2H V^n \\
&= \mathrm{div}_{tan} \mathbf{V}.
\end{aligned}
$$

In this expression, H denotes the mean curvature of the surface. Notice that we can then use Stokes thoerem and the conditions on $\mathbf{V}$ to write

$$(3.10) \qquad \int_\Gamma \mathrm{div}_{tan} \mathbf{F} \, d\Gamma = \int_\Gamma 2H (\mathbf{F} \cdot \mathbf{n}) \, d\Gamma.$$

We can now proceed to calculate the derivative of the functional. Before we consider the form of our specific functional 3.3, let us compute $\mathcal{J}'$ for

$$\mathcal{J}(\Gamma^\lambda) = \int_{\Gamma^\lambda} J(\mathbf{u}) d\Gamma^\lambda,$$

where J is defined on Γ^λ and $\mathbf{u} = \mathbf{u}_\lambda = \mathbf{u}|_{\Gamma^\lambda}$. Pulling back to Γ,

$$\mathcal{J}(\Gamma^\lambda) = \int_{\Gamma^\lambda} J(\mathbf{u}) d\Gamma^\lambda = \int_\Gamma J(\mathbf{u}_\lambda \circ F^\lambda) Jac(\lambda) \, d\Gamma,$$

then

$$
\begin{aligned}
\frac{d}{d\lambda} \mathcal{J}|_{\lambda=0} &= \lim_{\lambda \to 0} \lambda^{-1} (\mathcal{J}(\Gamma^\lambda) - \mathcal{J}(\Gamma)) \\
&= \lim_{\lambda \to 0} \lambda^{-1} \{ \int_\Gamma \{ J(\mathbf{u}^\lambda) - J(\mathbf{u}) \} + J(\mathbf{u}^\lambda)(Jac(\lambda) - 1) d\Gamma \}
\end{aligned}
$$

---

[1] Of course, it is easy to see what it is we are calculating here from the following simple consideration. As indicated by 3.9,

$$Jac(\lambda) = \det(F_*^\lambda) = \det(I + \lambda \nabla_{tan} \mathbf{V}),$$

and thus,

$$\frac{d}{d\lambda} Jac(\lambda)|_{\lambda=0} = \mathrm{tr}(\nabla_{tan} \mathbf{V}) = \mathrm{div}_{tan} \mathbf{V},$$

and the calculation in the text gives this function in terms of the components of $\mathbf{V}$.

$$= \int_\Gamma J(\mathbf{u})\dot{\mathbf{V}} + J(\mathbf{u})\mathrm{div}_{tan}\mathbf{V}\, d\Gamma$$

$$= \int_\Gamma J'(\mathbf{u})\mathbf{V} + V^n\frac{\partial}{\partial n}J(\mathbf{u}) + \mathbf{V}\cdot\nabla_{tan}J(\mathbf{u}) + J(\mathbf{u})\mathrm{div}_{tan}\mathbf{V}\, d\Gamma$$

$$(3.11)\qquad = \int_\Gamma J'(\mathbf{u})\mathbf{V}\, d\Gamma + \int_\Gamma (2HJ(\mathbf{u}) + \frac{\partial}{\partial n}J(\mathbf{u}))V^n\, d\Gamma,$$

having used (3.10). By use of the chain rule, the first integral can be seen to measure the way that the solution to (3.2,3.4-3.6) changes with respect to the boundary; we will deal with that term later. Our interest for the moment is on the second integral (we'll refer to the term in the integrand multiplying V^n as *boundary flux*), which for the functional (3.3) takes the form:

$$(3.12)\qquad \int_\Gamma \{H|\mathbf{u}\times\mathbf{n}|^2 + \frac{\partial}{\partial n}(\frac{|\mathbf{u}\times\mathbf{n}|^2}{2})\}V^n\, d\Gamma$$

Because of the non-penetrability condition, the first term is mean curvature times the square speed on the boundary which of course is zero if the fluid adheres to the boundary and there is no forcing. Both terms can be linked with the rate of vorticity generation at the bounding surface ([12]), but this doesn't seem to very useful in terms of a design sensitivity analysis, which is our concern here. Instead we will re-write the second term in terms of the square speed and the flux of the square speed and a third term which can be related to "effective" curvature.

If we compute the normal derivative in the expression (3.12),

$$\frac{\partial}{\partial n}\frac{|\mathbf{u}\times\mathbf{n}|^2}{2} = \mathbf{n}\cdot\nabla\frac{|\mathbf{u}\times\mathbf{n}|^2}{2}$$

$$= \mathbf{n}\cdot((\mathbf{n}\times\mathbf{u})\cdot\nabla(\mathbf{n}\times\mathbf{u})) + \mathbf{n}\cdot(\mathbf{n}\times\mathbf{u})\times(\nabla\times(\mathbf{n}\times\mathbf{u}))$$

we see that the first term on the right hand side simplifies because of the boundary condition of no-penetration:

$$\mathbf{n}\cdot((\mathbf{n}\times\mathbf{u})\ \cdot\nabla(\mathbf{n}\times\mathbf{u}))$$
$$= -(\mathbf{n}\times\mathbf{u})\cdot((\mathbf{n}\times\mathbf{u})\cdot\nabla\mathbf{n})$$
$$(3.13)\qquad = -(S(\mathbf{u}\times\mathbf{n}),\mathbf{n}\times\mathbf{u}).$$

This condition also simplifies the the second term :

$$\mathbf{n}\cdot(\mathbf{n}\times\mathbf{u})\ \times(\nabla\times(\mathbf{n}\times\mathbf{u}))$$
$$= \mathbf{n}\times(\mathbf{n}\times\mathbf{u})\cdot(\nabla\times(\mathbf{n}\times\mathbf{u}))$$
$$= [(\mathbf{n}\cdot\mathbf{u})\mathbf{n} - (\mathbf{n}\cdot\mathbf{n})\mathbf{u}]\cdot[\nabla\times(\mathbf{n}\times\mathbf{u})]$$
$$= -\mathbf{u}\cdot(\mathbf{u}\cdot\nabla\mathbf{n} - \mathbf{n}\cdot\nabla\mathbf{n} - 2H\mathbf{u})$$
$$(3.14)\qquad = 2H|\mathbf{u}|^2 + \nabla_n(|\mathbf{u}|^2) - (S\mathbf{u},\mathbf{u});$$

here $\mathbf{S}$ is the shape operator from surface theory [14]. If $\mathbf{u}$ were a unit vector on the surface then $(S\mathbf{u},\mathbf{u})$ would be normal curvature of the surface in the

334 THOMAS SVOBODNY

direction of $\mathbf{u}$. Thus in the case that $\mathbf{u}$ does not vanish on the surface, we can consider this tangential blowing as an enhancement to the curvature of the surface; the complete flux term for the boundary variation is then

$$H|\mathbf{u}|^2 + \frac{\partial}{\partial n}\left(\frac{|\mathbf{u}|^2}{2}\right) + ((H\mathbf{I} - \mathbf{S})\mathbf{u}, \mathbf{u}) + ((H\mathbf{I} - \mathbf{S})(\mathbf{u} \times \mathbf{n}), (\mathbf{u} \times \mathbf{n}))$$

If the surface streamlines are regular, it is seen that the last two terms cancel; however, we will keep them because they may be of some use in certain numerical maneovres.

The first integral in describes the variation of the integral due to the change of the solution on the changing domain. By the chain rule [21] we get (again we use to simplify)

$$\begin{aligned}
\int_\Gamma J'(\mathbf{u})\mathbf{V}\,d\Gamma &= \int_\Gamma (\mathbf{u} \times \mathbf{n}, \mathbf{w} \times \mathbf{n}) + (\mathbf{u} \times \mathbf{n}, \mathbf{u} \times \frac{d\mathbf{n}}{d\lambda}) \\
(3.15) &= \int_\Gamma \mathbf{u} \cdot \mathbf{w}.
\end{aligned}$$

The variable $\mathbf{w} = \mathbf{u}'$ is the shape derivative of $\mathbf{u}$ at $\lambda = 0$; it satisfies the system of differential equations and boundary conditions

$$(3.16) \qquad \mathbf{u}_t + \mathbf{u} \cdot \nabla \mathbf{w} + \mathbf{w} \cdot \nabla \mathbf{u} = -\operatorname{grad}(p) - \operatorname{curl}(\operatorname{curl}\mathbf{w}),$$
$$\operatorname{div}\mathbf{w} = 0,$$
$$\mathbf{w}|_{\partial\Omega\backslash\Gamma} = 0,$$
$$(3.17) \qquad \mathbf{w} \cdot \mathbf{n}|_\Gamma = -\operatorname{div}_{tan}(\mathbf{u})(\mathbf{V} \cdot \mathbf{n}),$$
$$(3.18) \qquad \nabla \times \mathbf{w} \times \mathbf{n}|_\Gamma = \Sigma(u),$$

where $\Sigma(u) = \frac{\partial}{\partial n}\boldsymbol{\omega} \times \mathbf{n}|_{\eta=0} = \frac{\partial}{\partial n}(\frac{\partial u}{\partial n}\mathbf{t}_1 + \frac{\partial v}{\partial n}\mathbf{t}_2)$, (cf. 2.2). This then is the description of the directional gradient of our functional. For computational purposes it is desirable to put this derivative in a variational framework. We will introduce this *weak* derivative later in the article. For the moment, we will discuss some technical matters.

4. Existence, uniqueness, and differentiability. We will gather here just a few facts regarding the well-posedness of our state equations. As our intent is to present a method by which we can construct a surface, we are interested in the conditions under which a minimum exists to our optimization and when we can define a gradient. The variational system of equations (3.7) has a solution if the surface field $\mathbf{f}$ is in $\mathbf{H}^{1/2}(\Gamma)$. This can be shown as in [1], as our system of equations falls into the class of equations studied there. Furthermore, one can show regularity results of the following form [20]: the solution map

$$(\Gamma, \mathbf{f}) \to \mathbf{u}$$

is continuous on

$$\operatorname{Lip} \times \mathbf{H}^{1/2}(\Gamma) \to \mathbf{H}^1(\Omega) \cap C^{0,\alpha}(\Omega)$$

We will always assume at least this much regularity with regard to the data (that is, the boundary and objects defined on the boundary).

It turns out that uniqueness of the state equation is necessary for the existence of a minimum; for this purpose, one can define a generalized Reynold's number $\mathcal{R}$, so that if $\mathcal{R} < \infty$ then the solution to (3.7) is unique. This number depends on the geometry, boundary data, and viscosity, being large for a large domain, large data, and small viscosity. For details on this construction, consult ([10],[1],[20]). Furthermore, we need a compactness condition in order to establish existence of a minimum; essentially what is need is some uniform control over the Lipshitzness of the boundary. The easiest way to see how to do this is to assume that Γ is given by a Monge patch [20]:

$$\Gamma^\phi = \{(\xi_1, \xi_2, \zeta) : \zeta = \phi(\xi_1, \xi_2), (\xi_1, \xi_2) \in D\}$$

Of course, we generalize to the case where Γ is given by a finite number of Monge patches. Now define

$$U_\beta = \{\Gamma^\phi : ||\phi||_{H^4(D)} \leq \beta\}$$

PROPOSITION 4.1. *Assume that* $\mathbf{h}$ *and* $\mathbf{f}$ *are in* $\mathbf{H}^2$*; suppose that* $\mathcal{R} < 1$*. There exists a* ϕ^* *such that*

$$\mathcal{J}(\Gamma^{\phi^*}) = \min_{\phi \in U_\beta} \mathcal{J}(\Gamma^\phi)$$

Proof: Let $\{\phi_n\} \subset U_\beta$ be a minimizing sequence, and let $\{\mathbf{u}_n\}$ be the corresponding solutions to the state equation. The assumptions are more than enough to guarantee that

$$||\mathbf{u}_n||_{H^1} + ||\mathbf{u}_n||_{C^{0,\alpha}} \leq C$$

And so (passing to subsequences):

$$\phi_n \to \phi^* \text{ in } H^3(D),$$

$$\mathbf{u}_n \rightharpoonup \mathbf{u}^* \text{ in } H^1(\Omega),$$

$$\mathbf{u}_n \to \mathbf{u}^* \text{ in } C(\bar{\Omega})$$

The equation for $\mathbf{u}_n$ is

$$\int_{\Omega_n} \nu(\nabla \times \mathbf{u}_n) \cdot (\nabla \times \boldsymbol{\tau}) \ + \ (\nabla \times \mathbf{u}_n) \times \mathbf{u}_n \cdot \boldsymbol{\tau} dx$$
$$= \int_{\Gamma_n} \nu(\mathbf{f} \times \mathbf{n}) \cdot \boldsymbol{\tau} d\Gamma, \quad \forall \boldsymbol{\tau} \in X.$$

If $\boldsymbol{\tau} \in C^1(\bar{\Omega})$ then clearly

$$\int_{\Gamma^{\phi_n}} (\mathbf{f} \times \mathbf{n}) \cdot \boldsymbol{\tau}\, d\Gamma \to \int_{\Gamma^{\phi^*}} (\mathbf{f} \times \mathbf{n}) \cdot \boldsymbol{\tau}\, d\Gamma;$$

and so,

$$\int_{\Omega_*} \nu(\nabla \times \mathbf{u}_*) \cdot (\nabla \times \boldsymbol{\tau}) \;+\; (\nabla \times \mathbf{u}_*) \times \mathbf{u}_* \cdot \boldsymbol{\tau}\, d\mathbf{x}$$

$$= \int_{\Gamma} \nu(\mathbf{f} \times \mathbf{n}) \cdot \boldsymbol{\tau}\, d\Gamma, \quad \forall \boldsymbol{\tau} \in X \cap C^1;$$

but C^1 is dense. Since, $\mathcal{R} < 1$, the solutions to this system are unique and we conclude that $\mathbf{u}_* = \mathbf{u}(\phi^*)$. Finally, we can pass to the limit

$$\lim_{n \to \infty} \mathcal{J}(\Gamma^{\phi_n}) = \mathcal{J}(\Gamma^*).$$

We proceed to establish a variational form for the directional derivative of the functional. The gradient in this weak form will be of interest in devising a computational method for computing Γ. The gradient of $\mathcal{J}$ at Γ is

$$(4.1) \qquad \int_{\Gamma} \mathbf{u} \cdot \mathbf{w} + \{H|\mathbf{u}|^2 + \frac{\partial}{\partial n}(\frac{|\mathbf{u}|^2}{2})\} V^n\, d\Gamma$$

and the equation for $\mathbf{w}$ is (3.16), which was derived by applying the chain rule whereby $\mathbf{u} = \mathbf{u}'$ is the shape derivative of $\mathbf{u}$ with respect to boundary variations. To make this expression for the gradient well-defined, one must show that this shape derivative exists. One would like to do this by computing the material derivative of the solution and applying (3.8), for example, as explained in [17] for the case of a linear equation: one pulls back to the $\lambda = 0$ domain and constructs the equation that the pull-back $\mathbf{u}^\lambda$ satisfies on Ω. This will be an equation

$$G(\lambda \mathbf{V}, \mathbf{u}^\lambda) = 0$$

of the form of the Navier-Stokes equations but with continuously varying coefficients (assuming $\mathbf{V} \in C^2$). One can get that $\dot{\mathbf{u}}$ exists by applying the implicit function theorem at the point $(0, \mathbf{u})$, i.e., $G(0, \mathbf{u})$ are the equations (3.2). In our case, however, we cannot express directly $\mathbf{u}^\lambda$ in powers of λ because the map $D_2 G(0, \mathbf{u})$ is not bijective. This operator is similar to that given by the left-hand side of (3.16). As pointed out in ([5],§3) this operator is semi-Fredholm and onto and thus one can apply the surjective version of the implicit function theorem [24] to conclude that the material derivative and thus the shape derivative is well-defined. Thus the gradient (4.1) is well-defined.

In formulating the weak version of this gradient by introducing an adjoint variable, the essential non-homogeneous boundary condition (3.17) presents us with a difficulty. One possibility for dealing with this is to first enforce this condition by introducing a multiplier as in [5] and then dualizing. However, since here it is only the normal component that is involved, it seems easiest to introduce into the derivative $\mathcal{J}'$ a potential term $\mathbf{u} \cdot \nabla \phi$ that will take care of the condition. More precisely, let $\mathbf{w} = \bar{\mathbf{w}} + \nabla \phi$, where

$$
(4.2) \qquad
\begin{aligned}
\Delta \phi &= 0, && \text{in } \Omega \\
\frac{\partial \phi}{\partial n} &= -\text{div}_{tan} \mathbf{u} \, V^n, && \text{on } \Gamma
\end{aligned}
$$

That this Neuman problem is solvable follows from the fact that $\mathbf{w}$ is solenoidal. It is clear that $\bar{\mathbf{w}}$ solves the same system as $\mathbf{w}$, but now with homogeneous boundary condition (3.17). At this point we introduce the adjoint variable $\xi \in X$ that satisfies

$$
(4.3) \qquad
\begin{aligned}
\int_\Omega (\nabla \times \xi) \cdot (\nabla \times \boldsymbol{\tau}) &\quad -(\nabla \times \mathbf{u}) \times \xi \cdot \boldsymbol{\tau} \, dx \\
-\nabla \times (\mathbf{u} \times \xi) \cdot \boldsymbol{\tau} &= \int_\Gamma \mathbf{u} \cdot \boldsymbol{\tau} \, d\Gamma, \quad \forall \boldsymbol{\tau} \in X.
\end{aligned}
$$

This is in fact dual to (3.16) as can be verified by an integration by parts, noting that

$$
\int_\Omega (\nabla \times \xi) \cdot (\nabla \times \mathbf{w}) \, dx = \int_\Omega \nabla \times (\mathbf{u} \times \xi) \cdot \mathbf{w} \, dx,
$$

since $\mathbf{u} \times \xi$ is a vector normal to Γ. Now choose, in the equation (4.3), $\boldsymbol{\tau} = \bar{\mathbf{w}}$. Then, by an integration by parts,

$$
\int_\Gamma \mathbf{u} \cdot \bar{\mathbf{w}} \, d\Gamma \int_\Omega (\nabla \times \xi) \cdot (\nabla \times \bar{\mathbf{w}}) \quad -(\nabla \times \mathbf{u}) \times \bar{\mathbf{w}} \cdot \xi \, dx
$$
$$
-\nabla \times \bar{\mathbf{w}} \times \mathbf{u} \cdot \xi = \int_\Gamma \Sigma(\mathbf{u}) \cdot \xi \, d\Gamma,
$$

where we have used the weak formulation of (3.16). Then we have the following

PROPOSITION 4.2. *The boundary functional $\mathcal{J}$ has a derivative at any $\Gamma \in H^4$ in any direction $\mathbf{V} \in C^2$ and*

$$
(4.4) \qquad
\begin{aligned}
\mathcal{J}'(\Gamma)\mathbf{V} &= \\
&\int_\Gamma \Sigma(\mathbf{u}) \cdot \xi + \mathbf{u} \cdot \nabla \phi + \{ H|\mathbf{u}|^2 + \frac{\partial}{\partial n}(\frac{|\mathbf{u}|^2}{2}) \} V^n \, d\Gamma
\end{aligned}
$$

where ξ and ϕ are defined as above.

5. Conclusion. Once one has an expression such as (4.4), the real work can begin. One would like a discrete version of the gradient to be able to use a gradient-like algorithm for minimizing the functional. We have in mind here what is apparently known in the trade as a design sensitivity analysis ([8],[15],[16],[19]). Simply, this name is descriptive of its origins in design. Subsequent to discretization, a direction in the approximation space is typically associated with the node of a triangulation of the design surface and a variation in that direction (in approximation space) is associated with the movement of that node in a particular direction (in physical space). Thus, with the idea that we associate minima with stationarity of the functional (local extrema are always a problem), we can push the physical nodes in different directions to see how "sensitive" the functional is to such movement. Of course, by the time one gets around to constructing an algorithm for this it resembles a gradient-type programming algorithm.

From the expression for the gradient of $\mathcal{J}$ it is seen that the it needs to have at each step current values for the state variable $\mathbf{u}$ as well as the auxiliary variables ξ and ϕ. A steady flow is found by computing the state equation in time until a suitable mean flow is achieved which is then fed to the adjoint eqn which is integrated backward in time until steady. The initial condition is always the steady condition from the previous optimization step. Thus the flow takes some time to settle down adjusting to the new boundary at each step in the optimization. The spatial discretization is donw through a finite-element approximation. The finite-element grid is dictated by the discretization of the surface. For example, if Γ is given by a superposition of height functions, then it is easy enough to align the grid with the outer glow; however if the surface is given by an orthogonal mesh of curvature lines, then clearly one wants to move out normally in the mesh. Although the computational requirements of this problem seem enormous, it does provide a one-shot method for constructing surfaces that do not allow separation.

REFERENCES

[1] C.Begue,C.Conca,F.Murat,and O.Pironneau, Les équations de Stokes et de Navier-Stokes avec des conditions aux limites sur la pression, *Nonlinear PDE and applications; Collège de France Seminar*, IX, 1988

[2] P.K.Chang, *Control of Flow Separation*, Hemisphere, 1976

[3] M. Gad-el-Hak and D. Bushnell, Separation Control: Review, Journal of Fluids Engineering, 113(1), 1991

[4] M. Gunzburger, *Finite Element Methods for Viscous Incompressible Fluids*, Academic Press, Boston, 1989

[5] M.Gunzburger, L.Hou, T.Svobodny, Analysis and Finite Element Approximation of Optimal Control Problems for the Stationary Navier-Stokes Equations with Dirichlet Controls, *Math. Model. Num. Anal.*, 25(6),

[6] M.Gunzburger, L.Hou, T.Svobodny, Analysis and Finite Element Approximation of Optimal Control Problems for the Stationary Navier-Stokes Equations with Distributed and Neumann Controls, *Math. Comp.*,57(195), 1991

[7] M.Gunzburger, L.Hou, T.Svobodny, Boundary Velocity Control of Incompressible Flow with an Application to Viscous Drag Reduction, SIAM J. Control & Optim., 30(1), 1992

[8] J.Haslinger and P.Neittaanmäki, *Finite Element Approximation for Optimal Shape Design*, Wiley, Chichester, 1988

[9] G. R. Hough, (Editor), *Viscous Flow Drag Reduction*, AIAA, New York, 1980

[10] O. Ladyzhenskaya, *The Mathematical Theory of Viscous Incompressible Flow*, Gordon and Breach, New York, 1963

[11] R.Legendre, Séparation de l'écoulement laminaire tridimensionnel, *Rech. Aéro.*, 54, 1956

[12] M.J.Lighthill, Theory of Boundary Layers in *Laminar Boundary Layers*, (Rosenhead, Ed.), Dover, 1963

[13] J. Nečas, *Les Methodes Directes en Theorie des Equations Elliptiques*, Masson, Paris, 1967.

[14] B.O'Neill, *Elementary Differential Geometry*, Academic Press, New York, 1966

[15] O.Pironneau, *Shape Design for Elliptic Systems*, Springer, New York, 1984

[16] H.Rabitz, Sensitivity Analysis of Combustion Systems, in *The Mathematics of Combustion*, SIAM, Philadelphia, 1985

[17] J.Simon, Differentiation with respect to the domain in boundary value problems, *Numer. Fun. Anal. Optim.* , 2 1981

[18] A.Skow and D. Peake, Control of the forebody vortex orientation by asymmetric air injection, in AGARD LS-121, 1982

[19] J.Sokolowski and J-P. Zolesio, Introduction to Shape Optimization, Springer, 1992

[20] S.Stojanovic and T. Svobodny, A Variational Approach to Shape Optimization for the Navier-Stokes Equations, to appear

[21] S.Stojanovic and T. Svobodny, Computation of the Surface of a Viscous Fluid, to appear

[22] C.Truesdell, *Kinematics of Vorticity*, Indian , 1954

[23] G.Werlé, Flow Visualization, in AGARD LS-121, 1982

[24] E.Zeidler, *Nonlinear Functional Analysis and its Applications*, Springer, 1986

RECENT ADVANCES IN STEADY COMPRESSIBLE AERODYNAMIC SENSITIVITY ANALYSIS

ARTHUR C. TAYLOR III[*], PERRY A. NEWMAN[†], GENE J.-W. HOU[‡], AND HENRY E. JONES[§]

1. Introduction. An overview is given of some recent accomplishments by different researchers in calculating gradient information of interest from modern flow-analysis codes. Of particular interest here is *advanced* computational fluid dynamics (CFD) software, which solves the *nonlinear* multidimensional Euler and/or Navier-Stokes equations. The accurate, efficient calculation of aerodynamic sensitivity derivatives is very important in design-oriented applications of these CFD codes to single discipline and multidisciplinary problems [1,2].

Sensitivity analysis methods are classified in this study as belonging to either of two categories: the discrete (quasianalytical) approach or the continuous approach. This roughly follows the classification presented in Ref. [3], where the two methods are referred to as the implicit gradient approach and the variational approach, respectively. These two broad categories in essence differ by the order in which discretization and differentiation of the governing equations and boundary conditions is undertaken; for the former approach, the discretization precedes differentiation. In the final analysis of either case, a large discrete system of linear equations must be solved when calculating the sensitivity derivatives.

The principal focus of the present discussion is the discrete approach, for which the basic equations are presented; the major difficulties, together with proposed solutions, are reviewed in some detail. However, advantages and disadvantages are associated with each of the two categories of methods. Thus, a brief discussion of some recent research activity that involves the continuous approach is also included.

2. The discrete approach.

2.1. Summary of basic equations. After discretization, the nonlinear, multidimentional steady-state governing equations of fluid flow and the boundary conditions are approximated as a large system of coupled nonlinear algebraic equations as

$$(2.1) \qquad R(Q(D), X(D), D) = 0$$

---

[*] Assistant Professor, Department of Mechanical Engineering, Old Dominion University, Norfolk, VA 23529.

[†] Senior Research Scientist, NASA Langley Research Center, Hampton, VA

[‡] Associate Professor, Department of Mechanical Engineering, Old Dominion University, Norfolk, VA 23681.

[§] Research Scientist, U.S. Army AeroFlightdynamics Directorate, Hampton, VA 23681.

where Q is the vector of field variables, X is the computational grid, and D is a vector of independent input (design) variables. Differentiation of Eq. (2.1) yields the matrix equation

$$(2.2) \qquad R' = \frac{\partial R}{\partial Q}Q' + \frac{\partial R}{\partial X}X' + \frac{\partial R}{\partial D} = 0$$

where $R' \equiv \frac{dR}{dD}$; $Q' \equiv \frac{dQ}{dD}$ (the sensitivity of the field variables); and $X' \equiv \frac{dX}{dD}$ (the "grid sensitivity"). This latter sensitivity will be discussed subsequently in greater detail. The linear Eq. (2.2) is first solved for Q', in order that the sensitivity derivatives of aerodynamic output functions, F, can be calculated subsequently. That is,

$$(2.3) \qquad F = F(Q(D), X(D), D)$$

and differentiation of Eq. (2.3) yields

$$(2.4) \qquad F' = \frac{\partial F}{\partial Q}Q' + \frac{\partial F}{\partial X}X' + \frac{\partial F}{\partial D}$$

where $F' \equiv \frac{dF}{dD}$, which are sensitivity derivatives of interest. Alternatively, the necessity of solving Eq. (2.2) for Q' is eliminated by first solving the linear equation

$$(2.5) \qquad \left(\frac{\partial R}{\partial Q}\right)^T A + \left(\frac{\partial F}{\partial Q}\right)^T = 0$$

where A is a discrete adjoint variable matrix associated with the functions F. Then F' is computed as

$$(2.6) \qquad F' = \frac{\partial F}{\partial X}X' + \frac{\partial F}{\partial D} + A^T\frac{\partial R}{\partial X}X' + A^T\frac{\partial R}{\partial D}$$

For maximum computational efficiency, Eq. (2.2) is solved for Q' if the dimension of F is greater than that of D; otherwise, Eq. (2.5) is solved for A if the dimension of D is greater than that of F.

A number of researchers have successfully pursued the preceding quasi-analytical approach to calculate sensitivity derivatives from nonlinear flow-analysis codes of varying degrees of complexity. For example, Elbanna and Carlson (Ref. [4]) have computed sensitivity derivatives for various airfoil flows from the transonic small-disturbance equation, and, more recently, for three-dimensional (3D) flow over a wing from the full potential flow equation (Ref [5]). Drela (Ref. [6]) has computed derivatives for airfoil flows from a streamline coordinate formulation of the two-dimensional (2D) Euler equations, coupled with the boundary-layer equations, to account for viscous effects.

The calculation of quasianalytical sensitivity derivatives is reported by Taylor et al. (Ref. [7,8]), Hou et al. (Ref. [9]), and Baysal et al. (Ref.

[10,11]) for interior channel flows from a conventional upwind finite-volume solution strategy applied to the 2D Euler equations in body-oriented coordinates. These researchers have subsequently extended this work to calculate sensitivity derivatives for 2D laminar flows from the thin-layer Navier-Stokes (TLNS) equations, including external flows over isolated airfoils (Refs. [12,13]). Calculation of quasianalytical aerodynamic sensitivity derivatives with an upwind finite-volume solution of the Euler equations has also been reported by Beux and Dervieux (Ref. [14]) for a 2D channel flow. In many of the references cited thus far, the quasianalytical sensitivity derivatives were not only shown to agree very well (as expected) with derivatives computed by the method of finite differences, but were obtained with significantly less computational effort.

Despite the success reported in these works, however, severe difficulties remain, and these must be overcome, so that efficient, accurate calculation of gradient information from large-scale modern CFD software can become routine, particularly for turbulent 3D flows over complex geometries. Three such major difficulties identified here are:

1. Solution of the extremely large system of linear equations (either Eq. (2.2) for Q' or Eq. (2.5) for A)
2. Accurate differentiation of all terms in the flow-analysis code (which can become an extremely complex task) to be used in computing the sensitivity derivatives
3. Evaluation of the "grid sensitivity" term X', in Eqs. (2.2) and (2.4), or in Eq. (2.6).

These three problems will be discussed subsequently in greater detail; included in this discussion will be some recent research efforts that have been undertaken to overcome these obstacles. Further discussion of these and other difficulties is also found in Ref. [15].

2.2. Methods for equation solution. If a strict application of Newton iteration is possible and applied in solving the nonlinear flow Eq. (2.1) for Q, then clearly the solution of the linear Eq. (2.2) for Q' (or Eq. (2.5) for A) becomes simply an efficient back-substitution procedure. This procedure has been demonstrated in the references cited thus far. However, the formal implementation of Newton iteration is not feasible for advanced CFD codes on current supercomputers because available memory does not permit direct LU factorization of the coefficient matrix when solving the Euler or Navier-Stokes equations for large $2D$ or practical $3D$ problems.

As an alternative to pure Newton iteration, typical CFD codes employ what is sometimes called "quasi-Newton" iteration which can be expressed as

$$(2.7) \qquad -\frac{\partial \widetilde{R^n}}{\partial Q}\Delta Q = R^n$$

$$(2.8) \qquad \begin{aligned} Q^{n+1} &= Q^n + \Delta Q \\ n &= 1,2,3... \end{aligned}$$

The left-hand-side coefficient matrix operator $\frac{\partial \widetilde{R^n}}{\partial Q}$ of Eq. (2.7) is, in many CFD codes, at best only a very rough approximation to the exact Jacobian matrix operator that is associated with true Newton iteration. Thus, Eqs. (2.7) and (2.8) are intended to represent a broad spectrum of implicit and explicit iterative algorithms that are common to CFD software.

Some important computational difficulties are associated with the linear sensitivity equations when they are iteratively solved in the standard form given by Eqs. (2.2) and (2.5). Most importantly, the coefficient matrix, $\frac{\partial R}{\partial Q}$, (and also $(\frac{\partial R}{\partial Q})^T$) is characterized by a lack of diagonal dominance (for spatially higher order accurate, standard CFD methods) and perhaps by poor overall conditioning. The result is poor performance, or even failure (divergence), of conventional iterative methods, when applied to the sensitivity equations in standard form (Refs. [5] and [12]). Furthermore, approximations of computational convenience cannot be introduced into any of the terms of these equations without affecting the accuracy of the sensitivity derivatives that are computed at convergence.

One approach that addresses these difficulties is given by Eleshaky and Baysal (Ref. [16]). In this work, a domain decomposition strategy, together with a preconditioned conjugate-gradient (CG) algorithm, is successfully applied to iteratively solve the sensitivity equations in standard form for an airfoil flow from the TLNS equations. An initial indication of the feasibility of this approach in 3D was recently demonstrated on an axisymmetric nacelle configuration (Ref. [17]). A CG technique was also introduced in Ref. [5] for obtaining sensitivity derivatives from the 3D full potential equation for a wing.

Another strategy has been developed by Korivi et al. (Ref. [18]) and Newman et al. (Ref. [19]), where the sensitivity equations are recast and solved in *incremental iterative form*; for Eq. (2.2), this form is

$$(2.9) \qquad -\frac{\partial \widetilde{R}}{\partial Q}\Delta Q' = R'^m = \frac{\partial R}{\partial Q}Q'^m + \frac{\partial R}{\partial X}X' + \frac{\partial R}{\partial D}$$

$$(2.10) \qquad \begin{aligned} Q'^{m+1} &= Q'^m + \Delta Q' \\ m &= 1,2,3,... \end{aligned}$$

In Eq. (2.9), the left-hand-side coefficient matrix, $\frac{\partial \widetilde{R}}{\partial Q}$, represents any convergent, computationally convenient approximation of the exact Jacobian matrix. In particular, the identical approximate left-hand-side operator and algorithm that are used to solve the nonlinear flow equations can also be used to solve the linear sensitivity equations. Comparisons of Eqs. (2.7) and (2.8) with Eqs. (2.9) and (2.10) reveal that the linear sensitivity equations (Eq. (2.2)) are solved by interchanging the right-hand side of Eq. (2.7) with that of Eq. (2.9) and "freezing" the left-hand-side operator. At convergence, the accuracy of the sensitivity derivatives is not compromised if the the the terms on the right-hand side of Eq. (2.9) are evaluated consistently. The use of the incremental iterative strategy is also applicable in solving Eq. (2.5); in this case, the left-hand-side operator, $\widetilde{\frac{\partial R}{\partial Q}}$, must be transposed.

Implementation of the incremental iterative strategy for solving Eqs. (2.2) and (2.5) has been successfully demonstrated in Ref. [18]. In this work, two airfoil problems using the TLNS equations were considered: low Reynolds number laminar flow and high Reynolds number turbulent flow. The well-known, spatially split, approximate factorization algorithm was used to solve the nonlinear flow and the linear sensitivity equations in incremental iterative form. Derivatives, with respect to geometric shape and nongeometric shape input variables, were accurately computed; they compared well with the method of finite differences, but were significantly less costly to obtain. For these two airfoil problems, attempts to solve the sensitivity equations in standard form with conventional iterative methods failed because of the lack of diagonal dominance, as discussed previously. Furthermore, use of an "in-core" direct solution of these equations was not feasible; the large number of points in the computational grid exceeded the storage allocation on the standard Cray-2 computer queues. Burgreen and Baysal (Ref. [20]) have recently extended their earlier work to combine the efficient preconditioned conjugate-gradient algorithm with the incremental iterative formulation to solve the sensitivity equations for an airfoil flow.

The incremental iterative formulation is very flexible. This formulation should allow the future development of algorithms which are specifically tailored for the highly efficient solution of these equations on advanced machines, including massively parallel architectures. Most significantly, the incremental iterative formulation increases the feasibility of solving the sensitivity equations for advanced 3D CFD codes. Korivi et al. (Ref. [21]) have demonstrated the use of this strategy to efficiently and accurately calculate quasianalytical sensitivity derivatives for a space-marching 3D Euler code with supersonic flow over a blended wing-body configuration.

2.3. Construction of complicated derivatives. Application of the quasianalytical methods that have been described requires the construction and evaluation of many derivatives (e.g. the Jacobian matrices, $\frac{\partial R}{\partial Q}$ and $\frac{\partial R}{\partial X}$), found in the preceding equations. For advanced CFD codes, the task of constructing exactly all of these required derivatives "by hand" and then building the software for evaluating these terms is extremely complex, error prone, and practically speaking, impossible. For example, the inclusion of even the most elementary turbulence model adds a tremendous level of complexity to the Jacobian matrices, $\frac{\partial R}{\partial Q}$ and $\frac{\partial R}{\partial X}$, even in $2D$. Reference [18] shows that failure to consistently differentiate the turbulence modeling terms can result in unexpectedly large errors in the sensitivity derivatives that are calculated. Other common features associated with advanced CFD software that are expected to severely increase the complexity of these terms include the use of multigrid for convergence acceleration, and/or either structured multiblock or unstructured grid capability for application to complex geometric configurations.

A promising possible solution to this problem may be found in the use of a technique known as automatic differentiation (AD), which involves application of a precompiler software tool that automatically differentiates the application program source code from which sensitivity derivatives are to be obtained. The output of the AD precompiler procedure is a new source code which, upon compilation and execution, will compute the numerical value(s) of the derivative(s) of any specified output function(s) with respect to any specified input parameter(s). In addition, this new program will perform the function evaluations of the original code. Computation of derivatives via AD should not be confused with the use of a mathematical symbolic manipulation software package (e.g., MACSYMA, Ref. [22]). This latter approach was employed extensively in Ref. [5], for example.

An AD precompilar software tool called ADIFOR (**A**utomatic **DI**fferentiation of **FOR**tran, Ref. [23]) has recently been tested by Bischof et al. (Ref. [24]) and Green et al. (Refs. [25,26]) in applications to an advanced CFD flow-analysis code called TLNS3D (Ref. [27]). The TLNS3D code solves the 3D TLNS equations using central difference approximations of all spatial derivatives and employs an explicit solution algorithm that includes a highly efficient, state-of-the-art multigrid convergence acceleration technique. In these studies, a high Reynolds number, turbulent, 3D transonic flow over the ONERA M6 wing was selected as the example problem.

The ADIFOR procedure generated a new version of the TLNS3D code that was augmented with the capability to calculate the derivatives of lift, drag, and pitching moment with respect to a variety of different types of input parameters (including parameters related to the geometric shape of the wing). The sensitivity derivatives that were calculated by AD compared very well with the same derivatives calculated by finite differences. The computational cost of generating the results was roughly the same for both methods; however, this cost was very high. Nevertheless, the results reported in Refs. [24,25,26] are encouraging in that they confirm the feasibility of applying AD to advanced 3D CFD codes. In particular, the AD procedure was proven to be capable of generating accurate derivatives, even for a complicated iterative solution algorithm such as multigrid and with the extra level of complexity due to turbulence modeling.

When AD is applied directly to a typical iterative CFD code, the resulting AD-enhanced CFD code must calculate the required sensitivity derivatives through a similar iterative process. From the discussion in Refs. [23,24], the process whereby sensitivity derivatives are iteratively calculated after the application of AD can be represented conceptually by combining Eqs. (2.7) and (2.8) (i.e., the basic CFD solution procedure) and differentiating with respect to D; the result is

$$Q'^{n+1} = Q'^n - P^n R'^n - P'^n R^n$$
$$(2.11) \qquad n = 1, 2, 3...$$

where $P \equiv \left(\widetilde{\frac{\partial R}{\partial Q}} \right)^{-1}$.

References [23,24] note that as an option for improved overall computational efficiency, the original CFD code can be used to first generate a well-converged numerical solution of the nonlinear flow equations *before* the AD-enhanced CFD code is executed to calculate the sensitivity derivatives. When implemented in this way, the derivative calculations via Eq. (2.11) and AD essentially reduce to the previously discussed incremental iterative formulation of Eqs. (2.9) and (2.10) because R^n is very small. Unfortunately, differentiation through the complete iterative CFD solution algorithm and repeated calculation of its derivatives (represented by P'^n in Eq. (2.11)), although unnecessary, is *not* avoided. The computationally wasteful, repeated calculation of P'^n is probably a very significant part of the total work represented by Eq. (2.11). Furthermore, the AD-enhanced CFD code will continue to iterate on the well-converged solution to the nonlinear flow equations. A concept for deactivation of the AD for some parts of the code or calculations was suggested in Ref. [24].

An alternative strategy has been proposed by Newman et al. (Ref. [19]) for applying AD to large-scale CFD codes. If successful, the method would circumvent the computationally wasteful aspects (previously discussed) associated with the conventional direct application of AD to CFD codes. Reference [19] proposes that AD be judiciously applied to differentiate only the right-hand side of Eq. (2.7); the resulting terms would be placed on the right-hand side of the incremental iterative formulation of Eq. (2.9). That is, AD would be used to assist in the accurate construction of the terms required on the right-hand side of Eq. (2.9); the original CFD code and solution algorithm would be used "as is" for the left-hand side of Eq. (2.9). The resulting method should effectively combine an existing, highly efficient, iterative solution algorithm with a fast, reliable procedure for constructing all of the required derivatives.

2.4. Discussion of grid sensitivity terms. Typical CFD calculations are performed on a computational mesh that is "body-oriented." Changes in the geometric shape result in the movement of grid points throughout the entire mesh – not just on the boundaries. Therefore, for design variables that are related to geometric shape, the grid sensitivity matrix, X', of Eqs. (2.2), (2.4), and (2.6) is nonzero, nonsparse, and requires special consideration to evaluate computationally.

One method for calculating these grid sensitivity terms is by finite differences. If forward finite difference approximations are selected, for example, the mesh generation code is used to produce one additional perturbed grid for a slightly perturbed value of each geometric shape design variable of interest. This approach has been successfully used in many of the references cited thus far. This procedure is generally expected to be reliable in producing accurate grid sensitivity terms because the relationships that are associated with the mesh generation process should be very smooth by design.

An analytical method for evaluating the grid sensitivity derivatives has been proposed by Taylor et al. (Ref. [8]), which involves the chain rule and direct differentiation of the relationships that are used by the mesh generation code to distribute the grid points throughout the interior of the computational domain. Computationally, the geometric shape of the domain is defined by the grid points that lie on the boundaries (i.e., on the body surfaces). These boundary grid points, X_B, can be viewed as the principal input variables of the mesh generation code, whereas the complete set of mesh points, X, are the output variables. Furthermore, the boundary grid points of interest are a function of the geometric shape design parameters. Thus, the function of the mesh generation code is expressed as

$$(2.12) \qquad\qquad X = X(X_B(D))$$

Differentiation of Eq. (2.12) with respect to D yields the working relationship for X'

$$(2.13) \qquad\qquad X' = \frac{\partial X}{\partial X_B} X'_B$$

where the matrix, $X'_B \equiv \frac{dX_B}{dD}$, is a very small subset of X'. Typically, the derivative X'_B can be evaluated analytically; it depends on the specific shape and particular parameterization of the body surface in terms of the design variables, D. The matrix $\frac{\partial X}{\partial X_B}$ is unique to the particular mesh generation program employed to distribute grid points throughout the domain and can be evaluated by a one-time direct differentiation of the relationships used. Smith and Sadrehaghighi (Ref. [28]) and Sadrehaghighi et al. (Refs. [29,30]) have pursued in some depth the analytical approach of Eq. (2.13) to efficiently calculate accurate grid sensitivity derivatives for airfoil flows. This method is also used by Burgreen et al. in Ref. [31].

Another approach for calculating grid sensitivity derivatives is proposed by Taylor et al. (Ref. [12]) and is also used by Korivi et al. (Ref. [18]). The method employs an elastic membrane analogy applied to the computational domain. That is, to remesh after a geometric shape change and to calculate grid sensitivity derivatives, the domain is assumed to obey the laws of linear elasticity. The procedure involves the use of a finite-element computer code for structural analysis to compute the required grid sensitivity information. A detailed explanation of this method is given in Ref. [12].

Green et al. (Ref [26]) have applied AD (i.e., ADIFOR) directly to the grid generation program to successfully calculate the grid sensitivity terms. These grid derivatives were subsequently coupled directly to the AD-enhanced TLNS3D flow code. As previously discussed, the final result is the successful calculation of aerodynamic sensitivity derivatives with respect to geometric design parameters for 3D turbulent flow over the ONERA M6 wing.

2.5. Comments on simultaneous analysis and optimization. Gradient information, whether it be sensitivity derivatives or adjoint variables, is required for design-oriented applications. In the approach of Sobieski (Ref. [2]), the global optimization, with its multidisciplinary objective function and constraints, is the outermost iteration loop and drives the various single discipline analyses and their corresponding sensitivity codes. Each discipline furnishes both functional and gradient information at each optimization iteration step. In the case of iterative single-discipline solutions, both the functional and gradient information should be well converged. Other formulations for the multidisciplinary design optimization problem have been proposed by Cramer et al. (Ref. [32]). These formulations involve the nature and extent of optimization and analysis partitioning or mixing.

When a single-discipline analysis code employs an iterative solution algorithm (i.e., CFD), then embedding the optimization iteration within the discipline solution iteration may have significant computational advantages. This approach, of course, is possible and has been done for single-discipline (optimization) design codes in both the discrete (discussed here) and continuous (discussed in the next section) approaches.

Rizk (Ref. [33]) proposed a simultaneous analysis and optimization technique called the single-cycle scheme and recently summarized several CFD applications of this technique (Ref. [34]). The design version of TRANAIR, as discussed by Huffman et al. (Ref. [35]), incorporates sensitivity analysis via both direct (such as Eq. (2.2)) and adjoint (such as Eq. (2.5)) techniques. For the adjoints, however, the output functions F (such as Eq. (2.3)) are the objective function and the constraints; these adjoint solutions are embedded in the flow-analysis solutions (i.e., simultaneous analysis and optimization).

Two other discrete approach techniques for simultaneous analysis and optimization have been reported by Orozco and Ghattas (Ref. [36]) and Hou et al. (Ref. [37]). These independent derivations essentially arrive at the same set of equations to solve for the flow and adjoint variables and are also closely related to the variational or control theory techniques discussed in the next section.

3. The continuous approach. An important advantage is associated with the continuous method for computing gradient information, where the governing equations and boundary conditions, or their corresponding weak variational form, are differentiated with respect to the design variables prior to the discretization and solution of the resulting sensitivity equations. The advantage is increased *flexibility*. For example, a completely different strategy might be selected to discretize and solve the sensitivity equations from the strategy used to solve the flow equations. A simpler governing equation or set of governing equations other than those used in the flow analysis might be selected and solved to estimate flow sensitivity information. With this increased flexibility comes the possibility that some of the major difficulties associated with the discrete approach (which were discussed earlier) might be mitigated, or completely avoided.

Shubin and Frank (Ref. [3]) have concluded that aerodynamic sensitivity derivatives used with gradient-based design optimization procedures should be *consistently discrete*. That is, they should be essentially the exact derivatives of the discrete algebraic system that approximately models the continuous problem. Shubin and Frank assert that the use of inconsistently discrete derivatives can cause significant slowdown or even complete failure of optimization procedures. Furthermore, they note that use of a continuous formulation can yield derivatives that are not consistently discrete; according to Ref. [3], derivative inconsistency traced to this source can create severe problems for the optimization algorithm. Generally, the continuous approach can yield consistently discrete derivatives (or at least very close approximations of the same) when a careful discretization of the sensitivity equations is selected that is compatible with the one used to discretize the flow equations; in addition, of course, the sensitivity equations must also be derived from the original flow equations. Therefore, the requirement that the sensitivity derivatives be consistently discrete will severely restrict the principal advantage of the continuous approach (i.e., flexibility). However, the necessity of always using consistently discrete sensitivity derivatives for gradient-based optimization is refuted in part by Borggaard (Ref. [38]), who examines the identical quasi-one-dimensional nozzle flow problem with a normal shock (as investigated previously in Ref. [3]). In Ref. [38], the judicious use of inconsistently discrete derivatives is shown in some cases to be beneficial, resulting in "successful" optimization, whereas use of the consistently discrete derivatives results in failure.

A continuous formulation by Yates (Ref. [39]) and Yates and Desmarias (Ref. [40]) has successfully demonstrated the accurate and efficient calculation of aerodynamic sensitivity derivatives from the integral-equation representation of the governing equations of aerodynamics and of the resulting aerodynamic sensitivity equations. Results to date have been reported only for linear aerodynamic theory, in which the method reduces to a conventional boundary element procedure. In principle, however, this strategy might be extended to efficiently compute aerodynamic sensitivity derivatives for 3D, nonlinear, viscous flow. An integral-equation representation results in solution procedures that are unique and that have advantages over standard finite-difference and/or finite-volume methods for solving the flow equations; these advantages then carry over in solution of the resulting sensitivity equations (Refs. [39,40]).

Continuous formulations for aerodynamic sensitivity derivatives applied to the 2D Euler equations are reported in Refs. [14] and [41]. Recall that Ref. [14] also gave results for the discrete approach. Borggaard et al. (Ref. [41]) successfully used the continuous approach to calculate sensitivity derivatives by direct differentiation of the 2D Euler equations together with the boundary conditions. With the methods of Ref. [41], the existing CFD software can be modified in a relatively straightforward manner to also efficiently solve the linear sensitivity equations. In particular, both the nonlinear flow and the linear sensitivity equations are solved in incremental iterative form using the identical approximate operator. The extension of this methodology to 3D viscous flow appears to be feasible in principle.

Another important consequence of the methods presented in Ref. [41] is the apparent absence of grid sensitivity terms of the type discussed previously for the discrete approach. Of course, with the complete lack of any grid sensitivity terms, the sensitivity derivatives that are calculated cannot be consistently discrete (for design parameters that are geometric shape related). In Ref. [41], the governing fluid equations and boundary conditions are first differentiated in physical (Cartesian) coordinates; thereafter, the resulting sensitivity equations are transformed to and then numerically solved in generalized computational coordinates (as are the nonlinear flow equations). However, if the governing fluid equations are first transformed to computational coordinates and are thereafter differentiated, then the resulting sensitivity equations are the same as those obtained in Ref. [41], with one important exception. For design variables related to the geometric shape, some additional terms appear that involve derivatives of the transformation from physical to computational coordinates. The discretization and subsequent solution of the sensitivity equations would then involve approximation of these terms as "grid sensitivity" terms.

Jameson (Refs. [42,43]) has demonstrated the use of control theory applied to aerodynamic optimization, wherein the 2D Euler equations are used. In this work, gradient information used in the optimization is obtained through numerical solution of a continuous adjoint variable problem. A similar technique has also been used in Ref. [44]. Ta'asan et al. (Ref. [45]) demonstrated the use of a continuous adjoint variable formulation for the gradient-based aerodynamic design optimization of an airfoil from the small-disturbance equation. More recently, this work has been extended to the 2D full potential equation (Ref. [46]). Of particular interest in Refs. [45,46] is that the optimization strategy features simultaneous minimization of the object function and solution of the discrete nonlinear flow equations, and includes a heavy dependence on the careful use of multigrid for efficiency. In Refs. [43,45,46], an incremental iterative formulation is used to solve the equations (i.e., the nonlinear flow and the linear adjoint equations), after discretization.

4. Summary. An overview has been presented of some recent research activities that have concentrated on the problem of efficient and accurate calculation of gradient information from advanced CFD codes. This review was not intended to be exhaustive (i.e., some important recent advances have likely been overlooked). In particular, some studies, which appear to be more related to optimization procedures, have also been omitted. For the discrete approach, the basic equations of aerodynamic sensitivity analysis were outlined, and three of the most important computational difficulties associated with solving the sensitivity equations were discussed. In addition some potential remedies for these problems were surveyed. Although significant advances using the discrete approach have been made, many obstacles remain that must be overcome before the calculation of quasianalytical sensitivity derivatives becomes routine for turbulent 3D flows.

In principle, the flexible nature of the continuous approach might possibly be exploited to overcome some of the computational difficulties that have been discussed for the discrete approach. At the same time, however, the consequences of this flexibility are typically sensitivity derivatives that are different in the sense that they are not consistently discrete. This result can have an impact on the performance of optimization algorithms; whether the impact is large or small, or even good or bad, is not yet clear. The continuous formulation can be applied in a careful manner to produce the consistently discrete derivatives (or very nearly these derivatives). However, then the advantage of flexibility for the continuous approach is sacrificed and the difference between the discrete and the continuous approaches becomes more an issue of philosophy than of substance.

5. Acknowledgements. The work of A.C.T., III and G.J.-W.H. was supported by NASA grant NAG-1-1265.

REFERENCES

[1]   J.S-. SOBIESKI, *The Case for Aerodynamic Sensitivity Analysis*, NASA CP-2457, pp. 77–96, January 1987.
[2]   J.S-. SOBIESKI, *Multidisciplinary Optimization for Engineering Systems: Achievements and Potential*, NASA TM 101566, March 1989.
[3]   G.R. SHUBIN AND P.D. FRANK, *A Comparison of Two Closely-Related Approaches to Aerodynamic Design Optimization*, in *Third International Conference on Inverse Design Concepts and Optimization in Engineering Sciences (ICIDES-III)*, pp. 67-78, Washington, D.C., October 1991. (Also Technical Report AMS-TR-163, Boeing Computer Services, April 1991.)
[4]   H.M. ELBANNA AND L.A. CARLSON, *Determination of Aerodynamic Sensitivity Coefficients Based on the Transonic Small Perturbation Formulation*, Journal of Aircraft, Vol. 27, No. 6, June 1990, pp. 507–515. (Also AIAA Paper 89-0532, January 1989).

[5]   H.M. ELBANNA AND L.A. CARLSON, *Determination of Aerodynamic Sensitivity Coefficients Based on the Three-Dimensional Full Potential Equation*, AIAA Paper 92-2670, June 1992.

[6]   M. DRELA, *Viscous and Inviscid Inverse Schemes Using Newton's Method*, In *Special Course on Inverse Methods for Airfoil Design for Aeronautical and Turbomachinery Applications*, AGARD Report No. 780, May 1990, pp. 9.1–9.16.

[7]   A.C. TAYLOR III, V.M. KORIVI, AND G.W. HOU, *Taylor Series Approximation of Geometric Shape Variation for the Euler Equations*, AIAA Journal, Vol. 30, No. 8, August 1992, pp. 2163–2165. (Also AIAA paper 91-0173, January 1991).

[8]   A.C. TAYLOR III, G.W. HOU AND V.M. KORIVI, *Methodology for Calculating Aerodynamic Sensitivity Derivatives*, AIAA Journal, Vol. 30, No. 10, October 1992, pp. 2411-2419. (Also in *AIAA/ASME/ASCE/AHS/ASC 32nd Structures, Structural Dynamics, and Materials Conference*, April 1991, pp. 477-489, AIAA Paper 91-1101-CP).

[9]   G.J-.W. HOU, A.C. TAYLOR III, AND V.M. KORIVI, *Discrete Shape Sensitivity Equations for Aerodynamic Problems*, AIAA Paper 91-2259, June 1991. (Also to appear, International Journal of Numerical Methods in Engineering).

[10]   O. BAYSAL AND M.E. ELESHAKY, *Aerodynamic Sensitivity Analysis for the Compressible Euler Equations*, ASME Journal of Fluids Engineering, Vol. 113, No. 4, December 1991. (Also in *Recent Advances and Applications in CFD*, ed. by O. Baysal, ASME-FED, Vol. 103, Winter Annual Meeting, November 1990, pp. 191–202).

[11]   O. BAYSAL AND M.E. ELESHAKY, *Aerodynamic Design Optimization Using Sensitivity Analysis and Computational Fluid Dynamics*, AIAA Journal, Vol. 30, No. 3, March 1992, pp. 718–725. (Also AIAA Paper 91-0471, January 1991).

[12]   A.C. TAYLOR III, G.W. HOU, AND V.M. KORIVI, *Sensitivity Analysis, Approximate Analysis, and Design Optimization for Internal and External Viscous Flows*, AIAA Paper 91-3083, September 1991. (Also to appear, International Journal of Numerical Methods in Fluids).

[13]   M.E. ELESHAKY AND O. BAYSAL, *Airfoil Shape Optimization Using Sensitivity Analysis on Viscous Flow Equations*, ASME Journal of Fluids Engineering, Vol 115, No. 1, March 1993, pp. 75–84. (Also in *Multidisciplinary Applications of Computational Fluid Dynamics*, ed. by O. Baysal, ASME-FED, Vol. 129, Winter Annual Meeting, December 1991, pp. 27–37).

[14]   F. BEUX AND A. DERVIEUX, *Exact-Gradient Shape Optimization of a 2-D Euler Flow*, Finite Elements in Analysis and Design, Vol. 12, 1992, pp. 281–302.

[15]   G.R. SHUBIN, *Obtaining 'Cheap' Optimization Gradients from Computational Aerodynamics Codes*, Technical Report AMS-TR-164, Boeing Computer Services, June 1991.

[16]   M.E. ELESHAKY AND O. BAYSAL, *Aerodynamic Shape Optimization Via Sensitivity Analysis on Decomposed Computational Domains*, in *AIAA CP-9213, Fourth AIAA/USAF/NASA/OAI Symposium on Multidisciplinary Analysis and Optimization*, September 1992, pp. 98–109 (AIAA paper 92–4698-CP).

[17]   M.E. ELESHAKY AND O. BAYSAL, *Preconditioned Domain Decomposition Scheme for Three-Dimensional Aerodynamics Sensitivity Analysis*, in *AIAA CP-933, AIAA 11th Computational Fluid Dynamic Conference*, July 1993, pp. 1055–1056.

[18]   V.M. KORIVI, A.C. TAYLOR III, P.A. NEWMAN, G.J-.W. HOU, AND H.E. JONES, *An Approximately Factored Incremental Strategy for Calculating Consistent Discrete Aerodynamic Sensitivity Derivatives*, in *AIAA CP-9213, Fourth AIAA/USAF/NASA/OAI Symposium on Multidisciplinary Analysis and Optimization*, September 1992, pp. 465–478, (AIAA paper 92-4746-CP, and also to appear, Journal of Computational Physics).

[19] P.A. NEWMAN, G.J-.W. HOU, H.E. JONES, A.C. TAYLOR III, AND V.M. KORIVI, *Observations on Computational Methodologies for use in Large-Scale, Gradient-Based, Multidisciplinary Design*, in *AIAA CP-9213, Fourth AIAA/USAF/NASA/OAI Symposium on Multidisciplinary Analysis and Optimization*, September 1992, pp. 531–542, (AIAA paper 92-4753-CP).

[20] G.W. BURGREEN AND O. BAYSAL, *Aerodynamic Shape Optimization Using Preconditioned Conjugate Gradient Methods*, in *AIAA CP-933, AIAA 11th Computational Fluid Dynamics Conference*, July 1993 pp. 278–288 (AIAA paper 93-3322-CP).

[21] V.M. KORIVI, A.C. TAYLOR III, G.W. HOU, P.A. NEWMAN, AND H.E. JONES, *Sensitivity Derivatives for Three-Dimensional Supersonic Euler Code Using Incremental Iterative Strategy*, in *AIAA CP-933, AIAA 11th Computational Fluid Dynamics Conference*, July 1993, pp. 1053–1054.

[22] *MACSYMA Reference Manual*, Version 13, Computer Aided Mathematics Group, Symbolics, Inc., 1988.

[23] C.BISCHOF AND A. GRIEWANK, *ADIFOR: A Fortran System for Portable Automatic Differentiation*, in *AIAA CP-9213, Fourth AIAA/USAF/NASA/OAI Symposium on Multidisciplinary Analysis and Optimization*, September 1992, pp. 433-441 (AIAA paper 92-4744-CP).

[24] C. BISCHOF, G. CORLISS, L. GREEN, A. GRIEWANK, K. HAIGLER, AND P. NEWMAN, *Automatic Differentiation of Advanced CFD Codes for Multidisciplinary Design*, presented at the Symposium on High-Performance Computing for Fligh Vehicles, December 1992, Arlington, VA. (To appear in Computing Systems in Engineering, Vol. 3, No. 6, 1993).

[25] L.L. GREEN, P.A. NEWMAN, AND K.J. HAIGLER, *Sensitivity Derivatives for Advanced CFD Algorithm and Viscous Modelling Parameters Via Automatic Differentiation*, in *AIAA CP-933 AIAA 11th Computational Fluid Dynamics Conference*, July 1993, pp. 260–277, (AIAA Paper 93-3321-CP).

[26] L. GREEN, C. BISCHOF, A. CARLE, A. GRIEWANK, K. HAIGLER, P. NEWMAN, *Automatic Differentiation of Advanced CFD Codes With Respect To Wing Geometry Parameters for MDO*, abstract in *Second U.S. National Congress on Computational Mechanics*, August 16-18, 1993, Washington, D.C.

[27] V.N. VATSA AND B.W. WEDAN, *Development of a Multigrid Code for 3-D Navier-Stokes Equations and Its Application to a Grid Refinement Study*, Journal of Computers and Fluids, Vol. 18, No. 4, pp. 391–403, 1990.

[28] R.E. SMITH AND I. SADREHAGHIGHI, *Grid Sensitivity In Airplane Design*, in *Proceedings of the Fourth International Symposium on Computational Fluid Dynamics*, Vol. 1, September 1991, Davis California, pp. 1071–1076.

[29] I. SADREHAGHIGHI, R.E. SMITH, AND S.N. TIWARI, *Grid and Design Variable Sensitivity Analysis for NACA Four-Digit Wing-Sections*, AIAA Paper 93-0195, January 1993.

[30] I. SADREHAGHIGHI, R.E. SMITH, AND S.N. TIWARI, *An Analytical Approach to Grid Sensitivity Analysis*, AIAA Paper 92-0660, January 1992.

[31] G.W. BURGREEN, O. BAYSAL, AND M.E. ELESHAKY, *Improving the Efficiency of Aerodynamic Shape Optimization Procedures*, in *AIAA CP-9213, Fourth AIAA/USAF/NASA/OAI Symposium on Multidisciplinary Analysis and Optimization*, September 1992, pp. 87–97, (AIAA paper 92-4697-CP and to appear, AIAA Journal).

[32] E. CRAMER, P. FRANK, G. SHUBIN, J. DENNIS, AND R. LEWIS, *On Alternative Problem Formulations for Multidisciplinary Design Optimization*, in *AIAA CP-9213, Fourth AIAA/USAF/NASA/OAI Symposium on Multidisciplinary Analysis and Optimization*, September 1992, pp. 518–530, (AIAA 92-4752-CP).

[33] M. RIZK, *The Single-Cycle Scheme: A New Approach to Numerical Optimizations*, AIAA Journal, Vol. 21, 1983, pp. 1640–1647.

[34] M. RIZK, *Optimization by Updating Design Parameters as CFD Iterative Flow Solutions Evolve*, in *Multidisciplinary Applications of Computational Fluid Dynamics*, ed. by O. Baysal, ASME-FED, Vol. 129, Winter Annual Meeting, December 1991, pp. 51–62.

[35] W.P. HUFFMAN, R.G. MELVIN, D.P. YOUNG, F.T. JOHNSON, J.E. BUSSOLETTI, M.B. BIETERMAN, AND C.L. HILMES, *Practical Design and Optimization in Computational Fluid Dynamics*, AIAA Paper 93-3111, July 1993.

[36] G. OROZCO AND O. GHATTAS, *Optimal Design of Systems Governed by Nonlinear Partial Differential Equations*, in *AIAA CP-9213, Fourth AIAA/USAF/NASA/OAI Symposium on Multidisciplinary Analysis and Optimization*, September 1992, pp. 1126–1140, (AIAA 92-4836-CP).

[37] G.W. HOU, A.C. TAYLOR III, S.V. MANI, AND P.A. NEWMAN, *Simultaneous Aerodynamic Analysis and Design Optimization*, abstract in *Second U.S. National Congress on Computational Mechanics*, August 16–18, 1993, Washington, D.C.

[38] J.T. BORGGAARD, *On the Presence of Shocks in Domain Optimization of Euler Flows*, this volume.

[39] E.C. YATES, JR., *Integral-Equation Methods in Steady and Unsteady Subsonic, Transonic, and Supersonic Aerodynamics for Interdisciplinary Design*, NASA TM 102577, May 1990.

[40] E.C. YATES, JR. AND R.N. DESMARIAS, *Boundary-Integral Method for Calculating Aerodynamic Sensitivities with Illustration for Lifting-Surface Theory*, Proceedings of the Third International Congress of Fluid Mechanics, Cairo, Egypt, January 1990.

[41] J. BORGGAARD, J.A. BURNS, E. CLIFF, AND M. GUNZBURGER, *Sensitivity Calculations for a 2D Inviscid, Supersonic Forebody Problem*, NASA CR-19144 and ICASE Report No. 93-13, March 1993.

[42] A. JAMESON, *Aerodynamic Design Via Control Theory*, Journal of Scientific Computing, Vol. 3, pp. 233-260, 1988. (Also NASA CR-181749 and ICASE Report No. 88-64, November 1988.)

[43] A. JAMESON, *Automatic Design of Transonic Airfoils to Reduce Induced Pressure Drag*, Princeton University MAE Report 1881, 1990. (Also 31st Israel Annual Conference in Aviation and Aeronautics, February 1990).

[44] J.R. LEWIS, G.R. PETERS, AND R.K. AGARWAL, *Airfoil Design Via Control Theory Using Euler Equations*, in *Multidisciplinary Applications of Computational Fluid Dynamics*, ed. by O. Baysal, ASME-FED, Vol. 129, Winter Annual Meeting, December 1991, pp. 39–49.

[45] S. TA'ASAN, G. KURUVILA AND M.D. SALAS, *Aerodynamic Design and Optimization in One Shot*, AIAA Paper 92–0025, January 1992.

[46] G. KURUVILA, private communication, written report in preparation.

REMARKS ON THE CONTROL OF TURBULENT FLOWS

ROGER TEMAM*

Introduction

Our aim in the article is to address some theoretical and computational questions related to the control of viscous incompressible flows governed by the Navier-Stokes equations or related equations.

This article comprises three parts where we study three types of problems which correspond to different preoccupations and utilize different tools for their solution.

In Section 1 we study some control problems where the objective is to minimize, in some sense, turbulence. Distributed control, boundary control problems for thermohydraulics and for a channel flow in space dimension two are considered. After modeling the problem, we show the existence of an optimal control, and derive the necessary conditions of optimality (NCO) for the problem, using the adjoint state.

In Section 2 we consider in space dimension three one of the problems from Section 1, namely the distributed control problem. The analysis of Section 1 does not apply here since the initial value problem for the Navier-Stokes equations is not known to be well posed in dimension three. The existence of an optimal control is established and, if the optimal state is sufficiently regular, we are able with appropriate methods, to derive the necessary conditions of optimality.

In Section 3 we study the optimal control of the Stochastic Burgers equations. It was shown that the Burgers equations forced by a white noise produce turbulence phenomena similar to those observed for fluid. The objectives and the methods are now different. Instead of looking for an optimal control, we only try to decrease the cost function by using a one step control procedure. Theoretical questions are not addressed here but we report on numerical results which show a very significant decrease of the cost function.

The results in Sections 1, 2, 3 are based on references [1], [2] and [7] where further details can be found.

1. Modeling of some control problems. We describe three model problems in control of fluids (control of turbulence).

1.1. A model distributed control problem: control by volume forces. We consider the incompressible Navier–Stokes equations in a smooth bounded two–dimensional domain Ω, on an interval of time $[0, T]$.

---

* Laboratoire d'Analyse Numérique, Bat. 425, Université Paris Sud, Orsay and Institute for Scientific Computing and Applied Mathematics, Indiana University, Bloomington, Indiana.

358   ROGER TEMAM

We set $Q_T = \Omega \times [0, T]$ and recall the equations

$$
\text{(1.1)} \quad
\begin{cases}
\dfrac{\partial u}{\partial t} - \nu \Delta u + (u \cdot \nabla)u + \nabla p = f & \text{in } Q_T, \\[2mm]
\nabla \cdot u = 0 & \text{in } Q_T, \\[2mm]
u = 0 & \text{on } \partial\Omega \times [0, T], \\[2mm]
u\big|_{t=0} = u_0.
\end{cases}
$$

Here $u = (u_1, u_2)$ is the velocity vector, p the pressure, $\nu > 0$ the kinematic viscosity; f, which will be the control, represents volume forces.

The mathematical setting of the equation is well known and we do not recall it with full details (see e.g. [13], [16]). Let

$$
V = \{v \in H_0^1(\Omega)^2, \operatorname{div} v = 0\},
$$

let H be the closure of V in $L^2(\Omega)^2$, we consider the linear unbounded operator A in H associated with the Stokes problem, and the nonlinear operator B defined by

$$
B(u) = B(u, u), \qquad \forall\, u \in V,
$$

$$
(B(\varphi, \psi), \theta) = \sum_{i,j=1}^{2} \int_\Omega \varphi_i \frac{\partial \psi_j}{\partial x_i} \theta_j \, dx.
$$

Then (1.1) is equivalent to a differential equation in H for $u = u(t)$, where $u(t) \in V$:

$$
\text{(1.2)} \quad
\begin{cases}
\dfrac{du}{dt} + \nu A u + B(u) = f \, in \, H, \quad t \in (0, T), \\[2mm]
u(0) = u_0.
\end{cases}
$$

We will use subsequently the Fréchet derivative of B and its adjoint:

$$
B'(\varphi) \cdot \psi = B(\varphi, \psi) + B(\psi, \varphi)
$$
$$
\langle B'(\varphi)^* \cdot \psi, \theta \rangle = \langle B(\varphi, \theta), \psi \rangle + \langle B(\theta, \varphi), \psi \rangle.
$$

The control problem

In the language of control theory (see [14]), f is the *control*, $u = u_f$ is the *state*. We want to reduce the turbulence as measured by

$$
\int_0^T \int_\Omega |\operatorname{curl} u_f(x, t)|^2 \, dx dt,
$$

and hence we introduce the cost function:

$$
J(f) = \frac{1}{2} \int_0^T \int_\Omega |f(x, t)|^2 dx dt + \frac{1}{2} \int_0^T \int_\Omega |\operatorname{curl} u_f(x, t)|^2 dx dt.
$$

The problem is then

$(\mathcal{P}_1)$ *To minimize $J(f)$, for $f \in L^2(0, T; H)$.*

Concerning the existence of an *optimal control* we have

THEOREM 1.1. *For u_0 given in H, there exists at least an element $\bar{f}$ in $L^2(0, T; H)$, and $\bar{u} \in \mathcal{C}([0, T]; H) \cap L^2(0, T; V)$, such that $J(f)$ attains its minimum at $\bar{f}$ and $\bar{u} = u_{\bar{f}}$.*

Remark 1.2. Of course, since $J(f)$ is not a convex function, we cannot assert that $\bar{f}$ is unique.

The necessary conditions of optimality (NCO)

Basically (see e.g. [10])., they consist in writing that

$$(1.3) \qquad\qquad\qquad J'(\bar{f}) = 0,$$

or

$$\langle J'(\bar{f}), f \rangle = 0, \qquad \forall f,$$

where J' is the Fréchet derivative of J.

A convenient expression of J' can be given by using the adjoint state which is defined by the adjoint of the linearized equations.

Equation (1.2) linearized around an orbit $u = u_f$ reads

$$(1.4) \qquad \begin{cases} \dfrac{\partial v}{\partial t} + \nu Av + B'(u_f) \cdot v = 0, & \text{in } H, \quad t \in (0, T), \\ v(t) \in V \quad \text{a.e.}, \\ v(0) = 0, \end{cases}$$

where simply $B'(u_f) \cdot v = B(u_f, v) + B(v, u_f)$.

The adjoint equation of (1.4) is

$$(1.5) \qquad \begin{cases} -\dfrac{\partial w}{\partial t} + \nu Aw + B'(u_f)^* w = h, & \text{in } H, \quad t \in (0, T), \\ w(t) \in V \quad \text{a.e.}, \\ w(T) = 0. \end{cases}$$

Here $B'(u_f)^*$ is the adjoint of $B'(u_f)$ in H and we have introduced h in the right–hand side of the first equation (1.5). Note that w depends on f (through u_f) and on h; for completeness we can write $w = w_f(h)$ or $w(h)$. Now the *adjoint state* for the present control problem is $\tilde{w} = w_f(h)$ for $h = \nabla \times \nabla \times \bar{u}$. Introducing the pressure like functions, we can interpret

(1.5) as the following set of partial differential equations and boundary conditions:

$$
(1.6) \quad
\begin{cases}
-\dfrac{\partial \tilde{w}}{\partial t} - \nu \Delta \tilde{w} + (\nabla \bar{u})^t \cdot \tilde{w} - (\bar{u} \cdot \nabla)\tilde{w} + \nabla \tilde{q} = \nabla \times \nabla \times \bar{u}, \\[2mm]
\qquad \text{in } Q_T = \Omega \times (0, T), \\[2mm]
\operatorname{div} \tilde{w} = 0 \quad \text{in } Q_T, \\[2mm]
\tilde{w} = 0 \quad \text{on } \partial\Omega \times (0, T), \\[2mm]
\tilde{w}(x, T) = 0, \quad x \in \Omega.
\end{cases}
$$

Using $\tilde{w} = \tilde{w}(\nabla \times \nabla \times u_f)$ one can prove that

$$
J'(f) = f + w_f(\nabla \times \nabla \times u_f),
$$

$$
\langle J'(f), f^* \rangle = \int_0^T \int_\Omega [f + w_f(\nabla \times \nabla \times u_f)]f^* \, dx\, dt, \quad \forall f^*.
$$

The following theorem follows then promptly from (1.3) (see [1,2]).

THEOREM 1.3. *Let $\{\bar{f}, \bar{u}\}$ be an optional pair for problem $(\mathcal{P}_1)$. Then the following equality holds*

$$
(1.7) \qquad \bar{f} + \tilde{w}_{\bar{f}}(\nabla \times \nabla \times \bar{u}) = 0
$$

where $\tilde{w}$ is the adjoint state, solution of the adjoint linearized problem.
Furthermore, we have the following regularity property for $\bar{f}$:

$$
\bar{f} \in L^\infty(0, T; V) \cap L^2(0, T; H^2(\Omega)^2).
$$

The NCO for problem $\mathcal{P}_1$ consist of

– equation (1.2) with $f = \bar{f}, u = \bar{u}$,

– equation (1.5) (or (1.6) with $w = \tilde{w}, h = \nabla \times \nabla \times \bar{u}, f = \bar{f}$,

– equation (1.7).

Of course this set of equations is not easy to solve; however one can compute (or hope to compute) $\bar{u}, \bar{f}, \tilde{w}$ by using optimization alogrithms such as the gradient or conjugate gradient algorithm.

Numerical Algorithms

The classical gradient algorithm for $(\mathcal{P}_1)$ consists in defining recursively a sequence of $f_n \in L^2(Q_T)$. Starting from an arbitrary $f_0 \in L^2(Q_T)$, we write ($w_n = w_{fn}(\nabla \times \nabla \times u_{fn})$):

$$
f_{n+1} = f_n - \rho_n J'(f_n),
$$

i.e.,

$$f_{n+1} = f_n - \rho_n(f_n + w_n).$$

The numbers $\rho_n > 0$ must be chosen properly.

The conjugate gradient algorithm for $(\mathcal{P}_1)$ consists in defining recursively two sequences f_n, k_n. Starting from an arbitrary $f_0 \in L^2(\mathcal{Q}_T)$ we write $(w_n = w_{fn}(\nabla \times \nabla \times u_{fn}))$:

$$k_0 = f_0 + w_0,$$

$$k_n = f_n + w_n$$

$$+ k_{n-1} \frac{\int_{\mathcal{Q}_T} (f_n - f_{n-1} + w_n - w_{n-1}) \cdot (f_n + w_n) dx dt}{\int_{\mathcal{Q}_T} |f_{n-1} + w_{n-1}|^2 dx dt}$$

$$f_{n+1} = f_n - \rho k_n,$$

where $\rho > 0$.

Both algorithms converge to $\bar{f}$ if f_0 is chosen close enough to $\bar{f}$ and the ρ_n are suitable. Unfortunately, for realistic flows, these algorithms necessitate a computing power beyond that presently available. In the rest of this section we describe, in a similar manner, the modeling of some related flow control problems. In Section 3 we will describe suboptimal procedure which are more feasible.

1.2. A boundary control problem in thermohydraulics. The "system"[1] that we consider here is a two–dimensional layer of fluid heated (or cooled) from above and below. The flow is periodic in direction x_1 (period L_1), at rest at the bottom of the layer $x_2 = 0$, driven at velocity $u = \varphi$ on top of the layer, $x_2 = L_2$. The boundary velocity φ is the *control*; the *state* of the system is given by the field of velocities $u = u_\varphi$ and the field of temperatures $\tau = \tau_\varphi$, solutions of the classical Boussinesq equations:

$$\begin{cases} \dfrac{\partial u}{\partial t} - \nu \Delta u + (u \cdot \nabla)u + \nabla p = -q(\tau - \tau_0)e_2 & \text{in } Q_T = \Omega \times (0, T), \\[2mm] \dfrac{\partial \tau}{\partial t} + (u \cdot \nabla)\tau - \kappa \Delta \tau = 0 & \text{in } Q_T, \\[2mm] \nabla \cdot u = 0 & \text{in } Q_T. \end{cases}$$

(1.8)

The density is one, ν, k, g are positive constants, $e_2 = (0, 1)$ is the unit vertical vector; Ω is the domain $(0, L_1) \times (0, L_2)$. These equations are supplemented by the boundary conditions described before, namely the periodicity in the x_1 direction and

(1.9)
$$\begin{cases} u = \varphi, \tau = \tau_2 \text{ at } x_2 = L_2, \\[2mm] u = 0, \tau = \tau_1 \text{ at } x_2 = 0, \end{cases}$$

---

[1] In the sense of control theory

and by initial conditions

$$(1.10) \qquad \{u(x,0) = u_0(x)\}, \quad \tau(x,0) = \tau_0(x), \quad x \in \Omega.$$

For φ, τ_2, τ_1 given in $L^2(0,T; H^{1/2}(0,L_1))$, one can show that equations (1.8)–(1.10) posses a unique solution $\{u, \tau\} = \{u_\varphi, \tau_\varphi\}$.

We encounter a regularity difficulty for the choice of the cost function. A simple choice would be

$$J_1(\varphi) = \frac{\ell}{2} \int_0^T \int_0^{L_1} |\varphi(x_1,t)|^2 dx_1 dt + \frac{m}{2} \int_0^T \int_\Omega |\nabla \times u_\varphi(x,t)|^2 dx dt,$$

$m, \ell > 0$. However it is not easy to obtain the existence and uniqueness of solution of (1.8)–(1.10) if we only assume that $\varphi \in L^2(0,T; H^{1/2}(0,L_1))$. This result may not be true if we require $u_\varphi \in L^2(0,T; H^1(\Omega)^2)$. On the other hand for $\varphi \in L^2(0,T; H^{1/2}(0,L_1)^2)$, the function J_1 above may not attain its infimum. Hence we choose the less convenient cost function

$$J_2(\varphi) = \frac{\ell}{2} \int_0^T |\varphi(\cdot,t)|^2_{H^{1/2}(0,L_1)^2} dt + \frac{m}{2} \int_0^T \int_\Omega |\nabla \times u_\varphi(x,t)|^2 dx dt.$$

It can be shown as for Theorem 1.1, that this function J_2 attains its minimum on $L^2(0,T; H^{1/2}(0,L_1)^2)$, at least at one point $\bar\varphi$ with corresponding state $\{\bar u, \bar\tau\} = \{u_{\bar\varphi}, \tau_{\bar\varphi}\}$.

The necessary condition of optimality (NCO) is obtained by writing

$$J_2'(\bar\varphi) = 0$$

It is not easy to make it explicit because of the space $H^{1/2}(0,L_1)^2$. However, if we do not emphasize the existence of the optimal control but only the NCO, then we can make the NCO explicit in the case of J_1.

Indeed the Fréchet derivative J_1' can be computed using the adjoint state $\tilde w, \tilde\sigma$ which is solution of the following problem with $\varphi = \bar\varphi$ and $h = \nabla \times \nabla \times u_{\bar\varphi}$:

$$(1.11) \quad \begin{cases} -\dfrac{\partial \hat w}{\partial t} - \nu \Delta \tilde w - (u_\varphi \cdot \nabla)\tilde w + (\nabla u_\varphi)^t w^2 \\[2mm] \qquad \tilde\sigma \nabla \tau_\varphi + \nabla q = h \quad \text{in} \quad Q_T, \\[2mm] -\dfrac{\partial \tilde\sigma}{\partial t} - (u_\varphi \cdot \nabla)\tilde\sigma - \kappa \Delta \tilde\sigma = 0 \quad \text{in} \quad Q_T, \\[2mm] \nabla \cdot \tilde w = 0 \quad \text{in} \quad Q_T. \end{cases}$$

The boundary conditions are (1.9) homogeneized, and the "initial" conditions at $t = T$ read

$$\tilde w(x,T) = 0, \quad \tilde\sigma(x,T) = 0, \quad x \in \Omega.$$

Then $J_1'(\bar{\varphi})$ is equal to $\bar{\varphi} - \nu \dfrac{\partial \tilde{w}}{\partial x_2}$ and the NCO is

$$\bar{\varphi} - \nu \frac{\partial \tilde{w}}{\partial x_2} = 0 \text{ on } (0, T) \times (0, L_1).$$

The (usual) gradient algorithm consists in constructing a sequence of functions φ_n such that

$$\varphi_{n+1} = \varphi_n - \rho_n J_1'(\varphi_n)$$
$$= \varphi_n - \rho \left(\varphi_n - \nu \frac{\partial \tilde{w}_n}{\partial x_2} \right),$$

where $\tilde{w}_n$ is the solution of (1.11) with φ replaced by φ_n, h by $\nabla \times \nabla \times u_{\varphi_n}$.

Remark 1.4. Of course we could have chosen the top and bottom heatings, τ_1, τ_2 as the control functions for this problem.

1.3. Boundary control of channel flows. . This control problem is a very important one. Large scale computations are being performed on this problem at this time with the methods of Section 3 (cf. [3]). In this section we present the modeling of the control problem and describe briefly the theoretical results similar to those of Sections 1.1 and 1.2.

The "system" is a channel, $\Omega = (0, L_1) \times (0, L_2)$. The flow is incompressible and driven by a given flux. Hence we write the incompressible Navier–Stokes equations in $Q_T = \Omega \times (0, T)$:

$$(1.12) \qquad \begin{cases} \dfrac{\partial u}{\partial t} - \nu \Delta u + (u \cdot \nabla)u + \nabla p = 0, & \text{in } Q_T, \\ \nabla \cdot u = 0, & \text{in } Q_T. \end{cases}$$

The flow F is prescribed , i.e., if $u = \{u_1, u_2\}$:

$$(1.13) \qquad \int_0^{L_2} u_1(0, x_2; t)dx_2 = F \text{ (given)}.$$

The pressure p is of the form

$$p = P(t)x_1 + p',$$

where $P = P(t)$ is an unknown pressure gradient determined by F.

The boundary conditions are the periodicity of u and p' in direction x_1 (period L_1), and on the walls $x_2 = 0, L_2$, we have

$$(1.14) \qquad u(x_1, 0; t) = \varphi e_2, \quad u(x_1, L_2; t) = \varphi e_2.$$

Here e_2 is the vector $(0,1)$ in the plane and φ is the *control*. Of course the *state* is the pair $\{u, p\}$ defined by (1.10)–(1.14) which we supplement with an initial condition

$$(1.15) \qquad\qquad u(x, 0) = u_0(x), \quad x \in \Omega.$$

More precisely the functional setting of (1.12)–(1.15) is as follows. Let V be the space of functions v in $H^1(\Omega)^2$ which are L_1-periodic in direction x_1 which vanish at $x_2 = 0$ and L_2, and such that $\operatorname{div} v = 0$. Let H be the closure of V in $L^2(\Omega)^2$.

For $F \in \mathbb{R}$ and $\varphi \in H^{1/2}(\Gamma_1)^2$ given, we denote by $V_{F,\varphi}$ the set of functions v in V such that

$$(1.16) \qquad\qquad \int_0^{L_2} v_1(0, x_2)dx_2 = F, \quad v\big|_{x_2=0,L_2} = \varphi.$$

In particular for $F = 0, \varphi = 0$, we write $V_0 = V_{0,0}$. Of course $V_{F,\varphi}$ is the affine space

$$V_{F,\varphi} = V_0 + \Phi,$$

where $\Phi \in V$ is any function of V satisfying (1.16). Let also $H_{F,\varphi}$ and H_0 be defined in a similar way: $H_0 = H_{0,0}$, and $H_{F,\varphi}$ is the space of functions v in H such that[2]

$$(1.17) \qquad\qquad \int_0^{L_2} v_1(0, x_2)dx_2 = F, \quad v_2\big|_{x_2=0,L_2} = \varphi.$$

Again

$$H_{F,\varphi} = H_0 + \Phi.$$

Now for Φ given in $L^2(0, T; V)$, for u_0 given in $H_{F,\varphi}$, there exists a unique pair $\{u = u_\varphi, p = p_\varphi\}$, solution of (1.12)–(1.15); in particular $u \in L^2(0, T; V_{F,\varphi})$. As usual p is only defined up to an additive constant; as indicated before the pair $\{u_\varphi, p_\varphi\}$ is the *state* associated with the *control* φ.

In the control problem we want to choose φ so as to reduce the magnitude of the drag on the wall

$$\int_{\Gamma_1} \frac{\partial u_1}{\partial x_2} dx_1.$$

A possible choice of the cost function J is ($\ell, m > 0$):

$$J_1(\varphi) = \frac{\ell}{2}\|\varphi\|_X^2 + \frac{m}{2} \int_0^T \left(\int_{\Gamma_1} \frac{\partial u_1}{\partial x_2} dx_1 \right)^2 dt,$$

---

[2] Note the difference with (1.16).

including a time average of the square of the drag. An alternate, seemingly more interesting choice is

$$J_2(\varphi) = \frac{\ell}{2}\|\varphi\|_X^2 + \frac{m}{2}\int_0^T \int_{\Gamma_1} \left(\frac{\partial u_1}{\partial x_2}\right)^2 dx_1 dt,$$

the integral term in J_2 corresponding to a time average of surface stresses. As in Section 1.2, we have to choose X properly. The simplest choice, $L^2(0,T;L^2(\Gamma_1))$ is not suitable; hence we take $X = L^2(0,T;H^{1/2}(\Gamma_1)^2)$; more precisely X is the space of traces on the wall Γ_1 of the functions v of $L^2(0,T;V)$, satisfying the first condition in (1.16) (the flux condition), and such that $v' = \partial v/\partial t$ belongs to $L^2(0,T;V')$ (see e.g. [T3])[3]

The control problem is then

($\mathcal{P}_3$) *To minimize J_1 (or J_2) on X.*

We obtain (see [1]), the same theoretical results as in Section 1.1 and 1.2, namely

– For $F \in \mathbb{R}$ given, for u_0 given in H satisfying the first condition in (1.16), there exists an optimal triplet $\{\bar{u},\bar{p},\bar{\varphi}\}$, where $\bar{\varphi}$ is an optimal control (solution of ($\mathcal{P}_3$) and $\bar{u} = u_{\bar{\varphi}}, \bar{p} = p_{\bar{\varphi}}$ is the corresponding distribution of velocities and pressures.)

– We can write the necessary conditions of optimality for ($\mathcal{P}_3$) but they are rather involved. As in Section 1.2, they are easier to write if X is a subspace of $L^2(0,T;L^2(\Gamma_1)^2)$; see [1].

– We can think at implementing a gradient type algorithm for the numerical solution of problem ($\mathcal{P}_3$), but we meet two difficulties:

  • If $X \subset L^2(0,T;H^{1/2}(\Gamma_1)^2)$, then the gradient algorithm is not easy to make explicit (even theoretically)

  • If $X \subset L^2(0,T;L^2(\Gamma_2)^2)$, we can write gradient algorithms similar to those in Section 1.1, but the amount of computing is beyond the capacity of the available computers at present time as well as in a foreseeable future. We refer the reader to Section 3 for more affordable computations.

Remark 1.5. For other control problems for the Navier-Stokes equations see [11,12].

2. The three-dimensional case. (NCO) The question addressed here is a purely theoretical one. Since the mathematical theory of the Navier-Stokes equations in dimension three is not complete, we cannot write the necessary conditions of optimality in a straightforward way as we did in Section 1. In fact the modeling of the control problem itself raises some difficulties; if we consider the three dimensional analog of the problem in Section 1.1, we are not able, for f given, to define a unique state u_f.

Our aim in this section is to consider the three-dimensional version of the problem in Section 1.1, and to derive some partial results following [2],

---

[3] Hence φ is prescribed at time $t = 0$, equal to $u_{2,1}$, the second component of u_0.

in particular the necessary conditions of optimality when the optimal state $\bar{u}$ is sufficiently regular.

The 3-D Navier-Stokes equations read as in (1.1):

$$(2.1) \quad \begin{cases} \dfrac{\partial u}{\partial t} - \nu \Delta u + (u \cdot \nabla)u + \nabla p = f \quad \text{in,} \quad Q_T = \Omega \times (0,T), \\[2mm] \nabla \cdot u = 0 \quad \text{in } Q_T, \\[2mm] u = 0 \quad \text{on } \partial\Omega, \\[2mm] u(x,0) = u_0(x), \quad x \in \Omega. \end{cases}$$

Here Ω is now an open bounded domain in $\mathbb{R}^3$. The functional form of the equation is similar to (1.2):

$$(2.2) \quad \begin{cases} \dfrac{du}{dt} + \nu Au + B(u) = f, \\[2mm] u(0) = u_0. \end{cases}$$

We know that for every f in $L^2(0,T;V')$, and for every u_0 in H^4, there exists a *nonnecessarily unique* solution u of (2.2) such that

$$u \in L^\infty(0,T;H) \cap L^2(0,T;V).$$

For the control problem, the *control* is f, the *state* $u = u_f$. The *cost function* is

$$J(f) = \frac{\ell}{2}\int_0^T |A^{1/2}f(A)|^2 ds + \frac{m}{2}\int_0^T |\nabla \times u_f|_H^2 ds,$$

where $\ell, m > 0$ and $f \in L^2(0,T;V')$.

The control problem $(\mathcal{P}_4)$ consists in minimizing $J(f)$ for all f in $L^2(0,T;V')$. Furthermore, for each f, if u_f is not unique we minimize as well with respect to all possible $u = u_f$ satisfying (2.2).

The lack of uniqueness of u_f does not prevent us from proving the existence of an optimal pair $\bar{f}, \bar{u}$, as in Theorem 1.1. More precisely there exists an optimal $\bar{f}$ and a corresponding state $\bar{u} = u_{\bar{f}}$ solution of (2.2).

We want to derive the NCO. This will be done under the assumption

$$(2.3) \qquad \bar{u} \in L^8(0,T;L^4(\Omega)^3) \text{ i.e.,}$$

$$\int_0^T \left(\int_\Omega |\bar{u}(x,t)|^4 dx \right)^2 dt < \infty.$$

We have

---

[4] Notations are the same as in Section 1.1

THEOREM 2.1. *Let $\bar{u}$ be an optimal state such that (2.3) holds. Then there exists an adjoint state $\tilde{w}$ solution of*

$$
(2.4) \qquad
\begin{cases}
\tilde{w} \in L^2(0,T;V) \cap L^\infty(0,T;H) \\[2mm]
-\dfrac{\partial \tilde{w}}{\partial t} + \nu A\tilde{w} + B'(\bar{u})^*\tilde{w} = -m\nabla \times (\nabla \times \bar{u}), \\[2mm]
\tilde{w}(T) = 0,
\end{cases}
$$

and we have

$$
(2.5) \qquad \ell A^{-1}\bar{f} + \tilde{w} = 0.
$$

Sketch of the proof

Consider the modified problem
$(\tilde{\mathcal{P}}_4)$ To minimize

$$
\tilde{J}(f) = \tfrac{\ell}{2} \int_0^T |A^{-1/2}f(s)|^2 ds + \tfrac{m}{2} \int_0^T |\nabla \times u(s)|^2 ds
$$
$$
+ \tfrac{1}{4} \int_0^T \left(\int_\Omega |u - \bar{u}|^4 dx \right)^2 ds,
$$

for

$$
(2.6) \qquad
\begin{cases}
u \in \mathcal{C}([0,T];H) \cap L^2(0,T;V) \cap L^8(0,T;L^4(\Omega)^3), \\[2mm]
u' \in L^2(0,T;V'), \quad f \in L^2(0,T;V') \text{ and} \\[2mm]
u' + \nu Au + B(u) = f, \quad u(0) = u_0.
\end{cases}
$$

The solution of $(\tilde{\mathcal{P}}_4)$ exists, is unique and it is obviously the pair $\{\bar{f}, \bar{u}\}$.
Writing the NCO for $(\tilde{\mathcal{P}}_4)$, we obtain

$$
\int_0^T \langle \ell A^{-1}\bar{f} + \tilde{w}, \hat{f} \rangle_{V,V'} ds = 0,
$$

for all

$$
\hat{f} = \hat{u}' + \nu A\hat{u} + B'(\bar{u})\hat{u},
$$

with $\{\hat{f}, \hat{u}\}$ as in (2.5), and $\hat{u}(0) = 0$. We then prove (2.4) by showing that all such $\hat{f}$'s are dense in $L^2(0,T;V')$ (see [2] for the details).

3. Control of the stochastic burgers equation. Some optimal control problems have been described in Sections 1 and 2. From the point of view of control theory, they correspond to open loop full information control problems.

From the practical, computational point of view, we have seen that they correspond to problems which are not feasible at this time. We now consider a different type of problems, from a more practical viewpoint.

We are interested in active control and feedback procedures. We are less demanding and do not look for optimal control anymore; instead we look for procedures which are feasible and which produce an effective reduction of the cost function.

The model problem that we consider is the Stochastic Burgers equation, and we follow [7]. At this time, in a progressing work [3] we try to develop similar procedures for the control of the channel flow problem considered in Section 1.3.

We consider the Burgers equations with a white noise forcing. These equations are an interesting model for the Navier–Stokes equations (NSE); they are simpler than the NSE but it was shwon that the white noise forcing produces a behavior close to that of turbulent flows (cf. [6]). Other work on the control of Burgers equations appear in [4,5]; see also the references therein and in [7].

After nondimensionalization the Stochastic Burgers equations read

$$
(3.1) \quad
\begin{cases}
\dfrac{\partial u}{\partial t} + \dfrac{\partial}{\partial x}\dfrac{u^2}{2} - \dfrac{1}{\mathrm{Re}}\dfrac{\partial^2 u}{\partial x^2} = f + \chi, \quad 0 < x < 1, \quad t > 0, \\[2mm]
u(x,0) = u_0(x), \quad 0 < x < 1, \\[2mm]
u(0,t) = \psi_0(t), \quad u(1,t) = \psi_2(t).
\end{cases}
$$

Here Re is the Reynolds (like) number, χ is after nondimensionalization, a white noise random process in x with zero mean and mean square value 1,

$$
\langle \chi \rangle_x = 0, \qquad \langle \chi^2 \rangle_x = 1.
$$

The control problem The *state* of the system is described by the function u. The *control* could be f, or $\psi = \{\psi_0, \psi_1\}$, or the pair $\{f, \psi\}$. Hereafter we emphasize the boundary control case (control $= \psi$), but we present numerical results on both the boundary control case and the distributed control case (control $= f$); further details appear in [7].

We would like to reduce the "turbulence" as measured by

$$
(3.2) \qquad \int_0^1 \left(\frac{\partial u}{\partial x}(x,t) \right)^2 dx.
$$

Hence we consider the instantaneous cost function

$$
(3.3) \qquad J_t(\psi) = \frac{\ell}{2}|\psi_0(t)|^2 + \frac{\ell}{2}|\psi_1(t)|^2 + \frac{m}{2}\int_0^1 \left(\frac{\partial u}{\partial x}(x,t) \right)^2 dx,
$$

which accounts for the cost (3.2) and for the cost of implementing the control $\psi(\ell, m > 0)$. By boundary layer effect, the main contributions to the integral in (3.2) are produced near the boundary. Hence, instead of (3.2) we could consider

$$
\left(\frac{\partial u}{\partial x}(0,t) \right)^2 + \left(\frac{\partial u}{\partial x}(1,t) \right)^2 ;
$$

however the integral (3.2) leads to simpler computations.

A control problem similar to those of Section 1 could be set as before, with cost function

$$J(\psi) = E \int_0^T J_t(\psi)dt$$

However as indicated before, simplicity is preferred now over optimality and we will look for suboptimal choices of the control function ψ which produce a substantial reduction of $J_t(\psi)$ over a period of time.

We use a marching procedure (one step optimal control) based on a time descretization of the equations.

Time discretized Burgers equations

The Burgers equations (3.1) are written as an abstract evolution equation ($\nu = \mathrm{Re}^{-1}$):

$$(3.4) \qquad \frac{du}{dt} + \nu Au + R(u, \psi) = 0,$$

and using a Cranck Nicholson time discretization scheme, we obtain

$$(3.5) \frac{u^n - u^{n-1}}{\Delta t} + \frac{\nu}{2}A(u^n + u^{n-1}) + \frac{1}{2}(R(u^n, \psi^n) + (R(u^{n-1}, \psi^{n-1}))) = 0.$$

For $u = u^n$, we write (3.5) as

$$(3.6) \qquad \mathcal{A}u^n + R^n(u^n, \psi^n) = 0,$$

where

$$\mathcal{A}u^n = \left(I + \tfrac{\nu \Delta t}{2}A\right)u^n,$$

$$R^n(u^n, \psi^n) = \tfrac{\Delta t}{2}R(u^n, \psi^n) + \tfrac{\Delta t}{2}R(u^{n-1}, \psi^{n-1})$$
$$- u^{n-1} + \tfrac{\nu \Delta t}{2}Au^{n-1}.$$

The problem, similar to those considered in Section 1 is now the following:

Assuming that u^{n-1} and ψ^{n-1} are known, how to best determine u^n and ψ^n so as to minimize

$$J^n(\psi^n) = \frac{\ell}{2}(|\psi_0^n|^2 + |\psi_1^n|^2) + \frac{m}{2}\int_0^1 \left(\frac{\partial u^n}{\partial x}(x)\right)^2 dx.$$

As in Section 1 we can show, at each step n, that this problem has an optimal solution ψ^n with corresponding state u^n (solution of (3.6)). Using the adjoint state we can describe the optimality conditions for this problem; we can also describe gradient algorithm which has been effectively implemented. At each step n, this algorithm produces ψ^n as the limit of a sequence $\psi^{n,k}$, $k \to \infty$. For the details see [7].

In conclusion we emphasize the fact that $J^n(\psi^{n,k})$ does not necessarily decrease as k increases. In fact, setting $\psi^n = \lim_{k\to\infty} \psi^{n,k}$, we do not even assert that

$$(3.7) \qquad\qquad J^n(\psi^n) \leq J^{n-1}(\psi^{n-1}).$$

The effective, large scale computations, reported in [7] show that (3.7) may not be true as t evolves (n increases). However, over a period of time, some significant or substantial decreases of $J^n(\psi^n)$ is observed.

Figures 3.1 to 3.7 are borrowed from [7]. Figures 3.1, 3.2, 3.3 give some characteristics of the flow[5]. Figures 3.4 and 3.5 are related to a control problem, Figures 3.6 to a boundary control problem. The decrease of the cost function J is always important.

Finally let us emphasize Figures 3.7 which are very instructive. We attempted here to plot $u(0,t) = \psi_0(t)$ vs $\dfrac{\partial u}{\partial n}(0,t)$ with the hope of finding some actual feedback law

$$(3.8) \qquad\qquad \psi_0(t) = F\left(\frac{\partial u}{\partial x}(0,t)\right).$$

It appears from Figure 3.7 that there is no single valued feedback law of type (3.8) in this case.

---

[5] Δt_1 not mentioned before is the time step discretization of the white noise χ; $\Delta t_r > \Delta t$ for all computations.

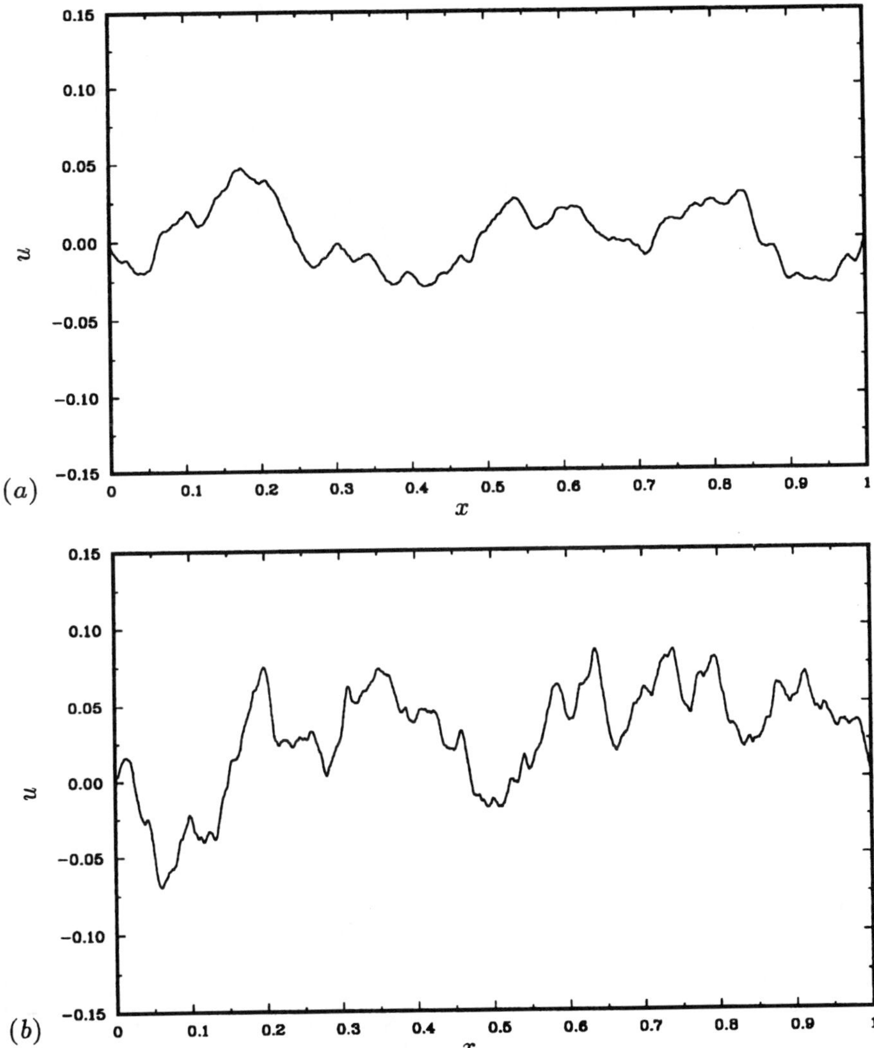

FIG. 3.1. *Instantaneous velocity profiles.* (a) $Re = 500, \Delta t_r = 0.1$; (b) $Re = 1500, \Delta t_r = 0.1$.

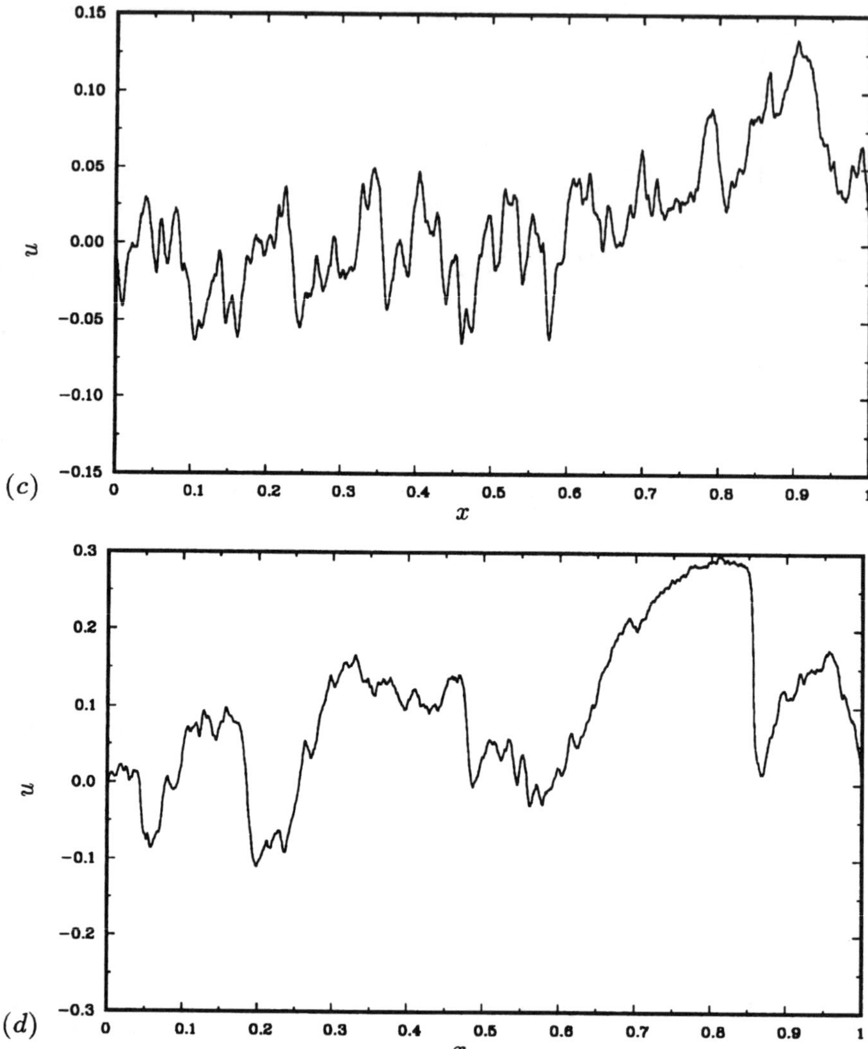

FIG. 3.1. *Instantaneous velocity profiles.(c)* $Re = 4500, \Delta t_r = 0.1$; *(d)* $Re = 4500, \Delta t_r = 1$.

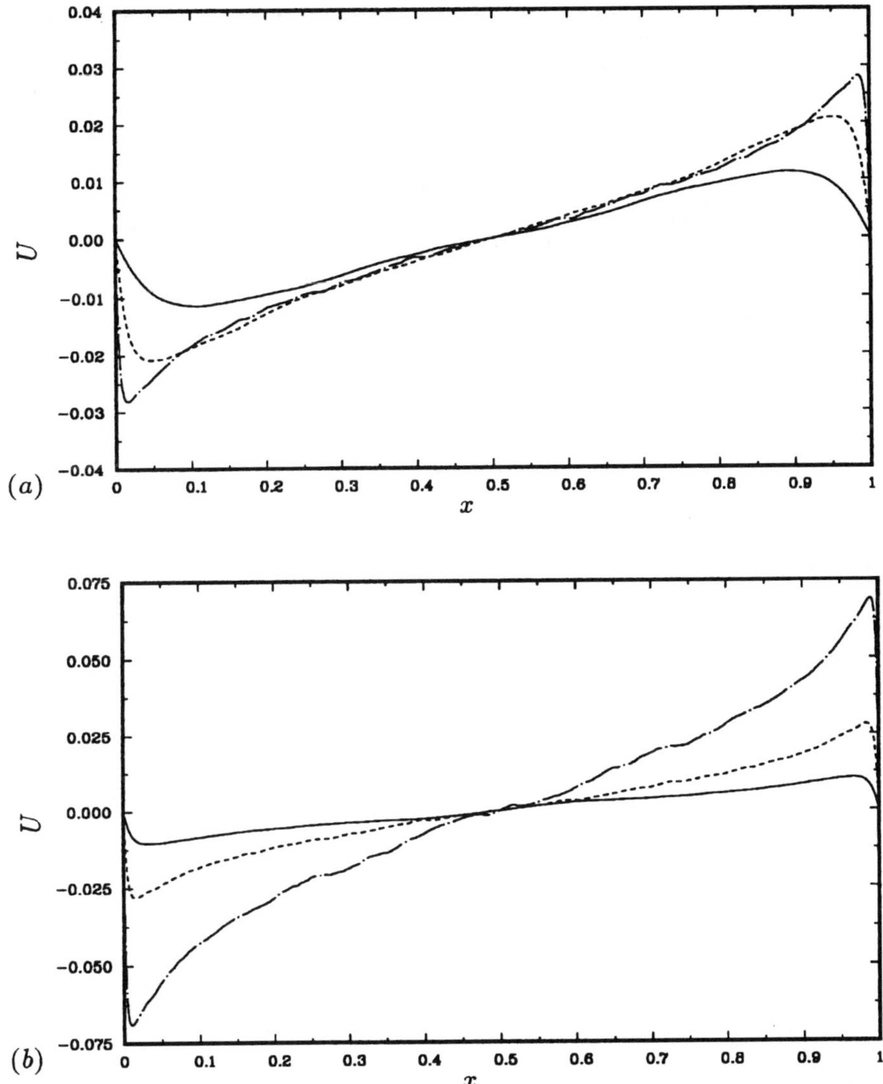

FIG. 3.2. *Mean velocity profiles.* *(a)* $\Delta t_r = 0.1$: —————, $Re = 500$; - - - -, $Re = 1500$; $Re = 4500$; *(b)* $Re = 4500$ —————, $\Delta t_r = 0.01$; - - - -, $\Delta t_r = 0.1$; - - - -, $\Delta t_r = 1$.

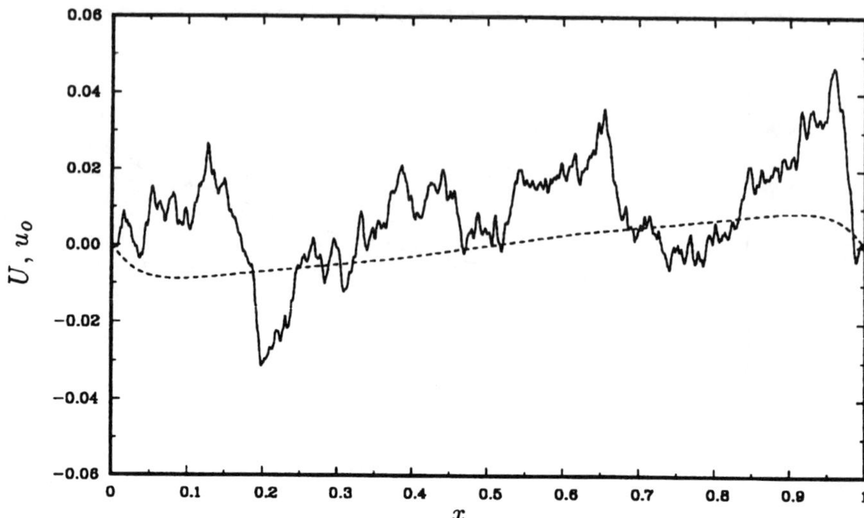

FIG. 3.3. *Velocity profiles:* —-, *time-averaged mean velocity U with no control and* $Re = 1500, \Delta t_r = 0.01;$ ————, *initial velocity* u_0 *for distributed and boundary controls.*

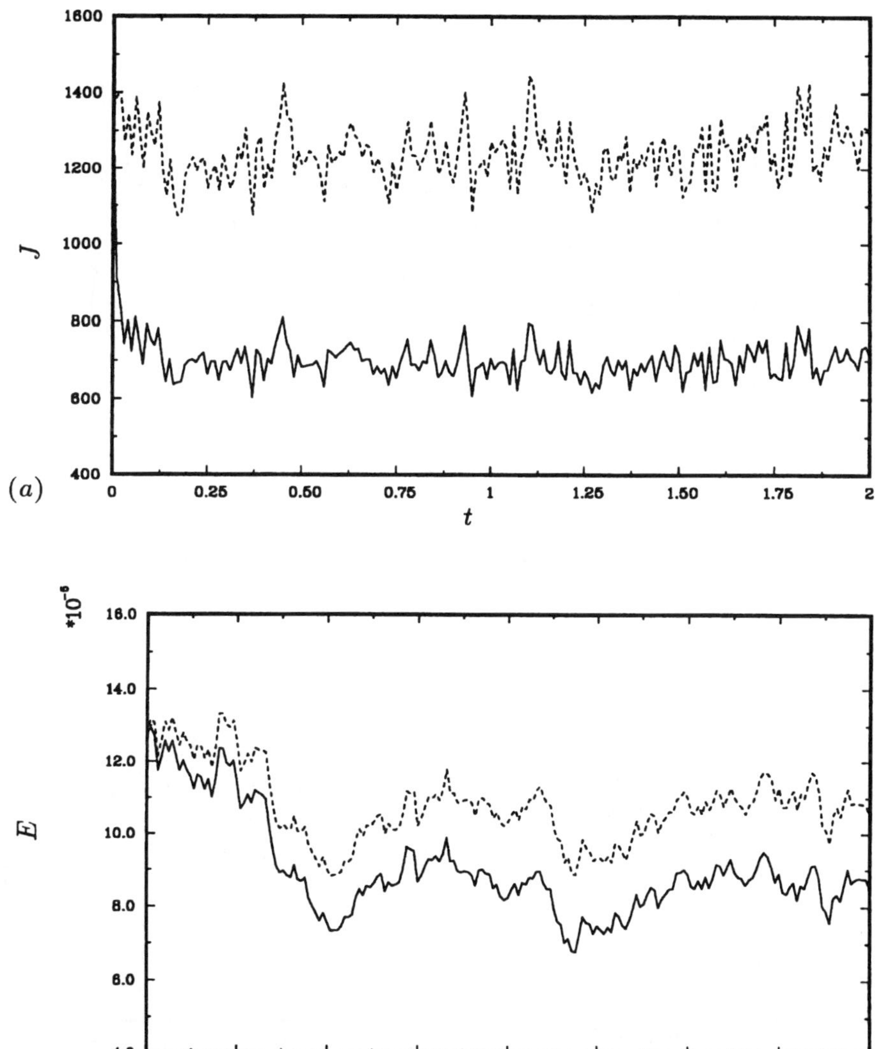

FIG. 3.4. Time history of flow parameters for a distributed control problem ($\ell = 1, m = 2047(= 1/\Delta x)$): ——————, with control; - - - -, without control. (a) Cost; (b) energy inside the domain;

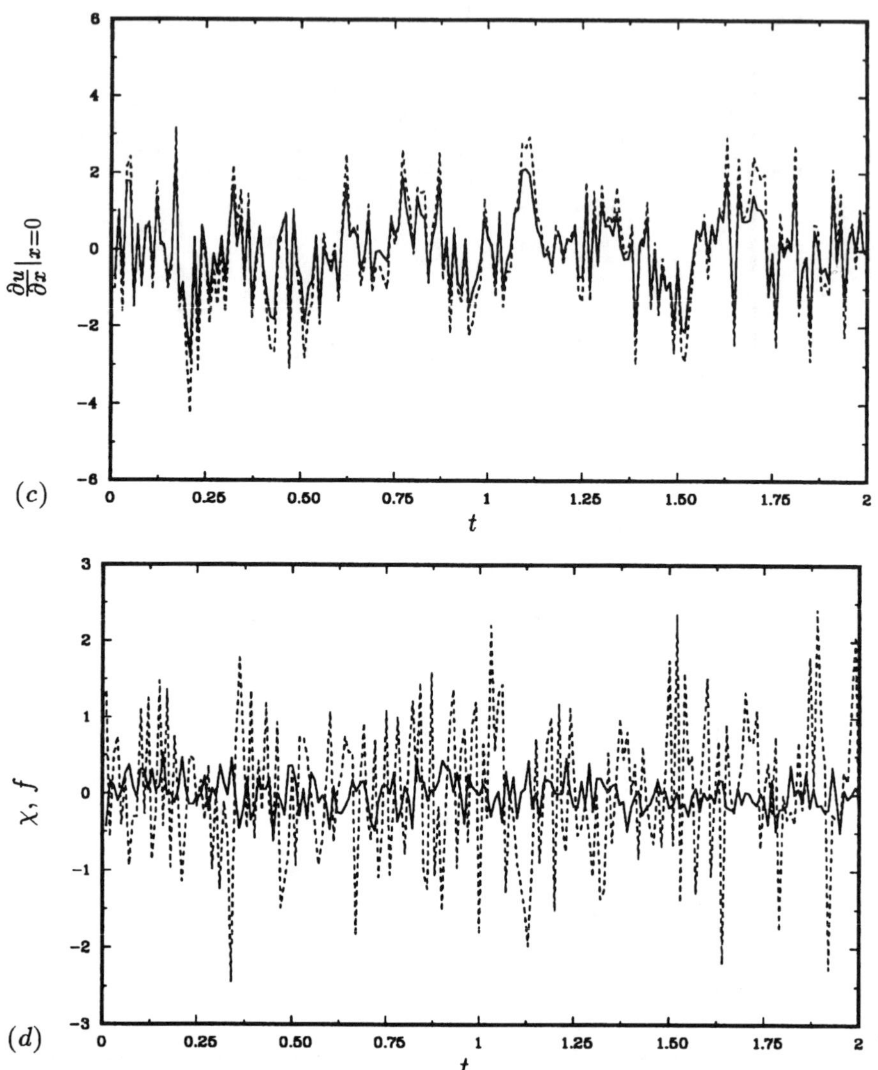

FIG. 3.4. *Time history of flow parameters for a distributed control problem ($\ell = 1, m = 2047(= 1/\Delta x)$):* ————, *with control;* - - - -, *without control.* *(c) wall velocity gradient $\partial u \partial x(x = 0$; (d) momentum forcings at $x = 0.5$;* ————, *control forcing f;* - - - -, *random forcing χ.*

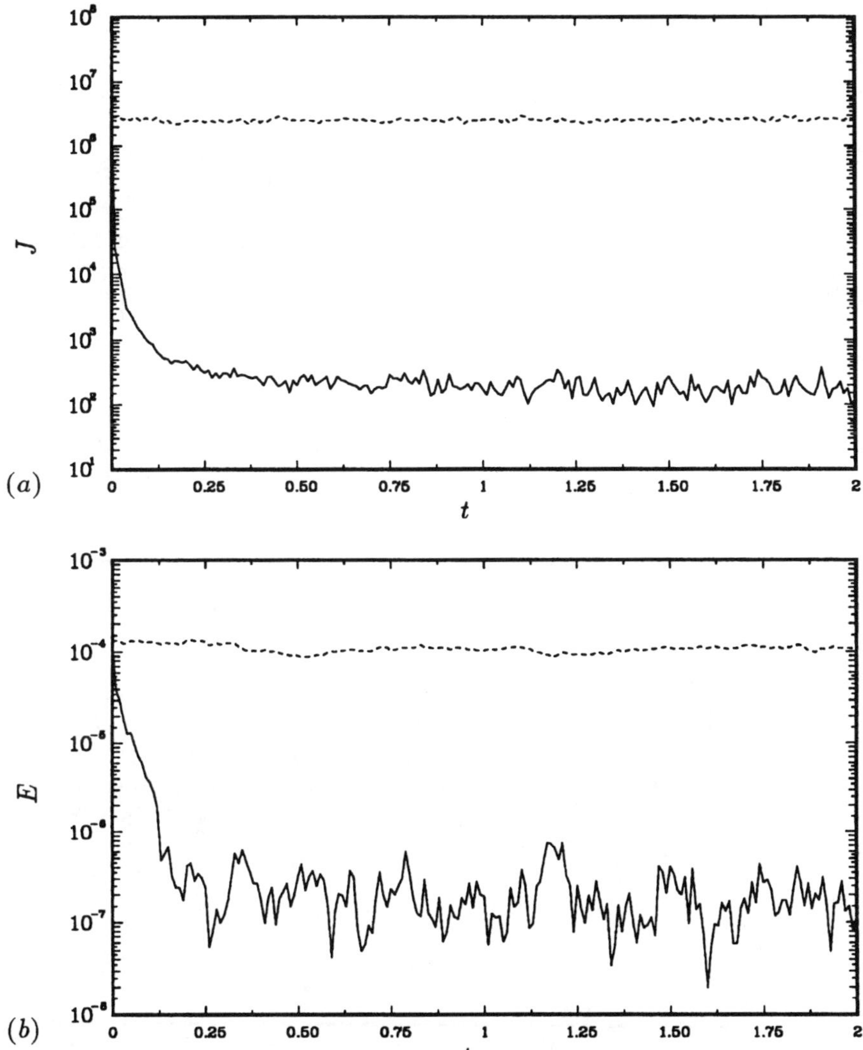

FIG. 3.5. *Time history of flow parameters for a distributed control problem* $\ell = 1, m = 4.2 \times 10^6 (= 1/\Delta x^2)$: ————, *with control;* - - - -, *without control.* (a) *Cost;* (b) *energy inside the domain.*

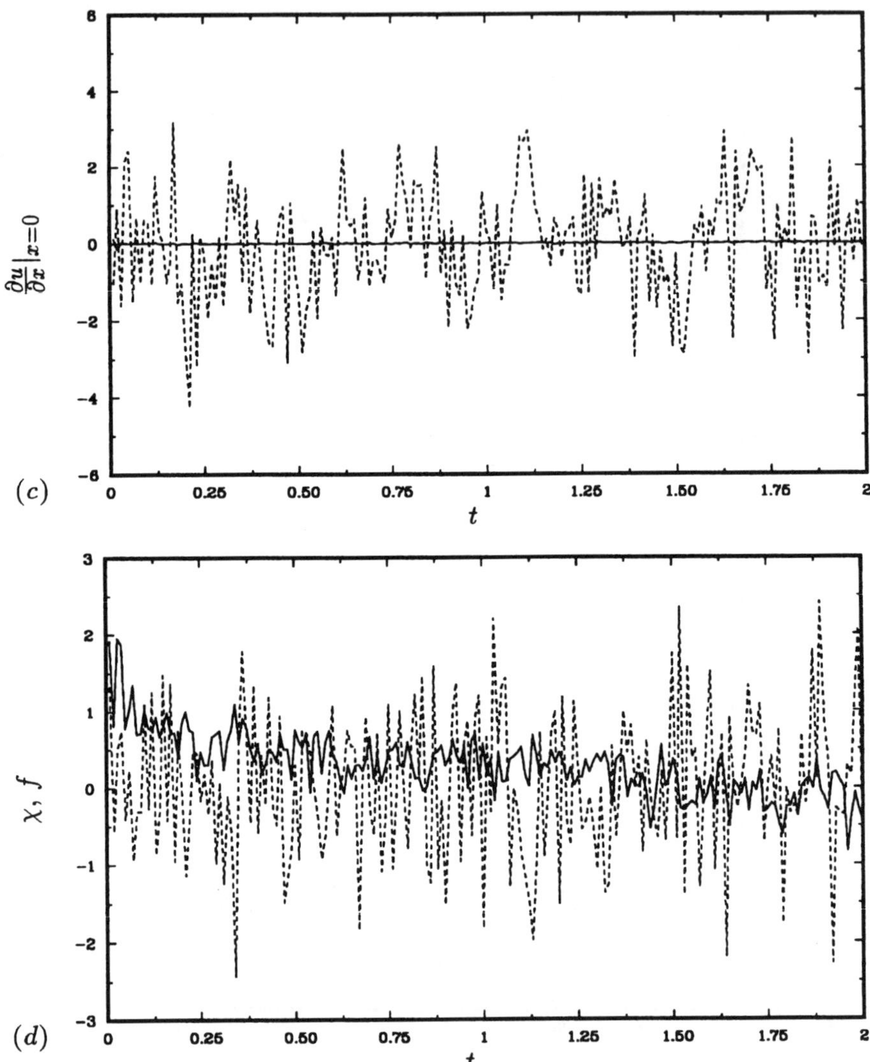

FIG. 3.5. *Time history of flow parameters for a distributed control problem* $\ell = 1, m = 4.2 \times 10^6 (= 1/\Delta x^2)$: —————, *with control;* - - - -, *without control. (c) wall velocity gradient* $\partial u \partial x (x = 0)$; *(d) momentum forcings at* $x = 0.5$; —————, *control forcing* f; - - - -, *random forcing* χ.

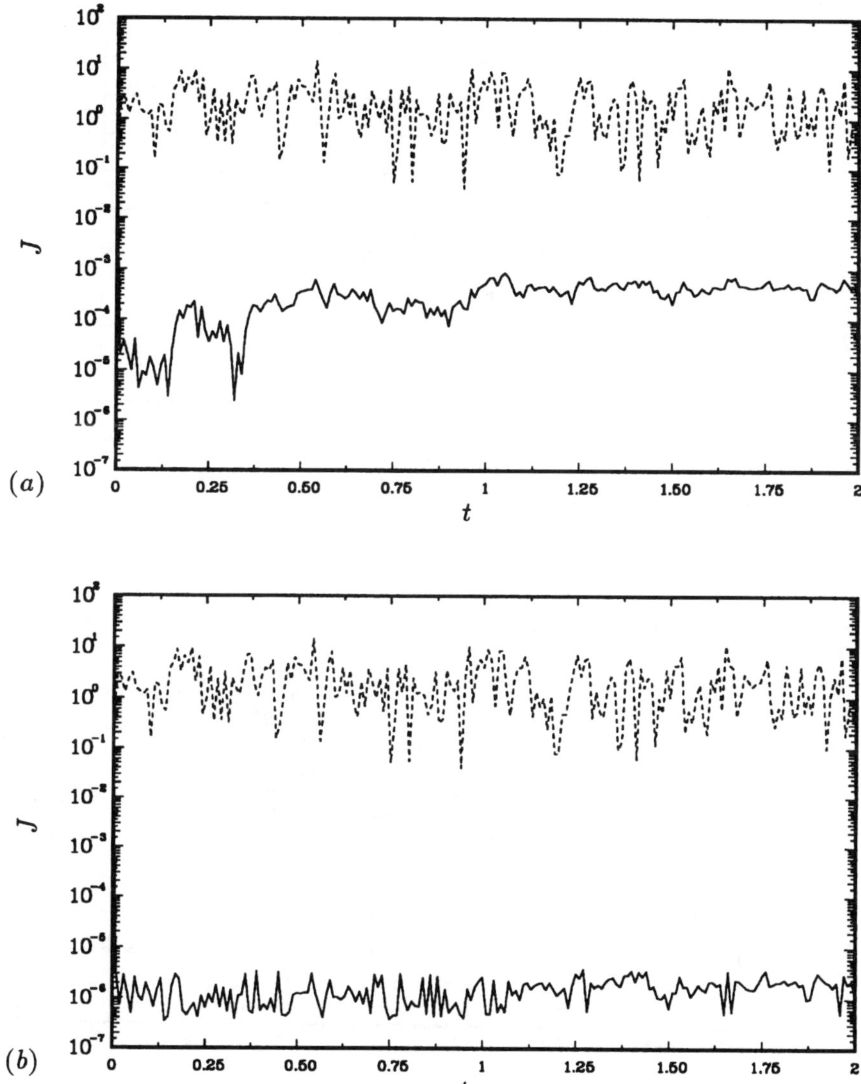

FIG. 3.6. *Time history of the cost for a boundary control problem (a) $\ell = m = 1$, (b) $\ell = 0, m = 1$.*

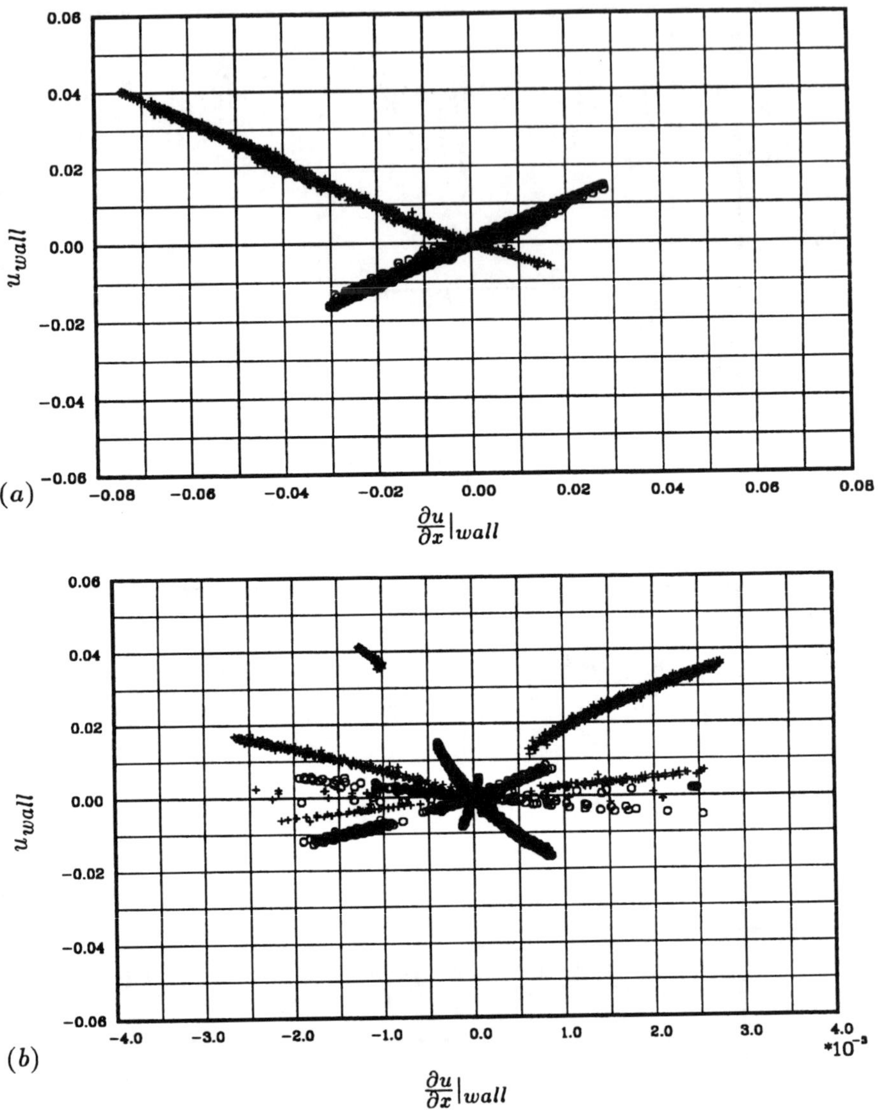

FIG. 3.7. *Phase diagram of control velocity and velocity gradient at the wall for the time interval* $0 < t < 2$: o, *at* $x = 0$, $+$, *at* $x = 1$. *(a) For* $\ell = 1, m = 5 \times 10^{-4}(= \Delta x)$, *(b) For* $\ell = m = 1$. *The phase diagram of case (c)* $\ell = 0, m = 1$ *is nearly same as that of case (b). Note that wall velocity gradients of case (b) are much smaller than those of case (a).*

REFERENCES

[1] F.ABERGEL, R.TEMAM, *On some control problems in fluid mechanics*, Theoret. Comput. Fluid Dynamics **1** (1990), 303–325.

[2] F.ABERGEL, R.TEMAM, *Optimal control of turbulent flows*, (in) [15].

[3] T.BEWLEY, H.CHOI, P.MOIN, R.TEMAM, *Control of channel flows*, (in preparation).

[4] J.A.BURNS, S.KANG, *A stabilization problem for Burgers' equation with unbounded control and observation*, Intern. Series of Num. Math. **100** Birkhauser Verlag, Basel 1991, 51–72.

[5] J.A.BURNS, S.KANG, *A control problem for Burgers' equation with bounded input (output)*, Nonlinear Dynamics **2** (1991), 235–262.

[6] D.H.CHAMBERS, R.J.ADRIANS, P.MOIN, D.S.STEWART, H.J.SUNG, *Karhunen-Loeve expansion of Burgers' model of turbulence*, Phys. Fluids **31** (1988), 2573.

[7] H.CHOI, R.TEMAM, P.MOIN, J.KIM, *Feedback control for unsteady flow and its application to the stochastic Burgers equations*, J. Fluid Mech. (to appear) (1993).

[8] G.DAPRATO, A.DEBUSSCHE, R.TEMAM, *Stochastic Burgers' equations* (to appear).

[9] G.DAPRATO, J.ZABCZYK, *Stochastic equations in infinite dimension*, Encyclopedia of Mathematics and Its Applications, Cambridge University Press, Cambridge 1992.

[10] I.EKELAND, R.TEMAM, Convex Analysis and Variational Problems, North Holland, Amsterdam 1976.

[11] M.D.GUNZBURGER, L.HOU, T.P.SVOBODNY, *Analysis and finite element approximations of optimal control problems for the stationary Navier-Stokes equations with Dirichlet controls*, Math. Model. Num. Anal. M2AN **25** (1991), 711.

[12] M.D.GUNZBURGER, L.HOU, T.P.SVOBODNY, *Boundary velocity control of incompressible flow with an application to viscous drag reduction*, SIAM J. Control Optim. **30** (1992), 167.

[13] J.L.LIONS, Quelques Méthodes de Résolution des Problèmes aux Limites Non-linéaires, Dunod, Paris, 1968.

[14] J.L.LIONS, Controle Optimal des Systèmes Gouvernés par des Équations aux Dérivées Partielles, Dunod, Paris 1969. (English translation) Springer-Verlag, New York.

[15] S.SRITHARAN, *Optimal control of viscous flows*. Frontiers in Applied Mathematics Series, SIAM, Philadelphia (to appear) 1993.

[16] R.TEMAM, Navier-Stokes Equations, North Holland Pub. Comp., Amsterdam 1977.

[17] R.TEMAM, Infinite Dimensional Dynamical Systems in Mechanics and Physics, Applied Mathematical Sciences Series **68** Springer-Verlag, New York 1988.

[18] R.TEMAM, *Navier-Stokes equations theory and approximation*, Handbook of Numerical Analysis, P.G.CIARLET, J.L.LIONS, (eds.) North Holland, Amsterdam (to appear).